LIQUID AND AMORPHOUS METALS

NATO ADVANCED STUDY INSTITUTES SERIES

Proceedings of the Advanced Study Institute Programme, which aims at the dissemination of advanced knowledge and the formation of contacts among scientists from different countries.

The series is published by an international board of publishers in conjunction with NATO Scientific Affairs Division

A Life Sciences B Physics	Plenum Publishing Corporation London and New York
C Mathematical and Physical Sciences	D. Reidel Publishing Company Dordrecht and Boston
D Behavioural and Social Sciences E Applied Sciences	Sijthoff & Noordhoff International Publishers B.V. Alphen aan den Rijn, The Netherlands and Germantown, Maryland, USA

Series E: Applied Sciences – No. 36

LIQUID AND AMORPHOUS METALS

edited by

E. LÜSCHER

and

H. COUFAL

Physics Department E13
Technical University of Munich
Federal Republic of Germany

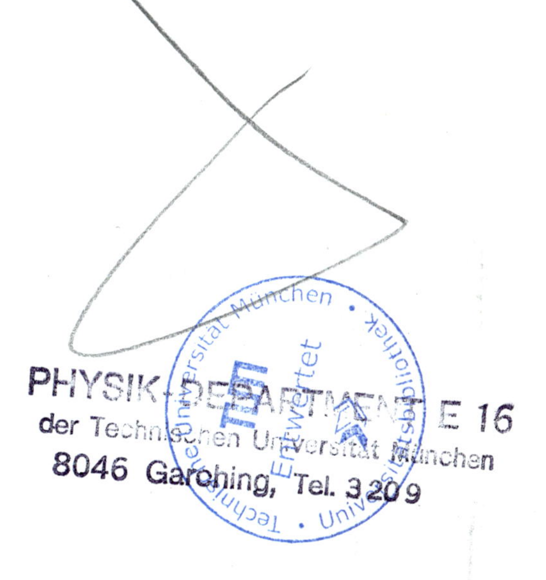

SIJTHOFF & NOORDHOFF 1980
Alphen aan den Rijn, The Netherlands
Germantown, Maryland, USA

Proceedings of the NATO Advanced Study Institute on
Liquid and Amorphous Metals
Zwiesel, Germany
September 11-22, 1979

ISBN 90 286 0680 7

Printed in The Netherlands.

PREFACE

Research on disordered condensed matter is rapidly expanding. In addition to the glassy state of insulators and semiconductors liquid and amorphous metals and alloys are of great interest. The understanding of these materials and their various properties is a tremendous challenge for experimental and theoretical physicists. Metallic glasses seem to offer an enormous potential in the development of new materials for technical applications.

The purpose for this Study Institute was to bring newcomers, insiders and peers from all over the world together to discuss basic principles and the state of affairs. The representation of latest results and the correlation of future research efforts was another important aim.

We would like to thank everybody for his contribution to the Institute and to its proceeding. As we all know the success of such a meeting is for a large part due to the efforts of all participants. Most of them presented lectures, contributed papers or posters. They were heavily involved in panel discussions on "Two Level Systems" and "The Stability of Metallic Amorphous Systems", in ad-hoc sessions on "Resistivity Minima" and "Melting" and in all types of informal meetings.

The help of the Organizing Committee in setting up the scientific programme and of our staff members that were instrumental in the organization of the Institute and of the social programme is gratefully acknowledged. Special thanks go to Mrs. M. Gerl as the good spirit and heart of the meeting, to Mr. K. Hackstein as the business manager and to Mr. H. Hacker as well as to the local authorities for their valuable help in the local organization. Last not least we are deeply indepted to the NATO Advanced Study Institute Programme and to the Schott-Zwiesel-Glaswerke AG for their financial support.

December 1979 E. Lüscher and H. Coufal

TABLE OF CONTENTS

1. INTRODUCTION

MACROSCOPIC AND MICROSCOPIC PROPERTIES OF LIQUID METALS AND ALLOYS: THE EXPERIMENTAL SITUATION

G. Fritsch

Hochschule der Bundeswehr München, ZWE Physik,
8014 Neubiberg, Germany

E. Lüscher

Physik-Department, Technische Universität München
8046 Garching, Germany

1. INTRODUCTORY REMARKS

This contribution is the introductory lecture to this NATO Advanced Study Institute. Therefore, the basic theoretical concepts are qualitatively treated in order to give the reader the background for the following lectures as well as for the experimental methods which will be discussed in the next sections. We intend to show how theory and experiment can be related at the various stages between first principles (microscopic point of view) and macroscopic properties. Hence, it will turn out that it is often useful to insert a measured quantity at a certain point in theory and to develop predictions therefrom which can be checked again by experiment.

In order to make our ideas more precise we have summarized some of the theoretical skeletons and a selection of the experimental ingredients in Table I. At various stages experimental results either from computer or from nature can be used to test the theoretical models.

Table I. Connections between experiments and theory

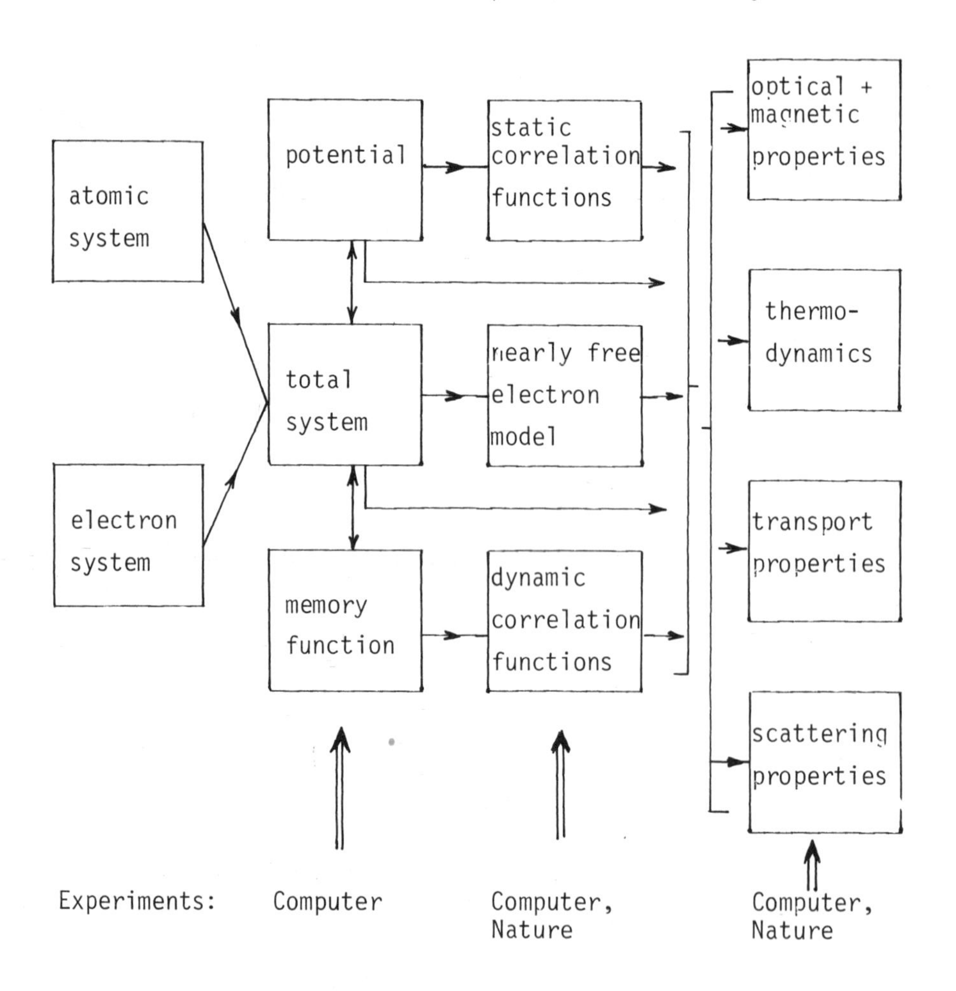

2. STRUCTURAL ASPECTS

2.1 Pure metallic liquids

2.1.1 Correlation functions

In order to describe the structure of liquids we introduce correlations of particles wihhin the system: the n-particle correlation functions: $g_n(\vec{r}_1,\vec{r}_2...\vec{r}_n,t)$.

One can ask how large the contributions of higher order correlation functions ($n > 2$) are! One will find that some properties can be described by using only a two particle correlation function $g_2(\vec{r}_1,\vec{r}_2)$ or in isotropic systems $g(r)$, where $r = |(\vec{r}_1-\vec{r}_2)|$. When we try to relate the forces to the structure of a liquid, the three body correlation function $g_3(\vec{r}_1,\vec{r}_2,\vec{r}_3)$ comes in.

Let us first summarize how $g(r)$ can be obtained theoretically. If n_0 describes the particle density than

$$dN = n_0 \; 4\pi r^2 \cdot g(r) \cdot dr$$

gives the number dN of particles in a spheric shell of thickness dr and radius r around a given particle. All theoretical models to calculate $g(r)$ start from the two particle interaction potential $v(\vec{r}_1,\vec{r}_2)$. In metallic systems the pseudopotential is used (see talk by R. Evans). From the configurational part of the partition sum one derives an hierarchy of equations (Bogoljubov, Born, Green) coupling the various correlation functions. The lowest order equation is /1/

$$-kT \nabla_{\vec{r}_1} \cdot [\ln g_2(\vec{r}_1,\vec{r}_2)] = \nabla_{\vec{r}_1} v(\vec{r}_1,\vec{r}_2) + n_0 \cdot$$

$$\int [\frac{g_3(\vec{r}_1,\vec{r}_2,\vec{r}_3)}{g_2(\vec{r}_1,\vec{r}_2)} \nabla_{\vec{r}_1} v(\vec{r}_1,\vec{r}_3)] \cdot d^3 r_3 \quad .$$

In order to get a closed system of equations one has to approximate g_3 by g_2, f.e. by the following equation (Kirkwood)

$$g_3(\vec{r}_1,\vec{r}_2,\vec{r}_3) = F(g_2(\vec{r}_1,\vec{r}_2),\; g_2(\vec{r}_1,\vec{r}_3),\; g_2(\vec{r}_2,\vec{r}_3)) \quad ,$$

where F means some appropriate function.

A second approach /1/ was derived by Percus and Yevick /2/ (PY) applying collective coordinates for the description of the system. The result can be formulated with the abbreviations:

$$h(r) = g(r) - 1 \quad , \quad f(r) = \exp\,[-v(r)/kT] - 1$$

and the Ornstein-Zernicke /1/ definition for the direct correlation function $c(r)$

$$h(r) = c(r) + n_0 \int_0^\infty c(|\vec{r}-\vec{r}'|) \cdot h(r') \cdot d^3 r' .$$

The idea behind this equation is, to split the total correlation function into a pair-term and one due to the remaining particles. It follows:

$$g(r) = [1+f(r)]\cdot[1+h(r)-c(r)] .$$

Using a systematic diagrammatic approach /1/, the hypernetted chain equation (HNC) was derived. It should be better than the PY-equation, since more diagrams are summed up:

$$g(r) = [1+f(r)] \cdot \exp [h(r)-c(r)] .$$

This can also be seen in the formal structure. However, it turns out that for short range potentials (hard spheres) PY gives best results, so one is forced to believe in some natural cancellation of diagrams, speaking in the language of the HNC-approximation.

It should be mentioned that knowing g(r) the potential v(r) could be derived in principle. We will discuss this important point later on.

Another way to relate potentials and g(r) uses thermodynamic perturbation theory starting from the properties of the so-called hard sphere systems.

2.1.2 Thermodynamics

In principle it is possible to calculate the thermodynamic properties of our metallic system by the following equations /1/:

$$E(T,V) = \frac{3}{2}kT + \frac{n_o}{2} \cdot N \cdot \int_0^\infty d^3r\; v(r)\cdot g(r) \quad ,$$

where E(T,V) is the internal energy,

$$P/kT = n_o\left[1 - \frac{n_o}{6RT} \cdot N \cdot \int_0^\infty d^3r\; v'(r)\cdot g(r)\right] \quad ,$$

the pressure equation, where v'(r) is the derivative of the potential as well as

$$kT \cdot \left(\frac{\partial n_o}{\partial P}\right)_T = 1 + n_o \cdot \int_0^\infty d^3r\; [g(r)-1] \quad .$$

This latter equation is called the compressibility equation, it does not contain the potential any more, explicitly.

All these equations have the disadvantage that the potential is not treated on equal footing in g(r) and in the quantities given above.

The three theoretical models discussed below overcome this difficulty. One starts from the idea, that the hard core (HC) of the potential is the most important part to define the structure. Hence, one tries to get exact results for the HC-part (repulsive) and to apply perturbation theory to the attractive part of the potential.

Let us first describe the properties of the hard sphere (HS) system: The HS-potential is given by

$$v(r) = \begin{cases} \infty & 0 < r < d \\ 0 & r \geq d \end{cases} \qquad \text{d: hard sphere diameter.}$$

The free energy can be calculated /3/:

$$F_{HS} = F_i\ (\eta) + RT \cdot f(\eta),$$

where F_i is the free energy of an ideal gas of the same density

$$F_i = \frac{3}{2} RT \cdot (1 - \ln T - f_o) - RT \cdot \ln \eta \quad ,$$

where f_o is a constant which is known. The parameter η is the packing density, defined by:

$\eta = \frac{\pi}{6} n_o d^3$. $f(\eta)$ must be calculated by some approximation:

Carnahan and Starling /4/ 1969 (CS) $f(\eta) = \dfrac{\eta(4-3\eta)}{(1-\eta)^2}$,

Frisch and Lebowitz /5/ 1964 (FL) $f(\eta) = \dfrac{3\eta(2-\eta)}{2(1-\eta)^2} - \ln(1-\eta)$.

The FL-solution yields directly to the PY-result for the compressibility. The latter expression comes from the fact that the PY-model can be solved exactly for the HS-potential /1/:

$$c_{HS}(r/d) = -\frac{1+2\eta}{(1-\eta)^4} \cdot \left[1+\frac{\eta}{2}(r/d)^3\right] + 6\eta \frac{(1+\eta/2)^2}{(1-\eta)^4} \cdot (r/d).$$

g(r) is calculated from the Ornstein-Zernicke relation.

The so-called static structure factor S(q) can be derived from the Fourier transform (FT): $c_{HS}(q \cdot d) = FT(c_{HS}(r/d))$ by:

$$S(qd) = \{1 - n_0 \cdot c_{HS}(q \cdot d)\}^{-1} \quad .$$

In order to compare how good these approximations are, consider Table II. Here the various virial expansions are compared to the exact one, calculated by cluster expansion.

Table II. Virialexpansions of various approximations to the hard sphere equation of state, P: pressure, n_0: number density, k_B: Boltzmann-constant and T: temperature.

$P/(n_0 k_B T)$	$= 1 + 4\eta + 10\,\eta^2 + 18.362\,\eta^3 \pm \ldots$	exact
	$= 1 + 4\eta + 10\,\eta^2 + 19\,\eta^3 \pm$	FL
	$= 1 + 4\eta + 10\,\eta^2 + 18\,\eta^3 \pm$	CS
	$= 1 + 4\eta + 16\,\eta^2 + 64\,\eta^3 \pm$	van der Waals. (a=0, b≠0)

As one can seen the PY-solution is not too bad when compared to the CS or the exact results. There are only deviations in the third virial coefficient.

The Barker-Henderson (BH) model /6,7/ uses the PY solution for the HS-potential as a reference system. The attractive part is added by perturbation theory. The HS-diameter d is determined by making the first order correction to the Helmholtz free energy F to disappear. So we get:

$$F(V,T) = F_{HS} + F_{BH} + Nu(n_0),$$

$$F_{BH} = 2\pi N n_0 \cdot \int_{\sigma}^{\infty} g_0(R) \cdot v(R) \cdot R^2 dR$$

$$- \pi N n_0 \cdot \int_{\sigma}^{\infty} [v(R)]^2 \cdot g_0(R) \cdot \left.\frac{dn_0}{dP}\right|_{HS} \cdot dR,$$

where $g_0(R)$ is the PY-solution of HS of diameter d. $dn_0/dP/_{HS}$ is derived from the compressibility equation for HS and $u(n_0)$ is the electronic contribution. σ describes the first zero of the interaction potential $v(R)$: $v(\sigma) = 0$

$$d = \int_0^{\sigma} [1 - \exp(-v(R)/kT)] \cdot dR = d(T) \quad .$$

As one can see, there's no improvement in $g(r)$ over the HS-solution with the PY-model, except for the use of $d(T)$. The weakness of the BH-model lies in the bad description of the real HC-behaviour of the potential

This point is improved in the Weeks - Chandler - Anderson /7,8/ (WCA)-model. The potential $v(R)$ is decomposed in a repulsive $v_r(R)$ and an attractive part $v_a(R)$ as follows:

$$v_r(R) = \begin{cases} v(R) - v_{Min} & R \leq R_{Min} \\ 0 & R > R_{Min} \end{cases}$$

$$v_a(R) = \begin{cases} v_{Min} & R \leq R_{Min} \\ v(R) & R > R_{Min} , \end{cases}$$

where

$$v_{Min} = v(R_{Min}) \quad .$$

The attractive part $v_a(R)$ is added as before as a perturbation to a reference system. But this reference system is now given by the exact potential $v_r(R)$. After doing this the reference system is approximated by the HS-system, which gives an equation for the quantity d. We get:

$$F(V,T) = F_{HS} + F_{WCA} + N \cdot u(n_0),$$

$$F_{WCA} = 2\pi N n_0 \cdot \int_0^{\infty} dR\, R^2 \cdot g(R) \cdot v_a(R)$$

with $g(r) = y_{HS}(R) \cdot \exp\left[- \frac{v_r(R)}{kT}\right]$,

which is an improved two particle correlation function as well as

$$y_{HS}(R) = g_{HS}(R) \cdot \exp\left[v_{HS}(R)/kT\right].$$

The HS-diameter d is determined by:

$$\int_0^{R_{Min}} R^2 B(R) \cdot dR = 0,$$

$$B(R) = y_{HS}(R)\cdot\left\{\exp\left(-\frac{v_r(R)}{kT}\right) - \exp\left(-\frac{v_{HS}(R)}{kT}\right)\right\}.$$

There are still other methods to calculate F and g(R) from the potential. One of them uses the Gibbs-Bogoljubov inequality /9/, which states that the true free energy is small or equal the free energy of a reference system plus a perturbation averaged with respect to the reference system. The HS-diameter is used as a variational parameter to minimize F.

2.1.3 Structure Factor

From the discussion given above, one sees that the two particle correlation function plays a vital role. Hence, we will give some further relations which will correlate g(r) with measurable quantities.

We define the static structure factor S(q) by

$$S(q) = 1 + n_o \int (g(r)-1)\cdot\exp\left[i\vec{q}\vec{r}\right]\cdot d^3r.$$

Its value at q = 0 is given by a sum rule to be

$$S(0) = 1 + n_o\cdot\int (g(r)-1)\cdot d^3r = n_o\cdot kT \cdot \varkappa_T \quad ,$$

where $\varkappa_T$ is the isothermal compressibility. One recovers directly the compressibility equation mentioned before.

2.2 Liquid Alloys

Let us consider a system with two components A and B. The analysis of g(r) can be generalized in principle by introducing the appropriate potentials and correlation functions /10/. For two particles i.e. the relevant indices are AA, BB and AB.

So we have for g(r)

$$g(r) \rightarrow g_{ij}(r) \qquad \text{where } i,j = A,B \qquad \text{and}$$

$$g_{ij}(r) = g_{ji}(r).$$

In addition we define:

$$S_{ij}(q) = 1 + n_o\cdot\int\left[g_{ij}(r)-1\right]\cdot\exp\left[i\vec{q}\vec{r}\right]\cdot d^3r.$$

Important information is hidden in the limits $S_{ij}(q\rightarrow 0)$ (compare the compressibility equation for a one component system) /11/:

$$S_{AA}(0) = n_o \cdot kT\,\varkappa_T - \frac{c_B}{c_A} + A(\delta_A - \frac{1}{c_A})^2 \qquad \text{and}$$

$$S_{AB}(0) = n_o \cdot kT\,\varkappa_T + 1 - A(\delta_A - \frac{1}{c_A})(\delta_B - \frac{1}{c_B}) \quad , \qquad \text{where}$$

$$A = RT\left(\frac{\partial^2 \Delta G}{\partial c_A{}^2}\right)^{-1}_{P,T,N} \quad , \quad c_A,\ c_B \text{ are the concentrations and}$$

$$\delta_A = \frac{1}{v_m}\left(\frac{\partial v_m}{\partial c_A}\right)_{P,T,N} = -\frac{1}{v_m}\left(\frac{\partial v_m}{\partial c_B}\right)_{P,T,N}$$

with v_m the molare volume of the alloy.

ΔG is an important quantity, the Gibbs free energy of mixing. It's intimately connected to the stability of the alloy and to the deviation from an ideal solution.

The situation can also be considered from a different point of view. Bhatia and Thornton 1970 /12/ started from fluctuation theory. They ended up with the following description for the structure of an alloy in terms of three correlation functions:

i) number density fluctuation correlation $S_{NN}(q)$,
ii) concentration fluctuation correlation $S_{CC}(q)$,
iii) number density-concentration fluctuation correlation $S_{NC}(q)$.

It can be shown /11/ that these quantities are linearly related to $S_{AA}(q)$, $S_{BB}(q)$ and $S_{AB}(q)$. However, their limits $q\rightarrow 0$ are easier to interpret:

$$S_{CC}(0) = RT\left(\frac{\partial^2 \Delta G}{\partial c_A{}^2}\right)^{-1}_{T,P,N} = A \quad ,$$

$$S_{NN}(0) = n_o \cdot kT\,\varkappa_T + \delta^2 \cdot S_{CC}(0) \qquad \text{and}$$

$$S_{NC}(0) = -\,\delta \cdot S_{CC}(0) \quad , \qquad \text{where}$$

$$\delta = (v_{mA} - v_{mB})/(c_A v_{mA} + c_B\, v_{mB}).$$

v_{mA} and v_{mB} are the molare volumes of the pure components.

The same argumentation, valid for liquid alloys, may be transferred directly to amorphous alloys. A word of caution should be in place here. Whereas liquids are in thermal equilibrium this is not true for amorphous alloys. They are, if at all, only in a metastable equilibrium. Hence, a thermodynamic description is only appropriate if the time average can be performed properly.

2.3 Experimental situation

2.3.1 Direct results

The first ingredient of theory to be discussed is the two particle potential. It can be calculated theoretically from first principles but not very precisely. If d-electrons are involved it's even questionable to define a pseudopotential. Therefore the experimental derivation of the potential is an important point. This task can be performed either by suitably parametrizing the potential function and by fitting the parameters to thermodynamic properties via the theories discussed or by deriving the potential directly from a measured $g(r)$.

The first possibility has been used extensively i.e. see Dagens et al. /13/ or Ashcroft /14/. These pseudopotentials seem to be the best ones available, especially if already some physics is included in the parametrisation procedure.

The second method suffers from the fact that $S(q)$ or $g(r)$ must be known to a very high accuracy (<1 %) in various regions of q or r in order to give meaningful results. Thus, potentials derived by PY- or HNC-methods should be considered with caution.

We are already discussing the second important structural quantity: The various $g_{ij}(r)$ functions. They can be measured since their Fourier-transform is related directly to the scattering cross section $d\sigma/d\Omega$ /15/:

$$d\sigma/d\Omega = A(q) \cdot S(q),$$

where the value of q is connected to the wave vectors of the incoming and outgoing particles (waves) and to scattering angles Ω. $S(q)$ is the static structure factor for a one-component system. $A(q)$ contains the scattering properties of the particles of the system as well as of the scattered particles (X-rays, neutrons etc.). It should be mentioned that corrections due to energy transfer, multiple scattering, absorption, container effects are by no means straight forward. Some results are given in Figs. 1 and 2.

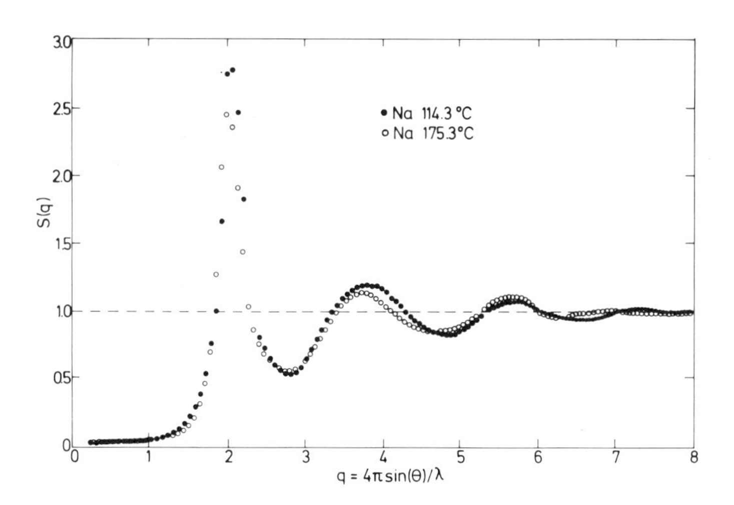

Fig.1. X-ray scattering results for liquid soldium at two temperatures (R.B. Schierbrock, Diploma work, 1970).

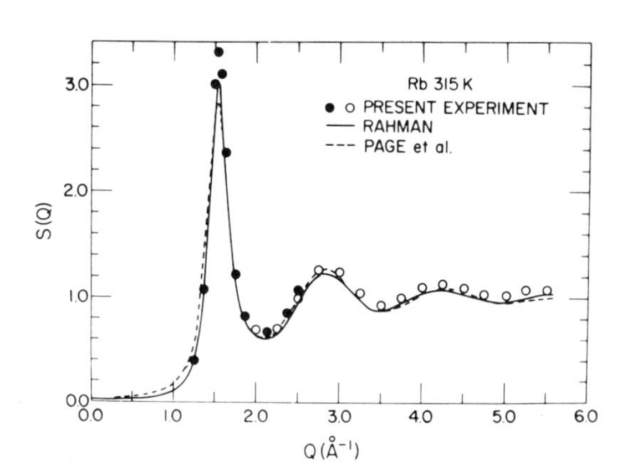

Fig.2. n-scattering results for liquid Rubidium (J.R.D. Copley, J.M. Rowe 1974).

In case of alloys S(q) is given either by /11/:

$$S(q) = \frac{1}{\langle a^2 \rangle} [c_A^2 \cdot a_A^2 \cdot S_{AA}(q) + c_B^2 \cdot a_B^2 \cdot S_{BB}(q) + 2c_A c_B a_A a_B (S_{AB}(q) + \frac{(\Delta a)^2}{2a_A a_B})] \quad \text{with}$$

$$\Delta a = a_A - a_B \ , \quad \langle a^n \rangle = c_A a_A^n + c_B a_B^n$$

or equivalently in terms of the Bhatia-Thornton-structure factors /11/:

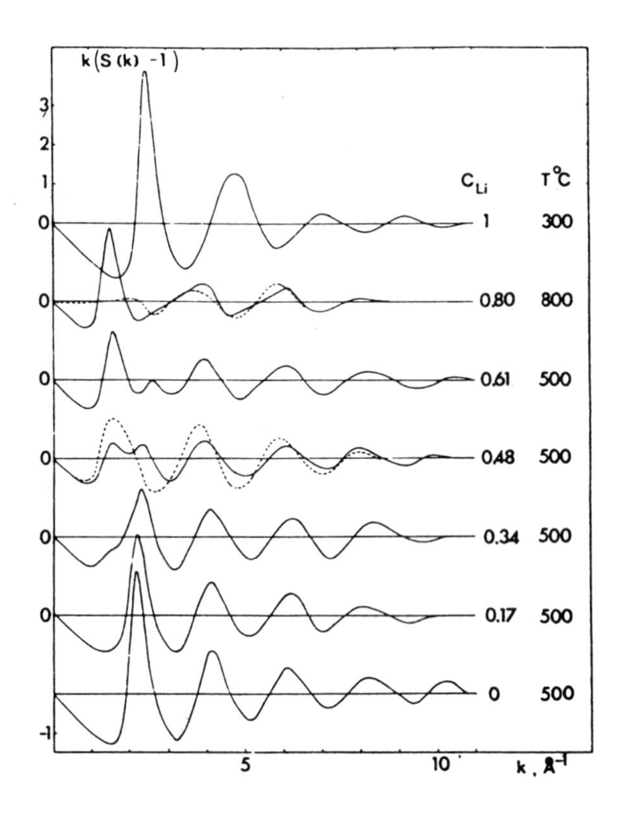

Fig.3. Experimental data for the quantity k(S(k)-1) versus k at various compositions of the liquid alloy ^{7}LiPb. The coherent scattering cross sections are $\sigma_{coh,Pb}$=11.11 barn and $\sigma_{coh,Li}$=-0.66 barn (99.05 % ^{7}Li). The dashed curves are theoretical estimates (H. Ruppersberg, H. Egger, J.Chem.Phys. 63, 4095 (1975)).

$$S(q) = \frac{1}{\langle a^2 \rangle} \left[\langle a \rangle^2 \cdot S_{NN}(q) + (\Delta a)^2 \cdot S_{CC}(q) + 2 \cdot \langle a \rangle \cdot \Delta a \cdot S_{NC}(q) \right] \quad .$$

The a_j $j = A,B$ are the scattering lengths for the two components. Since, for neutrons the scattering length a depends on the isotopes used and this quantity can be negative, one has a wider field of possibilities to separate the various structure factors. Fig. 3 reports some experimental results. Applying X-rays, one has to use the anomalous dispersion part of the form factor close to an absorption edge in order to vary the scattering properties /16/.

We would like to make the following comments:

i) If one compares a high temperature solid metal with a liquid metal the differences are not large. Certainly long range order is indicated by Bragg-reflexes (see Fig. 4).

ii) A comparison of S(q) of liquid alloys and amorphous alloys exhibits a close similarity, were it not for a shoulder in the second peak in the amorphous alloy S(q) (see talk of G. Jacucci).

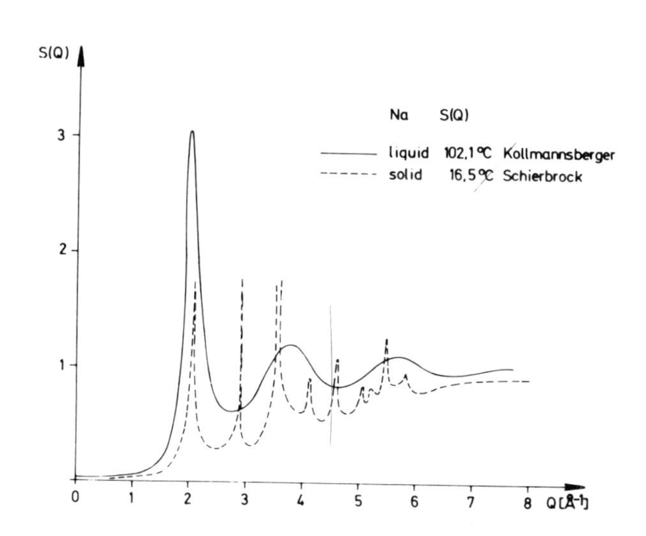

Fig.4. Comparison between the static structure factors of Na of a polycrystalline and a liquid sample. (G. Kollmannsberger, G. Fritsch, E. Lüscher Z.Physik 233, 308 (1970) and R.B. Schierbrock, G. Fritsch Phys.Letters 48A, 151 (1974)).

iii) A measurement of $S(q)$ yields the correlation function $g(r)$. The latter can in turn be calculated from various models of the liquid (Bernal, Hosemann etc. /17/) or compared with the results from BBG, PY, HNC or WCA etc. theories. It can also be used as an ingredient to calculate other properties, like electrical conductivity (see Sect. V) etc. (compare also Table I).

2.3.2 Computer experiments /1,18/

The invention of large and fast computers allows to evaluate the behaviour of a system of 400 to 1400 particles - made virtually infinite by periodic boundary conditions - microscopically. Hence, theories with given potential functions $v(r)$ can be checked directly. There's no uncertainty concerning the knowledge of the potential $v(r)$. Static properties are determined by two methods, the Monte Carlo (MC) and the Molecular Dynamics (MD) methods.

MC-method: Configurations of the systems are generated at random and weighted by a factor

$$\exp\left[-\frac{1}{kT}\sum_{i<j} v(R_{ij})\right].$$

From an average over many configurations weighted in such a way, $g(r)$ and other properties may be calculated.

MD-method: The equations of motion are solved numerically for time steps around 10^{-14} s. Hence, the full time dependence of the system is known. Performing a time and configurational average gives static properties.

Limitation due to the smallness of the basic cell (available range in q-space) and effects from super-periodicity from periodic boundary conditions must be considered carefully. Figs 5 and 6 illustrate some results.

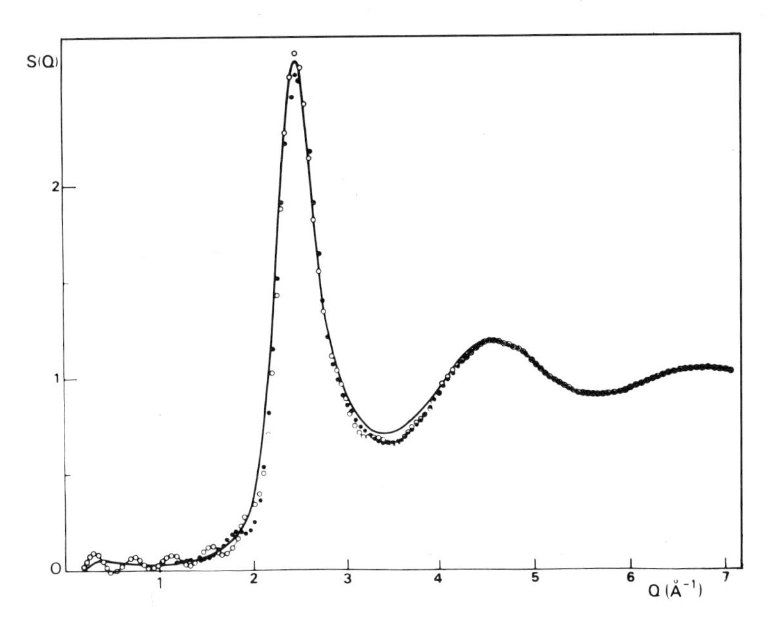

Fig.5. Molecular dynamics calculation as compared to experimental results for liquid Lithium (G. Jacucci, M.L. Klein, R. Taylor Solid States Comm. 19, 657 (1976)).

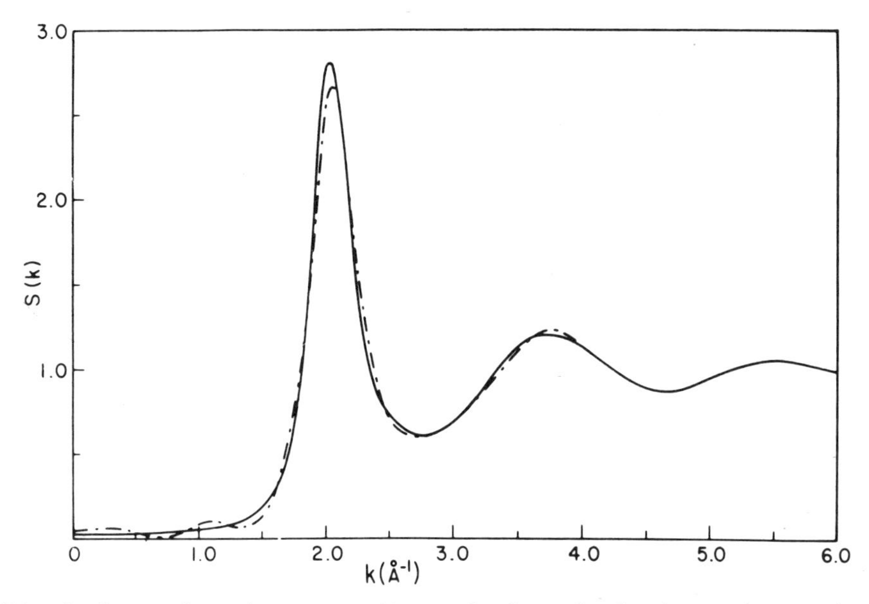

Fig.6. Comparison between a Monte Carlo calculation and experimental data for the structure factor of liquid Sodium at 100 °C (R.D.Murphy, M.L.Klein Phys.Rev. 18, 2640 (1973)).

3. DYNAMICAL ASPECTS

3.1 Theoretical concepts

3.1.1 Mean square displacements

In a liquid we expect diffusive as well as vibrational motion of the particles, the latter are reminescent of the solid state. Such a situation can best be illustrated by considering the function $\langle r^2(t)\rangle$ /19/. This quantity, the mean square displacement, describes, how far an average particle has moved from the origine in a time t. (The situation will be discussed in more detail in A. Sjölander's talk.) Schematically we expect the following behaviour (Fig. 7):

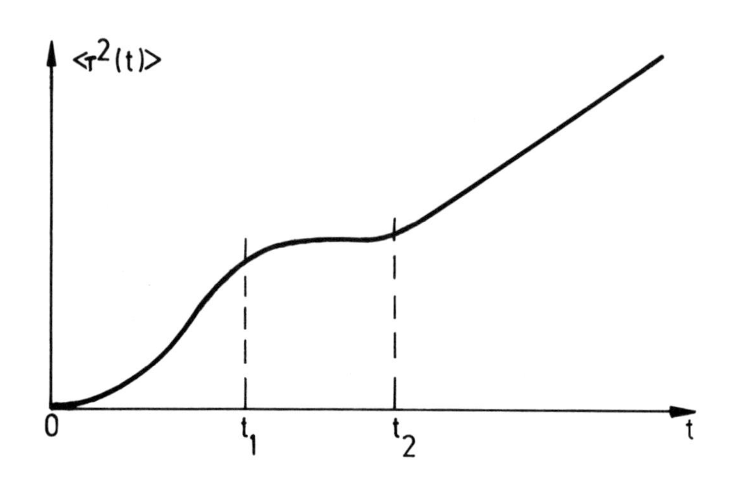

Fig.7. Principal behaviour of the mean square displacement function $\langle r^2(t)\rangle$.

We see, we must add a "free-flight"-region, where the particle is not influenced by others.

region: $0 < t < t_1$

$$\langle r^2(t)\rangle = \overline{v^2}\cdot t^2 = \frac{3kT}{M}\cdot t^2,$$

where $\overline{v^2}$ is the mean square velocity, M the mass of the particle, T the temperature and k the Boltzmann-constant.

region: $t_1 < t < t_2$

$$\langle r^2(t) \rangle \approx \text{constant},$$

region: $t > t_2$

$$\langle r^2(t) \rangle = 6 \cdot Dt \quad ,$$

according to Einstein with D: diffusion constant.

The interesting question arises, how $\langle r^2(t) \rangle$ is connected to the microscopic properties of a many particle system. Before giving some hints to the answer of this question let us discuss another interesting point: How about the "dispersion" curves in liquid and amorphous structures?

3.1.2 Dispersion curves

Concerning the vibrational part one could be tempted to introduce quantities similar to the phonons in the solid state. However, there are two important limitations:

i) The lifetimes τ of these excitations are very short: $\tau \lesssim t_2$,

ii) There is no periodicity in the system.

The conclusions are:

i) No sharp energy- $E = \hbar\omega$ and q-values can be assigned to these excitations,

ii) No Brillouinzone can be defined, i.e. we do have excitations with wave vectors $q > \frac{2\pi}{a}$, where a is the nearest neighbor distance.

The question, whether a kind of blurred Brillouinzone due to the first peak in S(q) does still exist is yet undecided. Nevertheless, we can draw the following picture (Fig. 8).

Here the slope of the longitudinal mode for small q is given by the sound velocity $c_S = (B_S/\rho)^{1/2}$, where ρ is the density and B_S the adiabatic bulk modulus. For high q values we expect free particle behaviour, described by

$$c_f = \sqrt{\overline{v^2}} = \sqrt{kT/M}$$

The deviation from this straight line indicates collective behaviour.

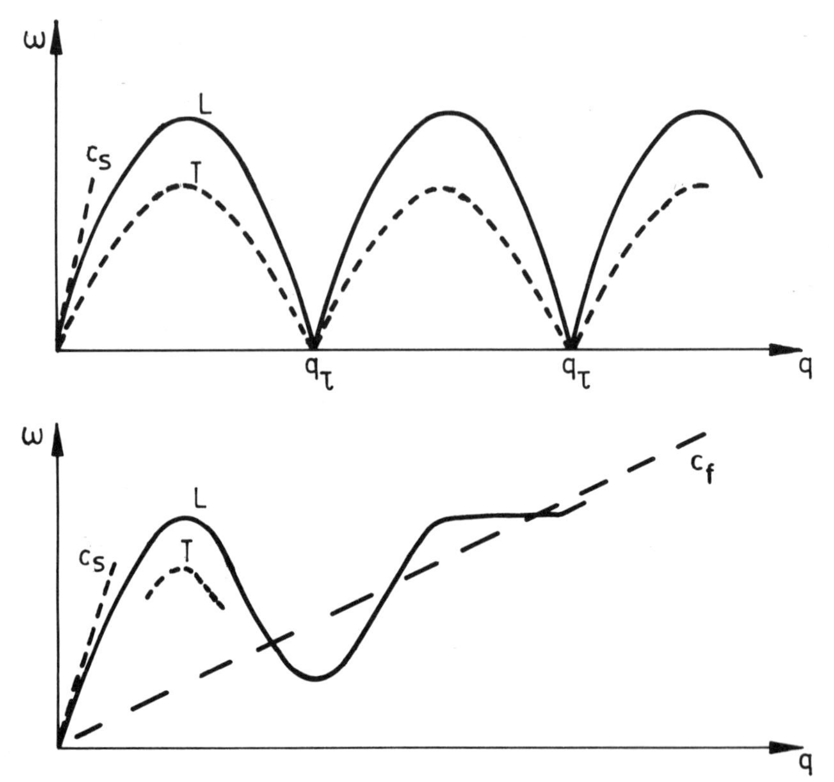

Fig.8. Dispersion curves in a crystalline solid and a liquid metal. L: longitudinal, T: transversal branch. q_τ: reciprocal lattice vector, c_s: velocity of sound, c_f: thermal velocity.

Concerning the transversal branch, we do not expect to find any collective excitations of this kind at low q, since the liquid supports no static shear. However, at higher q, there is no reason to believe that transversal modes will not be present.

The minimum in the $\omega(q)$ curve is reminescent of the solid state, but since long range order (LRO) doesn't exist any more there's no zero in $\omega(q)$. We do not attempt to discuss any theoretical model in this contribution (see talk by A. Sjölander and W. Götze).

3.1.3 Velocity-autocorrelation function /19/

We define a function $\Psi(t)$ as

$$\Psi(t) = \frac{\langle \vec{v}(o)\vec{v}(t)\rangle}{\langle v^2 \rangle} = \frac{M}{3kT} \langle \vec{v}(o)\vec{v}(t)\rangle .$$

The symbol $< >$ indicates an average over all particles of the system. $\psi(t)$ is called the velocity autocorrelation function because it correlates the velocities v of the same particle at times zero and t. From this definition the mean square displacement $< r^2(t)>$ is calculated by

$$<r^2(t)> = 2\cdot\int_0^t dt'\cdot(t-t')\cdot<\vec{v}(o)\vec{v}(t')> = \frac{6kT}{M}\int_0^t dt'\cdot(t-t')\Psi(t') \quad .$$

Next let us consider the Fourier-transform of $\psi(t)$ for the x-components of the velocity

$$\rho(\omega) = \frac{1}{2\pi}\cdot\int_0^{+\infty}<v_x(o)v_x(t)>\cdot\exp[-i\,\omega t]\cdot dt.$$

$\rho(\omega)$ is the analogue to the phonon-density of states in ω-space for a crystal. It can be shown /19/ using the behaviour of $<r^2(t)>$ for larger times and the above equation, that

$$D = \pi\cdot\rho(o) = \frac{1}{2}\cdot\int_0^{+\infty}<v_x(o)v_x(t)>\cdot dt = \frac{kT}{2M}\cdot\int_0^{\infty}\Psi_x(t)\,dt\,.$$

Hence, all the interesting quantities can be extracted, once $\psi(t)$ is known by some microscopic theory. Mori 1965 /21/ has given arguments how to include memory effects of the system under consideration into the autocorrelation function (see A. Sjölander's talk). He ends up with the equation

$$\dot{\psi}(t) = -\int_0^t \psi(t-t')\cdot K(t')\cdot dt'.$$

K(t) is called the memory function. Memories for diffusive motions are provided by longitudinal and transversal density fluctuations. We will not discuss the concept in any more detail here, but simply give a schematic picture of $\Psi(t)$. Without memory one would have simply $\Psi_x(t) = \exp[-t/\tau]$, with $\tau = 2MD/(kT)$. Hence: $K(t) = \delta(t)\cdot\frac{kT}{2MD}$. Deviations therefrom indicate memory (Fig.9).

Negative parts in $\Psi(t)$ can be interpreted as hard core effects, since the velocity at time t is just opposite to the one at time zero. Therefore oszillations around the average $\Psi(t)$ point towards collective vibrational excitations. The diffusion coefficient is determined by the net area under the curve (see equation above).

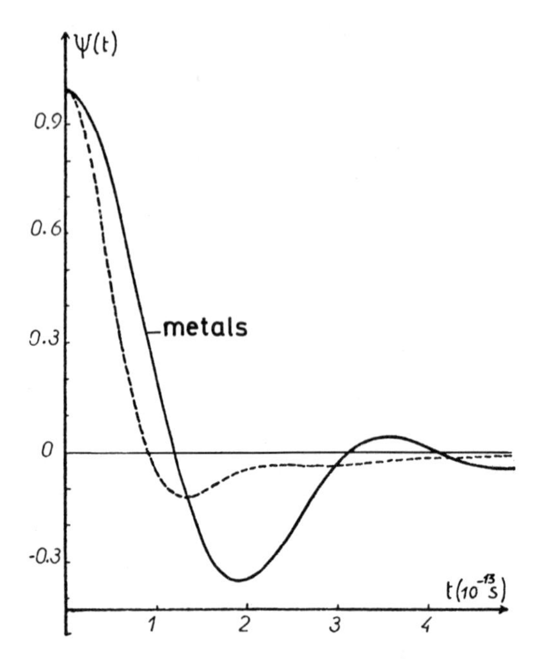

Fig.9. Schematical drawing of the normalized velocity auto-correlation function $\Psi(t)$ for metals and insulating liquids. (D.Schiff Phys.Rev. 106, 151 (1969)).

3.2 Experimental situation

3.2.1 Some remarks

Experimental information about the microscopic dynamics of metallic liquids is obtained mainly from inelastic neutron scattering. In addition valuable results can be extracted from molecular dynamics calculations. However, as will be shown, there is no way to evaluate the functions of interest $\langle r^2(t)\rangle$, $\Psi(t)$ or $K(t)$ directly from experiment. The procedure is as follows: Start from some first principle theory or some suitable parametrized physical model, predict the experimental results and compare them with the measurements. The advantage of the computer experiments is the fact, that those functions in question can be calculated directly from the computer "data". Hence, for testing theories they are of great importance. Not very much has been done concerning liquid alloys and amorphous metals. There are computer experiments on liquid Na-K alloys showing collective behaviour (see talk by G. Jacucci) and inelastic neutron scattering data from amorphous alloys (see talk by J.B. Suck).

In this section we will not discuss methods of measurements and data on the diffusion coefficient, this analysis is performed in Section VI. As can be seen from the behaviour of $\psi(t)$ the value of D is not very sensitive to the detailed microscopic dynamics. Some additional information on D is given in the talk by M. Gerl.

3.2.2 Incoherent neutron scattering

The scattering experiment is schematically performed in the following manner (Fig.10). A neutron beam is monochromatized by some means (crystal Bragg reflection, time of flight etc.) and then directed at the sample.

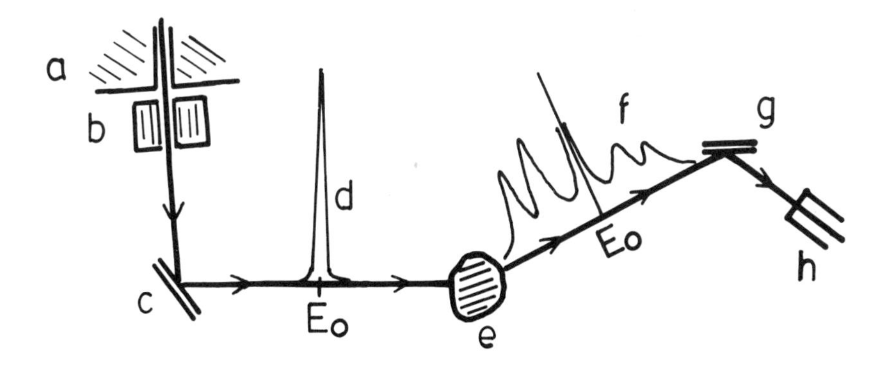

Fig.10 Principle set-up for inelastic neutron scattering. a: reactor, b: collimator, c: first monochromator, d: incoming neutron-energy distribution (E_0), e: specimen, f: scattered neutron-energy distribution, g: energy-analyser and h: counter.

The scattered neutrons are analysed in terms of scattering angle and energy. Energy transfer E and momentum transfer $\hbar q$ is defined by

$$E = \hbar\omega = |E_i - E_s| \qquad \hbar\vec{q} = \hbar\vec{k}_i - \hbar\vec{k}_s \quad ,$$

where the indices refer to: i: incoming and s: scattered beam, respectively. Detailed scattering geometry is much more complicated because of intensity and focussing considerations. The double differential scattering cross section $d^2\sigma/d\Omega d\omega$ is given by /15/:

$$\frac{\partial^2\sigma}{d\Omega d\omega} = \frac{1}{4\pi}\cdot\left(\frac{k_s}{k_i}\right)\{\sigma_{coh}\cdot S(q,\omega) + \sigma_{incoh}\cdot S_s(q,\omega)\} \quad ,$$

with σ_{coh}, σ_{incoh} the coherent and incoherent scattering cross-section /15/, respectively.

$S(q,\omega)$ is the total scattering factor, a generalization of the function $S(q)$:

$$S(q) = \int_{-\infty}^{+\infty} S(q,\omega)\cdot d\omega \quad .$$

It can be written:

$$S(q,\omega) = S_d(q,\omega) + S_s(q,\omega) \quad ,$$

where the index d indicates "distinct" and s "self". $S(q,\omega) = FT(g(r,t))$, (FT is the Fouriertransform), where $g(r,t)$ is the time dependent two particle correlation function. Making use of $g(r,t)$ a physical interpretation can be given for the meaning of the indices d and s. Whereas "d" describes the correlation of distinct particles within space and time, "s" correlates the movement of a given particle with its position at $t = 0$. Hence, s and d contribute to the collective excitations (coherent) and s only to the incoherent movements.

The physics of the system is burried in the behaviour of $S_d(q,\omega)$ and $S_s(q,\omega)$. Both contributions may be separated experimentally, since their relative magnitude depends critically on the location in q-space where the experiment is performed.

Let us first analyse the self-part of the scattering function $S_s(q,\omega)$. As only single particle motion is involved, it is easier to handle than the collective behaviour. Assuming for $g_s(r,t)$ a simple diffusive form, one can write immediately (Vineyard 1958) /22/:

$$g_s(r,t) = \{6\pi\cdot\langle r^2(t)\rangle\}^{-3/2}\cdot \exp\{-r^2/(6\langle r^2(t)\rangle)\} \quad ,$$

this expression being exact for $t = 0$ and for large t.

In the latter case we have: $\langle r^2(t)\rangle = 6\cdot Dt$.

However, this expression is assumed to be a good approximation in the whole time scale. Thus, we have:

$$S_s(q,\omega) = \frac{1}{\pi}\int_0^{\infty} dt\cdot\cos\omega t\cdot\exp\left[-q^2\cdot\langle r^2(t)\rangle /6\right] \quad .$$

Now we need a model for $\langle r^2(t)\rangle$ in order to evaluate $S_s(q,\omega)$. As a simple example we propose (Cocking 1969 /23/):

$$\langle r^2(t)\rangle = \langle r^2_{diff}(t)\rangle + \langle r^2_{bound}(t)\rangle \quad ,$$

$$\text{with} \quad \langle r^2_{diff}(t)\rangle \quad \begin{cases} \sim t^2 & \text{for } t \to 0 \\ \sim t & \text{for } t \to \infty , \end{cases}$$

$$\text{and} \quad \langle r^2_{bound}(t)\rangle \quad \begin{cases} \sim t^2 & \text{for } t \to 0 \\ = \text{const.} & \text{for } t \to \infty . \end{cases}$$

The model yields:

$$S_s(q,\omega) = \frac{C}{\pi} e^x \cdot \sum_{n=0}^{\infty} \frac{(-1)^n x^n (n+x)}{n! \{(n+x)^2 + (\omega C)^2\}} \quad ,$$

where $x = D \cdot C \cdot q^2$ and C is a constant.

If we take as a crude first approximation only the n = 0 term

$$S_s(q,\omega) = \frac{C}{\pi} \frac{x}{x^2 + \omega^2 C^2} \quad ,$$

$S_s(q,\omega)$ shows a peak in a scan q = const as a function of ω. The half width, $\Delta\omega$ at half height, of this peak is:

$$\Delta\omega = D \cdot q^2 \quad .$$

Measuring $S_s(\omega, q=\text{const})$ for various values of q gives the diffusion coefficient. An example is given in Fig.11.

For comparison we mention a recent work on solid sodium /24/. We see, that incoherent scattering yields $S_s(q,\omega)$ and allows a test of theoretical models for K(t), Ψ(t) and the diffusion coefficient D.

3.2.3 Coherent neutron scattering

Besides the function $S_s(q,\omega)$ in addition $S_d(q,\omega)$ is involved in the coherent scattering problem. On the other hand the full correlation within the system is available in principle.

We will not treat here the complicated theory, but simply give some results. Collective excitations show up as peaks in $S(q,\omega)$ /25/, f.e. when a ω-run at q=const is made. The width of such a peak is correlated with the life time of these excitations. It is however, always modified by instrumental resolution.

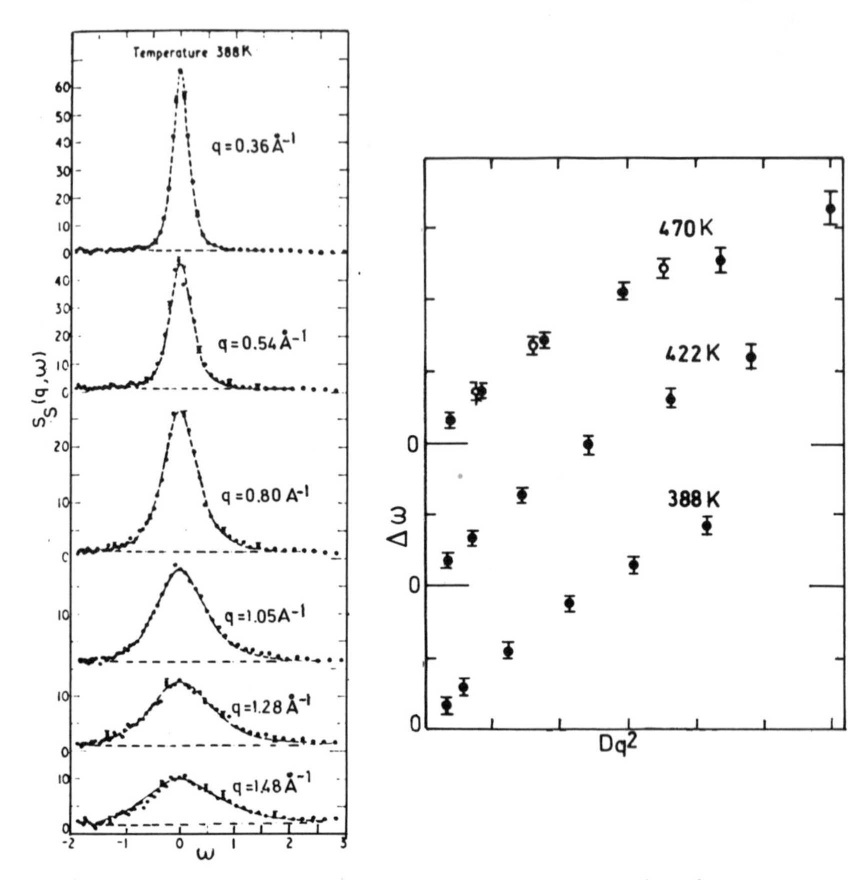

Fig.11. The self-dynamical structure factor $S_S(q,\omega)$ for Na at 388 K at several values of the wave vector q together with the half width of this quasi-elastic peak versus Dq^2. (S.J. Cocking J.Phys. C 2, 2047 (1969)).

It is essential to realize that only longitudinal excitations can be determined, because no reciprocal lattice vector can be defined (lack of LRO) and because of the factor $(\vec{q}\vec{e})^2$ in the scattering cross section /26/. $\vec{e}$ is the polarisation vector. There are some discussions going on, to whether a sharp first peak in S(q) may act as a kind of reciprocal lattice vector. This in turn could permit to get hold of transversal excitations in the "second Brillouin zone". Some results are given in Figs. 12 and 13.

As one can see, "dispersion curves" may be derived, if collective peaks exist. Metallic liquids like Na, K, Pb and Cs do show these excitations. They are not seen in substances like liquid Ar.

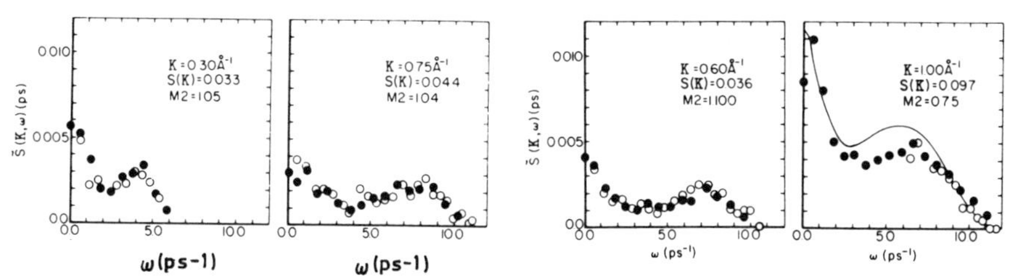

Fig.12 Comparison between experimental dynamical structure factors $S(q,\omega)$ and MD-computer calculations for liquid Rubidium (J.R.D. Copley, J.M. Rowe Phys.Rev.Letters 32, 49 (1974) and A. Rahman Phys.Rev.Letters 32, 52 (1974)).

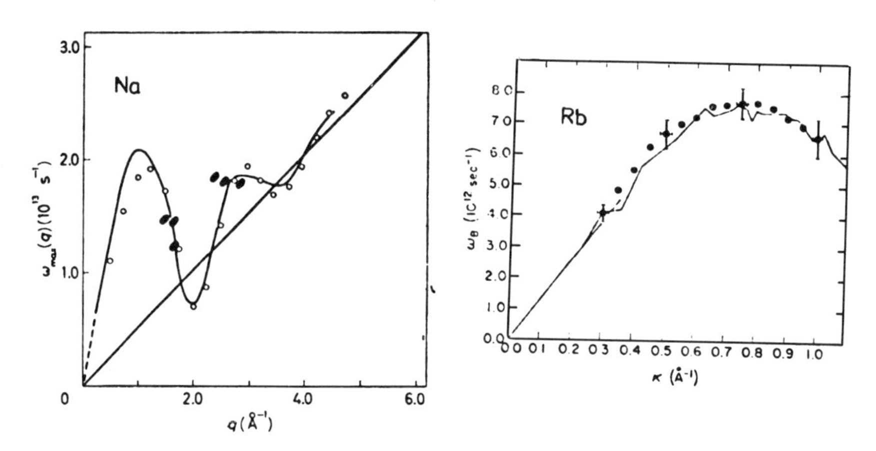

Fig.13 Dispersion-relation for liquid Rb and Na. The dashed line corresponds to the velocity of sound. Full lines: Na: direct calculation from a potential (also open dots) Rb: Computer calculation by Rahman. Full dots: experimental values. (A. Rahman Phys.Rev.Lett. 32, 52 (1974) and G. Cubiotti, K.N. Pathak, K.S. Singwi , M.P. Tosi, Lett. Nuov.Cim. 4, 799 (1970)).

3.4.2 Computer experiments

It has already been discussed, that time dependent properties can be derived from molecular dynamics (MD) calculations. From the positions of the particles a direct evaluation of $S(q,\omega)$ is possible. Hence, all the information contained in this function can be extracted and compared either with theoretical calculations or with experimental results. A. Rahman's results (1974) /27/ are also reproduced in Fig. 12.

4. ELECTRONIC PROPERTIES

4.1 Introductory remarks

Thus far, we have only considered the ionic system. The electrons entered the whole picture only through the potential and through the function $u(n_0)$ in the thermodynamic free energy. In crystalline systems the electrons can be described by Bloch waves with a well defined $\vec{k}$-number. If the pseudo-potential formalism is applied, plane waves and hence the free electron-gas model is a good description of the situation. However, in a disordered system it is by no means clear that plane waves should be a good approximation of the behaviour of the electrons. The quantity $\vec{k}$ possesses no longer a sharply defined value. In addition, the mean free path of an electron is extremely short. Values of only a few average atomic distances are common. Having this in mind it's very surprising, that the nearly free electron model (NFE) with effective masses represents a satisfactory description of the electron system of many metallic liquids. This conclusion was born out by experimental facts and awaits theoretical support.

4.2 Density of states

In the NFE-model the density of states $n(E) = dn_0/dE$ is given by the well known equation:

$$n(E) = \frac{4\pi}{h^3} \cdot (2\, m_{eff})^{3/2} \cdot \sqrt{E} \quad ,$$

where n_0 is the number of electrons per volume, h Planck constant, m_{eff} the effective mass of the electrons and E their energy ($n(E)$: density of states per energy and volume).

The picture is complicated by the following facts:

i) Many-electron effects /18/ must be included in this one-electron picture. This may change the E-behaviour as indicated in Fig. 14.

ii) The periodic structure of crystalline solids gives rise to the band structure of the electronic system, due to electron-ion interaction. Since the first peak in $S(q)$ is an indication of short range order (SRO), one might expect a dip in the density of states at k_P (position of the first peak in reciprocal space).

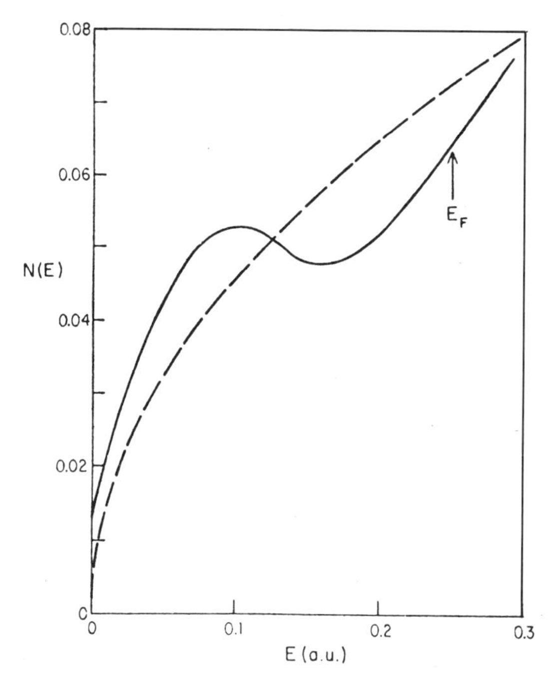

Fig.14. Comparison of calculated density of states with the nearly free electron model prediction (dashed) for liquid Hg at 20 ^{o}C. (K. Ichikawa Phil.Mag. 27, 177 (1973)).

iii) In many alloys and liquid metals d electrons are simultaneously present. These electrons are contained in narrow bands with high densities of states $n_d(E)$. Hence, an appreciable change in the total density of states will occur. If it happens that such a d-band coincides with the Fermi-energy, it is not clear at all how many d- or s-electrons are present (see talk by H. Beck).

4.3 Experimental situation

There is a variety of methods which permits in principle to measure directly the density of states. Since they all determine transitions between two electronic states, the observed intensity (f.e. electrons in photoemission studies, see below) is proportional to

$$\sum_{i,f} |\langle f|\hat{H}|i\rangle|^2 \cdot \delta(E_f - E_i - \hbar\omega) \cdot \delta(E - E_f) \quad ,$$

where the indices refer to i: initial state and f: final electron state, resp..

If no indirect transitions are present - a good approximation in disordered systems - the matrix element $|\langle f|H|i\rangle|^2$ is only weakly energy dependent. In this case, we get for the intensity

$$\sim n\,(E - \hbar\omega)\cdot n\,(E) \quad ,$$

where n(x) is the density of states as defined above.

Methods to be used are /29/:

i) Soft X-ray emission: Here, an electron of the conduction band jumps back into a core hole, created i.e. by electron impact or X-ray absorption. In many cases the L-states are used as final states. An example is given in Fig. 15.

ii) Photoemission: The number of emitted electrons is determined as a function of the energy of the incident light beam. Since the work function is important, a very clean surface is absolutely necessary. Fig. 16 gives some experimental evidence.

iii) Ultraviolett Photoemission Spectroscopy (UPS):
In this method monochromatic ultraviolet light is incident at the sample. The number of emitted electrons is measured as a function of $\hbar\omega$. An example is reported in Fig. 17. See also talk by P. Oelhafen. There the first results on metallic glasses at high temperatures are given.

4.4 Optical properties

The optical properties of a metallic system are defined by its complex refraction index $\hat{n}(\omega)$. This quantity can be related in turn to the dielectric constant $\varepsilon(\omega)$ and to the frequency dependent electrical conductivity $\sigma(\omega)$ by /30/:

$$\hat{n}(\omega) = (\varepsilon(\omega) + i\,\sigma(\omega)/(\omega\,\varepsilon_0))^{1/2} \quad ,$$

with ε_0 the absolute dielectric constant.

For the NFE-model $\sigma(\omega)$ can be derived to be (Drude, Sommerfeld)

$$\sigma(\omega) = \sigma_0 \cdot \frac{1+i\omega\tau}{1+\omega^2\tau^2} \, ,$$

where τ describes the collision-time of the electrons and σ_0 is the ordinary zero frequency conductivity (DC).

$$\varepsilon(\omega) = 1 + \frac{n_a\alpha}{\varepsilon_0} - n_0\, e^2/(m_{eff}\,\varepsilon_0\,\omega^2).$$

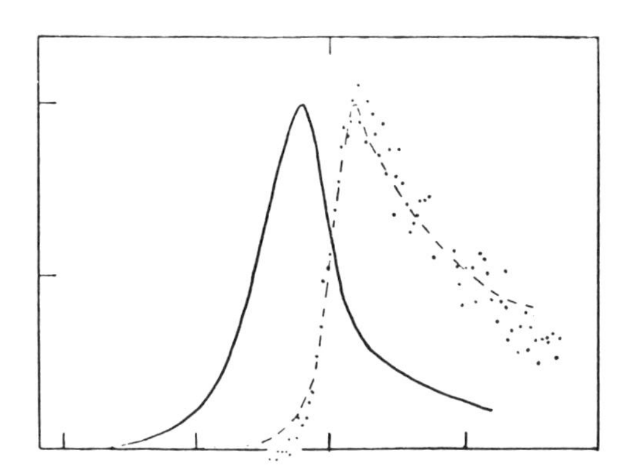

Fig.15. A comparison between various soft X-ray data for liquid Cobalt. Solid line: L_3-emission curve at 2.5 keV, dotted line: L_3 self-absorption data and dashed: thin film absorption curves. (N.H. March Can.J.Chem. 55, 2165 (1977)).

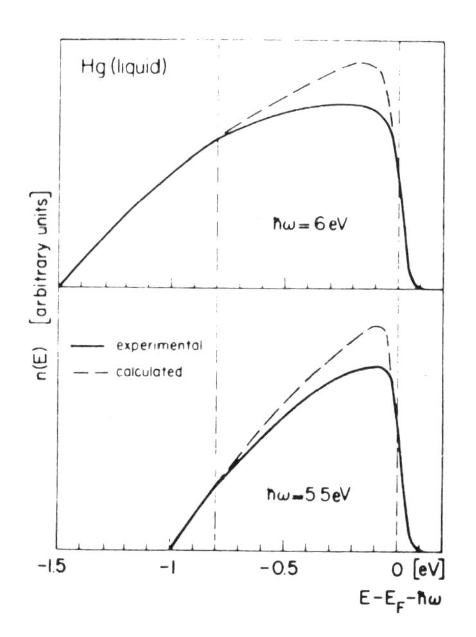

Fig.16. Comparison of calculated and measured electron energy distribution curves for liquid Hg. No dip is visible, compare Fig. 14! Solid line: experiment, dashed: calculated. (N.H. March Can.J.Chem. 55, 2165 (1977)).

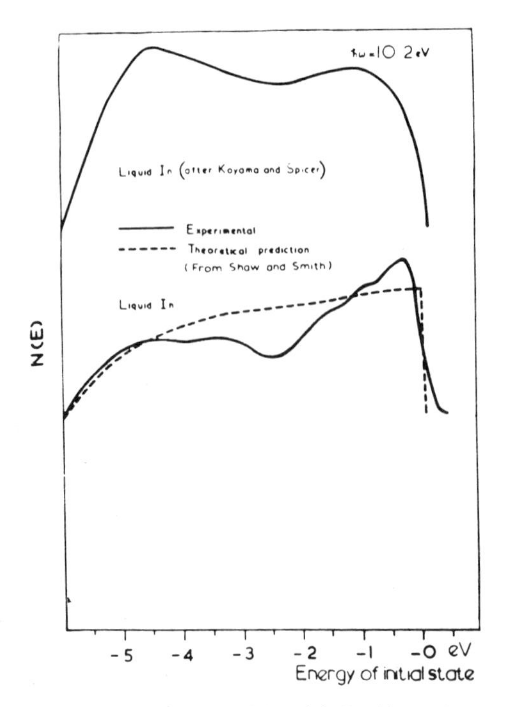

Fig.17. Density of states for liquid Indium from experiment and theory. Solid lines: experimental, dashed line: theoretical prediction. (N.H. March Can.J.Chem. 55, 2165 (1977)).

Here $n_a\alpha/\varepsilon_0$ is the ionic contribution, with n_a ionic density and α the core-polarizability of the ions. n_0 is the electron density. The third part of the equation above describes the behaviour of the NFE. It can also be written $-\omega_p^2/\omega^2$ with ω_p the plasma-frequency. Hence, $\varepsilon(\omega)$ doesn't contain very interesting information. Let us therefore concentrate on $\sigma(\omega)$.

In the spirit of the NFE-model the quantity τ in $\sigma(\omega)$ can be expressed in terms of the static structure factor S(q) and the imaginary part of the electronic screening function. Thus, $\sigma(\omega)$ can be calculated inserting the measured S(q) (or using the PY result) and compared with experiment. If however, the NFE-model is not adequate the situation is much more complicated /29/.

Experimental determinations of $\sigma(\omega)$ can be made by ellipsometric measurements at free metallic surfaces. Extremely clean surfaces are necessary. Some results for liquid Na are given in Fig. 18 together with the prediction of the simple NFE-model (m_{eff} = 1.17 m_0). The small deviation above 2.2 eV is attributed to some interband absorption.

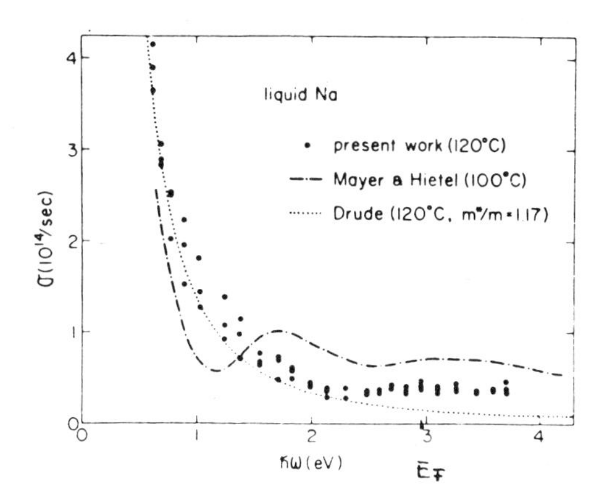

Fig.18. High frequency electrical conductivity $\sigma(\omega)$ of liquid Na. The Fermi-energy is indicated. (N.H. March Can.J.Chem. 55, 2165 (1977)).

4.5 Magnetic Properties

The magnetic properties of metallic systems may be described by the magnetic susceptibility χ. The total susceptibility is the sum of three contributions:

$$\chi = \chi_i + \chi_p + \chi_d \quad ,$$

where χ_i is the ion-core contribution, χ_p the electronic spin susceptibility and χ_d the electronic orbital susceptibility. Since one always measures the total susceptibility, the quantity χ_i has to be taken from some theoretical treatment and substracted from the data. Within the limits of the NFE-model one has (Pauli, Landau) /30/ for the volume susceptibility:

$$\chi_{po} = 3/2 \cdot n_o \mu_B^2 \cdot \mu_o/E_F \quad \text{and} \quad \chi_{do} = -\chi_p/3.$$

μ_o: absolute permeability, μ_B Bohr's magneton and E_F Fermi-energy.

Both of these quantities should be modified in order to take the electron-electron interaction into account /29/. Hence, it seems that the magnetic susceptibility is very sensible to small deviations from the NFE-model. χ_p can be measured directly by conduction electron spin resonance techniques, as indicated in Fig. 19.

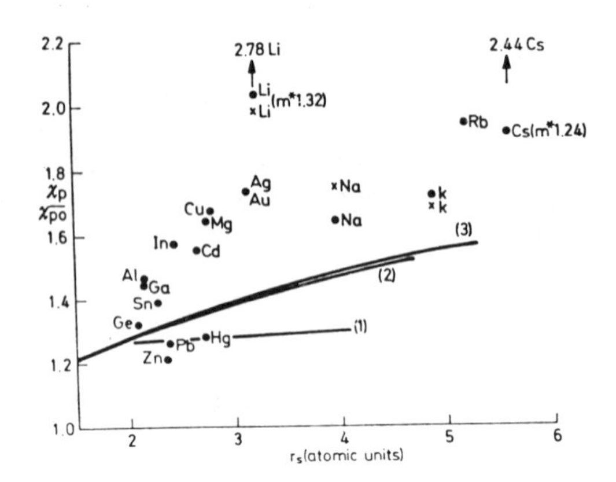

Fig. 19. Spin susceptibilities of liquid metals, normalized to the Pauli value χ_{po} with $m_{eff}=m_e$; crosses: ESR-technique, solid lines: calculated from Landau-Fermi-liquid model with different choices of the parameters. (N.H. March Can.J.Chem. 55, 2165 (1977)).

Table III. Experimental and theoretical single OPW Knight-shifts at the melting points /29/.

	$K_{exp.}$	$K_{theor.}$
Li	0.026	0.09
Na	0.116	0.15
Cu	0.264	0.52
Ga	0.449	0.53
In	0.786	0.79
Te	0.38	0.70
Bi	1.41	1.32

Another more indirect method is to determine the Knight shift K by nuclear magnetic resonance. Theoretically K is given by:

$$K = \frac{8\pi}{3} \cdot v_m \cdot P \cdot \chi_p \quad ,$$

where v_m is the atomic volume and P describes the density of conduction electrons at the atomic nucleus. From this expression χ_p can be derived, once $v_m \cdot P$ is known. There are several publications dealing with this problem /31/. However, one finds that whereas χ_p changes quite a bit at melting, K does not. "Unfortunately the Knight shifts remain difficult to interpret in a really convincing way" /29/. Results at the melting temperature for liquid metals are summarized in Table III.

4.6 Superconductivity

It is clear that liquid metals are no superconductors since the temperature is much to high. However, metallic glasses can be cooled down to low temperatures. Many of them show superconductivity. A summary of the transition temperatures is given in Table IV.

Table IV. Superconducting transition temperatures of some metallic glasses.

Liquid quenched	T_c/K
$La_{80}Au_{20}$	3.5
$La_{80}Ga_{20}$	3.8
$La_{70}Cu_{30}$	3.5
$La_{78}Ni_{22}$	3.0
$Zr_{70}Pd_{30}$	2.4
$Zr_{65}Cu_{35}$	2.0
$Zr_{60}Cu_{40}$	1.6
$Zr_{70}Co_{30}$	3.3
$Zr_{70}Be_{30}$	2.8
$Zr_{75}Rh_{25}$	4.5
$Nb_{60}Ni_{40}$	1.5
$Nb_{58}Rh_{42}$	4.7

Liquid quenched	T_c/K
$(Mo_{80}Re_{20})_{80}P_{10}B_{10}$	8.7
$Mo_{80}P_{10}B_{10}$	9.0
$(W_{60}Ru_{40})_{80}P_{20}$	4.3
sputtered, co-evaporated	
Nb_3Si	3.9
$Mo_{\approx 80}N_{\approx 20}$	8.3
$Mo_{68}Si_{32}$	6.7
Nb_3Ge	3.6
$Re_{70}Mo_{30}$	7.6

(P. Duwez, W.L. Johnson, J.Less-Common Met. 62, 215 (1978), W.C. Johnson J.Appl.Phys. 50, 1557 (1979), J. Willer, G. Fritsch, E. Lüscher, to be published).

Since these metallic glasses are alloys, the mean free paths of the electrons are very short. They are in the order of several mean atomic distances. Therefore the correlation length is also short and we are in the limit of a so-called dirty superconductor. Many of these alloys show pronounced conductivity fluctuations above T_c because of the short correlation length. As transition metals are often included, the role of the d-electrons has to be considered. More details are given by the talks of K. Bennemann and S. Methfessel.

5. THERMODYNAMIC PROPERTIES

5.1 Some remarks

In this section some experimental results for thermodynamic properties are discussed. In principle most of them can be calculated by the thermodynamic models discussed in Section 2.1.2.. Using the Helmholtz free energy F(T,V), the compressibility κ_T and the specific heat c_v may be evaluated by considering second derivatives with respect to V and T. The same is true for the sound velocities once the corresponding deformations are introduced in F. The density is related to the equation of state P(V,T) which in turn is the first derivative of F with respect to V. However, the situation is much more complicated in alloys, since SRO and changes of the binding types play an important role (see talk by J. Hafner).

5.2 Density and Expansivity

The density of amorphous alloys can be determined directly from geometry and weight. Densities of liquid metals or alloys are not so easily accessible, especially at higher temperatures. The maximum bubble pressure technique is often applied /32/. An example is given in Fig. 20. The upper curve of this Figure shows the thermal expansivity $\alpha_p = 1/V \cdot (dV/dT)_p$. Whereas the expansivity is a straight line as a function of the composition, the specific volume $v = 1/\rho$ (ρ: density) indicates clearly the nonideal character of the alloy.

Empirically /33/ the density can be fitted to a power series:

$$\rho = \rho_0 + AT + BT^2 + CT^3 + \dots \quad ,$$

ρ_0 being some normalization value (here at T = 273 K).

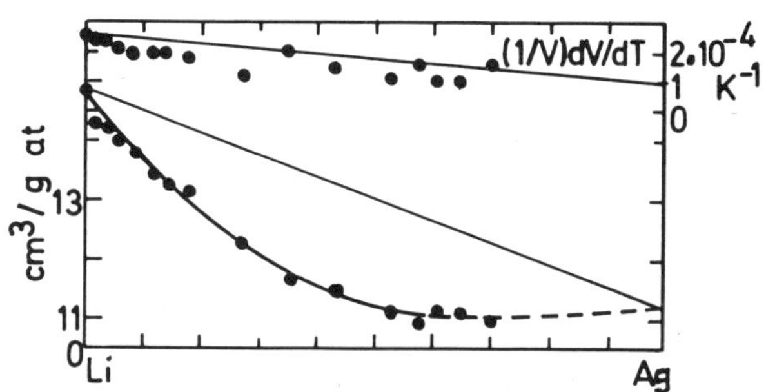

Fig.20. Molare volume and expansivity as a function of composition for a liquid LiAg alloy at 600 ^{o}C. (H. Reiter, H. Ruppersberg, W. Speicher, Z.Naturf. 31A, 47 (1976)).

Therefrom one gets:

$$\alpha_P \approx \frac{1}{V_o}\cdot\left(\frac{dV}{dT}\right)_P = -\frac{A}{\rho_o} - 2\left(\frac{B}{\rho_o} - \frac{A^2}{\rho_o^{\,2}}\right)T - 3\left(\frac{C}{\rho_o} - \frac{AB}{\rho_o^{\,2}} + \frac{A^3}{\rho_o^{\,3}}\right.T^3 \ldots =$$

$$= a_o + a_1 T + a_2 T^2 + \ldots \quad .$$

Table V gives some results for liquid Na and K, according to Hoch et al /33/:

Table V. Empirical constants for a power series equation of the thermal expansivity.

	a_o/K^{-1}	a_1/K^{-2}	a_2/K^{-3}
Na	$2.3211\cdot10^{-4}$	$4.0465\cdot10^{-8}$	$-1.7802\cdot10^{-11}$
K	$2.3930\cdot10^{-4}$	$7.3294\cdot10^{-8}$	$-1.6588\cdot10^{-11}$

5.3 Specific heat

At high temperatures the isobaric specific heat may be written for a crystalline solid

$$c_{P,solid} = 3R\cdot D\left(\frac{\Theta_D}{T}\right) + bT + A(T) \quad .$$

The first term is the harmonic lattice contribution with R: gas constant, D(x): Debye function and Θ_D: Debye temperature. The second one describes the electronic contribution and the third corresponds to anharmonic effects.

Having in mind the picture of a liquid we developped in Section 3, one is tempted to try a similar expression for the liquid state. The third contribution however, has to be modified then, since anharmonic effects are large. In addition specific heat will be needed to populate the new degree of freedom in a liquid: translation. As less energy is necessary for translation ($\frac{3}{2}$ RT for free particles) than for vibration (3 RT in the harmonic case), one has to take into account a complicated balance. Empirically we may write /34/:

$$c_{P,\text{liquid}} = 3\ R\cdot D(\frac{\Theta_D}{T}) + b'T + d/T^2 \quad .$$

In checking this equation let us plot $(c_{P,\text{liquid}}-3RD)/T$ versus $1/T^3$ in order to get a straight line. The result /35/ is reproduced in Fig. 21.

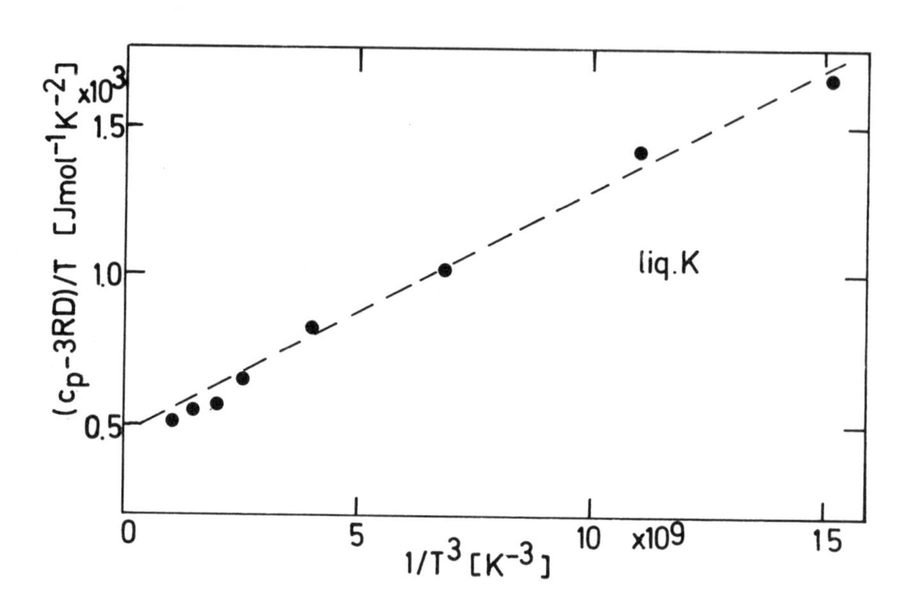

Fig.21. The quantity $(c_p-3RD(\Theta_D/T))/T$ is plotted versus $1/T^3$ in order to check the semi-empirical equation given in the text. cp: isobaric specific heat of liquid potassium. (T.B. Douglas, A.F. Ball, D.C. Ginnings, W.D. Davis, J.Am.Chem.Soc. 74, 2472 (1952)).

Table VI contains some data derived from similar fits to experimental data.

Table VI. Some constants derived from a semi-empirical equation, fitted to experimental results (symbols see text).

	Θ_D/K	T_M/K	b'/(J/(mole·K^2))	d/(J·K/mole)
Li	344	453.7	$3.1\cdot10^{-3}$	$1.08\cdot10^{6}$
Na	158	371	$3.1\cdot10^{-3}$	$1.08\cdot10^{6}$
K	91	336.4	$4.7\cdot10^{-3}$	$0.87\cdot10^{6}$
Pb	105	600.4	$1.8\cdot10^{-3}$	$2.66\cdot10^{6}$

-Fig. 22 should give the reader an impression of the large variety of the specific heat behaviour.

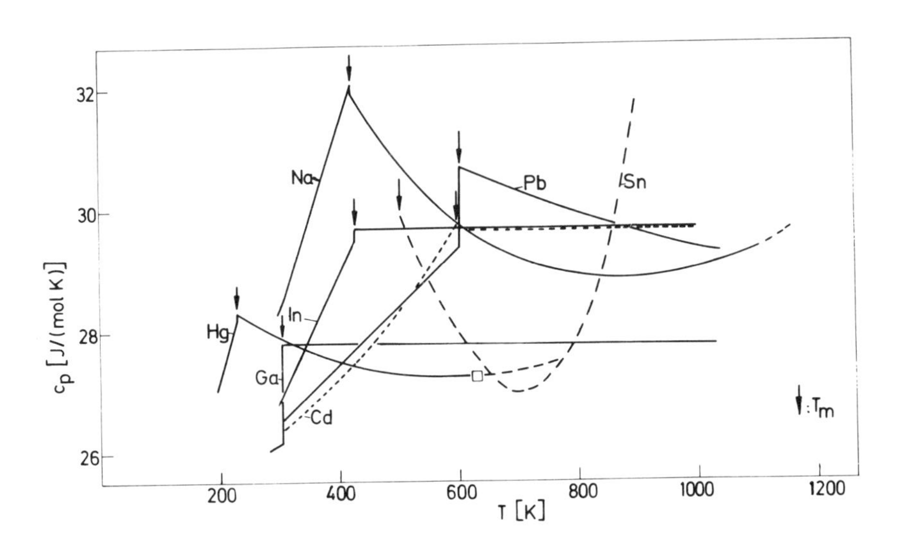

Fig.22. Isobaric specific heat for several pure molten metals. The melting point is indicated by an arrow. The rectangle refers to the boiling point.

Experimental set-ups for the determination of the specific heat are described in detail in literature: For a compilation see Komarek 1978 /36/. In addition enthalpies of mixing of alloys can be measured up to 1800 K. This quantity has been mentioned in Section 2.2 and will be discussed further in Section 5.5.

5.4 Compressibilities and sound-velocities

The compressibilities can be measured either isothermally, yielding $\varkappa_T$ or dynamically, giving $\varkappa_S$ the adiabatic compressibility. Both are related by an expression which can be calculated in principle from the equation of state. Since static measurements are difficult (container problems etc.) mostly sound velocity measurements are used to derive $\varkappa_S$. The relevant equation for an isotropic system is:

$$\varkappa_S = 1/(\varrho v_s^2) \quad ,$$

where ϱ is the mass density and v_S the longitudinal velocity of sound. Transversal sound waves are only found at very high frequencies ($\nu \gg 1$ GHz) in liquids. Not very much is known about the behaviour of the sound velocities in amorphous alloys.

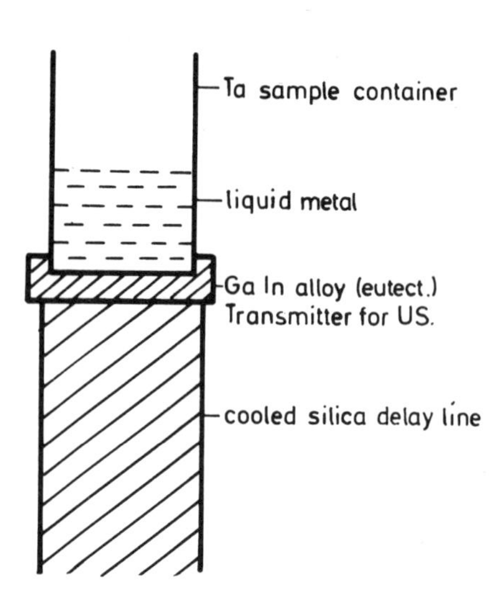

Fig.23. Experimental set-up for measuring velocities of sound of liquid metals at high temperatures.

Before discussing the compressibilities in more detail, let us examine how the measurement of the sound velocities can be performed. Normally an ultrasonic technique is used (pulse-echo-method). The set-up /37/ is shown schematically in Fig. 23. Since ultrasonic transducers fail at temperatures above 500 °C, a cooled delay line must be inserted.

Some experimental results for liquid Ba, Sr, Mg and Ca are reproduced in Fig. 24.

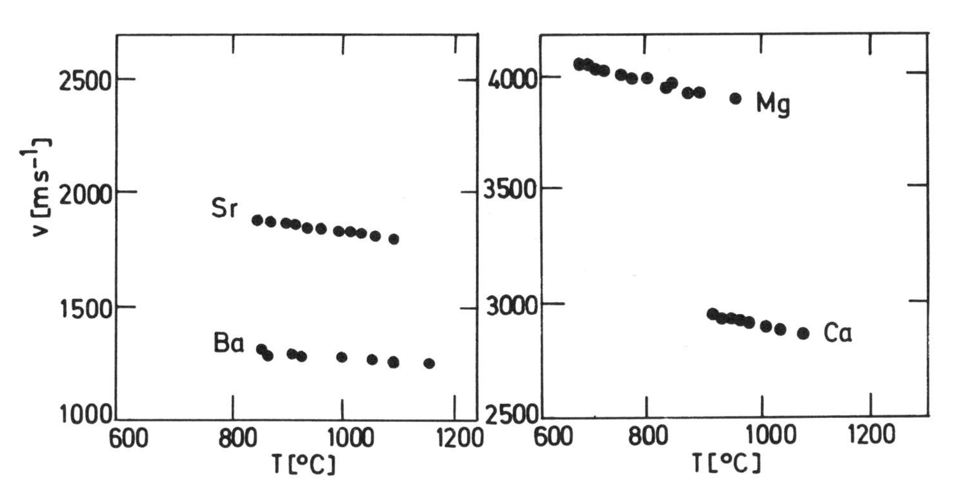

Fig.24. Temperature variation of the sound velocity in liquid Magnesium, Calcium, Strontium and Barium. (S.P. McAlister, E.B. Crozier, J.F. Cochran, Can.J. Physics 52, 1847 (1974)).

In an effort to explain the compressibility, quite a lot of theoretical models have been proposed in literature /37/. We would like to offer some of them and to compare their predictions with experimental results (see also talk by R. Evans):

i) In the spirit of the jellium model and the NFE model Bohm and Staver derived:

$$\varkappa_{BS} = 3/(2n \cdot E_F) \quad ,$$

where jellium means a positive smeared-out background, stemming from the ions, n is the electron number density and E_F the Fermi energy.

ii) Including effects from electron exchange and correlation in this model, yields (Overhauser /38/):

$$\varkappa_0 = \varkappa_{BS} \cdot (1 - (1+\alpha)/(\pi a_0 k_F))^{-1}$$

with $\alpha \approx 0.10$ for metallic densities, a_0: Bohr radius and k_F the Fermi wave vector. Models i) and ii) do only contain the electronic contributions.

iii) From a hard sphere fluid (see Sect. 2.1.2 using the PY-solution) one derives:

$$\varkappa_{HS} = \frac{v_m}{RT} \cdot (1-\eta)^4/(1+2\eta)^2 \quad ,$$

where η is the packing density ($\eta \approx 0.45$ at T_M) and v_m the molare volume.

iv) Improvement over $\varkappa_{HS}$ can be gained by using the Carnahan-Starling /4/ equation of state:

$$\varkappa_{CS} = \frac{v_m}{RT} \cdot (1-\eta)^4/[(1+2\eta)^2 - \eta^3(4-\eta)] \quad .$$

v) In versions iii) and iv) no direct effects from the electron gas are included, this draw-back is overcome in the next two expressions given by Ascarelli /39/:

$$\varkappa_{A,1} = \left[\frac{RT}{v_m} \cdot \left(\frac{(1+2\eta)^2}{(1-\eta)^4} - \frac{4}{3}\frac{1+\eta+\eta^2}{(1-\eta)^3} + \frac{2}{15}\frac{ZE_F}{k_B T}\right)\right]^{-1} \quad ,$$

where Z is the valence, E_F the Fermi energy and k_B the Boltzmann-constant. This equation was calculated in applying the PY-result for hard spheres.

The following expression for $\varkappa$ uses instead of the Carnahan-Starling modification:

$$\varkappa_{A,2} = \left[\frac{RT}{v_m} \cdot \left(\frac{(1+2\eta)^2 - \eta^3(4-\eta)}{(1-\eta)^4} - \frac{4}{3}\frac{1+\eta+\eta^2-\eta^3}{(1-\eta)^3} + \frac{2}{15}\frac{ZE_F}{k_B T}\right)\right]^{-1} .$$

Table VII. Comparison between the compressibilities ($\cdot 10^{-11}$ m^3/J) calculated from various approximations. Experimental values of the adiabatic and the isothermal compressibility, the Grüneisenparamter γ, the density and the isobaric expansivity are included (at T_M).

	$\rho/\cdot 10^3 (kg/m^3)$	$\alpha_p/\cdot 10^{-5} K^{-1}$	γ	κ_{BS}	κ_0	κ_{HS}	κ_{CS}	$\kappa_{A,1}$	$\kappa_{A,2}$	κ_S^{Exp}	κ_T^{Exp}
Mg	1.590	1.665	1.33	1.78	3.54	5.05	5.59	4.10	4.35	3.81	5.06
Ca	1.369	1.62	1.33	5.26	13.8	8.05	8.84	8.30	8.95	8.27	11.0
Sr	2.375	1.10	1.13	7.71	23.3	10.4	11.4	11.0	11.9	11.6	13.1
Ba	3.23	0.825	1.05	9.32	30.6	12.8	14.0	13.6	14.7	17.0	17.9

Table VII gives a summary of calculated values and experimental results. Expansivity and density are included for completeness. All data refer to the melting point. Since the bulk moduli $1/\varkappa$ have to be added, the electron and the hard core part contribute to $\varkappa$ in equal manner.

In order to illustrate, how the compressibility varies, when the composition is changed, we give results /40/ for the adiabatic compressibility of LiAg alloys. Fig. 25 shows the relevant data, taken at 600 °C by sound velocity measurements.

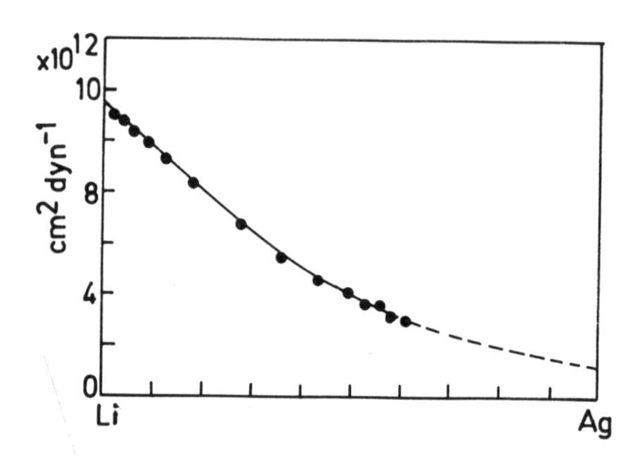

Fig. 25. Variation of the adiabatic compressibility of a liquid LiAg alloy at 600 °C with composition.
(H. Reiter, H. Ruppersberg, W. Speicher Liquid Metals 1976, edts. R. Evans, D.A. Greenwood).

5.5 EMF-Measurements

Important thermodynamic information on alloy systems can be obtained by measuring electro-motoric forces (EMF's) in especially deviced concentration cells.These cells use an electrolyt, a pure electrode - constructed from a component of the alloy - as well as an alloy electrode. Then the behaviour of the pure element in the alloy can be derived. Depending on temperature- and concentration-range, the situation may be more complicated, because of chemical reactions of parts of the cells with one another.

The chemical potential μ_i of a pure component in an alloy is given by /41/

$$\mu_i = RT \cdot \ln a_i + \mu_i^o \quad ,$$

where a_i is the activity (an effective concentration, equal to the ratio of the partial pressure of the i-th component to the total pressure), R the gas constant and μ_i^0 the chemical potential of the pure substance.

The activity a_i can be written as:

$$a_i = \gamma_i \cdot c_i$$

with γ_i the activity coefficient and c_i the mole per mole concentration. Since the equality $a_i = c_i$ is true for a regular solution, a deviation of γ_i from 1 describes non-ideal behaviour.

The activity can be obtained from EMF measurements as follows:

$$E_i = E_0 - \frac{RT}{F_0} \cdot \ln a_i \quad ,$$

where E_0 is a constant and F_0 indicates the Faraday-constant.

Knowing the concentrations of the components (i.e. in a two component system) as well as the activity coefficients, the Gibbs free enthalpy of mixing ΔG may be calculated by:

$$\Delta G = RT \cdot (c_1 \cdot \ln \gamma_1 + c_2 \cdot \ln \gamma_2) \quad .$$

Therefrom the entropy of mixing can be evaluated together with the quantity $A = RT \cdot (\partial^2 \Delta G / \partial c_i^2)^{-1}$.

Some experimental data on the system LiPb are reproduced in Figs. 26 and 27 /42/.

Large changes in those quantities around $c_{Pb} \approx 0.2$ to 0.3 are clearly recognized. Since the electrical resistivity also shows a deep minimum in this range, these data indicate ionic or covalent character of the alloy (charge transfer) (see also talk by R. Hensel).

It is interesting to compare these findings with the results of neutron scattering /11/ in the limit $q \to 0$ (see Sect. 2.2). Fig. 28 gives this comparison. The agreement of the two methods is quite convincing.

One can ask the question: How many two-component systems do show such anomalies, i.e. deviations from regular solution behaviour? Fig. 29, taken from Komarek's work shows what is known up to now.

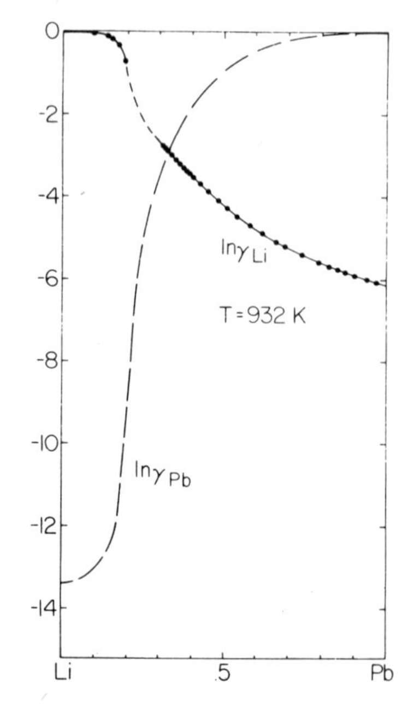

Fig.26. Variation of the logarithm of the activity coefficients with atomic composition of a LiPb alloy at T=932 K. (M.-L.Saboungi, J.Marr, M. Blander J.Chem.Phys.68, 1375 (1978)).

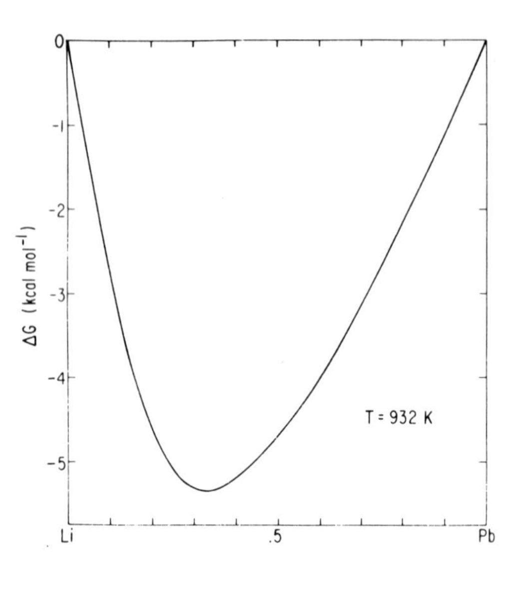

Fig.27. Variation of the excess Gibbs free energy of mixing ΔG with the atomic composition at T=932 K. (M.-L.Saboungi, J.Marr, M.Blander J.Chem.Phys. 68, 1375 (1978)).

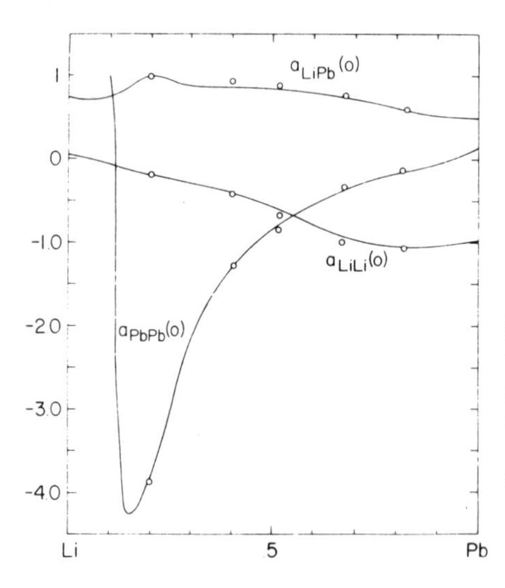

Fig.28. Calculated variations of the partial structure factors at q=0. Circles: Experimental results from neutron scattering. (M.-L. Saboungi. J.Marr, M. Blunder J.Chem.Phys. 68, 1375 (1978)).

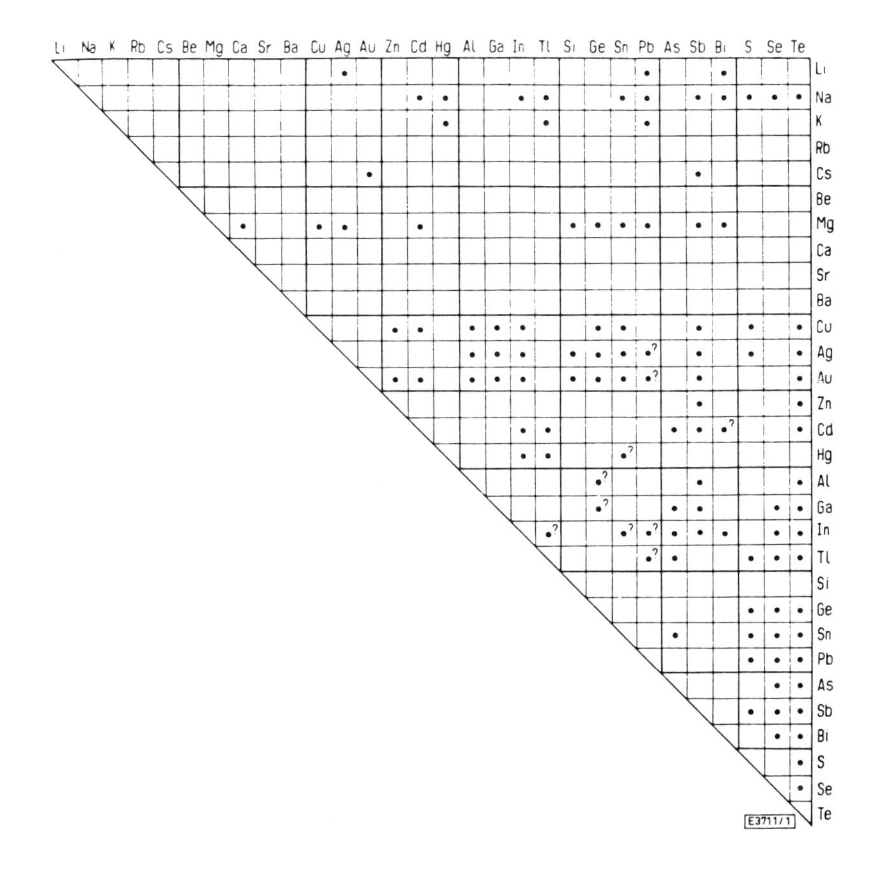

Fig.29. Review for binary alloy systems showing non-ideal behaviour in the liquid state.
(K.L. Komarek Ber.d.Bunsen-Ges. 81, 936 (1977)).

6. TRANSPORT PROPERTIES

6.1 Theoretical aspects

Transport properties include transport mediated by the ionic as well as by the electronic system. In liquid metals and alloys we may define the viscosity η, thermal conductivity λ_i and diffusion coefficient D of the ions. In metallic glasses viscous flow will occur only if the materials are highly strained (see talk by H.U. Künzi). The electronic system contributes to electrical σ and thermal λ_e conductivities.

Concerning the thermal conductivity λ_i of the ions, not very much is known for disordered systems. The only possibility available is to apply the Leibfried-Schlömann /43/ equation for the crystalline state to liquid and amorphous systems as well, using f.e. a Debye temperature Θ_D determined from specific heat data (Sect. 5.3):

$$\lambda_i \approx 3.6 \cdot 10^7 \cdot A \cdot a \cdot \Theta_D^{\,2} \cdot (\gamma + 0.5)^{-2} \cdot \Theta_D/T \quad ,$$

with A the atomic weight in kg, a the mean atomic distance in Å and γ the Grüneisenparameter, defined by

$$\gamma = - \partial \ln \Theta_D / \partial \ln V \quad .$$

Viscosity η and diffusion coefficient D will be discussed later.

Since the mean free path of the electron is so short, there is the question, if the NFE-model is still appropriate. One is tempted to apply a hopping (tight binding approximation) type model to describe the conductivity (see talk by F. Brouers).

However, it turned out that the Faber-Ziman theory is able to describe the electrical conductivity quite well. This is also true for its temperature dependence /44/.

In the spirit of the NFE model the thermal conductivity λ_e is coupled to the electrical conductivity σ by the Lorentz number L_0. This will be discussed in Section 6.3. It must be emphasized once more: There is no sound foundation for the application of the NFE-model despite its great success.

6.2 Electrical conductivity

Within the weak scattering model (NFE) (Faber, Ziman /10/) the scattering of an electron by an ion can be described by a weak electron-ion pseudopotential v(q). Such a potential gives the same scattering behaviour for a plane electron wave as does the true potential to the true electron wave function.

For pure metals the electrical resistivity $\rho_e(T) = 1/\sigma$ is given by /10/

$$\rho_e(T) = \frac{3\pi m_e^2}{\hbar^3 e^2} \cdot \frac{n_o^{-1}}{k_F^2} \cdot <|v(q)|^2 \cdot S(q)> \quad ,$$

with

$$<f(q)> = 4 \int_0^{2k_F} f(q)\, q^3\, dq \,/\, (2k_F)^4 \quad ,$$

where m_e and e are the electronic mass and charge respectively, k_F the Fermi wave vector, n_o the electron number density and $\hbar$ Planck constant divided by 2π.

Since the integral is heavily weighted towards the upper boundary $2k_F$ by the factor q^3, it is often enough to discuss the behaviour of the integrant at $q = 2\, k_F$. Supposing v(q) doesn't depend on volume or temperature - an assumption which is not very sound - we may judge the behaviour of ρ_e from the behaviour of S(q). As the preintegral factors depend on temperature and pressure to a very good approximation only through the volume change, a dimensional analysis gives:

$$\Delta\rho_e/\rho_e \Big|_{\text{preintegral}} = \frac{1}{3}\frac{\Delta V}{V} = \begin{cases} \frac{1}{3}\alpha_P \cdot \Delta T \\ -\frac{1}{3}\kappa_T \cdot \Delta P \end{cases} \quad ,$$

where α_P and κ_T is the isobaric expansivity and the isothermal compressibility resp.. ΔT and ΔP indicate the changes in temperature and pressure resp..

The peaks in S(q) increase with decreasing volume and vice versa. The temperature effect at constant volume shows the same behaviour. This reflects the ordering or disordering of the system with changing volume (reminescent of the Bragg peaks). The wings of the peaks in S(q) behave oppositely to the peaks, since they are dominated by the dynamical excitations in the system.

Hence, it follows that one-electron metals (Na,K) should have low resistivity and positive temperature coefficient (TC), as $2\, k_F$ is half way to the first maximum in S(q). Two electron metals (Ca, Ba etc.) should have high resistivity and negative TC's. A full discussion is given in the literature /44/.

The whole formulation of the theory can also be carried over to the single site t-matrix t(q), replacing v(q). For elements having the d-band at the Fermi-surface, i.e. transition metals, the d-electrons dominate and we may write (see, however, talk by H. Beck):

$$\varrho_e(T) \approx \frac{30\pi^2\hbar^3}{m_e e^2 E_F v_m k_F^2} \cdot S(2k_F) \cdot \frac{\Gamma^2}{\Gamma^2 + 4(E_d - E_F)^2} ,$$

where E_F is the Fermi energy, v_m the molare volume, Γ the width of the d-band and E_d its position.

There are some additional comments for the liquid rare earth metals at place /44/. Localized spins have to be considered as well as the break-down of Mathjessen's rule. The first point is illustrated in Fig. 30, where T_P describes a crystalline phase transition and T_M indicates the melting point.

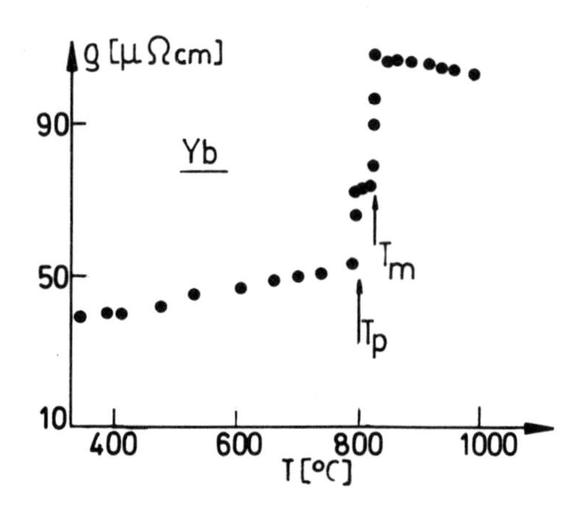

Fig.30. Electrical resistivity of solid and liquid Ytterbium. T_P: phase transition fcc-bcc, T_M: Melting point. (H.-J. Güntherodt, E. Hauser, H.U. Künzi, liquid Metals 1976, edt. R. Evans, D.A. Greenwood).

In both cases the rearrangement of the spins has an influence on the properties in the solid state.

Fe, Co and Ni have a lower electrical resistivity /44/ than the rare earth metals. The interpretation of the data requires contributions from spin disorder scattering.

In liquid alloys and metallic glasses the formulation has to be extended in the following way /10/:

$$\varrho_e(T) = \frac{3\pi^2 m_e^2}{\hbar^3 e^2} \cdot \frac{n_o^{-1}}{k_F^2} \cdot \langle c_A \cdot |t_A|^2 \cdot (1 - c_A + c_A \cdot S_{AA}) +$$

$$+ c_B \cdot |t_B|^2 \cdot (1 - c_B + c_B \cdot S_{BB}) + c_A \cdot c_B \cdot (t_A^* t_B +$$

$$+ t_A t_B^*) \cdot (S_{AB} - 1) \rangle \quad ,$$

where t_A, t_B are the t-matrices for the two components (the star indicates the complex conjugate), c_A, c_B are the mole/mole concentrations and S_{AA}, S_{BB}, S_{AB} the partial structure factors as introduced in Section 2.2.

Since almost no data are available concerning S_{AA}, S_{AB} and S_{BB}, it is difficult to discuss their variation with T and P. For metallic glasses very often a simple-minded picture is used. One assumes only one average t-matrix and only one average structure factor together with the stability condition $2k_F = q_P$ (see talks by H. Beck and J. Hafner). q_P is the position of the first peak in S(q). Then, besides a high resistivity, a negative TC results. Many metallic glasses follow this pattern.

In order to give some impression of the accuracy of the theory for liquid alloys, a calculation /45/ using PY-structure factors and pseudo-potentials is reproduced in Fig. 31. The overall agreement seems to be good.

6.3 The Lorentz-number L_o

Within the context for the NFE-model, there is a connection between λ_e and ϱ_e:

$$\frac{\lambda_e \cdot \varrho_e}{T} = L_o = \frac{\pi^2}{3} \frac{k_B^2}{e^2} \quad ,$$

where the symbols have the usual meaning and L_o is the Lorentz-number.

Since the NFE-model is believed to be a good approximation for liquid metals and alloys, the product $\lambda_e \varrho_e / T$ should be roughly equal to L_o. This is not exactly true, since one has a contribution to λ and ϱ from the ions. The product $(\lambda_i + \lambda_e)/\varrho_e$ can be measured directly by the Kohlrausch method /46/. These experiments give in most cases very good agreement with the equation stated above, supporting the NFE-model.

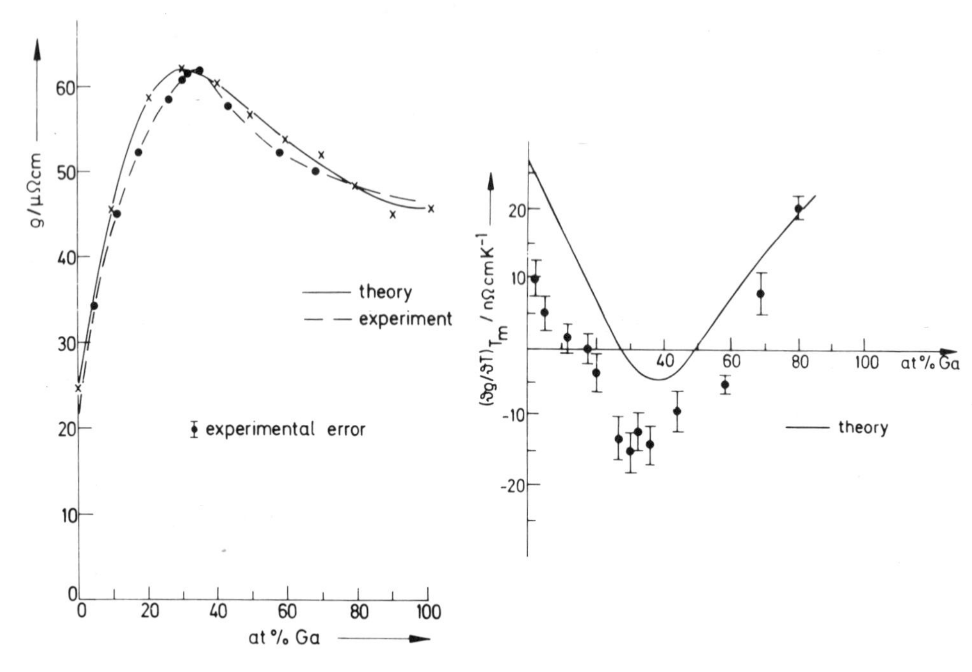

Fig.31. Calculated electrical resistivity ϱ and calculated temperature coefficient versus atomic composition as compared to experiment for a liquid Cu-Ga-Alloy at 1350 K. Solid line: theory, dots: experiment. (Ch. Holzhey, H. Coufal, G. Schubert, J. Brunnhuber, S. Sotier, Z.f.Physik B 33, 25 (1979)).

However, there has been some discussion recently /47/ to whether the effects of convection have been considered correctly. It may be that convection is not driven by density changes but by electromigration. Such an effect can have a 5 - 10 % influence on the data /47/.

6.4 Diffusion

The diffusion coefficient of the ionic system is an important parameter of its dynamics. Theoretical aspects have already been described in Sect. 3.1.3.

In this Section we would like to concentrate on experimental methods. In addition we will give some data. A recent review was published by Nachtrieb 1976 /48/.

One way to measure the diffusion coefficient D is by means of inelastic neutron scattering, mentioned already in Section 2.2.2. Another way is to use the nuclear magnetic resonance (NMR) technique. The diffusion coefficient is burried in the spin lattice relaxation time T_2. Hence, pulsed NMR is to be recommended.

Besides these more or less microscopic methods there are macroscopic ones, relying on Fick's laws of diffusion:

$$\vec{n} = - D \cdot \text{grad}\, c \qquad \frac{\partial c(\vec{r},t)}{\partial t} = D \cdot \Delta c(\vec{r},t) \quad ,$$

where $\vec{n}$ is the vector of the particle current density and c the concentration of certain test particles. Such probes can be different isotop species, radioactive tracers or other elements (impurity diffusion). Given a well specified distribution at $t = 0$, one studies its spatial variation at any later time. The accuracy claimed is around 3 %.

Amongst the various methods we will mention the capillary reservoir tracer technique, where the test substance is introduced into the liquid sample by a capillary; the long capillary methods /49/, where the starting distribution is as follows: Half of a capillary is filled with the normal sample, half of it with the sample enriched with test particles; the shear cell technique /50/, which consists of a staple of misaligned discs containing capillary holes: Diffusion is started by aligning the discs and stopped by misaligning them again. Finally we have the diaphragm cell method /51/, where the pure material and the alloy to be diffused are separated by a semipermeable membrane. The principles of the experimental set-ups are explained in Fig. 32. Some results for liquid metals are reported in Table VIII.

Table VIII. Some diffusion data for liquid metals. D_m value at the melting point.

	T_M/K	range fitted/K	$D_o/cm^2/s$	Q/kJ/mole	$D_m/cm^2/s$
Ga	303.0	303- 372	$1.1 \cdot 10^{-4}$	4.73	$1.71 \cdot 10^{-5}$
Na	370.7	372- 500	$0.92 \cdot 10^{-3}$	9.87	$3.82 \cdot 10^{-5}$
K	335.5	371- 557	$0.76 \cdot 10^{-3}$	8.52	$3.67 \cdot 10^{-5}$
Rb	311.7	330- 510	$0.66 \cdot 10^{-3}$	8.35	$2.70 \cdot 10^{-5}$
Ag	1233.7	1248-1623	$5.8 \cdot 10^{-4}$	32.3	$2.55 \cdot 10^{-5}$

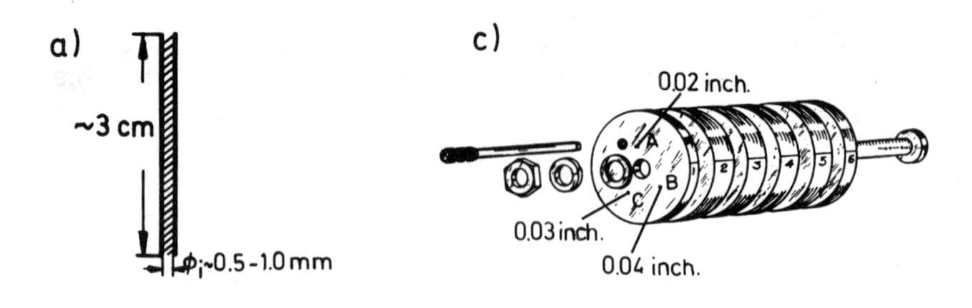

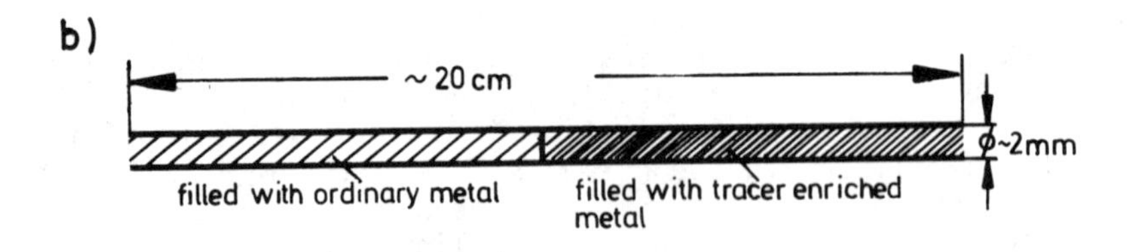

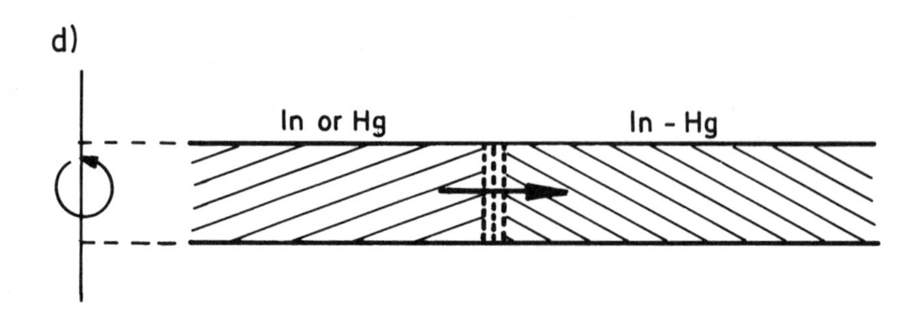

Fig.32. Comparison of various experimental set-ups for determination of the diffusion coefficient of liquid metals and alloys.
a) Capillary reservoir tracertechnique,
b) Long capillary methods, also with radioactive tracer techniques,
c) Shear cell technique,
d) Diaphragm cell method.

The quantities given are related by the following equation (Arrhenius-form):

$$D(T) = D_0 \cdot \exp\left[-Q/(RT)\right] \quad .$$

D_0 is a pre-exponential constant and Q indicates the activation energy for diffusion.

For the alkali - metals the values of $D(T_M) = D_m$ scale accurately with M, as is shown in Fig. 33.

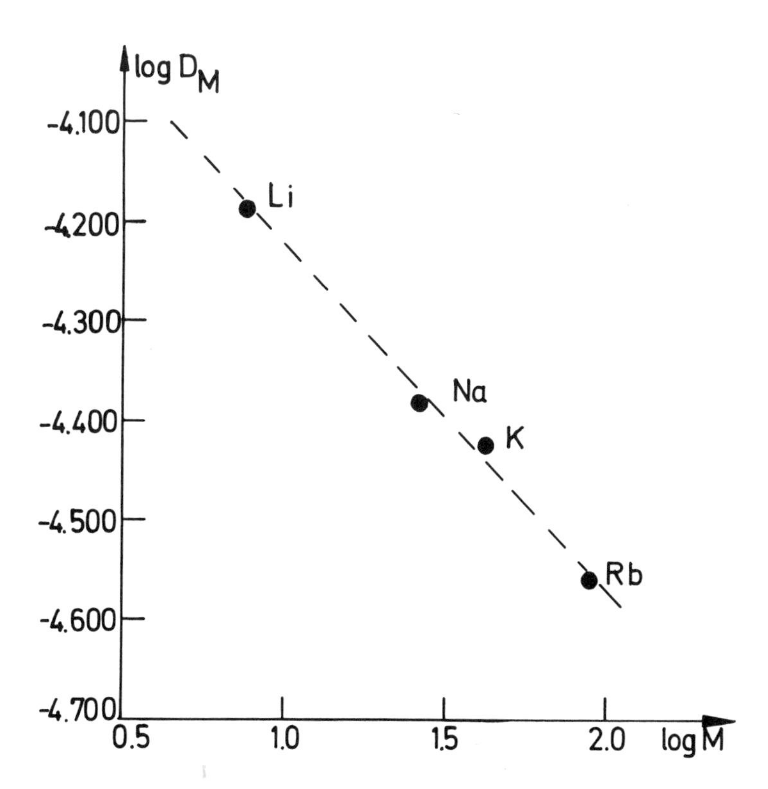

Fig. 33. Logarithm of the diffusion coefficient at melting D_m versus logarithm of the atomic weight M for the alkali metals. (N.H. Nachtrieb Berichte d.Bunsengesell. 80, 678 (1976)).

The general behaviour of the diffusion coefficient with temperature is illustrated in Fig. 34.

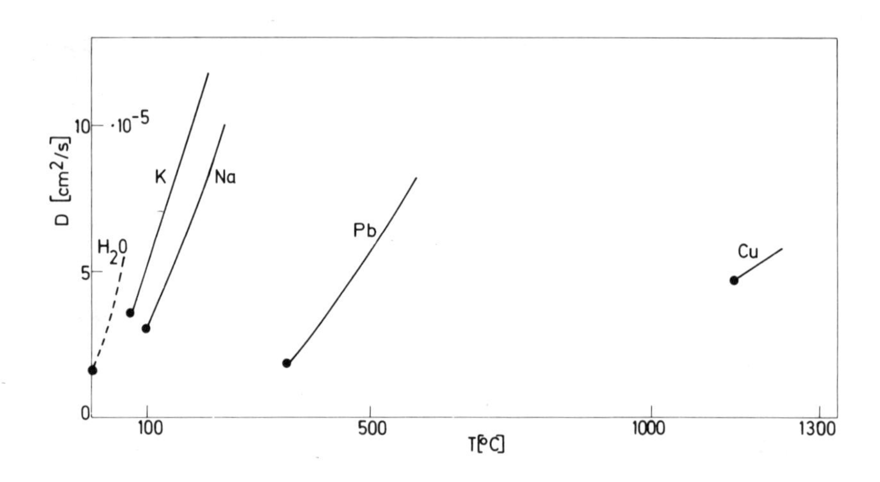

Fig.34. Diffusion coefficient D for some liquid metals and for water as a function of temperature. The dots indicate the melting temperatures.

6.5 Viscosity

The viscosity η of a liquid system is intimately related to its diffusion coefficient. Using the Einstein-Smoluchowski equation for the mean square displacement due to diffusion and the definition of the diffusion coefficient yields /52/:

$$D \cdot \eta = C \cdot T \quad ,$$

where C should be a constant, being inversely proportional to some effective radius of the diffusing particle (Stokes-Einstein relation).

Since D rises exponentially with temperature, η should fall with increasing T. This fact is also born out by experiment, illustrated in Fig. 35.

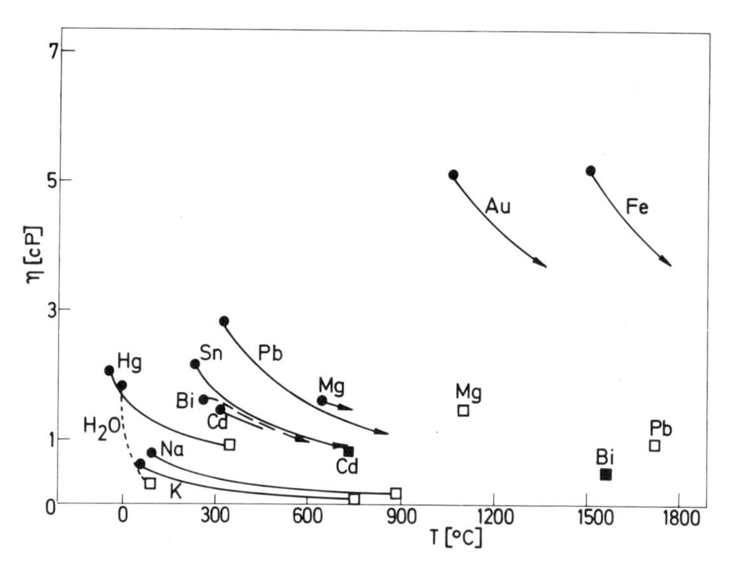

Fig.35. Viscosities η for some liquid metals and for water as a function of temperature. Dots indicate the melting points and quadrangles the boiling temperatures.

Inserting the Arrhenius-equation for D in the Stokes-Einstein relation gives:

$$\ln \eta = a + b/T + c \cdot \ln T \quad ,$$

where c should be equal to one. However, it turns out, that c is around -0.5. Hence, c depends on T, indicating that the effective radius changes with T. Some data are collected in Table IX:

Table IX. Some viscosity data as fitted to the equation given in the text (η/centipoise and T/K).

	range fitted/K	a	b	c
Na[1]	371-1089	0.4353	226.933	-0.4694
K[2]	342-1407	0.6402	181.269	-0.5624
NaK[1]	373-923	1.3535	136.785	-0.7748

[1] Argonne Nat.Lab. Rep. ANL-7323, 1967: G.H. Golden, J.G. Tokar,

[2] Nucl.Syst.Mat. Handbook, Vol.1/Pt.V, Hanford Eng. Devel. Lab. Richland 1971.

7. CONCLUDING REMARKS

Since this contribution was intended to be an introduction to the subjects discussed in detail in this NATO Advanced Study Institute, it is neither complete nor explaining the details. There are certain subjects which are only mentioned in passing. This is especially true for the properties of metallic glasses. Also certain special conditions, like dilute systems, closeness to critical points or phase transitions to ionic or covalent states contain important information to structure and dynamics of metallic systems. We hope that the lack of data especially for liquid alloys and metallic glasses will stimulate further research.

There are also very interesting new experimental methods like i.e. extended X-ray absorption fine structure (EXAFS) or positron annihilation (see talks by R. Hänsel and W. Triftshäuser) which promise great potential powers concerning the determination of SRO. More extensive use should also be made in applying pressure as a variable in experiments. Measurements of partial structure factors are highly welcome.

Finally it's worth mentioning that the concepts of dislocations /53/ may be significant even in amorphous or liquid systems (see talk by R. Cotterill). If such ideas help to explain "two-level"-systems, mechanical problems as well as relaxation phenomena have to await further confirmation.

Acknowledgement: The authors would like to thank Mrs. Gerl for typing the manuscript and Mrs. Harlandt for making the drawings.

REFERENCES

1. J.P. Hansen, J.R. McDonald, Theory of Simple Liquids Academic Press, London 1976
   N.H. March, Liquid Metals, Pergamon Press, Oxford 1968
   F. Kohler, Liquid State, Verlag Chemie, Weinheim 1972
2. J.K. Percus, G.J. Yevick, Phys.Rev. 110, 1 (1958)
3. R. Kumaravadivel, R. Evans, J.Phys. C 9, 3877 (1976)
4. N.F. Carnahan, K.E. Starling, J.Chem.Phys. 51, 635 (1969)
5. H.L. Frisch, J.L. Lebowitz, The Equilibrium Theory of Classical Fluids, Benjamin New York 1964
6. J.A. Barker, D. Henderson, J.Chem.Phys. 47, 2856, 4714 (1967)
7. J.A. Barker, D. Henderson, Rev.Mod.Phys. 48, 587 (1976)
8. J.D. Weeks, D. Chandler, H.C. Anderson, J.Chem.Phys. 54, 5237 (1971)
9. D. Stroud, N.W. Ashcroft, Phys.Rev. 5, 371 (1972)
10. T.E. Faber, J. Ziman, Phil.Mag. 8, 153 (1965)
11. H. Ruppersberg, H. Egger, J.Chem.Phys. 63, 4095 (1975)
12. A.B. Bhatia, D.E. Thornton, Phys.Rev. B2, 3004 (1970)
13. L. Dagens, M. Rasolt, R. Taylor, Phys.Rev. 11, 2726 (1975)
14. N.W. Ashcroft, J. Lekner, Phys.Rev. 145, 83 (1966)
15. L. van Hove, Phys.Rev. 95, 249 (1954)
16. H.S. Chen, Y. Waseda, Phys.Stat.Sol. a51, 593 (1979)
17. R. Hosemann, Phys. Blätter 34, 511 (1978)
    J.D. Bernal, Proc.Roy.Soc. A280, 299 (1964)
18. D. Schiff, Phys.Rev. 186, 151 (1969)
19. S.C. Jain, R.C. Bhandari, Physica 52, 393 (1971)
20. J. Hubbard, J.L. Beeby, Preprint 1968
21. H. Mori, Progr.Theoret.Phys. 33, 423 (1965), 34, 399 (1965)
22. G.H. Vineyard, Phys.Rev. 110, 99 (1958)
23. S.J. Cocking, J.Phys. C 2, 2047 (1969)
24. M. Ait-Salem, T. Springer, A. Heidemann, B. Alefeld, Phil.Mag. 39, 797 (1979)
25. P.D. Randolph, K.S.Singwi, Phys.Rev. 152, 99 (1966)
26. M.A. Krivoglaz, Theory of X-ray and Thermal Neutron Scattering by Real Crystals, Plenum Press New York 1969

27. A. Rahman, Phys.Rev. A9, 1667 (1974), Phys.Rev.Letters 32, 52 (1974)

28. K. Ichikawa, Phil.Mag. 27, 177 (1973)

29. N.H. March, Can.J.Chemistry, 55, 2165 (1977)

30. J. Ziman, Principles of the Theory of Solids, Cambridge Univ.Press London 1972

31. T.E. Faber, Adv. Physics 16, 637 (1967)
F.Cyrot-Lackmann, Phys.kond.Materie 3, 75 (1964)

32. H. Ruppersberg, W. Speicher, Z.Naturf. 31a, 47 (1976)

33. M. Hoch, T. Vernardakis, Ber.d.Bunsenges. 80, 770 (1976)

34. M. Hoch, High-Temp.-High Press. 4, 493, 659 (1972)

35. T.B. Douglas, A.F. Ball, D.C. Ginnings, W.D. Davis, J.Am. Chem.Soc. 24, 2472 (1952)

36. K.L. Komarek, Ber.d.Bunsenges. 81, 936 (1977)

37. S.P. McAlister, E.B. Crozier, J.F. Cochran, Can.J.Physics 52, 1847 (1974)

38. A.W. Overhauser, Phys.Rev. B3, 1888 (1971)

39. P. Ascarelli, Phys.Rev. 173, 271 (1968)

40. H. Reiter, H. Ruppersberg, W. Speicher, Liquid Metal 1976, Institute of Physics, Bristol 1977

41. A. Münster, Chemische Thermodynamik, Verlag Chemie Weinheim 1969; R.E. Smallman, Mod.Phys.Metallurgy, Butterworths, London 1970

42. M.-L. Saboungi, J. Marr, M. Blander, J.Chem.Phys. 68, 1375 (1978)

43. G. Leibfried, E. Schlömann, Math.Phys.Chem. Abt.4, 71 (1954)

44. G. Busch, H.-J. Güntherodt, Sol.State Phys. Vol.29, eds. F. Seitz, D. Turnbull, H. Ehrenreich, Academic N.Y. 1974

45. Ch. Holzhey, H. Coufal, G. Schubert, J. Brunnhuber, S. Sotier, Z.f.Physik B 33, 25 (1979)

46. W. Haller, H.-J. Güntherodt, G. Busch, edts: R. Evans, D.A. Greenwood, Liquid Metals 1976, Institute of Physics, Bristol, 1977

47. M.J. Laubitz, J.G. Cook, 16th Int.Thermal Conductivity Conf. Nov. 7-9, 1979, Chicago/USA

48. N.H. Nachtrieb, Ber.d.Bunsenges. 80, 678 (1976)

49. S. Larsson, L. Broman, C. Roxbergh, A. Lodding, Z.f.Naturf. 25a, 1472 (1970)

50. J. Petit, N.H. Nachtrieb, J.Chem.Phys. 24, 1027 (1956)
51. F.E. Butler, F.O. Shuck, Trans.AIME 245,3 (1969)
52. G. Fritsch, Transport, Akad.Verlagsanstalt Wiesbaden 1979
53. R.M.J. Cotterill, Phys.Rev.Letters 42, 1541 (1970).

ATOMIC THEORY OF LIQUID DYNAMICS

Alf Sjölander

Institute of Theoretical Physics, S-412 96 Göteborg, Sweden, and Solid State Scence Division, Argonne National Laboratory, Argonne, Illinois 60439, USA

I. INTRODUCTION

Matter can appear either as a gas, a liquid or a solid, and we can also distinguish a fourth phase, a plasma. The atoms are then ionized and the constituents are charged particles, which interact through the long range Coulomb interaction. An often utilized model system is the one component plasma, consisting of only one kind of charged particles and with a uniform neutralizing background. This is a good first approximation for the conduction electrons in simple metals and the positive ions are then assumed to be smeared out uniformly. In astrophysical objects, like white dwarfs, the positive ions are the basic constituents and the electrons form here the neutralizing background.

The dynamics of a system depends strongly on which phase it appears. In a gas of reasonably low density the atoms are moving freely with sudden interruptions of binary collisions, where only two atoms are involved at a time. These collisions provide the mechanism for transfer of energy and momenta from one atom to another and for the evolution towards equilibrium of non-uniformities in the gas. This is well described through the famous Boltzmann equation, introduced more than hundred years ago. It governs the time evolution of the one-particle phase-space function $f(\vec{r}\vec{p}t)$, which specifies the density as a function of position and time for atoms of definite momenta, and reads

$$(\frac{\partial}{\partial t} + \frac{1}{m}\vec{p}\cdot\vec{\nabla}_r)f(\vec{r}\vec{p}t) = C\{f,f\}. \quad (I.1)$$

Here, m is the atomic mass and $C\{f,f\}$ is a certain bilinear integral operator on $f(\vec{r}\vec{p}t)$, which takes care of the binary collisions.

The conservation of the total number of particles, the total momentum, and the total kinetic energy in each binary collision is built into the form of $C\{f,f\}$ and it causes the system to approach local thermal equilibrium after a number of collisions. The dynamics is thereafter described entirely through the hydrodynamic variables, the density $n(\vec{r}t)$, the current density $\vec{j}(\vec{r}t)$, and the local temperature $T(\vec{r}t)$. The hydrodynamic equations, which are obtained from (I.1), determine the evolution of these variables and explicit expressions are obtained for the hydrodynamic transport coefficients; the shear viscosity η, the bulk viscosity ζ, and the thermal diffusion constant D_T. Actually, the bulk viscosity vanishes in this case. The full dynamics is within hydrodynamics expressed in terms of collective modes; density waves, longitudinal and transverse currents, and heat diffusion. It is obvious that this cannot be appropriate for small time intervals, where the dynamics is governed by free particle motions. Depending on whether we consider slow and long wavelength phenomena or rapid and short wavelength ones, we have to focus our attention either on the collective aspects or on the single particle ones.

When considering plasmas, particular attention has to be given to the long range Coulomb interaction. This introduces a macroscopic electric field on each charge particle, arising from the surrounding particles, and and extra term has to be added to (I.1), changing this to

$$(\frac{\partial}{\partial t} + \frac{1}{m}\vec{p}\cdot\vec{\nabla}_r)f(\vec{r}\vec{p}t) - \{\int d\vec{r}'d\vec{p}'\vec{\nabla}_r v_c(\vec{r}-\vec{r}')f(\vec{r}'\vec{p}'t)\}\vec{\nabla}_p f(\vec{r}\vec{p}t) = C\{f,f\}. \tag{I.2}$$

Here, $v_c(r)$ is the Coulomb potential and $C\{f,f\}$ is the appropriate collision term. The added term introduces a screening mechanism where each particle repells the surrounding ones and introduces through the background a local charge distribution of the opposite sign. This screens out the Coulomb field from the primary particle. The efficiency of the screening depends, however, on the velocity of the particle. For a slow one the surrounding has time enough to adjust and makes the screening very effective. For a rapid particle, on the other hand, the opposite is true and the characteristic time scale for screening is the inverse plasma frequency $\omega_p^{-1} = (4\pi ne^2/m)^{-1}$, e being the charge of each particle. The collision term is essential in order to bring the system to proper local equilibrium.

For solids, particularly crystalline ones, a completely different dynamical description is normally used. The atoms are here localized and make only small excursions $\vec{u}(\vec{R}t)$ from their equilibrium positions $\vec{R}$. The harmonic approximation implies that each atom (ion) feels an effective harmonic force from the surrounding atoms (ions) and it depends linearly on their displacements. Newton's equation takes the form

$$\frac{d^2}{dt^2} u^{\alpha}(\vec{R}t) = \sum_{R'\beta} \Gamma^{\alpha\beta}(\vec{R}-\vec{R}')u^{\beta}(\vec{R}'t), \qquad (I.3)$$

where the force constants $\Gamma^{\alpha\beta}(\vec{R})$ are related to the second spatial derivative of the interaction potential. The general solution is a superposition of displacement waves, called phonons, and each one is characterized by its wavevector $\vec{q}$, frequency $\omega(\vec{q})$, and polarization direction. The dynamics of the system is here specified completely in terms of undamped phonons, which form the elementary excitations. Anharmonic corrections become significant at higher temperatures and they introduce scattering between the phonons and create a damping of the displacement waves. At the same time a shift of the phonon frequencies occurs in accordance with Kramers-Kronig relation. All motions are still described through collective modes and there is no need to introduce the concept of single particle motions separately, as long as we ignore diffusion of atoms.

A metal consists of localized ions, which are more or less rigid objects, and of highly mobile conduction electrons. The latter adjust quickly to the instantaneous positions of these ions and screen out the long range part of the bare ion-ion interaction. Particularly for monovalent metals the adjustment of the electrons can be calculated in linear response and one arrives at an effective screened ion-ion potential to be used in (I.3). Its Fourier transform is of the form[1]

$$v^{ii}_{eff}(q) = V^{ii}(q) + \frac{\{V^{ie}(q)\}^2}{v_c(q)} \left[\frac{1}{\varepsilon(q)} - 1\right], \qquad (I.4)$$

where V^{ii} is the bare ion-ion interaction, V^{ie} is the interaction between an ion and a single electron, $v_c(q) = 4\pi e^2/q^2$ is the Coulomb potential, and $\varepsilon(q)$ is the static dielectric function of the electron gas. Somewhat different forms for the above quantities have been given in the literature and they seem to provide through (I.4) and (I.3) quite good phonon dispersion curves, when tested against experiments[2]. There are, however, some few points to notice. By eliminating the electrons and replacing them by a static effective ion-ion potential we ignore the fact that the ions can lose energy to the conduction electrons and this can then diffuse out very rapidly. It will in the hydrodynamic limit result in a heat diffusion constant which is two orders of magnitude too small. Fortunately, this does not have any significant effect on the phonons in the wavevector region of main interest.

For liquids and also for disordered solids the situation is considerably more complicated and we have still no fully satisfactory quantitative model for these systems. In comparison with crystals the regular order is lost and for liquids the displacements are no more small quantities. Considering liquid metals, the ions are still moving very slowly relative to the electrons and we may incorporate their effects into an effective interaction between the ions. This, however, does not necessarily imply a two-body interaction. A key issue at present is whether we may for all practical purposes approximate such a complicated multi-ion interaction by an effective pair-potential and whether this can be extracted from the

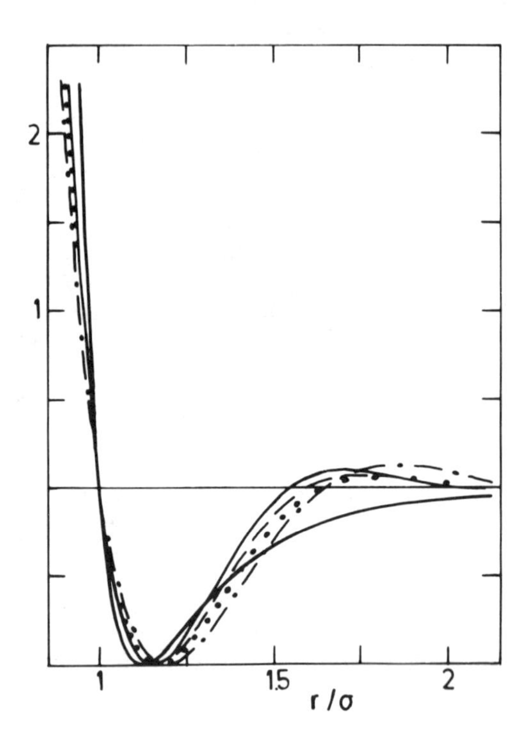

Fig. 1: Effective interionic potential for the alkali metals; sodium (full curve), potasium (dashed curve), rubidium (dotted curve), and lithium (dot-dashed curve). A corresponding Lennard-Jones potential is plotted for comparison (full curve). All curves are scaled to have the same value at their minima and to cross the zero-axis at the same relative distance (taken from L. Sjögren, J. Phys. C 12, 425 (1979); original data are from G. Jacucci (unpublished)).

static structure factor, for instance. Monovalent metals should be favourable in this respect and some recent results are indeed very encouraging. This question is particularly pertinant to computer simulations on liquid metals. A proper theory should, of course, treat the ions and the conduction electrons on equal footing and efforts in that direction have been made in the past[3]. However, most of the quantitative calculations are so far based on the simpler approach with an effective pair-potential between the ions. The present lectures will be based on the same assumption. This implies that the motion of the electrons enter only through the ion-ion potential and explicit electronic properties are ignored. Figure 1 shows some calculated potentials for monovalent metals and they are compared with the Lennard-Jones potential appropriate for liquid argon. We notice that the steep repulsive part is somewhat softer for the metals and their potentials are also more harmonic around the minimum. It has been found that the shape of the static structure factor S(q) is strongly determined by the steep part of the potential and quite good values have been obtained by simply assuming a hard core interaction with the radius as an adjustable parameter[4]. However, it is not possible to reproduce S(q) both for large and small q-values by such a simple approach and Weeks, Chandler and Anderson[5] developed a method for calculating S(q), where the hard core system is used as a reference system and the attractive part of the interaction potential is added as a perturbation. The inversion of their procedure has been used later in order to obtain an effective pair-potential from the measured S(q) and this has given very promising results[6].

II. EXPERIMENTAL AND MOLECULAR DYNAMICS RESULTS

When considering metals, we should distinguish experimental data referring primarily to the motion of the conduction electrons, such as electric conductivity, from those referring to the ionic motion, such as mass transport and thermal properties. We shall here be concerned only with the latter ones and these are also of two different kinds. We have those connected with the static properties of the system as various thermodynamic quantities and particularly the static structure factor and the corresponding pair correlation function g(r). Figure 2 shows S(q) for liquid rubidium. We shall here assume that these are known and concentrate on the dynamical aspects. Such properties enter in the measurements of the hydrodynamic transport coefficients, but most detailed information is obtained through inelastic neutron scattering. Figure 3 illustrates the experimental set up. A well collimated and monochromatic beam of neutrons with incident energy E_o and wavevector $\vec{k}_o$ is sent through the liquid sample and one measures the intensity of the scattered neutrons for various outgoing energies (E) and directions ($\vec{k}$). The result is

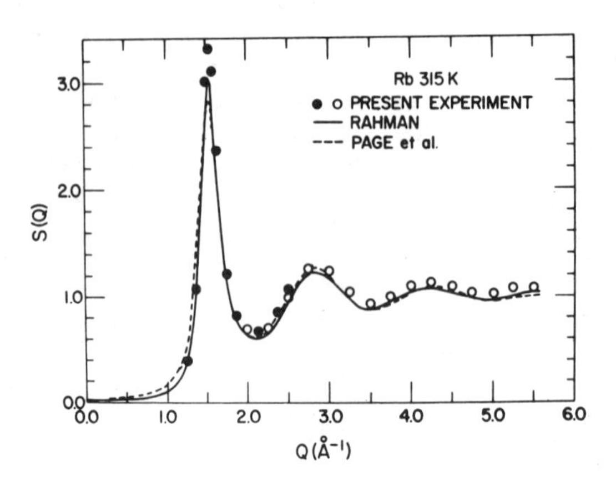

Fig 2: Static structure factor for liquid rubidium (from ref. 11).

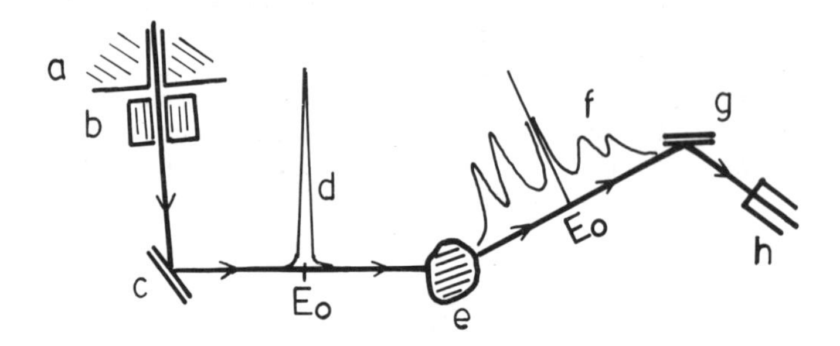

Fig. 3: Experimental set up for inelastic neutron scattering experiment: a) reactor b) shielding c) monochromator crystal d) energy spectrum of incident neutron beam e) sample f) energy spectrum of scattered neutrons (typical for single crystal sample) g) analyser crystal h) neutron counter.

usually expressed in terms of a dynamical structure factor $S(q,\omega)$, first introduced by Van Hove[7], where

$$\text{Intensity} \propto S(q\omega) \qquad \text{(II.1)}$$

with $\hbar\omega = \hbar^2(k_o^2-k^2)/2m$ and $\hbar\vec{q} = \hbar(\vec{k}_o-\vec{k})$ being the energy and the momentum transferred to the liquid in the scattering event. The proportionality factor contains certain trivial kinematical factors and the nuclear scattering length. Depending on the experimental arrangements one can measure the intensity at fixed scattering angle, where q becomes a function of ω and of the angle, or one measures $S(q,\omega)$ with q and ω varying independently. In time of flight measurements it is actually $\omega^2 S(q,\omega)$ which is measured. Neutron scattering experiments for liquids are in reality very difficult and time consuming and they involve various kinds of corrections to be made for counter efficiency, container scattering, resolution broadening, and multiple scattering. The last correction is particularly severe for q-values below the main peak in S(q) due to reduction of scattering intensity. This and other restrictions imply that one cannot always reach all (q,ω)-values of interest and it is particularly difficult to reach small (q,ω)-values, where hydrodynamics begins to be valid. Here, however, other experimental methods are available, like ultrasonic measurements and others. Due to insufficient energy resolution and to improper or no correction for multiple scattering or other reasons, the interpretation of experimental data before 1965 have been rather uncertain[8]. These measurements did, however, indicate that the motions of the ions have several features in common with what is found in solids and they created an intense activity among the theoreticians, mainly in developing simple phenomenological models to explain the experimental results[9]. Particularly early experiments on liquid and polycrystalline aluminum[10] was thought to show that both short wavelength longitudinal and transverse "phonons" were seen, analogous to what occurs in the crystalline phase. The individual ions were believed to execute a combination of vibratory and diffusive motions and models were constructed to take into account this kind of physics. Even though these models are now outdated for liquid metals, they have more recently been applied in other connections as, for instance, in fast ion conductors and supercooled liquids.

Recent high resolution measurements have been carried out on liquid rubidium[11] (Fig. 4), nickel[12], lead[13] and aluminum[14], and the old data on aluminum have been reanalysed in light of more recent theories[15]. For rubidium a clear resonance was found in $S(q,\omega)$ for wavenumbers up to 1 $Å^{-1}$ and they disappear rapidly as we move up on the lower side of the main peak of S(q). This is quite remarkable for it means that very short wavelength collective density oscillations can propagate, corresponding to longitudinal phonons in the crystal. The main peak in the static structure factor corresponds to the first reciprocal lattice vector in the solid and 1 $Å^{-1}$ lies therefore beyond the first Brillouin zone. Recent experiments on liquid lead have also revealed similar short wavelength resonances. Liquid insulators, like argon[16] (Fig. 5) and neon[17], have shown resonances in $S(q,\omega)$ only for small q-values.

A very useful technique for studying dynamics of liquids was introduced through computer simulations[18], so called molecular

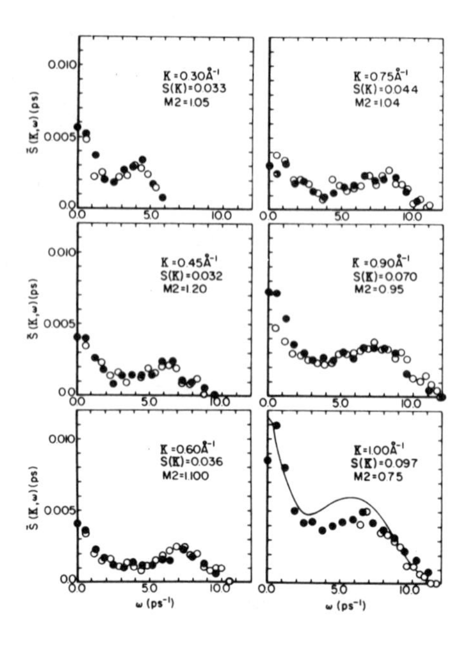

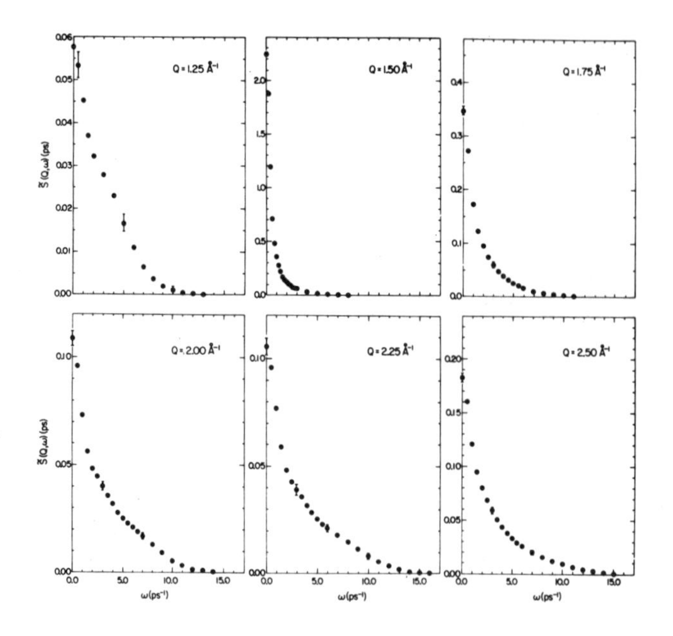

Fig. 4: Measured $S(q,\omega)$ for liquid rubidium (taken from ref. 11)

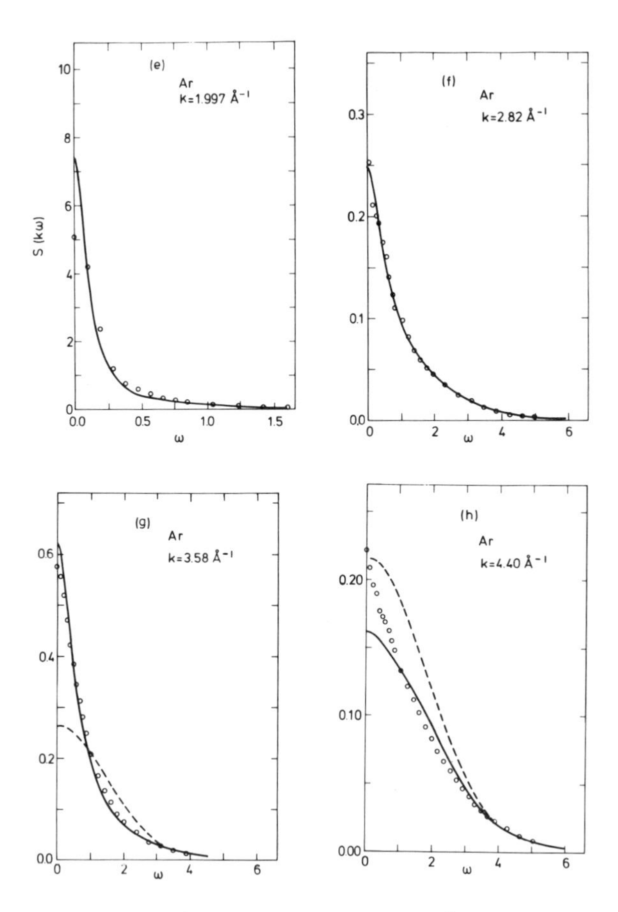

Fig. 5: Measured $S(q,\omega)$ for liquid argon (taken from ref. 20). The open rings are the experimental points from ref. 16.

dynamics. Given an interaction potential, one simulates the ionic motions by solving Newton's equation for a set of particles, usually from a few hundred up to a few thousand. This has been found to give good quantitative values for various quantities of interest, among them $S(q,\omega)$, provided an accurate interparticle potential is used. The method is by the theoreticians viewed as an experiment and a very useful aspect of this method is that one can here vary different input parameters at will and in this way increase the possibility of testing the theoretical model calculations. Also, one may determine quantities which are not directly accessible to real experiments, such as the single particle velocity correlation function. For supercooled liquids and amorphous solids this seems even more important. By experimentalists molecular dynamics is viewed as a theoretical calculation and by comparing the results with experimental data one can draw conclusions concerning the effective ion-ion interaction and other matters. Figures 6 and 7 show some results for $S(q,\omega)$ from molecular dynamics, using theoretically calculated ion-ion potentials. Comparison between figure 4 and 6 show the kind of agreement one obtains between molecular

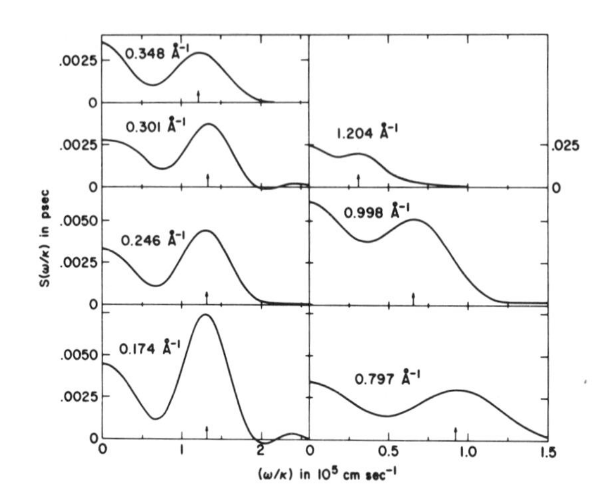

Fig. 6: Calculated $S(q,\omega)$ from MD for liquid rubidium (taken from A. Rahman, Phys. Rev. Letters 32, 52 (1974)).

dynamics and real experiments. The differences arise here mostly from an inaccurate potential, used as input in molecular dynamics, and from experimental uncertainties. The velocity correlation function, obtained for rubidium[19], is shown in Figure 8. This quantity is of considerable interest and it enters directly in NMR and Mössbauer measurements and its time integral gives the self-diffusion constant

$$D_s = 1/3 \int_0^\infty dt \, \langle \vec{v}(t) \cdot \vec{v}(o) \rangle \quad . \tag{II.2}$$

It is a great challenge to the theoreticians to connect the above experimental results to the atomic motions in the liquid and to answer such questions as how collective and single particle motions enter in $S(q,\omega)$. A point of particular interest is to

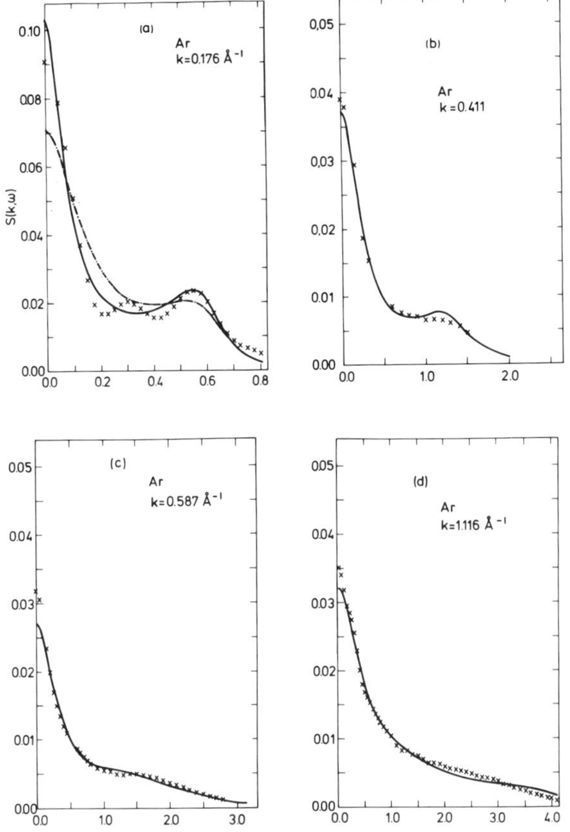

Fig. 7: Calculated $S(q,\omega)$ from MD for liquid argon (taken from ref. 20). The crosses are MD data from D. Levesque, L. Verlet and J. Kürkijärvi, Phys. Rev. A7, 1690 (1973).

clarify when hydrodynamics becomes applicable. It was shown by Van Hove that in the neutron scattering experiments one measures precisely the spatial and temporal Fourier transform of the density correlation function. One defines therefore

$$S(q\omega) = \frac{1}{2\pi n} \int d\vec{r} \int dt \, \exp[i(\vec{q}\cdot\vec{r}-\omega t)] \, \langle n(\vec{r}t)n(oo)\rangle \qquad (II.3)$$

where $n(\vec{r}t) = \Sigma\delta(\vec{r}-\vec{r}_\ell(t))$ is the microscopic particle density and $\vec{r}_\ell(t)$ denotes the position of the ℓ:th atom. n is the mean density and the bracket $\langle\ldots\rangle$ means an ordinary thermal average. A direct consequence of the definition is that for $2\pi/q$ much less than the

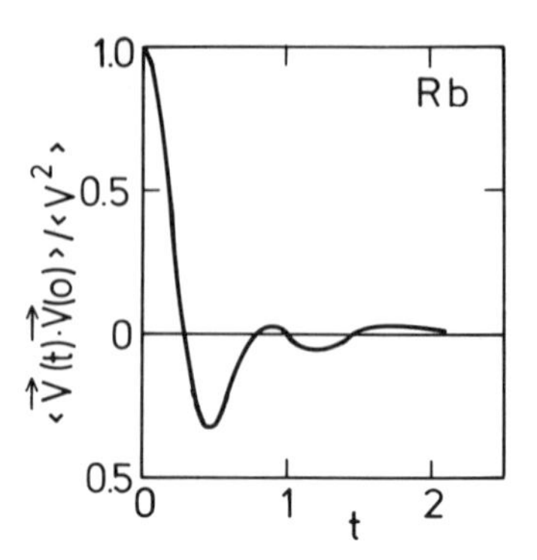

Fig 8: Calculated velocity correlation function from MD for liquid rubidium (taken from ref. 20 with unpublished data of A. Rahman).

interparticle distance the main information hidden in $S(q,\omega)$ refers to the motion of individual ions and collective aspects are of minor importance. In the extreme short wavelength limit only the free particle motion is revealed irrespective of the interaction. The opposite is true for small values of q and ω, where the phase factor in (II.3) varies only slowly over microscopic distances and times. The integrations imply then a course graining and ordinary hydrodynamics determines the time evolution of the density fluctuations. The density now couples to pressure fluctuations, which propagate as sound waves, and to entropy fluctuations, which are diffusive. This implies that $S(q,\omega)$ shows one peak around $\omega=0$, the heat diffusion peak, and two displaced peaks due to sound waves (Fig. 9). The width of the central peak is determined by the heat diffusion constant and that of the sound wave peaks mainly by the viscosity coefficients. The area under the central peak relative to that under the side peaks is within hydrodynamics given by (C_p/C_v-1), where C_p and C_v are, respectively, the specific heat at constant pressure and volume. Opposite to liquid argon, this ratio is for the simple metals rather small. This together with the fact that the heat diffusion constant is larger by a factor of hundred makes the effect on $S(q,\omega)$ from temperature fluctuations small compared to that for insulators[20]. Conventional molecular dynamics would, if used in this very long wavelength region, distort the physics on this point.

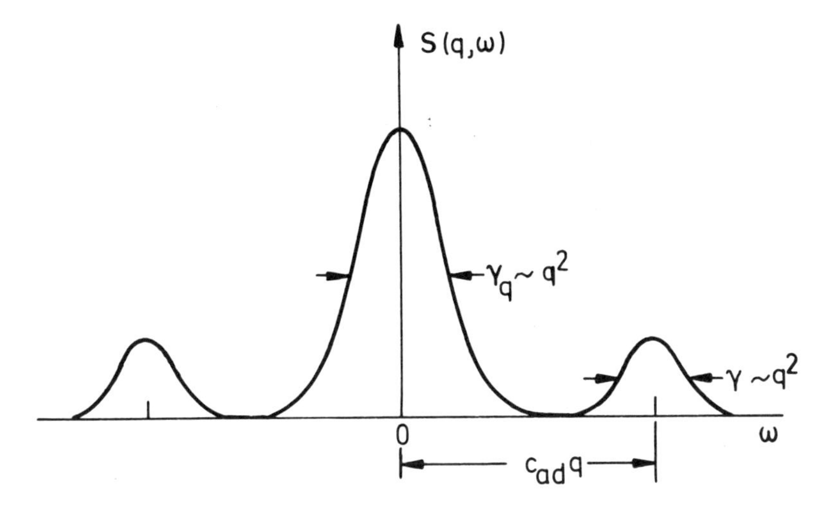

Fig. 9: S(q,ω) in the hydrodynamic regime.

The following rewriting of (II.3), using the definition of $n(\vec{r}t)$, reveals more directly the microscopic aspects:

$$S(q\omega) = \int dt\ e^{-i\omega t}\{\sum_{\ell=1}^{N} \langle \exp[i\vec{q}.\vec{r}_\ell(t)]\exp[-i\vec{q}.\vec{r}_1(o)]\rangle\}$$

$$= S_s(q\omega) + S_d(q\omega). \qquad (II.4)$$

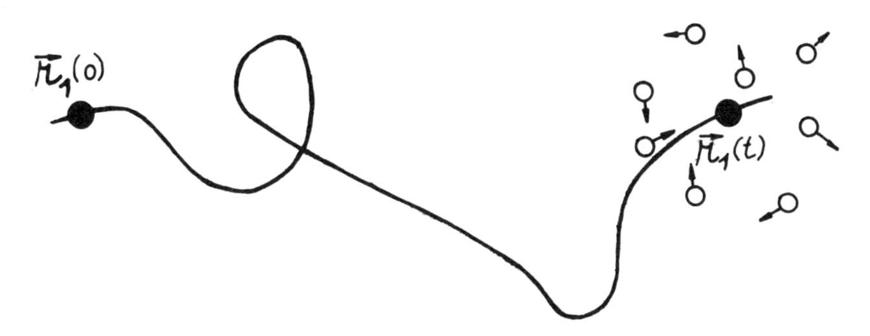

Fig. 10: Illustration of self motion and the accompanying backflow of the other ions.

The motion of the individual ions appear here explicitly and the term with $\ell=1$ and denoted by $S_s(q,\omega)$ contains information on the motion of one single ion. We may imagine how this ion performs some kind of Brownian motion and diffuses way from its initial position at t=0. $S_d(q,\omega)$, called the distinct part and containing all terms with $\ell\neq1$, gives information on how the ions around the first one move and how they are correlated to the position of the first ion. Figure 10 is meant to illustrate this, showing how the first ion moves from its position at t=0 to another position at a later time and how the other ions move and build up some kind of backflow around the first ion.

A basic aim of the theoretical analysis is to clarify to which extent and how the above aspects show up in $S(q,\omega)$. The same kind of picture is useful also for understanding the single particle velocity correlation function. Here, an ion with a certain initial velocity interacts with its surrounding, creating various kinds of disturbances. These spread out in the medium and react back on the first ion at a later time, whereever that ion may appear. So for instance, if the disturbances in the surrounding would show strong oscillatory behaviour, we expect that the force on the first ion also varies periodically. This should then show up as oscillations in $\langle\vec{v}(t)\cdot\vec{v}(0)\rangle$, which is indeed found in dense one-component plasmas[21]. From this argument we may expect that the velocity correlation function contains significant information on the collective motions in the liquid. It is quite understandable that the theoretical calculations have mostly concentrated on $S(q,\omega)$, for it is through this quantity a direct contact with experiments is obtained The computer simulations have in this respect broadened the view in an essential way.

III. SUM RULES

For convenience we introduce a separate notation for

$$F(q,t) = \sum_{\ell=1}^{N} \langle\exp[i\vec{q}\cdot(\vec{r}_\ell(t) - \vec{r}_1(o))]\rangle, \qquad (III.1)$$

usually called the intermediate scattering function, and we denote by $F_s(q,t)$ and $F_d(q,t)$ its self and distinct parts. We may here, by expanding the exponent in powers of t and then using Newton's equation, obtain values for the first few time derivatives of F(q,t) for t=0. Thus we have[9]

$$F(q,o) = S(q) , \qquad (III.2)$$

$$\left[\frac{d^2}{dt^2} F(q,t)\right]_{t=o} = \frac{k_B T}{m} q^2, \tag{III.3}$$

$$\left[\frac{d^4}{dt^4} F(q,t)\right]_{t=o} = \frac{3k_B T}{m} q^2 + \frac{n}{3m} \int d\vec{r}\; g(r)\nabla^2 V(r)$$

$$- \frac{n}{m} \int d\vec{r}\; g(r)\cos(qz) \frac{\partial^2 V(r)}{\partial z^2}, \tag{III.4}$$

and all the odd derivatives vanish. These relations hold only if we assume a pair-potential between the ions and treat their motion classically. The first relation reads in ordinary space

$$F(\vec{r},o) = \delta(\vec{r}) + ng(\vec{r}), \tag{III.5}$$

where the first term originates from $F_s(q,0)$, and it tells us that one ion is at t=0 located at the origin with the other ions being in equilibrium around the first one. Eq. (III.3) comes entirely from $F_s(q,t)$ and it simply reflects the fact that to begin with the first ion starts to move out from the origin, while the distribution of the other ions is still unchanged. Such changes enter later and show up for the first time in the last term in (III.4). The first two terms originate from $F_s(q,t)$. Also some higher order derivatives are known explicitly, and they are expressed in terms of the static two-, three- and four-particle correlation functions and contain further information on how the self-ion diffuses and how the surrounding responds to that. So for instance, it begins to fill the hole left behind by the first ion. This creates a strong disturbance of the surrounding, which will in due time spread through the medium. Due to lack of precise knowledge of the form of the higher order static correlation functions we have limited practical use of the relations for these higher order time derivatives.

The relations above imply for $S(q,\omega)$ the following:

$$\int \frac{d\omega}{2\pi} S(q\omega) = S(q), \tag{III.6}$$

$$\int \frac{d\omega}{2\pi} \omega^2 S(q\omega) = \frac{k_B T}{m} q^2, \tag{III.7}$$

$$\int \frac{d\omega}{2\pi} \omega^4 S(q\omega) = \frac{3k_B T}{m} q^2 + \frac{n}{m} \int d\vec{r}[1 - \cos(qz)]$$

$$\times\; g(r) \frac{\partial^2 V(r)}{\partial z^2}. \tag{III.8}$$

These are the zeroth, second, and fourth moments of $S(q,\omega)$ and they put certain restrictions on the dynamical structure factor. They have been used extensively in order to determine wavevector dependent parameters, entering in various model calculations.

IV. SOME EARLIER MODELS

Vineyard approximations[22]: In the early days of neutron scattering one was mostly concentrating the attention on the self motion and an obvious approximation was to assume that a single particle in a liquid performs a simple Brownian motion, following Langevin's equation. This implies that

$$F_s(q,t) = \exp[-\frac{q^2}{3}\int_0^t dt'(t-t')\langle\vec{v}(t').\vec{v}(o)\rangle] \tag{IV.1}$$

with

$$\langle\vec{v}(t).\vec{v}(o)\rangle = \frac{3k_BT}{m}\,e^{-\zeta t}, \tag{IV.2}$$

where the friction constant ζ can be determined from the self diffusion constant through eq. (II.2). $F_s(\vec{r}t)$ gives the probability of finding the self particle at a position $\vec{r}$ at time t, provided it was located at the origin at t=0, and $F_s(\vec{r},t)$ satisfies for large times the simple diffusion equation

$$\frac{\partial}{\partial t}F_s(\vec{r},t) = D_s\nabla^2 F_s(\vec{r},t). \tag{IV.3}$$

Vineyard then assumed that the surrounding medium is all the time in equilibrium around the above particle. This implies that

$$F_d(q,t) = [S(q)-1]\;F_s(q,t) \tag{IV.4}$$

and hence

$$F(q,t) = S(q)F_s(q,t). \tag{IV.5}$$

It was not a particularly good approximation and all the collective aspects, which should appear in F(q,t), were lost. Furthermore, it violated severely the second and fourth moments. Nevertheless, this was the first effort to understand what goes into neutron scattering. The shortcoming of the approximation is obviously due to our neglect of the finite time it takes the surrounding particles to reach equilibrium around the first particle.

Mean field approximation[23]: Here one takes into account in an approximate way the dynamics of the particles surrounding the first one. Let us use linear response in this connection. We assume that we all the time know the position of the particle which was originally at the origin. The induced density of the surrounding particles is easily calculated in linear response, and the response function relates to the full F(qt) or rather its Laplace transform $\tilde{F}(qz)$. This yields

$$\tilde{F}_d(qz) = [S(q)-1]\tilde{F}_s(qz) - \tilde{F}(qz)[1 - \frac{1}{S(q)}] [z \tilde{F}_s(qz)-1]. \quad (IV.6)$$

The first term is that from the Vineyard approximation and the second term contains the dynamics of the surrounding, where possible collective motions appear explicitly in $\tilde{F}(qz)$. One here recognizes an effective potential

$$v_{eff}(q) = - k_B T \, c(q), \qquad (IV.7)$$

where $c(q) = (1/n)\,[1 - 1/S(q)]$ is the direct correlation function. It has often in different contexts been argued that for a strongly interacting system, like an ordinary liquid, the bare interaction is renormalized and should be replaced by that above. For weak interaction it goes over to the bare potential.

Adding to (IV.6) the self part and solving for $\tilde{F}(qz)$ we obtain

$$\tilde{F}(qz) = S(q) \frac{\tilde{F}_s(qz)}{1 + nc(q)[z\tilde{F}_s(qz)-1]}, \qquad (IV.8)$$

and $S(q,\omega)$ is then found from

$$S(q,\omega) = \frac{1}{\pi} \text{ Re } \tilde{F}(q,z=i\omega). \qquad (IV.9)$$

This is the mean field result of Kerr and of Singwi et al. One must here make a separate calculation of $F_s(qt)$ and an often used approximation is the so called Gaussian one in (IV.1) with an appropriate form for the velocity correlation function.

This theory satisfies correctly the zeroth and second moments but gives only an approximate value for the fourth one. For long wavelengths it gives with a minor approximation

$$F(qz) = S(q) \frac{z + q^2\tilde{\alpha}(qz)}{z^2 + [k_B T q^2/mS(q)] + zq^2\tilde{\alpha}(qz)} \qquad \text{(IV.10)}$$

where $\tilde{\alpha}(qz)$ here relates to the memory function of the velocity correlation function, which will be introduced later. Eq. (IV.10) is the proper hydrodynamic form for $\tilde{F}(qz)$ and $[k_B T/mS(q=0)]$ is the isothermal sound velocity and $\tilde{\alpha}(qz)$ should for (q,z) tending to zero approach $(4\eta/3 + \zeta)$, if coupling to temperature is ignored. The above mean field theory gives, however, a $1/q^2$ singularity for $q \to 0$ and thus predicts an infinite value for the viscosity coefficient and it causes the hydrodynamic sound waves to disappear. This is not unexpected, since our result is analogous to the mean field result for plasmas and that is known to be incorrect in the hydrodynamic limit. What is missing is the collision term in (I.2).

From (IV.8-9) follows

$$S(q,\omega=o) = \frac{1}{\pi} S^2(q) \tilde{F}_s(q,z = o), \qquad \text{(IV.11)}$$

and here again one did not find particularly satisfactory values, when compared with experiments. However, the mean field approximation has on the whole given quite acceptable agreement with neutron scattering experiments and molecular dynamics for wavenumbers beyond 1 Å^{-1} or so, and it has been very useful in providing some insight into the dynamics. What the deficiencies tell us is that the disturbances in the medium around the first particle should be treated beyond linear response and be expressed not only in terms of the induced density but also through other modes.

Generalized hydrodynamics[24]: Here one begins from the hydrodynamic limit and generalizes this to larger q and ω values by introducing wavevector and frequency dependent transport coefficients. Some simple functional forms are assumed for these, consistent with both hydrodynamics and with known sum rules. It is not surprising that one can through this achieve good agreement with experiments for wavevectors below the main peak in S(q). A shortcoming of this treatment is that it describes all the motions in terms of only hydrodynamic modes, ignoring entirely the single particle aspects. This makes it inapplicable for large q-values and seems to lead to some misinterpretation[17] of the experimental data for intermediate wavenumbers. Another critical point against this approach is that it does not seem to give much of insight into the underlying physics, unless it is combined with a proper theory for how the transport coefficients vary with q and ω. A refinement in this direction was provided by Mori and we shall come back to this in the next section.

V. THE MORI APPROACH AND THE MODE COUPLING THEORY

Two dynamical variables, say A and B, are statistically uncorrelated if the equilibrium thermal average $\langle AB \rangle$ vanishes. We shall then say that A and B are orthogonal to each other. Pressure and temperature fluctuations are in this sense orthogonal, and it is often convenient to describe the dynamics through variables, which at equal time are orthogonal. Given a certain quantity $A(\vec{x}_1\vec{p}_1, \ldots, \vec{x}_N\vec{p}_N)$, being a function of the position and momenta of the particles in the system, its time evolution is governed by the equation

$$\frac{d}{dt} A(t) = L\, A(t) \; ; \tag{V.1}$$

where the Liouville operator L is

$$L = \sum_{\ell=1}^{N} \{\dot{\vec{r}}_\ell \cdot \vec{\nabla}_{r_\ell} + \dot{\vec{p}}_\ell \cdot \vec{\nabla}_{p_\ell}\} \, , \tag{V.2}$$

where $\dot{\vec{r}}_\ell = \vec{p}_\ell/m$, $\dot{\vec{p}}_\ell = -\sum_{\ell'} V(\vec{r}_\ell - \vec{r}_{\ell'})$, and V(r) being the pair interaction potential.

Let us split A(t) into a part, which is directly correlated to its initial value A(0) and the rest, writing

$$A(t) = \frac{\langle A(t)A(o)\rangle}{\langle A(o)A(o)\rangle} A(o) + \bar{A}(t) \, . \tag{V.3}$$

This is analogous to what we do in vector algebra, splitting an arbitrary vector, A(t), into its component along a specified direction A(0) and its orthogonal component $\bar{A}(t)$. Here, it is a projection in a functional space, where

$$PA(t) = \frac{\langle A(t)A(o)\rangle}{\langle A(o)A(o)\rangle} A(o) \; , \quad QA(t)\;(1-P)\bar{A}(t) = A(t) \; , \tag{V.4}$$

and we may split up (V.1) into two coupled equations

$$\frac{d}{dt} PA(t) = PLPA(t) + PLQA(t) ,$$

$$\frac{d}{dt} QA(t) = QLPA(t) + QLQA(t) . \tag{V.5}$$

If A is the velocity of a single atom, for instance, its time derivative would directly depend on the positions of the surrounding atoms. These positions are in the above sense orthogonal to the velocity of the first particle at equal time and we may consider the first particle and the surrounding as two different but interacting systems. The dynamics of the surrounding is then governed by $QA(t)$ and the coupling to the first particle enters through $QLPA(t)$. Solving formally the second equation above and inserting the result into the last term of the first equation, we eliminate all the degrees of freedom of the surrounding and we obtain an equation of the form[25]

$$\frac{d}{dt} A(t) - \Omega A(t) + \int_0^t dt' \Gamma(t-t')A(t') = f(t) , \tag{V.6}$$

where

$$\Omega = \langle \dot{A}(o)A(o)\rangle / \langle A(o)A(o)\rangle ,$$

$$\Gamma(t) = \langle \dot{A}(o)Q\exp[QLQt]Q\dot{A}(o)\rangle / \langle A(o)A(o)\rangle , \tag{V.7}$$

and

$$f(t) = \exp[QLQt]Q\dot{A}(o) . \tag{V.8}$$

This is a generalization of Langevin's equation for a stochastic variable and $f(t)$, being all the time orthogonal to $A(0)$, is here the random force. Multiplying (V.6) with $A(0)$ and taking the thermal average, we obtain the Mori equation for the auto-correlation function $\langle A(t)A(0)\rangle$;

$$\frac{d}{dt}\langle A(t)A(o)\rangle - \Omega \langle A(t)A(o)\rangle + \int_o^t dt'\Gamma(t-t')\langle A(t')A(o)\rangle = 0 \ . \tag{V.9}$$

Applied to the velocity correlation function one finds that $\Omega=0$ and, further assuming that $\Gamma(t) = \zeta\delta(t)$, we recover the result of Langevin's equation. Equation (V.9) is an exact formulation and the burden is now shifted to calculating Ω and particularly the memory function $\Gamma(t)$. This approach with some phenomenological ansatz for the memory function has been used extensively and with quite a lot of success[26].

Generalizations in various directions can easily be made. So for instance, we may introduce several variables $A_1(t), A_2(t), .. A_n(t)$ as the basic ones and it would then be convenient to have them orthogonal at equal time. Carrying through the arguments above, one arrives at a set of equations

$$\frac{d}{dt}\langle A_\mu(t)A_\nu(o)\rangle + \sum_\rho \Omega_{\mu\rho} \langle A_\rho(t)A_\nu(o)\rangle +$$

$$+ \sum_\rho \int_o^t dt' \ \Gamma_{\mu\rho}(t-t')\langle A_\rho(t')A_\nu(o)\rangle = 0 \ , \tag{V.10}$$

where $\Omega_{\mu\rho}$ and $\Gamma_{\mu\rho}(t)$ are generalizations of (V.7). Choosing as basic variables $n(qt)$, $j(qt)$, and $T(qt)$ - the spatial Fourier transform of the density, the current and the temperature -eq.(V.10) becomes analogous to the hydrodynamic equations with $\Gamma_{\mu\rho}(q\omega)$ being wavevector and frequency dependent transport coefficients[27]. In this way we obtain explicit, although rather formal, expressions for these, but they can be used as a starting point for approximations.

Another way of proceeding is to begin with just one variable, the one which is of interest, and to consider $(\exp[QLQt]\dot{A}(0))$ as a new one for which we write an equation analogous to (V.6). This introduces a new Ω and a new Γ with a modified projection operator. We may continue in this way until it is appropriate to stop. It leads for the Laplace transform of $\langle A(t)A(0)\rangle$ to a partial fraction expension of the form

$$\langle A(t)A(o)\rangle_z = \cfrac{\langle A(o)A(o)\rangle}{z - \Omega_1 + \cfrac{B_1}{z - \Omega_2 + \cfrac{B_2}{z - \Omega_3 + \Gamma_3(z)}}} \ . \tag{V.11}$$

A frequency dependent memory function enters only in the last step and, provided we can determine the constants Ω_1, $\Omega_2 \ldots$ and B_1, $B_2 \ldots$ the approximation enters only in calculating this function. Applying this approach to $S(q,\omega)$, we choose $A(t) = n(\vec{q},t)$. The various moments determine then uniquely the wavevector dependent constants above and the main effort lies in obtaining the memory function.

An approach of this kind has been exploited extensively by the Munich group[28], led by Götze. Let us consider some of their steps in calculating $S(q,\omega)$. They first write

$$\langle n(q,t)n(-q,o)\rangle_z = \cfrac{S(q)}{z + \cfrac{B_1(q)}{z + \cfrac{B_2(q)}{z + \tilde{\Gamma}(qz)}}} , \qquad \text{(V.12)}$$

where $B_1(q) \equiv \Omega_o^2(q) = k_B T q^2/mS(q)$ and $B_2(q) = \Omega_L^2(q) - \Omega_o^2(q)$ with

$$\Omega_L^2(q) = \frac{3k_B T}{m} q^2 + \frac{n}{m} \int d\vec{r}[1-\cos(qz)]g(r) \frac{\partial^2 V(r)}{\partial z^2} . \qquad \text{(V.13)}$$

$\Omega_L(q)$ is the frequency that is related to the fourth moment of $S(q,\omega)$ and the zeroth, second, and fourth moments are here satisfied irrespected of what approximation we make for $\tilde{\Gamma}(qz)$. The memory function is formally given by

$$\Gamma(q,t) = \frac{n}{m} \frac{1}{\Omega_L^2(q)-\Omega_o^2(q)} \langle B(q)Qe^{QLQt}QB(-q)\rangle , \qquad \text{(V.14)}$$

where $B(q)$ is of the form

$$B(q) = \int dq' [t_L(\vec{q},\vec{q}')\ j_L(\vec{q}')n(\vec{q}-\vec{q}') +$$

$$+ t_T(\vec{q},\vec{q}')\ j_T(\vec{q}')n(\vec{q}-\vec{q}')] , \qquad \text{(V.15)}$$

and the operator Q projects out all contributions proportional to the density and the current. Through $\Gamma(qt)$ enters a coupling between the original density mode $n(qt)$ and two combined modes, where one is the density mode with another wavevector and the other is either the longitudinal current or the transverse one. This is analogous to what happens in solids, where one phonon may decay into two other phonons due to anharmonicity. The coupling constants $t_L(\vec{q},\vec{q}')$ and $t_T(\vec{q},\vec{q}')$, which are frequency independent, are known and contain explicitly the interaction potential. Inserting B(q) in (V.14), Götze and collaborators make the plausible approximation of factorizing the resulting correlation function $\langle j_{L,T}(\vec{q}')n(\vec{q}-\vec{q}')Q\times \exp[QLQt]Qn(-\vec{q}+\vec{q}'')\ j_{L,T}(-\vec{q}'')\rangle$ and obtain an explicit expression for $\Gamma(qt)$, containing mode-mode coupling terms of the form

$$\langle j_{L,T}(\vec{q}',t)j_{L,T}(-\vec{q}',o)\rangle\ \langle n(\vec{q}-\vec{q}',t)n(-\vec{q}+\vec{q}',o)\rangle$$

and

$$\langle j_L(\vec{q}',t)n(-\vec{q}'\,o)\rangle\ \langle n(\vec{q}-\vec{q}',t)j_L(-\vec{q}+\vec{q}',o)\rangle\ .$$

It would lead us too far to continue here and we refer to the original articles for further details. Through eq. (V.12) and the corresponding ones for the longitudinal and transverse currents they obtain a closed set of equations, which besides the factorization approximation does not contain any further assumptions. It is true that the static coupling constants, entering in $\Gamma(qt)$, contain the three particle correlation function and this has to be approximated in an explicit numerical calculation. Götze et al have carried out a self consistent solution of their equations and typical results for $S(q,\omega)$ are shown in Figure 11. It should be added that coupling to temperature fluctuations are ignored here, and this leads in the hydrodynamics to a somewhat incomplete result.

A treatment similar in content but differing in form from that of the Munich group has been presented by Munakata and Igarashi[29]. They begin from the Mori equation (V.9), considering either the density or the current correlation functions, and for $Q\dot{A}(0)$, entering in $\Gamma(q,t)$, they project onto the single modes n_q and $\vec{j}_q$ and the double modes $n_{q'}n_{q-q'}$ and $\vec{j}_{q'}\vec{j}_{q-q'}$ and finally make use of the factorization approximation. They wanted to avoid using this mode coupling approach for the very short wavelength motions in $\Gamma(q,t)$ and introduced therefore a cut off wavevector, which was adjusted to give correct value for $S(q,\omega=o)$. In order to correct for the missing short wavelength part in $\Gamma(q,t)$ they renormalized the coupling constants in a phenomenological way so that the known initial value $\Gamma(q,t=o)$ is correct for all q. The numerical agreement with the experimental data seem to be of essentially the same quality as that of the Munich group. However, since Götze et al used the mode coupling approach for all wavevectors and secondly used systematically the partial fraction expansion, they did not introduce any phenomenological assumptions.

A similar treatment as that of Götze et al could be done for $F_s(qt)$ as well and it would be of great interest to test

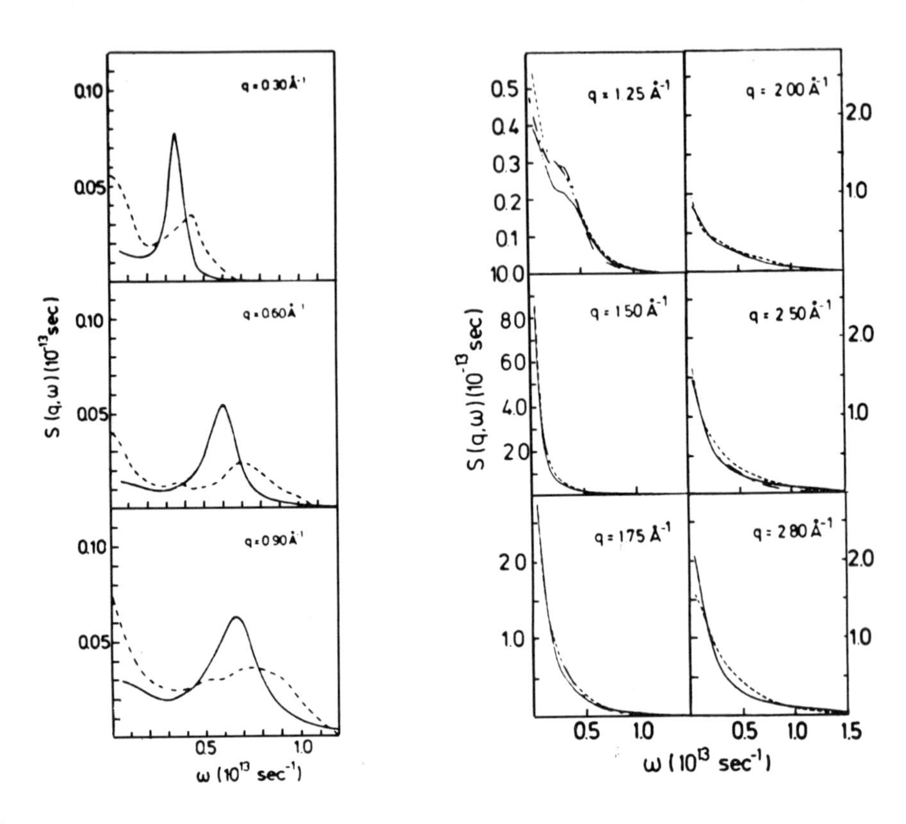

Fig. 11: $S(q,\omega)$ for liquid rubidium from first principle mode-mode coupling theory of the Munich group (taken from J. Bosse, W. Götze and M. Lücke, Phys. Rev. 18, 1176 (1978). The dashed curves are the experimental data from ref. 11.

the validity of their mode coupling approach for the single particle motion. We know that it does not give the correct free particle form

$$F_s(q,t) = \exp\left[-k_BT\, q^2t^2/2m\right] \qquad (V.16)$$

for large q-values but the main concern is how well it can reproduce available molecular dynamics data for intermediate wavevectors. The above deficiency leads to wrong results also for F(qt) for large q, but this was by Götze et al corrected for by adding to the mode-mode coupling term a free particle contribution. It seems, however, to be a somewhat artificial way of correcting an error in $F_s(qt)$. One may question whether the mode-mode coupling approach is particularly suitable for handling $F_s(qt)$ for large and intermediate wavenumbers. Another point of concern is whether in this way one is able to handle properly the rapid binary collisions entering also in liquids through the strong repulsive part of the interaction potential. This could be checked by considering dense gases, where these collisions dominate. We leave this interesting and very promising approach by these comments and hope that further work will be done along this line.

VI. THE KINETIC THEORY

Here one considers the full phase-space functions and tries to generalize the Boltzmann equation. We introduce the microscopic Klimontovich function

$$f(\vec{r}\vec{p}t) = \sum_{\ell=1}^{N} \delta(\vec{r}-\vec{r}_\ell(t)\ \delta(\vec{p}-\vec{p}_\ell(t)) \qquad \text{(VI.1)}$$

and the corresponding one for a single particle

$$f_s(\vec{r}\vec{p}t) = \delta(\vec{r}-\vec{r}_1(t))\ \delta(\vec{p}-\vec{p}_1(t)). \qquad \text{(VI.2)}$$

Akcazu and Duderstadt[30] were the first to apply the Mori formalism to these quantities but the detailed formal treatment, including the mode-mode coupling approximation, was given by Mazenko and others with applications to a system of hard spheres[31]. The extension to more realistic systems with a continuous interaction potential together with extensive numerical calculations have been carried out more recently by Sjögren and Sjölander[32].

The Mori equation for the phase-space correlation function $F(\vec{p}\vec{p}';\vec{r}t) \equiv \langle \delta f(\vec{r}\vec{p}t)\ \delta f(\vec{0}\vec{p}'0)\rangle$ takes the form

$$\left(\frac{\partial}{\partial t}+\frac{1}{m}\vec{p}.\vec{\nabla}_r\right)F(\vec{p}\vec{p}';\vec{r}t)+nk_BT\,\vec{\nabla}_p\phi_M(p).\int d\vec{r}''d\vec{p}''\,\vec{\nabla}_r c(\vec{r}-\vec{r}'')$$

$$\times\,F(\vec{p}''\vec{p}';\vec{r}''t)-\int_0^t dt''\int d\vec{r}''d\vec{p}''\Gamma(\vec{p}\vec{p}'';\vec{r}-\vec{r}'',t-t'')F(\vec{p}''\vec{p}';\vec{r}''t'')$$

$$=0, \tag{VI.3}$$

where $\delta f(\vec{r}\vec{p}t)$ is the deviation from the equilibrium value $n\phi_M(p)$, $\phi_M(p)$ being the Maxwellian distribution. $F(\vec{p}\vec{p}';\vec{r}t)$ contains information on how a disturbance in the six-dimensional phase-space evolves in time and the density and current correlations, among others, are easily obtained by integrating F with respect to the momenta after multiplying with appropriate functions of $\vec{p}$ and $\vec{p}'$.

Equation (VI.3) is a proper transport equation, which can be compared with the linearized version of (I.2). We recognize the third term above as the mean field contribution with $[-k_BTc(r)]$ being the effective potential. The last term, containing a memory function Γ, corresponds to the Boltzmann collision term $C\{f,f\}$. Non-linearities in (I.2) are here built into Γ and are found to cause long memory effects.

It is useful to compare (VI.3) with the original equation

$$\left(\frac{\partial}{\partial t}+\frac{1}{m}\vec{p}.\vec{\nabla}_r\right)F(\vec{p}\vec{p}';\vec{r}t)-\int d\vec{r}''d\vec{p}''\,\vec{\nabla}_r V(\vec{r}-\vec{r}'').\vec{\nabla}_p$$

$$\times\,\langle\delta f_2(\vec{r}\vec{p},\vec{r}''\vec{p}'',t)\,\delta f(o\vec{p}'o)\rangle=0\ , \tag{VI.4}$$

where here $V(r)$ is the bare interaction potential and $\delta f_2(\vec{r}\vec{p},\vec{r}''\vec{p}'',t)$ is the fluctuation in the two-particle phase-space function. This contains information both of how that particle moves which at t=0 is at the origin with the momentum $\vec{p}'$, and of how the other particles build up a surrounding backflow. Obviously, the same information is contained in Γ. In a similar way as we split up $F(q,t)$ in (III.1) into its self and distint parts, we can do the same for the phase-space function above, and for the self part we have the equation

$$(\frac{\partial}{\partial t} + \frac{1}{m}\vec{p}.\vec{\nabla}_r)F_s(\vec{p}\vec{p}';\vec{r}t) - \int_0^t dt'' \int d\vec{r}''d\vec{p}''\Gamma_s(\vec{p}\vec{p}'';\vec{r}-\vec{r}'';t-t'')$$

$$\times F_s(\vec{p}''\vec{p}';\vec{r}''t'') = 0 , \qquad \text{(VI.5)}$$

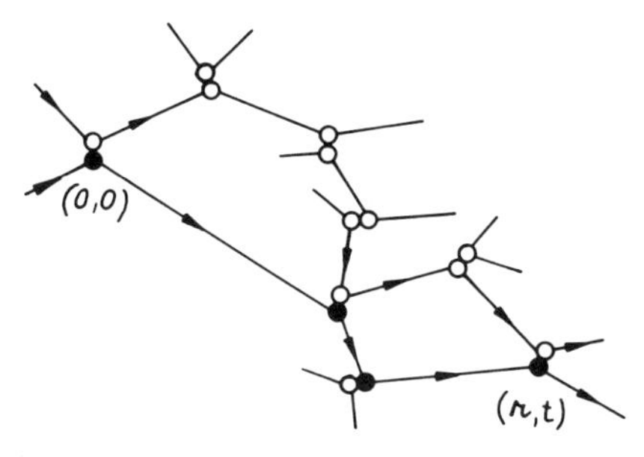

Fig. 12: Illustration of the dynamical events entering in $F_s(q,t)$ for a gas.

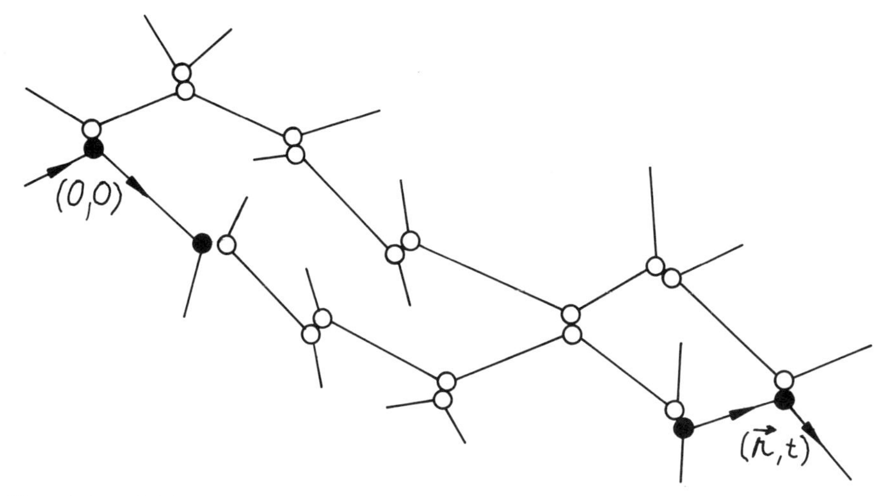

Fig. 13: Illustration of the dynamical events entering in F(q,t) for a gas

where another memory function enters. We notice that no mean field term appears here. Figures 12 and 13 are meant to illustrate the dynamics involved and this is most easily done for a dilute gas, where the elementary events are binary collisions. For F_s we follow all the time one particular particle, the one starting at the origin at t=0, and the other particles form a surrounding fluid where the collisions between these particles build up the collective motions we call the backflow. Except for a short initial time period, during which essentially a single binary collision occurs, we may describe the motion of the surrounding through collective variables, like density and current. This is consistent with available molecular dynamics results, which seem to show that the memory functions involved decrease rapidly in the beginning and later on have a much slower time dependence. The slow hydrodynamic modes in the backflow could explain the slow part, whereas the binary collision, connected with the steep repulsive part of the interaction potential, would be responsible for the initial rapid decay of the memory function. If so, we should split up the dynamics into at least three steps; the motion of the self particle, the binary collision of a surrounding particle with the self particle, and the collective motions in the surrounding at later times. Assuming that we have solved eq. (VI.5) for the self motion, we can use this in (VI.3) and in the Fourier-Laplace space write

$$\tilde{F}(\vec{p}\vec{p}';\vec{q}z) + ink_BTc(q) \int d\vec{p}_1 \; \tilde{F}_s(\vec{p}\vec{p}_1;\vec{q}z)[\vec{q}.\vec{\nabla}_{p_1} \phi_M(p_1)]\tilde{F}(\vec{p}_1\vec{p}';\vec{q}z)$$

$$- \int d\vec{p}_1 d\vec{p}_2 \; \tilde{F}_s(\vec{p}\vec{p}_1;\vec{q}z) \; \tilde{\Gamma}_d(\vec{p}_1\vec{p}_2;\vec{q}z) \; \tilde{F}(\vec{p}_2\vec{p}';\vec{q}z) = \tilde{F}_s(\vec{p}\vec{p}';\vec{q}z) \; , \quad \text{(VI.6)}$$

where

$$\Gamma_d = \Gamma - \Gamma_s \quad \text{(VI.7)}$$

A part of the original memory function Γ refers to the self-particle and, when this is split off, the remaining part Γ_d is connected entirely with the motion of the surrounding.

The functions Γ and Γ_s can be calculated exactly in the low density limit and eq. (VI.3) goes then over to the Boltzmann equation. Recollisions, which were ignored by Boltzmann, are found to be important also in a gas for explaining certain non-analytic behaviour of the transport coefficients. This is connected with a long time tail in the memory functions. A full treatment of the dynamical structure factor, based on (VI.3), has not been worked out as yet but is underway by Sjögren, and we shall instead illustrate the calculational procedure for the velocity correlation function[33]. More phenomenological treatments have been given by several authors with quite successful results[34]. Particularly the calculations of Sjögren and Sjölander[32] have a firm basis and the phenomenological ingredient concerned the treatment of the rapid binary collision. Eq. (VI.6) was the starting point and the Gaussian approximation was used for $F_s(q,t)$, and the backflow was described only through density and longitudinal current modes. This resulted in the following expression for the Laplace transform of F(q,t):

$$\tilde{F}(qz) = S(q)\,\frac{\tilde{F}_s(qz) - (1/k_BTq^2)\tilde{\Gamma}_d(qz)[z\tilde{F}_s(qz)-1]}{1 + [nc(q) - (z/k_BTq^2)\tilde{\Gamma}_d(qz)][z\tilde{F}_s(qz)-1]}\,, \qquad \text{(VI.8)}$$

where $\tilde{\Gamma}_d(qz)$ follows from $\tilde{\Gamma}_d$ in (VI.6). Putting $\tilde{\Gamma}_d(qz)=0$, we recover the mean field result in (IV.8). It leads in the hydrodynamic limit to (IV.10) with a finite value for $(4\eta/3 + \zeta)$. This was for liquid argon within 30 percent of the molecular dynamics value and for liquid rubidium it came out too large by roughly a factor two, resulting in a too broad sound wave resonance. Correct results are obtained for the zeroth, second, and fourth moments of $S(q,\omega)$ and the value at $\omega=0$ is

$$S(q,\omega=o) = \frac{1}{\pi}\, S^2(q)[\tilde{F}_s(q,z=o) + (m/k_BTq^2)\tilde{\Gamma}_d(q,z=o)]\,, \qquad \text{(VI.9)}$$

which should be compared with (IV.11). Figures 14 and 15 show some typical results and the corresponding mean field values are given for comparison. We notice a systematic improvement and this is particularly true for $S(q,\omega=0)$. The discrepancy we have in rubidium for 1 Å^{-1} is not due to our failure of reproducing hydrodynamics well, but must have some other origin. The self part happens to give a very significant contribution for this wavenumber and part of the discrepancy could be due to the use of the Gaussian approximation for $\tilde{F}_s(qz)$. We know that this can give a significant error in the width of $S_s(q,\omega)$ for intermediate wavenumbers[35]. An interesting analysis of the old experimental data on liquid aluminum has recently been done by Larsson[15] (Figure 16), where he compares his data with the predictions of the Vineyard treatment, of the mean field theory, and of eq. (VI.8). Here, he has added properly calculated multiple scattering contributions to the theoretical cross sections before comparing with the experiments. It clearly shows the significance of the last improvement and it does encourage further refinements along the same line.

In order to present the most recent and most advanced treatment within this kinetic theory, we shall consider the velocity correlation function $\phi(t)$[33]. The corresponding equation of

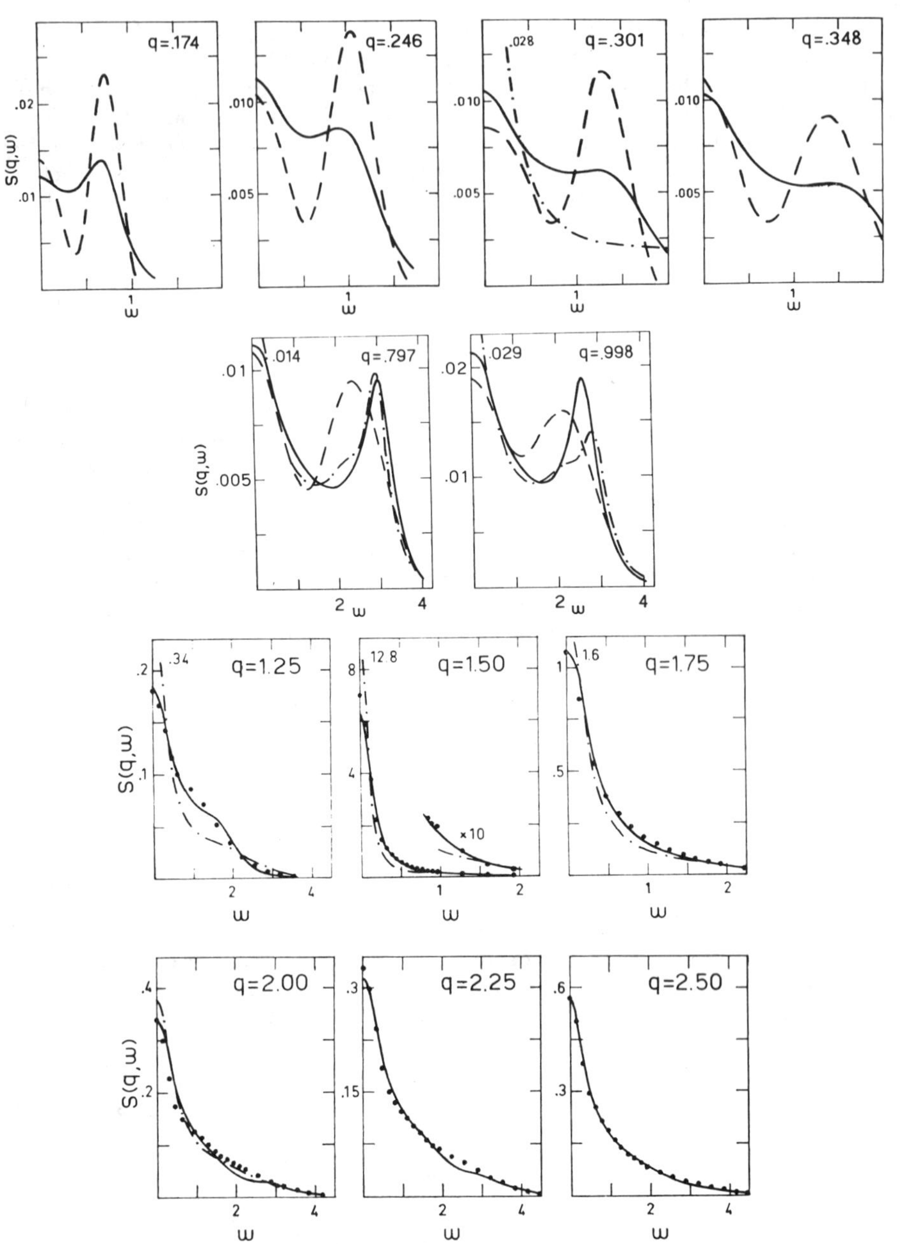

Fig. 14: S(q,ω) for liquid rubidium from the kinetic theory of the Gothenburg group (taken from L. Sjögren and A. Sjölander, Ann. Phys. 110, 421 (1978)). The dots are experimental data from ref. 11, the dashed curves are MD data from A. Rahman, Phys. Rev. Letters 32, 52 (1974), and the dot-dashed curves are from the mean field theory.

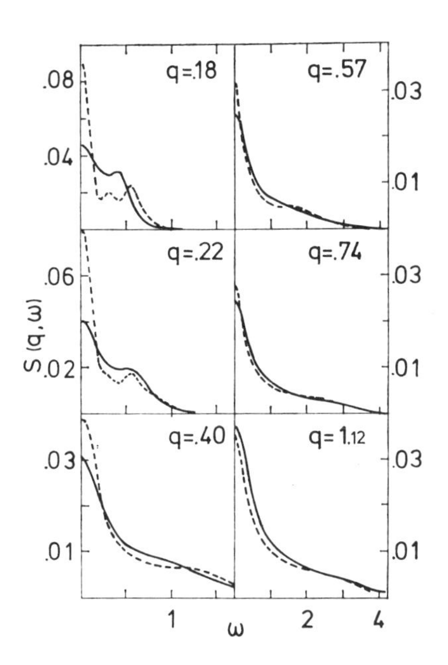

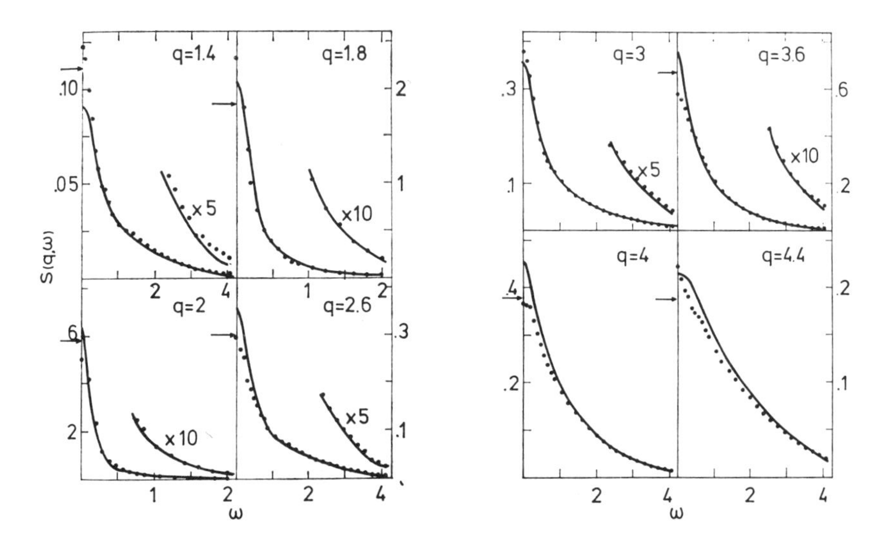

Fig. 15: S(q,ω) for liquid argon from the kinetic theory of the Gothenburg group (taken from L. Sjögren, Ann. Phys. 110, 156 (1978)). The dots are experimental data from ref. 16 and the dashed curves are MD data from D. Levesque, L. Verlet and J. Kürkijärvi, Phys. Rev. A7, 1690 (1973).

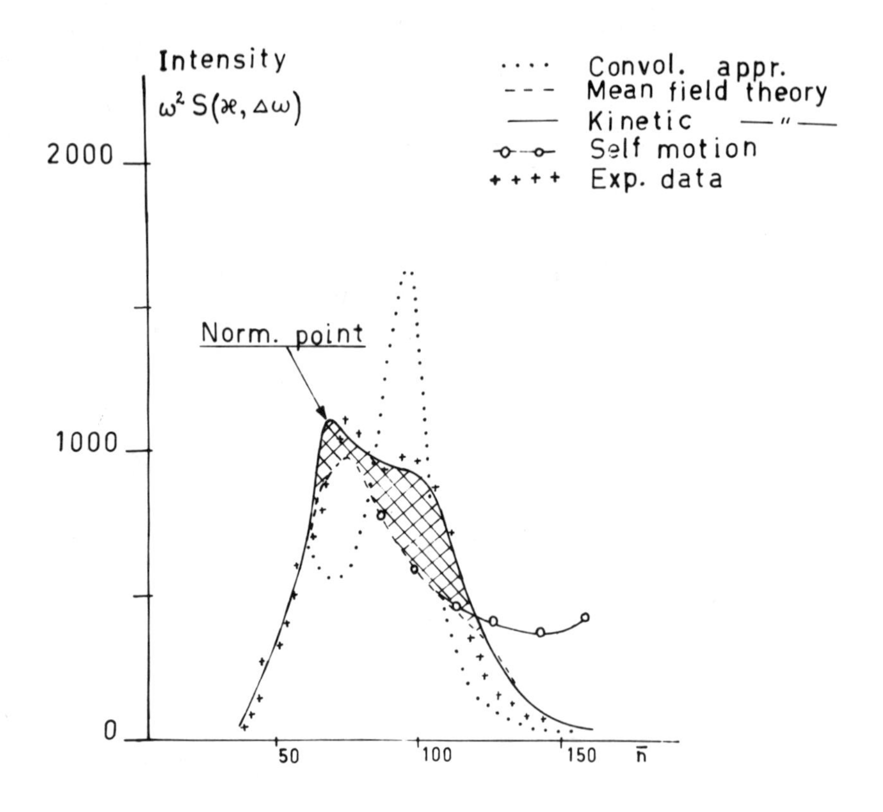

Fig. 16: New analysis of the old experiments on liquid aluminium (taken from ref. 15).

motion follows quite directly from (VI.5) and we obtain

$$\frac{d}{dt}\,\phi(t) + \int_0^t dt'\;\Gamma(t-t')\;\phi(t') = 0 \qquad \text{(VI.10)}$$

with the initial value $\phi(0) = \langle v^2 \rangle = 3k_BT/m$. The formal expression for $\Gamma(t)$ is known. It contains processes where the self particle interacts with a surrounding particle at time t', then continues to move through the medium, all the time interacting with this, up to the time t (see figure 17). The whole process can be split up into steps, where we first have one single binary collision with a subsequent free motion of the self particle. The surrounding particles are permitted to collide among themselves and in this way building up a collective backflow. All this is a

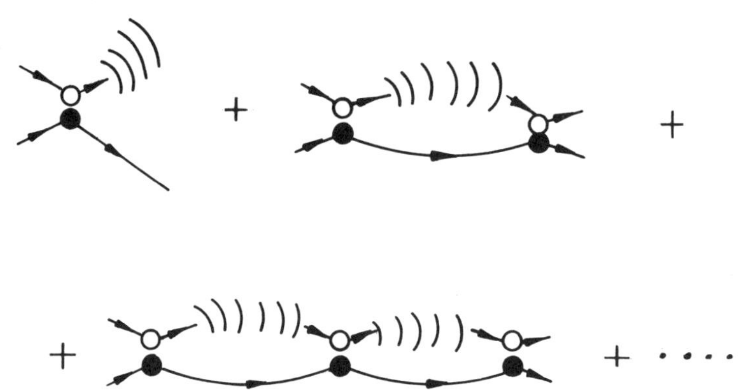

Fig. 17: Illustration of binary collision and recollisions entering in the velocity correlation function.

part of $\Gamma(t)$ and will be split off and denoted by $\Gamma^B(t)$. In the second step the self particle interacts once more with its surrounding, giving what is called the ring term in $\Gamma(t)$. This continues with more and more recollisions of the self particle as illustrated in the figure. It would not be appropriate here to go into the rather technical mathematical details, for which we refer to a forthcoming paper in Journal of Physics C, but state only that the full memory function splits into two parts, the binary part Γ^B and a collective part of a mode-mode coupling type with an essential frequency dependence in the coupling con-

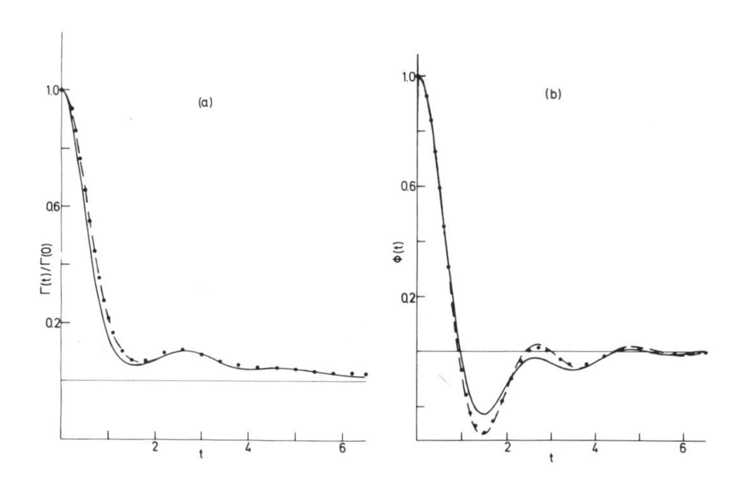

Fig. 18: Molecular Dynamics (dots) and theoretical (full curve) results for the velocity correlation function of liquid rubidium and the corresponding memory function (taken from ref. 33). The MD data are from unpublished work of A. Rahman.

stants. Figures 18 and 19 show comparisons with molecular dynamics data. The slight oscillations we find in the tail of $\Gamma(t)$ for rubidium is due to the short wavelength resonances we have in $S(q,\omega)$. Liquid argon shows no such resonances and there are no oscillations in $\Gamma(t)$ either. The oscillations in rubidium are analogous to those found in dense one-component plasmas[36], where they originate from the plasma oscillations in the backflow.

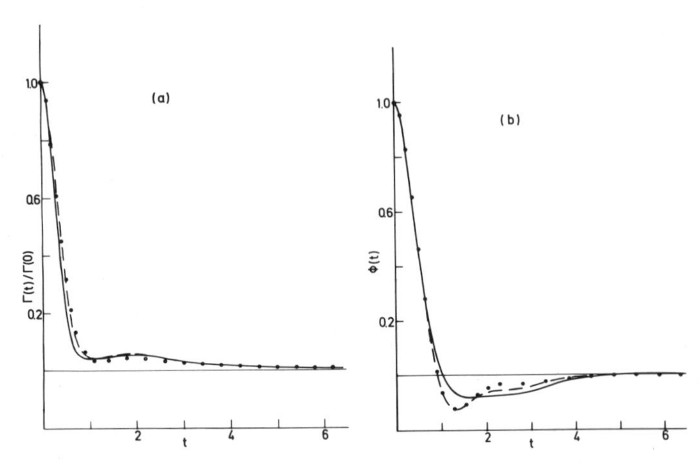

Fig. 19: Molecular Dynamics (dots) and theoretical (full curve) results for the velocity correlation function of liquid argon and the corresponding memory function (taken from ref. 33). The MD data are from D. Levesque and L. Verlet, Phys. Rev. A2, 2514 (1970).

VII. CONCLUDING REMARKS

We have here tried to give an up to date presentation of our understanding of the dynamics of simple liquids, including free electron like metals. Particularly the most recent developments, either along the lines used by Götze and collaborators or along those used by the Gothenburg group, have in a significant way deepened our understanding and brought out the most essential aspects of the dynamics involved. The earlier parametrized models have been replaced by first principle approaches without any adjustable parameters to be fitted to the experimental data. This makes that agreement between theory and experiments can be seriously evaluated and concerning discrepancies one can more easily trace their origin. It is obvious that we still have to make improvements, may be even some conceptual ones, before we have a true quantitative theory. In the approach of Götze, further tests of the applicability of their mode-mode decoupling seem necessary. Similarly, the Gothenburg group has so far always relied on the Gaussian approximation for the self motion and this certainly needs improvement. One has either to try to solve directly eq. (VI.5) or to include higher order terms beyond the Gaussian approximation, possibly based on the formally exact expression of Tokuyama and Mori[37]. All the calculations so far have ignored coupling to temperature fluctuations and this causes an important defect in the hydrodynamic limit and possibly also for intermediate wavevectors. It is no conceptual difficulty to extend the theory in this respect. The difficulties lie in finding good quantitative approximations. The same is true, when we want to extend the theory to handling the ions and the conduction electrons on equal footing.

The basic concepts, which have been introduced, are general enough to make the present approaches useful also for supercooled liquids and for amorphous solids. Here, however, considerable modification of the approximations are necessary. A qualitative analysis of available data on supercooled gallium[38] and argon[39] has been carried out recently along these lines[40] and it certainly encourages further studies of similar kind.

A characteristic feature of all the approaches, based on the Mori formalism, is that the static properties of the system have to be known and are not obtained on the same footing as the dynamics. For liquids this is an important point since several of the qualitative features of the dynamics are to a large extent determined by the local static structure. In order to have a complete theory for liquids one would demand that both the statics and the dynamics should come out from the theory and only the interaction potential being the input. Only then would one be sure that both parts are treated consistently. A good illustration of this point is critical dynamics near the gas-liquid critical point. $S(q)$ shows here a strong peak around $q=0$ due to the slow, long wavelength critical fluctuations. There is here an obvious connection between the statics and the dynamics and one could ask oneself, when making a particular mode-mode coupling approximation on the dynamics, whether this is consistent with the static structure factor one feeds in.

If this review could stimulate others to try to make further improvements on the recent approximate theories of liquids and further widen their applicability, it has served two purposes.

ACKNOWLEDGEMENTS

It is a great pleasure to acknowledge enlightening discussions with Professor Sidney Yip and with Professor Wolfgang Götze and other members of the Munich group. I also want to thank my collaborators over the recent years, Drs. Staffan Sjödin, Swapan Mitra, Lucasz Turski and above all Dr. Lennart Sjögren, who has contributed in the most significant way to the results of the Gothenburg group.

REFERENCES

1. T. Toya, J. Res. Inst. Catalysis, Hokkaido Univ. 6, 161, 183 (1958); 7, 60 (1959)
   G. Baym, Ann. Phys. 14, 1 (1961).
   L. J. Sham and J. M. Ziman, Solid State Phys. 15, 221 (1963)
2. D. J. W. Geldart and T. Taylor, Can. J. Phys. 48, 167 (1970).
   P. V. S. Rao, J. Phys. Chem. Solids 35, 669 (1974).
3. M. Watabe and M. Hasegawa, The Properties of Liquids Metals, ed. S. Takenchi, p. 133 (1973) (Taylor and Frances).
   J. Chihara, The Properties of Liquid Metals, p. 137 (1973) (Taylor and Frances).
   N. M. March and M. P. Tosi, Ann. Phys. 81, 414 (1973).
4. W. H. Young, Proc. Third Int. Conf. on Liquid Metals, p. 1,1977 (Conf. series no. 30, The Institute of Physics, Bristol and London).
5. J. D. Weeks, D. Chandler and H. C. Anderson, J. Chem. Phys. 54, 5237 (1971).
6. S. K. Mitra and M. J. Gillan, J. Phys. C 9, L515 (1976)
   S. K. Mitra, J. Phys. C 10, 2033 (1977).
7. L. Van Hove, Phys. Rev. 95, 249 (1954).
8. J. R. D. Copley and S. W. Lovesey, Repts. Prog. Phys. 38, 461 (1975).
9. A. Sjölander, in Thermal Neutron Scattering p. 291, ed. P. A. Egelstaff, 1965, (Academic Press, N.Y.).
   P. A. Egelstaff, An Introduction to the Liquid State, 1967 (Academic Press, N.Y.)
10. K. E. Larsson, U. Dahlborg and D. Jovic, Proc. Int. Conf. on Inelastic Scattering of Neutrons, vol. II, p. 117 (Vienna 1965)
    U. Dahlborg and K. E. Larsson, Arkiv f. Fysik 33, 271 (1966).
11. J. R. D. Copley and J. M. Rowe, Phys. Rev. Letters 32, 49 (1974); Phys. Rev. A9, 1656 (1974).
    J. B. Such, "Strengesetze des Polykristallinen und des flüssigen Rubidiums", KFK 2231 (1975), Gesellschaft für Kernforschung M. B. H., Karlsruhe (Ph.D. Thesis).
12. M. W. Johnson, B. McCoy, N. H. March and D. I. Page, Phys. Chem. Liquids, vol. 6, p. 243 (1977).
13. O. Söderström, M. Davidovic, U. Dahlborg and K. E. Larsson, Proc. Int. Symposium on Neutron Inelastic Scattering, vol. II, p. 67, Vienna 1977.
    O. Söderström, J. R. D. Copley, B. Dorner and J. B. Suck (private communications).
14. O. Eder, B. Kunsch, J. B. Suck, unpublished.
15. K. E. Larsson, Phys. Chem. Liquids, 1979 (in press).
16. K. Sköld, J. M. Rowe, G. Ostrowski, P. D. Randolph, Phys. Rev. A6, 1107 (1972).
17. H. Bell, H. Moeller-Wenghoffer, A. Kollmar, R. Stockmeyer, T. Springer and H. Stiller, Phys. Rev. A11, 316 (1974).
18. B. J. Alder and T. E. Wainwright, J. Chem. Phys. 31, 459 (1959)
    A. Rahman, Phys. Rev. 136, A405 (1964).

19. A. Rahman, unpublished (see ref. 32).
20. S. Sjödin and A. Sjölander, Phys. Rev. 18, 1723 (1978).
21. J. P. Hansen, I. R. McDonald and E. L. Pollock, Phys. Rev. A11, 1025 (1975).
22. G. H. Vineyard, Phys. Rev. 110, 999 (1958).
23. M. Nelkin and S. Ranganathan, Phys. Rev. 164, 222 (1967).
W. C. Kerr, Phys. Rev. 174, 316 (1968).
K. S. Singwi, K. Sköld and M. P. Tosi, Phys. Rev. A1, 454 (1970)
P. Ortoleva and M. Nelkin, Phys. Rev. A2, 187 (1970).
K. N. Pathak and K. S. Singwi, Phys. Rev. A2, 2427 (1970).
24. L. Kadanoff and P. C. Martin, Ann. Phys. 24, 419 (1963)
D. Forster, P. C. Martin and S. Yip, Phys. Rev. 170, 160 (1968)
C. H. Chung and S. Yip, Phys. Rev. 182, 323 (1969).
A. Z. Akcasu and E. Daniels, Phys. Rev. A2, 962 (1970).
25. H. Mori, Progr. Theor. Physics 33, 423 (1965).
26. V. F. Sears, Can. J. Phys. 48, 616 (1970)
J. Kurkijarvi, Ann. Acad. Sci. Fenn. A346, 1 (1970)
N. K. Ailawadi, A. Rahman and R. Zwanzig, Phys. Rev. A4, 1616 (1971).
K. Kim and M. Nelkin, Phys. Rev. A4, 2065 (1971)
K. N. Pathak and R. Bansal, J. Phys. C6, 1989 (1973)
P. K. Kahol, R. Bansal and K. N. Pathak, Phys. Rev. A14, 408 (1976)
F. Yoshida and S. Takeno, Progr. Theor. Phys. 53, 293 (1975); 58, 15 (1977).
F. Yoshida, Progr. Theor. Phys. 54, 1009 (1975); 56, 1374 (1976)
S. W. Lovesey, J. Phys. C6, 1856 (1973)
L. Sjögren, J. Phys. C11, 1493 (1978).
27. H. Mori, Progr. Theor. Phys. 34, 399 (1965).
28. W. Götze and M. Lücke, Phys. Rev. A11, 2173 (1975)
J. Bosse, W. Götze and M. Lücke, Phys. Rev. A17, 434 (1978); 17, 447 (1978); 18, 1176 (1978)
J. Bosse, W. Götze and A. Zippelius, Phys. Rev. 18, 1214 (1978).
29. T. Munakata and A. Igarashi, Progr. Theor. Phys. 58, 1345 (1977); 60, 45 (1978).
30. A. Z. Akcasu and J. J. Duderstadt, Phys. Rev. 188, 479 (1969).
31. G. F. Mazenko, Phys. Rev. 7, 209 (1973); 7, 222 (1973); A9, 360 (1974)
C. D. Boley, Ann. Phys. 86, 91 (1974); Phys. Rev. A11, 328 (1975)
E. P. Gross, J. Stat. Phys. 11, 503 (1974); 15, 181 (1976)
P. Resibois and J. L. Lebowitz, J. Stat. Phys. 12, 483 (1975)
G. F. Mazenko and S. Yip in Modern Theoretical Chemistry, vol. 6, 1977 ed. B. J. Berne (Plenum Press, N.Y.)
P. M. Furtado, G. F. Mazenko and S. Yip, Phys. Rev. A12, 1653 (1975); 14, 869 (1976)
J. R. Mehaffey, R. C. Desai and R. Kapral, J. Chem. Phys. 66, 1665 (1977).
J. R. Mehaffey and R. I. Cukier, Phys. Rev. A17, 1181, 1978.

32. L. Sjögren and A. Sjölander, Ann. Phys. 110, 122, 421 (1978)
L. Sjögren, Ann. Phys. 110, 156, 173 (1978); 113, 304 (1978),
J. Phys. C11, 1493 (1978).
33. L. Sjögren and A. Sjölander, J. Phys. C, 1979 (in press)
L. Sjögren, J. Phys. C, 1979 (in press)
34. M. S. John and D. Forster, Phys. Rev. A12, 254 (1975)
W. Götze and A. Zippelius, Phys. Rev. A14, 1842 (1976)
A. Zippelius and W. Götze, A17, 414 (1976)
S. Sjödin and A. Sjölander, Phys. Rev. A18, 1723 (1978).
35. A. Rahman, Phys. Rev. 136, A405 (1964)
D. Levesque and L. Verlet, Phys. Rev. A2, 2514 (1970)
B. R. A. Nijboer and A. Rahman, Physica 32, 415 (1966)
W. Götze and A. Zippelius, Phys. Rev. A14, 1842 (1976)
36. H. Gould and G. F. Mazenko, Phys. Rev. Lett. 35, 1455 (1975)
Phys. Rev. A15, 1274 (1977)
S. Sjödin and S. K. Mitra, J. Phys. A10, L163 (1977);
J. Phys. C11, 2655 (1978).
37. M. Tokuyama and H. Mori, Progr. Theor. Phys. 55, 411 (1976).
38. D. I. Page, D. H. Sanderson and C. G. Windsor, J. Phys. C6,
212 (1973).
L. Bosio and C. G. Windsor, Phys. Rev. Lett. 35, 1652 (1975).
39. A. Rahman, M. J. Mandell, J. P. McTague, J. Chem. Phys. 64,
1564 (1976)
M. J. Mandell, J. P. Tague and A. Rahman, J. Chem. Phys. 64,
3699 (1976).
40. A. Sjölander and L. A. Turski, J. Phys. C11, 1973 (1978).

THE THERMODYNAMICS AND STRUCTURE OF LIQUID METALS AND THEIR ALLOYS.

R. Evans

H. H. Wills Physics Laboratory, University of Bristol,

Bristol, BS8 1TL, U. K.

During the last decade or so considerable experimental and theoretical effort has gone into understanding the structure and the thermodynamic properties of liquid metals and their alloys. X-ray and neutron diffraction data is now available for many pure metals and several binary alloys and this complements a large body of thermodynamic data. Progress on the theoretical side has been significant. Indeed it is possible to argue that the simple (non-transition or rare-earth) pure metals are now well-understood - albeit not in the same quantitative detail as the argon-like insulating liquids. This progress has been achieved by combining results from two different branches of condensed matter physics. The pseudopotential theory of electrons in metals provides a realistic prescription for the effective interionic forces which can then be employed in the various schemes (Monte Carlo, molecular dynamics, diagrammatic expansions, perturbation theories) originally developed for treating the statistical mechanics of argon-like liquids. Transition and rare-earth metals are much less understood; for such systems there is, as yet, no tractable theory for the electronic structure which permits the calculation of effective interionic forces. Binary liquid alloys of simple metals exhibit an interesting diversity of behaviour; most systems show complete miscibility while others have miscibility gaps; in most alloys the electronic properties vary smoothly with concentration while in others 'compound' formation occurs at certain stoichiometric compositions with accompanying changes in electronic, structural and thermodynamic properties. Some progress has been made in understanding such phenomena.

These lectures will describe some of the work in this

general area. Since reviews of the subject by the present author (1) and by Ashcroftand Stroud (2) have recently been published, we do not publish the lecture notes for this school. Useful articles and references to recent work are listed below.

REFERENCES

1. R. Evans in 'Microscopic Structure and Dynamics of Liquids' ed. J. Dupuy and A. J. Dianoux (Nato ASI Series B. Vol 33 : Plenum) p. 153 (1978). This article discusses effective pairwise potentials, the application of thermodynamic perturbation theories to liquid metals, various results for the free energy, entropy and compressibility and schemes for calculating liquid structure factors. Some discussion of surface tension and methods for inverting experimental structure factor data to obtain interionic potentials is also given. The emphasis throughout is on pure metals.

2. N.W. Ashcroft and D. Stroud, Sol. St. Phys. 33 ed. H. Ehrenreich, F. Seitz and D. Turnbull (Academic) p. 1 (1978). This review covers many of the topics in ref. 1 and also gives useful background material on the theory of liquids and pseudo-potential theory. Liquid alloys are considered in more detail; theories of phase separation are described. Work on melting in the pure alkalis is reviewed.

3. A short review of the hard-sphere description of liquid metals and alloys is given by W.H. Young in 'Liquid Metals 1976' ed. R. Evans and D.A. Greenwood (Inst. of Phys. Conf. Series, 30) p. 1 (1977). The same volume contains many experimental and theoretical papers in this general area.

4. An instructive summary of work on concentration fluctuations and partial structure factors in binary alloys is given by A.B. Bhatia in 'Liquid Metals 1976' (see ref. 3) p. 21.

5. The recent book by M. Shimoji 'Liquid Metals' (Academic) (1977) provides a good introduction to the whole subject and has useful sections on thermodynamic and structural properties.

6. A general review of the electronic and cohesive properties of disordered simple metals (including work on disordered crystalline alloys, point defects and metallic surfaces) is given by R. Evans in 'Electrons in Disordered Metals and at Metallic Surfaces' ed. P. Phariseau, B.L. Gyorffy and L. Scheire (Nato ASI Series B Vol. 42 : Plenum) p. 417 (1979).

7. The results of ab-initio calculations of structure factors, excess volume, enthalpy and entropy for some binary alloys are presented by J. Hafner, Phys. Rev. A, 16, 351 (1977). This

paper shows how a rather simple alloy theory can give reasonable quantitative results even for the very small energy differences which occur in alloy properties.

8. M. Hasegawa and W.H. Young, J. Phys. F : Metal Phys., 7, 2271 (1977) have carried out first principles calculations of melting in Cd-Mg alloys.

9. P. Minchin, M. Watabe and W.H. Young, J. Phys. F : Metal Phys., 7, 563 (1977) have calculated the liquid-vapour co-existence characteristics of Na at low pressures.

10. R.D. Mountain, J. Phys. F : Metal Phys., 8, 1637 (1978) has performed Monte Carlo simulations of liquid Rb at high temperatures and low densities. The calculated structure factors differ from the measured data.

11. R. Evans and W. Schirmacher, J. Phys. C : Sol. St. Phys., 11, 2437 (1978) have discussed the long-wavelength behaviour of liquid structure factors and developed simple theories for the compressibility of liquid rare-gases and metals.

12. G. Jacucci, I. R. McDonald and R. Taylor, J. Phys. F : Metal Phys., 8, L121 (1978) have computed partial structure factors for an equimolar Na K alloy using realistic pairwise potentials and the molecular dynamics technique.

13. K.K. Mon, N.W. Ashcroft and G.V. Chester, Phys. Rev. B, 19, 5103 (1979) have modified the conventional pseudo-potential theory of the total energy of a metal to take into account core polarization effects. The inclusion of the fluctuating dipole interactions between the ion cores leads to substantial modifications of the usual pairwise potentials at relatively short separations for those metals with large ionic polariz-abilities. The authors show that the modified potentials, which now exhibit a secondary minimum at small separation, can account for the weak 'shoulders' that are observed in the measured structure factors of liquid such as Ga and Sn.

14. J.E. Enderby in 'The Metal Non-Metal Transition in Disordered Systems' ed. L.R. Friedman and D.P. Tunstall (Proc. 19th Scottish Univ. Summer School : Univ. of Edinburgh Press) p. 425 (1978) has reviewed recent work on the structure of 'compound' forming alloys.

15. F. Hensel, Advances in Physics (to appear) has reviewed the thermodynamic, magnetic and electrical properties of various 'compound' forming alloys. This article contains many useful references.

2. THEORY OF THE DISORDERED STATE

2.1 Liquid metals and alloys

CALCULATION OF ELECTRICAL RESISTIVITY OF LIQUID METALS

H. Beck

Institut de Physique de l'Université,
CH - 2000 Neuchâtel, Switzerland

We describe the liquid metal by the one-electron Hamiltonian

$$H = \frac{p^2}{2m} + \sum_n V(\vec{r} - \vec{R}(n)) + \sum_n I(\vec{r} - \vec{R}(n))\vec{\sigma} \cdot \vec{S}(n), \qquad (1)$$

$\vec{r}$, $\vec{p}$ and $\vec{\sigma}$ denoting position, momentum and spin of an electron. The electron-ion interaction includes a potential V and an exchange function I. Ionic positions $\vec{R}(n)$ and spins $\vec{S}(n)$ will be treated as given parameters (the latter assumption precludes the treatment of the Kondo problem) and will be subject to suitable configurational averages.

A rigorous evaluation of the electrical resistivity ρ might proceed in two steps :

(i) First the equilibrium electronic structure is determined by calculating the one-electron Green function G. This difficult task is not discussed here. Unless the electron-ion interaction is very strong we expect that the configurationally averaged G for a liquid metal could be interpreted in terms of some "average band structure". The band energies, $E(k) = \text{Re } E(k) + i\Gamma(k)$ will be somewhat smeared out by the effect of disorder (the "band states", labelled by k, should clearly be distinguished from the true eigenstates of (1)), but we expect that even in a transition metal (TM) or a rare earth metal (RE) one should be able to distinguish between (s-p)- and d-like bands (there will of course be hybridization). Approximate schemes leading to such results have been proposed [1, 2], but a really satisfactory calculation seems still to be missing.

(ii) In linear response theory Kubo's formula gives an exact formulation of the conductivity in terms of a current correlation function. (We shall not consider alternative approaches based on force correlation functions [3]). Various authors [4 - 6] have shown that, to first order in the energy uncertainty $\Gamma(k_F)/E_F$ at the Fermi energy, the resulting expression for ρ has the usual form, as derived in the framework of a Boltzmann equation [7] :

$$\rho = \frac{m}{ne^2} \int d^3k \; W(\vec{k},\vec{k}')(1 - \cos\theta_{\vec{k}\vec{k}'}) \; \delta(E_F - E_{\vec{k}}) \; \delta(E_F - E_{\vec{k}'}), \qquad (2)$$

with n = number density of electrons and $\theta_{\vec{k}\vec{k}'}$ = scattering angle. The transition probability W is an infinite series in V and G (usually the spin dependent part of (1) is not considered). If V is a small pseudopotential, appropriate for simple metals, the lowest order in V and the free-electron form G_0 yields Ziman's result [7] involving the structure factor S(q). Summing single-site scattering terms introduces the atomic t-matrix [6, 8], and suitable decoupling of higher order ionic structure factors allows for taking into account some multiple scattering effects [6, 9].

For treating TM and RE systems the usual single-site t-matrix approach [8, 10] is not really satisfactory, since it does not take into account the "true band structure", mentioned above : the practical calculation of the phase shifts is done for free electron boundary conditions and no distinction can be made between the different mobilities of s- and d-electrons. A suitable framework for calculating ρ should

(a) take into account the electronic equilibrium structure, but

(b) still treat the elementary electron-ion scattering event on the basis of free electron boundary conditions for the sake of numerical simplicity (as, e.g. KKR-method in crystalline band structure calculations only needs some atomic phase shifts to characterize electron-ion interactions), and

(c) allow for simple estimates for multiple scattering.

In what follows some of the ideas of this program will be taken into account in a very crude way.

The basic formula used in most calculations reads :

$$\rho = \frac{3\pi\,\Omega_0}{4e^2\hbar\, v_F^2\, k_F^4} \int_0^{2k_F} dq\; q^3\; S(q)\; |W(q,E_F)|^2 \qquad (3)$$

(Ω_0 = atomic volume, v_F = Fermi velocity). The generalization for alloys involves partial structure factors and transition probabilities [11], and including exchange interactions leads to the extension (9) at the end of this article.

Simple metals have quite successfully been treated by replacing W by a pseudopotential. Some calculations for various systems are mentioned in the review [12]. The famous "agreement between theory and experiment" is not so easy to judge, since very often model pseudopotentials are used the parameters of which are not well known and are in fact sometimes adjusted in order to yield the right ρ-value. Moreover (3) depends quite strongly on the details of S(q), which varies very rapidly in the vicinity of the first peak at $q = k_p$. Even using known partial structure factors for an alloy $A_x B_{1-x}$ and adjusting ρ to experiment at x = 0 and x = 1 does not always produce the right behavior for intermediate x. The occurence of negative temperature coefficients (TC) of ρ, specially in liquid alloys, is successfully explained by the T-dependence of S(q) : when $2k_F \approx k_p$ the integral (3) will decrease with rising T, since heating reduces the peak height of S(q).

Here we want to concentrate on phase-shift based calculations for TM's and RE's, where V is a muffin-tin potential V_{MT} and W the corresponding ionic t-matrix $t(q,E_F)$. We shall not deal with computational details but rather stress some delicate general problems.

Already the construction of V_{MT} is conceptually more questionable than in a crystal, since the neighbors of a given ion (contributing charge overlap) are not arranged regularly. In the spirit of the adiabatic approximation one should treat each electron-ion scattering event for a given specific ionic arrangement and do the configurational average at the end. Clearly such a procedure is not realizable, so one constructs an average $V_{MT}(r)$. Seemingly the best way to do this is to arrange neighbors on spherical shells, the probability distribution being given by the pair correlation function g(R), but then one usually has to admit nearest neighbor distances shorter than twice the muffin-tin radius R_o, unless the latter is chosen unreasonably small.

The phase shifts $\delta_\ell(E)$ usually show the following behavior : s- and p-phases vary rather monotonically with E, d-phases go through some kind of resonance (sharp for late TM, less so for RE) and for $\ell \geq 3$ they are negligible in the interesting domain of energy ($E \leq E_F$), even for RE's.

The crucial problem is to find E_F and $2k_F$ (in a theory which does not really care about the real electronic structure !) for systems where s-d hybridization creates strong deviations from quasi-free electrons in the vicinity of E_F. The Fermi energy E_F would be determined by G. Some authors [13 - 15] used a single site approximation to the latter, respectively to the integrated density of states (DS) :

$$N_{SS}(E) = N_0(E) + \frac{2}{\pi} \cdot \sum_{e}(2\ell+1)\ (\delta_\ell(E) - \delta_\ell(0)), \qquad (4)$$

given by the free electron value N_0 plus a Friedel-type correction. This estimate yielded Fermi energies for the RE's of the same order as in the crystalline states, producing an increasing trend through the series [15], as in the solid. In the late TM's E_F successively moves away from the d-resonance [10], which explains the decrease of ρ from Mn to Cu. A different method to determine E_F by using crystalline data and estimates for the bottom of the s-band in the liquid was used in ref. 10 and elsewhere.

The problem of k_F is more embarassing, since it clearly reveals the defects of not accounting for s- and d-states and their hybridization. Usually one assumes that the current is predominantly carried by s-electrons at the Fermi surface. For the electron-ion scattering they are treated as plane waves and $\delta_\ell(E)$ is the phase-shift of the angular momentum component ℓ of such a wave impinging on an ion. In many systems $\delta_2(E_F)$ is the largest contribution to the t-matrix in (3). It is, however, not quite clear to what extent such a treatment yields a realization of Mott's idea of strong s-d scattering [16]. In fact, he argues that the resonance can only build up if the lifetime of a d-electron (localized on an ion) with respect to a jump to a neighboring ion is long enough. This may be realized in an alloy (TM - Ge, etc.) with a large average TM-ion distance, but may fail for pure TM or RE liquids.

Thus, in our simplified picture we take into account only s-electrons, populating a free-electron-like Fermi sphere of radius k_F and giving rise to phase-shifts $\delta_\ell(E)$. Now we need some guide lines to determine the number n_c of such electrons per ion, and thus k_F itself. For noble metals the choice $n_c = 1$ seems obvious [10]. For the late 3-d TM band structure and partial DS calculations for the solid phase prompted Dreirach et al [10] to choose $n_c = 1$, whereas ref. [17], yielding a number n_s of s-like electrons for the crystalline RE series, increasing from 0.5 for La to about 1.3 for Lu, was the base of ref. [15]. Obviously the choice of n_c and thus of $2k_F$ has its direct consequences also on the TC of ρ. Alternatively one can obtain n_c in the same framework as E_F [13 - 15] : the partial DS for s- and p-states,

$$N_{sp}(E) \equiv N_0(E) + \frac{2}{\pi} \cdot \sum_{\ell=0}^{1} (2\ell+1)\ (\delta_\ell(E) - \delta_\ell(0)), \qquad (5)$$

yields n_c by

$$n_c = N_{sp}\ (E_F). \qquad (6)$$

For the trivalent RE's this produced the same variation from La to Lu as that of n_s in the crystal, which seems to demonstrate the usefulness of the approach.

At this stage one is faced with a final dilemma concerning $2k_F$ and E_F : the free electron relation yields

$$k_F^3 = 3\pi^2 n_c, \tag{7}$$

but the only choice which is consistent with the free electron boundary conditions used to calculate phase-shifts [18] is $k_F^2 = 2mE_F$. Sometimes, as for the divalent RE's, Eu and Yb, the latter yields the same value as (6) and (7) [14], but in many cases k_F would be too large, which yields too high values for ρ [18] and wrong TC's. While earlier work often used E_F and k_F values not consistent in this sense [18], the authors of ref. [15] tried to remedy this problem for the trivalent RE's in the following way : V_{MT} and its phase shifts were used to evaluate E_F and k_F by virtue of (4) to (7). The dispersion relation of the s-band was then approximated by

$$E(k) = E_B + k^2/2m \tag{8}$$

with the bottom energy $E_B = E_F - k_F^2/2m$. The fact that the band "takes off" at E_B instead of $E = 0$ reflects ion-electron interaction. We stress again, that a more complete theory would tell us how to deal with the scattering properties of the true quasi-particles of the liquid with energy-momentum relations like (8). As a crude approximation for the effect of the surroundings on the incoming electron a new muffin-tin potential $\tilde{V}_{MT}$ was built, equal to V_{MT} for $r < R_0$, but augmented by the amount E_B outside. Thus, with respect to $\tilde{V}_{MT}$, the incoming electrons moved like really free particles ($\tilde{E} = k^2/2m$), corresponding to the usual boundary conditions. New phase-shifts $\tilde{\delta}_\ell$ were then calculated at $\tilde{E}_F = E_F - E_B$ and used to evaluate ρ.

Details of resistivity calculations done along these lines are found in the quoted papers and further references therein. Structure factors are either taken from scattering experiments, if available, or modeled by hard spheres. Generally the agreement with experiment is reasonable, considering all the uncertainties and approximations. Already differing methods of handling the charge overlap or the exchange in V_{MT} can lead to drastically different ρ-values [13 - 15]. In order to demonstrate the usefulness of the procedure described in this contribution (which may indeed be surprising on a purely formal basis, since all these free-electron-like arguments are expected to break down when the mean free path is not much longer than typical ionic distances), I would like to summarize some facts and results concerning the RE's [14, 15, 19] :

(i) The numbers n_c of conducting electrons as determined by (6), (7) are in good agreement with the results for n_s in the crystal [17]. For Eu and Yb, which behave as divalent in the solid, $n_c \approx 1.8$, indicating only a small effect of s-d hybridization for $E \leq E_F$.

(ii) The monotonic increase of ρ from La to Lu is quantitatively reproduced. It is due to the increase of n_c and thus of the upper limit $2k_F$ in (3). The "exceptional" values of Eu (highest ρ) and Yb (lowest) are traced back to the potentials of these elements [14].

(iii) The most important fact pointing to an internal consistency in the whole approach is the relatively high value $n_c \approx 1.3$ for Lu. This leads to a $2k_F$-value for which a negative TC is predicted. Indeed, the TC of ρ, starting at slightly positive at La decreases through the series and is slightly negative for Lu [20]. The negative TC of Eu and Yb (for $n_c \approx 1.8$, $2k_F \geq k_p$) were also reproduced.

(iv) The experimental results for the TC of alloys confirm again the consistency of the n_c-values : alloys of RE's with polyvalent metals (Sn, Al, ..) show negative TC's in some concentration range (where $2k_F \approx k_p$), whereas those with monovalent elements in general do not [20]. Here one has to admit that GdCo is an interesting exception : it has a negative TC both in the liquid and glassy state ! This is a hint that the electronic structure of this d-band alloy is more complicated, so that a simple summation of n_c for both components is not allowed : it would yield a net $n_c < 1$ and thus a positive TC.

It is interesting to note that a recent first principles calculation of ρ for crystalline Pd [21] confirms that most of the current is indeed carried by "s-like" electrons, which are strongly scattered into d-states. This seems to provide strong support to Mott's [16] s-d scattering model and, at the same time, to our neglect of d-electrons in calculating ρ. In fact the latter may be even less mobile in a liquid, since the disorder will tend to reduce the "hopping opportunity".

Before concluding we briefly present a generalization of (3) including the spin-dependent part of (1). As proposed by Parrinello et al [22] one calculates two t-matrices t_J for total spin $J = S \pm \frac{1}{2}$ of the combined electron-ion system (S is the ionic spin). After a configurational average over ionic positions and spins [19] one ends up with an integral like (3) :

$$\rho = \frac{3\pi\Omega_0}{4e^2\hbar v_F^2 k_F^4} \int_0^{2k_F} dq\, q^3 [S(q)R_1(q) + (1+M(q))(R_2(q)-R_1(q))]. \quad (9)$$

Here

$$R_1(q) = \left| \sum_J \frac{2J+1}{2(2S+1)} \quad t_J(q,E_F) \right|^2 \quad (10)$$

$$R_2(q) = \sum_J \frac{2J+1}{2(2S+1)} \left| t_J(q,E_F) \right|^2, \quad (11)$$

and M(q) is the Fourier transform of the ionic spin correlation function (which is unimportant in the disordered liquid). Using this approach it has been shown [19] that even for Gd (with the highest 4f-spin of S = 7/2) the effect of including (s-f)-exchange in the liquid enhances ρ only by about 10 $\mu\Omega$ cm. This is also indicated by experiments [23]. It is still unclear, however, why the "magnetic" part of ρ seems to decrease continuously from room temperature (about 100 $\mu\Omega$cm for Gd) up to the melting temperature.

REFERENCES

1. G.J. Morgan, J. Phys. C 2, 1454 (1969).
2. A. Bansil, H.K. Peterson, L. Schwartz, Liquid Metals, 1976, Conference Series No. 30, Inst. of Physics, Bristol and London, p. 313.
3. M. Huberman, G.V. Chester, Adv. Physics 24, 489 (1975).
4. J. Rubio, J. Phys. C 2, 288 (1969).
5. L.E. Ballentine, J. Phys. C 3, L16 (1970).
6. N.W. Ashcroft, W. Schaich, Phys. Rev. B 1, 1370 (1970).
7. J.M. Ziman, Phil. Mag. 6, 1013 (1961).
8. R. Evans, D.A. Greenwood, P. Lloyd, Phys. Lett. A 35, 57 (1971).
9. H.N. Dunleavy, W. Jones, J. Phys. F 8, 1477 (1978).
10. O. Dreirach, R. Evans, H.-J. Güntherodt, H.-U. Künzi, J. Phys. F 2, 709 (1972).
11. T.E. Faber, J.M. Ziman, Phil. Mag. 11, 153 (1965).
12. G. Busch, H.-J. Güntherodt, Sol. State Phys. 29, 235 (1974).
13. A.H.M. Lopez-Escobar, J.S. Brown, Phil. Mag. 35, 1609 (1977).
14. B. Delley, H. Beck. D. Trautmann, F. Rösel, J. Phys. F 9, 505 (1979).
15. B. Delley, H. Beck. J. Phys. F 9 517 (1979).
16. N.F. Mott, Phil. Mag. 26, 1249 (1972).
17. J.C. Duthie, D.G Pettifor, Phys. Rev. Lett. 38, 564 (1977).
18. E. Esposito, H. Ehrenreich, C.D. Gelatt Jr. Phys. Rev. B 18, 3913 (1978).
19. B. Delley, H. Beck. J. Phys. F, to appear.

20. B. Delley, H. Beck, H.-U. Künzi, H.-J. Güntherodt, Phys. Rev. Lett. 40, 193 (1978).
21. F.J. Pinski, P.B. Allen, W.H. Butler, Phys. Rev. Lett. 41, 431 (1978).
22. M. Parrinello, N.H. March, M.P. Tosi, Nuovo Cimento 39 B, 233 (1977).
23. H.-J. Güntherodt, E. Hauser, H.-U. Künzi, p. 324 of ref. 2.

This work has been supported by the Swiss National Science Foundation.

THEORY OF THE CONDUCTIVITY OF LIQUID ALLOYS

F. Brouers

Institut für Theoretische Physik, Freie Universität,
1000 Berlin 33

1. INTRODUCTION

The electronic properties of many materials such as mixed crystals, glasses and metal alloys change dramatically with chemical composition. In the last few years a number of liquid alloys have been discovered which, although formed from two metallic components, exhibit non-metallic behaviour in a well-defined concentration range near stochiometry. In addition to their small electrical conductivity, these alloys often are found to have anomalous shifts in their NMR frequencies, densities, magnetic susceptibilities and thermopowers. One of the most thoroughly investigated examples is the liquid caesium-gold system.

Near the equiatomic composition a marked metal-non-metal transition is observed /1/ and simultaneously the magnetic susceptibility exhibits a pronounced diamagnetic minimum /2/. These authors have postulated that these effects are caused by a high degree of ionicity in the bonding of the system which is consistent with the large difference between the electronegativities of Cs atom. (Pauling electronegativity 0.7) and Au atom (electronegativity 2.1). Similarly for Li_xPb_y a dip in the conductivity occurs when the crucial concentration is such that $x = 4y$; this corresponding to four Li atoms each contributing one electron to an incomplete shell of the Pb atom. Neutron diffraction experiments at this concentration /3/ have suggested that there is a preference for unlike nearest neighbours in the liquid metal alloy. Another important result from neutron diffraction measurements is obtained from the radial distribution function. It is found that there is a well-defined short range order in the liquid metal alloy and that at the crucial concentration the co-ordination is very nearly equal to that present in the solid state phase of the alloy. The well-defined number of nearest neighbours is not necessarily indicative that all the nearest neighbours are of unlike types.

It is this rather well-defined short range order that has implied that a theory for the structure of disordered system may be applicable to this problem. In order to interpret and clarify the increasing number of experimental data for these liquid metal alloys, simple physical models and tight binding Green function technique have been used recently /4, 5, 6/. These papers are based upon the cluster Bethe-lattice method introduced by Brouers et al /7/ and developed by Falicov et al /8/ to describe the short range order. Although the model was developped for the solid state, it should give an idea of the liquid density of states. Using that model, Franz et al /6/ have been able to interpret the experimental data of liquid Cs_xAu_{1-x}, i.e. the variation of conductivity with concentration and temperature and the variation of magnetic susceptibility and Knight shift with concentration. The reason for the success of the model comes from the fact that a metal-insulator transition in CsAu is such a dominant phenomenon that details of the band structure are of secondary importance. The large amount of charge transfer, i.e. ionic character, and the resulting strong short range order are the decisive factors governing the properties of the system. Liquid CsAu can be considered as a reference system to which similar liquid alloy systems can be compared. A systematic study of how the physical parameters such as conductivity, magnetic susceptibility and Knight shift depend on the model parameters such as difference in electronegativity, relative bandwidths and size of Coulomb interactions and an improvement of the model to make contact with typical liquid properties such as partial structure factors and short range order parameters require a reformulation of the theory developped by Franz et al /6/ in the language of liquid physics using both model and experimental radial distribution functions.

In these lectures we want to review the tight binding formulation of liquid alloy conductivity based on the renormalized perturbation theory of Watson /9/, Anderson /10/ and Brouers et al /7/. A detailed bibliography of papers dealing with this approach can be found in Movaghar et al /11/.

2. THE LIQUID DENSITY OF STATES

We consider the following tight binding Hamiltonian in the Wannier basis:

$$H = H_0 + T \quad (1)$$

$$= \sum_i \varepsilon_i \; |i><i| \; + \sum_{i \neq j} t_{ij} \; |i><j| \quad (2)$$

Here $|i>$ is the localized state on site i and the t_{ij}, non zero only when i and j are nearest neighbours, are the hopping integrals responsible for the propagation of electrons. In the alloy A_xB_{1-x}, the random atomic levels assume one of the two possible values E_A and E_B, depending on whether an atom of type A or B occupies the site i. The hopping integrals can take the three possible values t_{AA}, t_{AB} and t_{BB} according to the occupation of site i and j.

For a given configuration, the site diagonal Green function may be defined by

$$G_{ii}(z) = <i| \frac{1}{z-H} |i> = g_{ii} + g_{ii} \sum_{j \neq i} t_{ij} \, G_{ji} \quad (2)$$

where

$$g_{ii}(z) = <i| \frac{1}{z-H_0} |i> = (z-\varepsilon_i)^{-1}$$

and

$$G_{ij}(z) = <i| \frac{1}{z-H} |j> = g_{ii} \, \delta_{ij} + g_{ii} \sum_{k \neq i} t_{ik} G_{kj} \quad (3)$$

By iterating equation (2) using equation (3), G_{ii} can be rewritten

$$G_{ii}(z) = (z-\varepsilon_i-\Delta_i)^{-1} \quad (4)$$

where Δ_i is the self-energy for state i and is given by

$$\Delta_i(z) = \sum_{j \neq i} \frac{t_{ij}^2}{z-\varepsilon_j} + \sum_{j \neq i} \sum_{k \neq i,j} \frac{t_{ij} \, t_{jk} \, t_{ki}}{(z-\varepsilon_j)(z-\varepsilon_k)} + \ldots \quad (5)$$

One can partially carry out the sums in equation (5) by including, in the energy denominators of (5), self-energies written as a sum of terms each of which corresponds to non-repeating path /9, 10, 7/

$$\Delta_i(z) = \sum_{j\neq i} t_{ij}\,(z-\varepsilon_j-\Delta^{(i)}(z))^{-1}\,t_{ji}$$

$$+\sum_{j\neq i}\sum_{k\neq i,j} t_{ik}\,(z-\varepsilon_k-\Delta_k{}^{(i,j)})^{-1} t_{kj}(z-\varepsilon_j-\Delta_j{}^{(i)})^{-1} t_{ji}$$

$$+\sum_{j\neq i}\sum_{k\neq i,j}\sum_{l\neq i,j,k} t_{il}(z-\varepsilon_l-\Delta_l{}^{(i,j,k)})^{-1}\,t_{lk}\cdots \qquad (6)$$

A term such as $\Delta_l{}^{(i,j,k)}$ denotes the self-energy on site l for a path in which the sites i,j,k have been removed and is defined by a series similar to equation (6) but with these sites excluded from all sums.

For example

$$\Delta_j{}^{(i)}(z) = \sum_{k\neq i,j} t_{jk}(z-\varepsilon_k-\Delta_k{}^{(i,j)})^{-1}\,t_{kj} +$$

$$\sum_{k\neq i,j}\sum_{l\neq i,j,k} t_{il}\ldots \qquad (7)$$

The local Green function

$$G_{ii} = G_{ii}\,(\vec{R}_1,\,\vec{R}_2,\,\ldots\,\vec{R}_N,\,E) = \frac{1}{E-\varepsilon_i-\Delta_i} \qquad (8)$$

depends on all other atomic sites.

The physically interesting quantity is the Green function averaged over all liquid and alloy configurations

$$\langle G_{ii}\rangle_c = \left\langle\frac{1}{E-\varepsilon_i-\Delta_i}\right\rangle_c \qquad (9)$$

The idea of the single site approximation is to treat exactly a particle at site i and to make an average over the (N-1) particles. The self-energy Δ_i does not depend on site i and the orbital energy E_i must be considered exactly. So one can make the average directly on Δ_i. One can therefore write

$$\langle G_{ii}\rangle_i = \frac{1}{E-\varepsilon_i-\langle\Delta_i\rangle_i} \qquad (10)$$

We shall first consider the liquid average.

One can see from (5) that the self-energy $\langle\Delta_i\rangle_i$ is independent of the orbital energy E_i and therefore the average self-energy $\langle\Delta_i\rangle_i$ will be calculated using the conditional distribution function defined by

$$P(\bar{R}_1,\bar{R}_2,\ldots\bar{R}_N) = P(\bar{R}_i)\, P(\bar{R}_i|\bar{R}_1,\bar{R}_2,\ldots,\bar{R}_N$$

i.e.
$$<\Delta_i>_i = \int P(\bar{R}_i|\bar{R}_1,\bar{R}_2,\ldots\bar{R}_N)\,\Delta_i \prod_{\substack{j=1\\(j\neq i)}}^{N} d\vec{R}_j \qquad (11)$$

for a given $\bar{R}_i$.

If we extend the single site approximation to all sites, we can write

$$<\Delta_i>_i = <\sum_{j\neq i} t_{ij} <G^i_{jj}>_j t_{ji} + \sum_{k\neq j\neq i} t_{ij} <G^i_{jj}>_j t_{jk} <G^{ij}_{kk}>_k t_{ki} + \ldots>_c \qquad (12)$$

with
$$<G^i_{jj}>_j = \frac{1}{E-\varepsilon_j-<\Delta^i_j>_j} \qquad (13)$$

$$<G^{ij}_{kk}>_k = \frac{1}{E-\varepsilon_k-<\Delta^{ij}_k>_k} \qquad (14)$$

The higher indices in the self-energy $\langle\Delta_k{}^{ij}\rangle_k$ indicate which sites are excluded from the average.

For high densites it is allowed to set

$$<\Delta_i>_i = <\Delta^i_j>_j = <\Delta^{ij}_k>_k = \ldots = \Sigma(E) \qquad (15)$$

This means that the quantities $\langle\Delta_i\rangle_i, \langle\Delta^i_j\rangle_j, \ldots$ are no longer different. Since in this approximation one averages also on j in $\langle\Delta^i_j\rangle_j$, one considers hops from k to j in (5). This gives a forbidden contribution of order N^{-1} which can be considered as negligible if the density is high enough.

With assumption (15), we can write the self-energy as

$$\Sigma(E) = <\sum_{j\neq i} t_{ij} <G_{ii}>_i t_{ji} + \sum_{\substack{k\neq j,i\\ j\neq i}} t_{ij} <G_{ii}>_i t_{jk} <G_{ii}>_i t_{ki} + \ldots \qquad (16)$$

and the local Green function is therefore given by

$$<G_{ii}>_i = \frac{1}{E-\varepsilon_i-<\Delta_i>_i} = \frac{1}{E-\varepsilon_i-\Sigma(E)} \qquad (17)$$

To calculate the liquid average we introduce the correlation function $g_N(\bar{R}_I,\ldots\bar{R}_N)$:

$$P_N(\bar{R}_1,\ldots,\bar{R}_N) = n^N g(\vec{R}_1,\ldots\vec{R}_N) \qquad (18)$$

$n = N/\Omega$ is the particle density.

The N-particle correlation function cannot be calculated and approximation have to be derived. The first approximation consists in expressing the N-particle correlation function in terms of pair-correlation functions.

For the pure liquid, the approximation of Matsubara et al /12/ consists simply in treating the system as completely random, i.e. $g(R_i, R_j) = 1$ for all pairs (i, j).

The approximation of Ishida and Yonezawa for the correlated case consists in decoupling the atomic correlation function into products of pairs but keeping only those joining atoms in successive hopping processes as illustrated in Fig. 1,

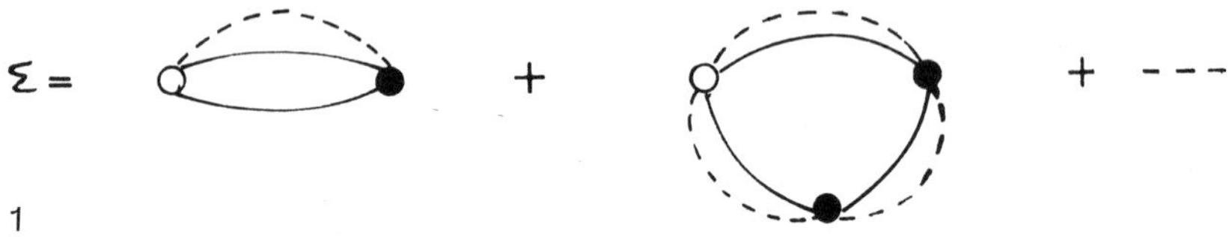

$\Sigma =$ + + ---

Fig. 1

i.e.
$$\Sigma = n \int g(R_i-R_j)t(R_i-R_j)\langle G\rangle t(R_j-R_i)dR_j + \tag{19}$$
$$+ n^2 \int g(R_i-R_j)t(R_i-R_j)\langle G\rangle t(R_j-R_k)g(R_j-R_k)\langle G\rangle t(R_k-R_i)g(R_k-R_i)dR_j dR_k$$

This expansion can be Fourier transformed yielding

$$\Sigma = n\int\frac{dk}{(2\pi)^3}\tilde{t}(k)\langle G\rangle t(k)+n^2\int\frac{dk}{(2\pi)^3}\tilde{t}(k)\langle G\rangle\tilde{t}(k)\langle G\rangle\tilde{t}(k)+\dots \tag{20}$$

with $\tilde{t}(k) = \int d^3r\, t(r)g(r)\exp(ikr)$ (21)

if $g(r) = 1, \quad \tilde{t}(k) = t(k)$.

One can sum (20) and the result given by

$$\Sigma(z) = \int\frac{d^3k}{(2\pi)^3}\frac{n\tilde{t}(k)\,\tilde{t}(k)}{z-\Sigma(z)-n\tilde{t}(k)} + n\int\frac{d^3k}{(2\pi)^3}\frac{(\tilde{t}_k t_k-\tilde{t}_k\tilde{t}_k)}{z-\Sigma(z)} \tag{22}$$

since here

$$\langle G(z)\rangle = (z-\Sigma)^{-1} \tag{23}$$

This equation has to be solved self-consistently yielding $\Sigma, \langle G\rangle$ and then the density of states since

$$n(E) = -\frac{1}{\pi}\,\mathrm{Im}\,\langle G(E+io)\rangle \tag{24}$$

This result is equivalent to Ishida et al /13/ self-energy and has been derived here in a simple and elegant way. It is an improvement over previous works which neglected the second term of (22). Numerical results can be found in Movaghar et al /14, 15/.

3. LIQUID ALLOY DENSITY OF STATES

The formalism of §2 can easily be generalized to liquid alloys with constituent depending hopping. The chain decoupling is generalized to a distribution of N atoms of N different types denoted by α_i and becomes

$$P(R_1^{\alpha_1},R_2^{\alpha_2})\ldots R_N^{\alpha_N}) = P(R_1^{\alpha_1},R_2^{\alpha_2})\ldots P(R_N^{\alpha_N}R_1^{\alpha_1}) \qquad (25)$$

The probability P(R) of finding an ion of type ß at a distance R from a given atom of type α within a volume element dR is thus

$$P(R)dR = g^{\alpha\beta}(R)\; n\; c_\beta\; dR \qquad (26)$$

Using the chain approximation, one can obtain for a binary alloy A_xB_{1-x} the following expansion

$$\Sigma_\alpha = \sum_\beta c_\beta n \int \sum_{j\neq i} t^{\alpha\beta}(\vec{R}_i-\vec{R}_j)\langle G\rangle_\beta g^{\alpha\beta}(\vec{R}_i-\vec{R}_j)t^{\beta\alpha}(\vec{R}_j-\vec{R}_i)d\vec{R}_j\; +$$

$$+n^2\sum_{\beta,\gamma} c_\gamma c_\beta \iint \sum_{j\neq i\neq k} \tilde{t}^{\alpha\beta}(\vec{R}_i-\vec{R}_j)\langle G\rangle_\beta \tilde{t}^{\beta\gamma}(\vec{R}_j-\vec{R}_k)\langle G\rangle_\gamma \tilde{t}^{\gamma\alpha}(\bar{R}_k-\bar{R}_i)d\bar{R}_j d\bar{R}_k+\ldots \qquad (27)$$

where

$$\tilde{t}^{\alpha\beta}(R_i-R_j) = g^{\alpha\beta}(R_i-R_j)t^{\alpha\beta}(R_i-R_j) \qquad (28)$$

and

$$\langle G\rangle_\nu = (E-\varepsilon_\nu-\Sigma_\nu)^{-1} \qquad (29)$$

Equation (27) can be Fourier transformed in the usual way and we obtain

$$\Sigma_\alpha(E)=n\int\frac{d^3k}{(2\pi)^3}\sum_\beta \tilde{t}_k^{\alpha\beta}t_k^{\beta\alpha}c_\beta\langle G\rangle_\beta+n^2\int\frac{d^3k}{(2\pi)^3}\sum_{\beta\gamma}\tilde{t}_k^{\alpha\beta}c_\beta\langle G\rangle_\beta\tilde{t}_k^{\beta\gamma}c_\gamma\langle G\rangle\,\tilde{t}_k^{\gamma\alpha}+\ldots$$

where $\tilde{t}_k^{\alpha\beta} = \int \tilde{t}^{\alpha\beta}(R)\exp(ikR)\, d\vec{R}$ (30)

One can sum the series $\Sigma\,\alpha$. It can be written as a geometric series in matrices A with A given by (for a binary alloy):

$$A = \begin{pmatrix} \tilde{t}_k^{AA} \langle G_A\rangle C_A & \tilde{t}_k^{AB} \langle G_B\rangle C_B \\ \tilde{t}_k^{BA} \langle G_A\rangle C_A & \tilde{t}_k^{BB} \langle G_B\rangle C_B \end{pmatrix} \tag{31}$$

We must calculate the summation

$$C = A^2 + A^3 + \ldots = 1 + A + A^2 + \ldots - 1 - A = \frac{1}{1-A} - 1 - A$$

For instance $C_{11} = \dfrac{A_{22}}{\det(1-A)} - 1 - A_{11}$ (32)

Therefore one gets

$$\Sigma_A(E) = \frac{n}{(2\pi)^3}\int d^3k \frac{1}{\det(E-nA)}\left[C_B\langle G\rangle_B \tilde{t}_k^{AB}\tilde{t}_k^{BA} + C_A\langle G\rangle_A(\tilde{t}_k^{AA})^2 - n\tilde{t}_k^{AA}\det A\right]$$

$$+ \frac{n}{(2\pi)^3}\int d^3k \left[C_A\langle G_A\rangle \tilde{t}_k^{AA} t_k^{AA} + C_B\langle G_B\rangle \tilde{t}_k^{AB}\tilde{t}_k^{BA} - C_A\langle G_A\rangle(\tilde{t}_k^{AA})^2\right.$$

$$\left. - C_B\langle G_B\rangle \tilde{t}_k^{AB}\,\tilde{t}_k^{BA}\right] \tag{33}$$

with

$$\tilde{t}_k^{AB} = \int t_{ij}^{AB}\, g_{ij}^{AB}\,(R_i - R_j)\, e^{-ikR_{ij}}\, dR_{ij}$$

$$t^{AB} = \tilde{t}^{AB} \quad (g_{ij}^{AB} = 1)$$

The Green function

$$G_\alpha = (E - \varepsilon_\alpha - \Sigma_\alpha(E))^{-1} \qquad \alpha = A,B \tag{34}$$

the self-energy $\Sigma_B(E)$ is obtained by changing the index A into B. The alloy density of states is therefore given by the average

$$\langle G_{ii}\rangle_{A_xB_{1-x}} = \sum_{\alpha=A,B} C_\alpha \langle G_\alpha\rangle = \frac{C_A}{E-\varepsilon_A-\Sigma_A(E)} + \frac{C_B}{E-\varepsilon_B-\Sigma_B(E)} \tag{35}$$

Numerical results can be found in Movaghar et al /15/.

The simplest case consists in assuming s-wave functions. In that case

$$\varphi_\nu(\vec{R}-\vec{R}_i) = \sqrt{\frac{\alpha_\nu^3}{\pi}}\; e^{-\alpha_\nu(\bar{R}-\bar{R}_i)} \qquad \nu = A,B \tag{36}$$

and

$$t^{\alpha\beta}(R_{ij}) = -\int \varphi_\alpha^*(\bar{R}-\bar{R}_i)\,\frac{e^2}{|\vec{R}-\vec{R}_i|}\,\varphi_\beta(\bar{R}-\bar{R}_j)\,d\bar{R}$$

$$\tilde{t}^{\alpha\beta} = \int t^{\alpha\beta}(R_{ij})\,g^{\alpha\beta}(R_{ij})\,e^{-ik\vec{R}_{ij}}\,d\vec{R}_{ij} \tag{37}$$

For the partial pair correlation functions one can use the modified-hard-core liquid pair correlation function

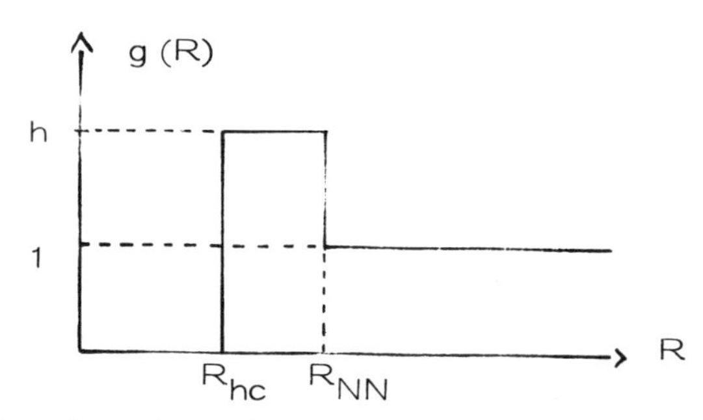

This model function has the essential characteristics of a realistic pair correlation function g(R). The surface between R_{hc} (hard core) and R_{NN} (next neighbours) is related to the number of ß atom neighbours of an α atom $Z_{\alpha\beta}$ by the expression

$$Z_{\alpha\beta} = n\int_{R_{NN}^{\alpha\beta}}^{R_{hc}^{\alpha\beta}} 4\pi\, g_{\alpha\beta}(R)\, R^2\, dR \qquad g_{\alpha\beta}(R) = h_{\alpha\beta} \tag{38}$$

4. THE KUBO-GREENWOOD FORMULA FOR LIQUID METAL ALLOYS

a) The linear response theory

To calculate the conductivity we start from the Kubo formula in the form established by Velicky /16/. This formula is derived in the framework of the linear response theory. One considers a time dependent perturbation

$$V = -A\, f(t) \tag{39}$$

where A is a one electron operator and f(t) a scalar function. One wants to know the variation of the thermodynamic averaged value of an observable B, $T\tau\, \rho B$, where ρ is the density matrix of the system. One considers only the first order terms and one assumes that the perturbation is applied adiabatically. This means that for $t = -\infty$ one has thermodynamic equilibrium $\rho(t = -\infty) = \rho_0$.

In the interaction representation one has

$$i\frac{d\rho}{dt} = [v(t), \rho(t)] \qquad (\hbar = 1) \tag{40}$$

The integration gives to first order in V

$$i(\rho-\rho_0) = \int_{-\infty}^{t} [v(t), \rho_0]\, d\tau \tag{41}$$

We assume that the mean value of B at equilibrium is zero. This gives

$$i\ T\tau\rho B = \int_{-\infty}^{t} d\tau\ T\tau\rho_0\ [B(t), v(\tau)] \tag{42}$$

We replace V by $-Af(t)$. As the commutator depends only on $(t-\tau)$ the integral has the form of a convolution. If we consider a perturbation of frequency ω

$$f(t) = \lambda e^{-i(\omega+io)t} \tag{43}$$

where the imaginary part io guarantees that the perturbation is applied adiabatically. The linear response will have the same time dependence. One gets

$$T\tau\rho B = \lambda e^{-i(\omega+io)t} \chi_{BA}(\omega^+) \tag{44}$$

with

$$\chi_{BA}(\omega^+) = i \int_0^{\infty} d\tau\ T\tau\rho_0\ [B(\tau), A(o)]\ e^{i(\omega+io)\tau} \tag{45}$$

We explicit this formula using the eigenstates $|\mu\rangle, |\nu\rangle$ of the non-perturbed system.

To make the thermodynamic average, one has to use for ρ_μ the Fermi function $f(\mu)$. This gives

$$\chi_{BA}(\omega^+) = -\sum_{\mu,\nu} \frac{f_\mu - f_\nu}{E_\mu - E_\nu + \omega}\ B_{\mu\nu} A_{\nu\mu} \tag{46}$$

This function can be extended in the complex plane. As usual for this type of functions, there is a cut on the real axis and $\chi_{BA}(z)$ is analytic in the two half-planes Im $z \gtrless 0$. This in particular yields the usual Kramers-Kronig relations.

The function $\chi_{BA}(z)$ can be rewritten as

$$\chi_{BA}(z) = -\int dE_1 dE_2 \frac{f(E_1)-f(E_2)}{E_1-E_2+z} \, \mathrm{Tr} \, \delta(E_1-H)B\,\delta(E_2-H)A \qquad (47)$$

and since

$$\delta(E-H) = -\frac{I_m}{\pi} G(E^+) = G^i(E) = \frac{i}{2\pi}[G^+(E)-G^-(E)] \qquad (48)$$

$$\chi_{BA}(z) = -\int dE_1 dE_2 \frac{f(E_1)-f(E_2)}{E_1-E_2+z} \, \mathrm{Tr} \, G^i(E_1)B\,G^i(E_2)A \qquad (49)$$

b) Static conductivity

We apply a uniform field $\vec{E}$. The perturbation is therefore $V = -e\vec{E}\cdot\vec{r}$. We want to know the induced current. The generalized conductivity $\sigma_{\alpha\beta}(\omega)$ is a tensor which depends on the components E_α and j_β.

For our one-electron Hamiltonian the A operator is given by

$$A = -e\,\hat{R}_\alpha \qquad (50)$$

whereas B is given by

$$B = -\frac{e}{\Omega}\,\hat{v}_\beta \qquad (51)$$

Ω is the volume of the crystal and $\hat{v}$ is the velocity operator. We limit ourselves to the space corresponding to the atomic states. The general commutation rules between $\hat{\vec{R}}$ and $\hat{\vec{v}}$ are no longer valid in this case. What is given are the matrix elements of $\hat{\vec{R}}$. To have the matrix element of $\hat{\vec{v}}$ one starts from

$$\hat{\vec{v}} = i\,[H,\vec{R}] \qquad (52)$$

and therefore

$$\hat{\vec{v}}_{nm} = i\,(\vec{R}_n-\vec{R}_m)\,H_{nm} \qquad (53)$$

or since the diagonal elements do not play any role

$$\hat{\vec{v}}_{nm} = i\,(\vec{R}_m-\vec{R}_n)\,t_{nm} \qquad (54)$$

Introducing these results and making the limit $\omega \to 0$, one finally obtains the Kubo-Greenwood formula in the form derived by Matsubara and Toyozawa assuming its isotropy in space

$$\sigma = \frac{2\pi e^2}{3\Omega} \int F(E) \left(-\frac{\partial f}{\partial E}\right) dE \tag{55}$$

$$F(E) = -\sum_{mnlk} R_{mn}\, t_{mn}\, G^i_{nl}(E)\, R_{lk}\, t_{lk}\, G^i_{km}(E) \tag{56}$$

In order to calculate F(E) it is useful to separate F(E) into contributions where the summations exclude equal indices. One therefore gets contributions where four, three or two sites are involved.

F(E) is given as

$$F(E) = -\left[F_4(E) + F_3(E) + F_2(E)\right] \tag{57}$$

These terms can be described by diagrams if we represent

$\vec{R}_{mn}\, t_{mn}$ by m ——>—— n

G^i_{mn} by m ——o—— n

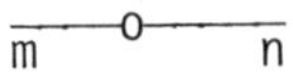

and G^i_{mm} by ⊚ m

$$F_4 = \sum_{m \neq n \neq l \neq k} \vec{R}_{mn} t_{mn} G^i_{nl} \vec{R}_{lk} t_{lk} G^i_{kn} \tag{58}$$

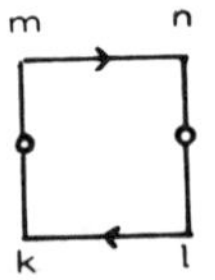

$$F_3 = F_3^a + F_3^b + F_3^c + F_3^d$$

$$F_3^a = \sum_{\substack{m \neq n \neq l \\ k=m}} \vec{R}_{mn} t_{mn} G^i_{nl} \vec{R}_{lm} t_{lm} G^i_{mn} \tag{59}$$

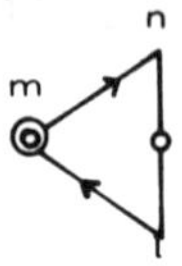

F_3^b 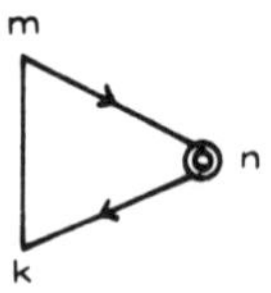 (60)

F_3^c 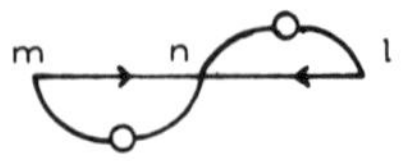 (61)

F_3^d 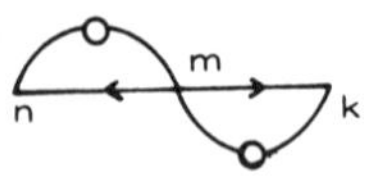 (62)

$$F_2 = F_2^a + F_2^b$$

$$F_2^a = \sum_{\substack{m \neq n \\ k=n, l=m}} \vec{R}_{mn}\, t_{mn}\, G^i_{nn}\, \bar{R}_{nm}\, t_{nm}\, G^i_{mm} \qquad (63)$$

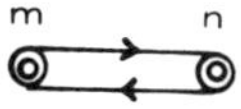

$$F_2^b = \sum_{\substack{m \neq n \\ k=n, l=m}} \vec{R}_{mn}\, t_{mn}\, G^i_{nm}\, \bar{R}_{mn}\, t_{mn}\, G^i_{nm} \qquad (64)$$

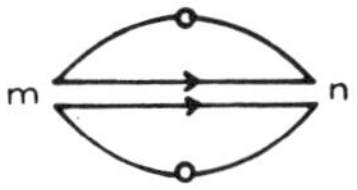

To calculate the conductivity one has to average

$$\langle F(E)\rangle = -\langle F_4(E) + F_3(E) + F_2(E)\rangle \qquad (65)$$

One is faced with the problem of calculating the average of products of Green functions

$$\begin{aligned}\langle G_{ln}(z_1)\, G_{km}(z_2)\rangle &= \langle G_{ln}(z_1)\rangle\langle G_{km}(z_1)\rangle \\ &+ \mathcal{L}(z_1,z_2) \sum_p \langle G_{lp}(z_1)\rangle\langle G_{kp}(z_2)\rangle\langle G_{pn}(z_1) G_{kp}(z_2)\rangle\end{aligned} \qquad (66)$$

The last term being called "vertex corrections". In some circumstances they can vanish. For instance, in the single-site approximation the vertex corrections vanish for $F_4(\mathfrak{z})$, $F_3(\mathfrak{z})$ and $F_2^a(\mathfrak{z})$.

The dependence of the various terms with density is the following

$$\begin{aligned} F_4(z) &\sim p^4 \\ F_3(z) &\sim p^3 \\ F_2(z) &\sim p^2 \end{aligned} \qquad (67)$$

where $p = 32\pi \frac{n}{\alpha^3}$ and α^{-1} the effective Bohr radius.

For $p > 1$ the dominant terms are $F_4(\mathfrak{z})$ and then $F_3(\mathfrak{z})$. The neglect of $F_2^b(\mathfrak{z})$ and of vertex corrections made by Matsubara and Toyozawa /12/ is justified for $p \gg 1$.

For $p < 1$ the dominant terms are $F_2(\mathfrak{z})$. In this region, as shown by Movaghar and Sauer, the neglect of $F_2^b(\mathfrak{z})$ is not justified. It can even give negative contributions to the conductivity. This is a manifestation of the breakdown of the single-site approximation. The theory has to be extended to consider cluster contributions and vertex corrections. It can be shown that the term F_2^a which is proportional to the square of the density of states corresponds to the random phase approximation of Hindley /17/.

Following Mott et al /18/ the conductivity

$$\sigma = \frac{2\pi}{3\Omega}\int dE \sum_{\mu,\nu} J_{\mu\nu} J_{\nu\mu}\, \delta(E-E_\nu)\, \delta(E-E_\mu)\, [-\frac{df}{dE}] \tag{68}$$

can be written if one decouples Green function and current operators averages

$$\sigma \propto \frac{1}{\Omega} n^2(E_F) \langle |\langle\mu|J|\nu\rangle|^2 \rangle_{Av} \tag{69}$$

If we apply the random phase approximation to the current average, one obtains the contribution F_2^a. This can easily be seen. One develops the eigenstates $|\mu\rangle$, $|\nu\rangle$ in terms of the atomic states $|n\rangle$

$$|\mu\rangle = \sum_n a_{n\mu} |n\rangle \tag{70}$$

In the case of a periodic system the coefficients are the plane waves $\exp(i\vec{k}\vec{R}_n)$ and the states $|n\rangle$ the Wannier functions. In a disordered system the amplitudes $a_{n\mu}$ vary from site to site. In the case where the mean free path is of the order of the interatomic distance phase correlation has disappeared and the motion of electrons is completely uncorrelated.

Substituting equation (70) into (69) one obtains

$$\langle |\langle\mu|J|\nu\rangle|^2 \rangle_{Av} = \langle \sum_{\substack{nm \\ n'm'}} a^*_{n\mu} a_{n'\nu} a^*_{m\nu} a_{m'\mu} \langle n|J|n'\rangle\langle m|J|m'\rangle \rangle_{Av} \tag{71}$$

The RPA consists in the summations to keep only the terms $n' = m$ and $n = m'$. Since

$$\langle |a_{n\mu}|^2 \rangle_{av} = \frac{1}{N} \tag{72}$$

one has

$$\langle |\langle\mu|J|\nu\rangle|^2 \rangle_{Av} = \frac{1}{N^2} \langle \sum_{nm} |\langle n|\, J\, |m\rangle|^2 \rangle_{Av} \tag{73}$$

Using the definition (53) of the velocity operator, one can see that the Mott formula

$$\sigma \sim \frac{1}{\pi} n^2(E_p)\, \bar{D}^2$$

with
$$\bar{D}^2 = \alpha \sum_{nm} R^2_{nm}\, t^2_{nm} > A_v$$

is equivalent to the term $\langle F_2^a \rangle_{av}$ of equation (63).

The Hindley approximation is valid only for L ↸ a and the neglect of F_3 and F_4 only for $p < 1$.

The contribution F_2^b is proportional to the non-local Green function G_{nm} which contains the phase memory necessary for an electron to go from site to n. If there is no phase coherence at all, this term does not contribute.

The present formalism can be generalized to liquid metal alloys by introducing indices and matrices as it was done for the density of states. We refer to the paper of Movaghar and Sauer /11/. This formalism was used by Franz et al /6/ to calculate the conductivity of liquid CsAu. In the intermediate regime between conductivity and percolation or localized regime it was assumed that the Mott - Hindley formula can be used. However, recent computer calculations /19/ have shown that the random phase approximation can be in some cases a bad approximation to describe the intermediate (diffusion) regime of random models. This fact together with the failure of the single-site approximation for some concentration and energy range, as pointed out by Movaghar and Sauer, shows that new progress in the theory has to be made before a good description of the conductivity of systems like AuCs liquid can be made.

REFERENCES

1. H. Hoshino, R.W. Schmutzler and F. Hensel, Phys. Lett. 51A (1975) 7
2. W. Freyland, G. Steinleitner, Ber. Bunsenges. Physik. Chem. 80 (1976) 810
3. H. Ruppersberg and H. Egger, J. Chem. Phys. 63 (1975) 4095
4. R.C. Kittler and L.M. Falicov, Phys. Rev. B18 (1976) 2506
5. A. Ten Bosch, J.L. Moran-Lopez and K.H. Bennemann, J. Phys. C. (1978) 2959

6. J. Franz, F. Brouers, C. Holzhey, J. Phys. F to be published 1979
7. F. Brouers, M. Cyrot, F. Cyrot-Lackmann, Phys. Rev. B7 (1973) 4370
8. L.M. Falicov and F. Yudurain, Phys. Rev. B12 (1975) 5664
9. K.M. Watson, Phys. Rev. 105 (1957) 1388
10. P.W. Anderson, Phys. Rev. 109 (1958) 1492
11. B. Movaghar and G.W. Sauer, J. Phys. F 9 (1979) 867
12. T. Matsubara and Y. Toyozawa, Prog. Theor. Phys. 26 (1961) 739
13. Y. Ishida and F. Yonezawa, Prog. Theor. Phys. 40 (1973) 731
14. B. Movaghar, D.E. Miller and K.H. Bennemann, J. Phys. F 4 (1974) 687
15. B. Movaghar, D.E. Miller, J. Phys. F 5 (1975) 261
16. B. Velicky, Phys. Rev. B184 (1969) 614
17. N.K. Hindley, J. Non-Cryst. Solids 5 (1970) 17
18. N.F. Mott and E.H. Davis, Electronic Processes in Non-Crystalline Materials, Claredon Press Oxford 1971
19. Prelovšek, Phys. Rev. Lett. 40 (1978) 1596

STRUCTURE AND DYNAMICS OF METALS AT HIGH TEMPERATURES

Gianni Jacucci and Michael L. Klein*

Instituto per la Ricerche Scientifiche e Technologiche, Libra Universita, degli Studi di Trento, Povo, Trento 38050, Italy. *Chemistry Division, National Research Council of Canada, Ottawa, Canada K1A 0R6

ABSTRACT. The nature of collective modes in metals at high temperatures is examined by computer simulation. Particular attention is given to bcc Na, K, and fcc Al and the alloy Rb_{71} K_{29}. Where possible comparisons with neutron and X-ray scattering data are presented.

1. COMPUTER SIMULATION

The Monte Carlo and molecular dynamics (MD) techniques are now well established methods for studying the properties of many body systems [1]. Basically, the Monte Carlo (MC) method is restricted to study equilibrium properties such as the structure and thermodynamics. The MD method, on the other hand, can be used to study time dependent correlation functions $f(t)$ and their associated power spectra $f(\omega)$ in addition to thermodynamic quantities. Typically, one studies systems composed of a few hundred particles with periodic boundary conditions being used to simulate an infinite system. One can study solids or liquids with equal facility. In the case of metals the interatomic forces are relatively long ranged and some care is needed to ensure that this is handled correctly [2]. In the work described in this talk we have used exclusively the interatomic potentials of Taylor and his coworkers [3]. In a typical MD "experiment" one follows the time evolution of the system for several thousand time steps of order 10^{-14} sec so that the macroscopic time corresponds only to about 10^{-10} secs. This is a short time but nonetheless long enough to study many interesting properties. In a Monte Carlo "experiment" properties are determined by averaging over about 10^6 configurations generated according to an appropriate prescription [1].

2. THEORY

The quantities we are primarily interested in are those obtained from neutron (or X-ray) scattering experiments. A systems response to a neturon (or X-ray) probe that transfers momentum $\hbar\vec{Q}$ and energy $\hbar\omega$ is embodied in the Van Hove function $S(\vec{Q},\omega)$. This in turn is related to the intermediate scattering function $F(\vec{Q},t)$ by the equation

$$S(\vec{Q},\omega) = \int dt\ \exp(i\omega t)F(\vec{Q},t) \tag{2.1}$$

where

$$F(\vec{Q},t) = <\rho(\vec{Q},t)\rho(-\vec{Q},0)>/N \tag{2.2}$$

and

$$\rho(\vec{Q},t) = \sum_i \exp(-i\vec{Q}\cdot\vec{r}_i(t)) \ . \tag{2.3}$$

Here, N is the number of particles in the system whose positions at time t are given by $\vec{r}_i(t)$. The latter are stored during a MD "experiment" so that quantities such as $F(\vec{Q},t)$ can be evaluated. The allowed values of $\vec{Q}$ are determined by the periodic boundary conditions and take the form

$$\vec{Q} = 2\pi(\ell,m,n)/L \tag{2.4}$$

where ℓ, m, and n are integers and L is the edge length of the (cubic) box containing the N particles.

The static structure factor $S(\vec{Q})$ can be obtained from the dynamical structure factor $S(\vec{Q},\omega)$ since

$$S(\vec{Q}) = \int d\omega S(\vec{Q},\omega)/2\pi \ . \tag{2.5}$$

For a liquid $S(\vec{Q})$ is related to the pair distribution function g(r) by the equation

$$S(\vec{Q}) = 1 + \frac{N}{V}\int d\vec{r}(g(r)-1)\exp(i\vec{Q}\cdot\vec{r}) \tag{2.6}$$

In the following sections we will present results for $S(\vec{Q})$, g(r), and $S(\vec{Q},\omega)$ obtained by computer simulation using what are believed to be realistic interatomic force models [3] for typical bcc and fcc metals.

3. STRUCTURE FACTORS

Before discussing the dynamics of solids at high temperatures it is worth spending a little time reviewing briefly known results on the pair distribution functions g(r) and the structure factor $S(\vec{Q})$ for simple metals. Figure 1 shows the spherically averaged pair distribution function g(r) for solid K calculated by MD. The solid line corresponds to a temperature of approximately one half the melting point T_m while the dashed line is for $T \sim 0.9\ T_m$ [4]. The main point to note is that even at one half the melting point the vibrational motion of the atoms in the bcc lattice issufficient to cause several of the shells of neighbours to appear to have merged. Thus although there should be eight distinct shells of neighbours (shown in figure 1) only four peaks are seen in g(r). Morever, close to the melting point the peaks around r/a = 1.5 are only barely separable.

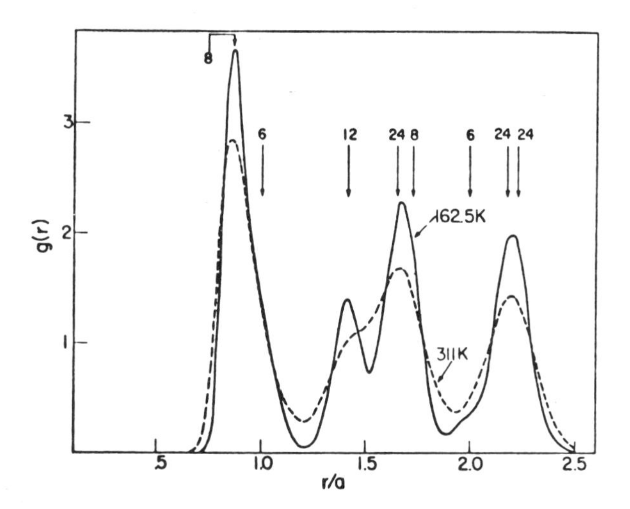

Fig. 1. Spherically averaged pair distribution functions g(r) for solid K at 162.5 K (solid line) and 311 K (dashed line). The position and number of neighbours at a given distance r in abcc lattice is indicated by the arrows and numbers respectively.

It is natural to wonder what happens to g(r) upon melting? The answer is shown in figure 2. Here we show the results of a MC calculation carried out for liquid Na [5] at 100°C. The starting configuration for this calculation was a bcc lattice, with lattice constant a = 4.37 Å since this yields the correct liquid density. The dotted curve shows the g(r) obtained from the initial part of the MC run, before the system had melted. One sees that this is very similar to result shown for K in figure 1. The solid line in figure 2 is the g(r) for liquid Na, obtained from the final portion of the MC experiment. Upon melting the main peak is g(r) shifts to smaller values of r and the second peak loses its double peaked structure. The overall differences between the solid and liquid g(r) are not so large as one might have thought.

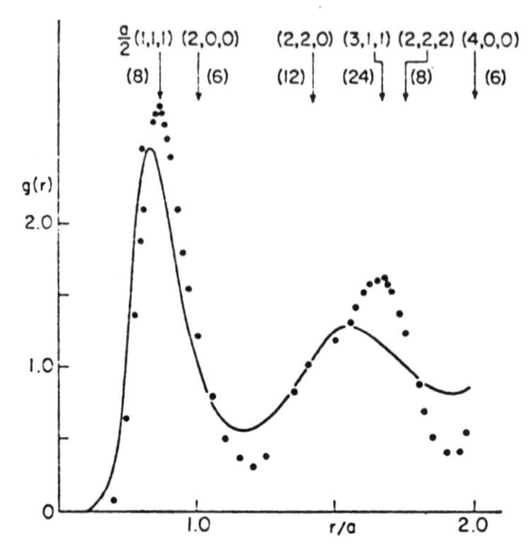

Fig. 2. Spherically averaged pair distribution functions g(r) for solid and liquid Na.

Figure 3 shows the S(Q) for both "solid" and liquid Na at 100°C. The liquid result agrees very well with experimental X-ray data

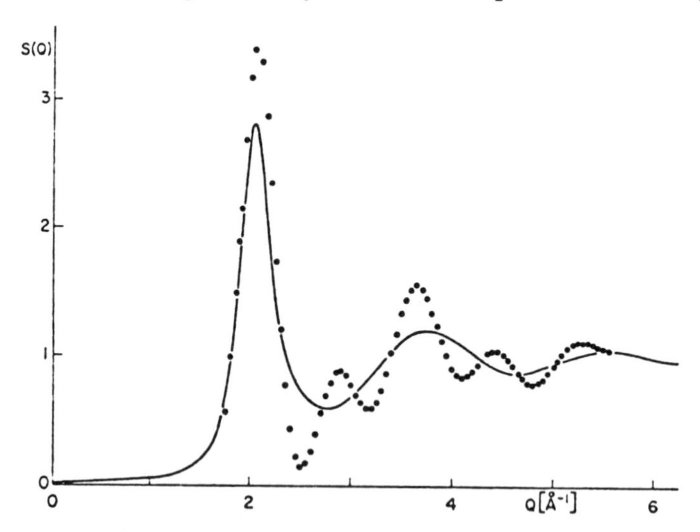

Fig. 3. The averaged static structure factor S(Q) for "solid" and liquid Na at 100°C and density corresponding to a = 4.37 Å (see text).

(shown in figure 4) although no explicit comparision is given here [5]. S(Q) for solid Na has been measured by the Munich group using thermal diffuse X-ray scattering from a polycrystalline sample [6]. These data are shown in figure 4 along with data for the liquid [7]. The agreement between the computer simulation and experiment is striking.

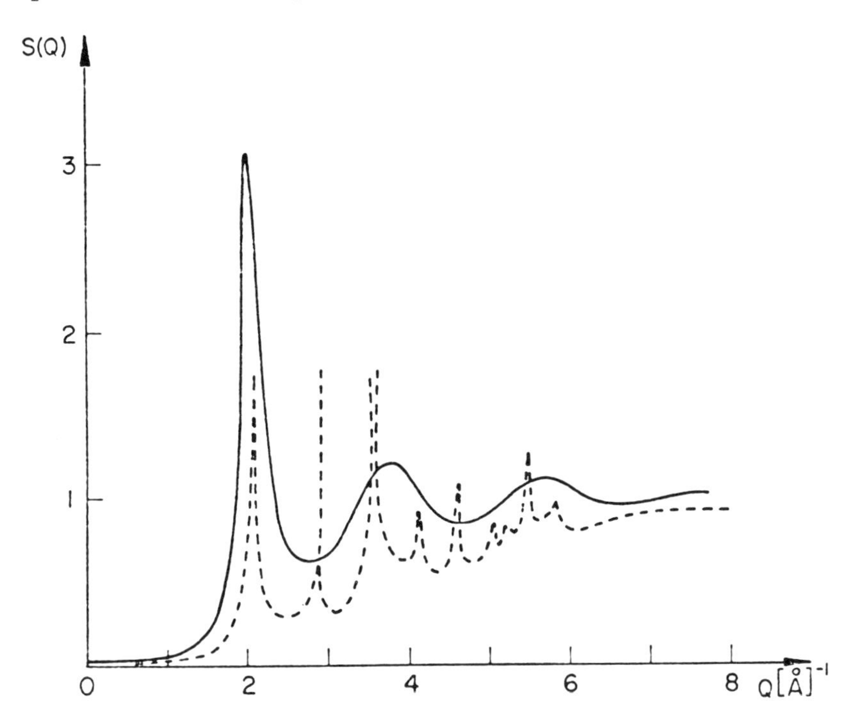

Fig. 4. The averaged static structure function S(Q) for solid Na at 16.5°C taken from ref [6] and the liquid S(Q) at 100°C from reference [7].

4. DYNAMICS OF HIGH TEMPERATURE SOLIDS

We now turn to a discussion of the dynamical properties of metals at high temperatures. By high temperature we mean $T \gg \theta$, where θ is the Debye temperature of the solid in question. Some time ago Buyers and Cowley [8] reported a detailed study of the temperature dependence of phonon frequencies and lifetimes in solid K. One such phonon is shown in figure 5. As the temperature is raised towards the melting point there is a significant softening of the mode frequency as well as appreciable change in linewidth accompanied by an increased multiphonon background. A detailed MD study has been carried out in an attempt to reproduce this behaviour [4]. The results are shown in figure 6. There one

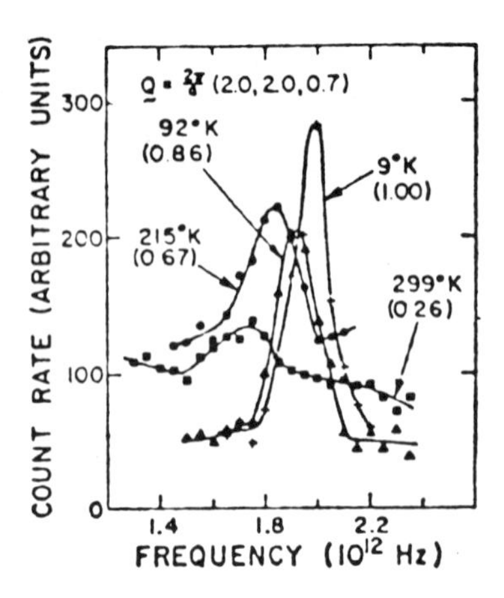

Fig. 5. Effect of temperature on a neutron group in K. The data are taken from ref. [8].

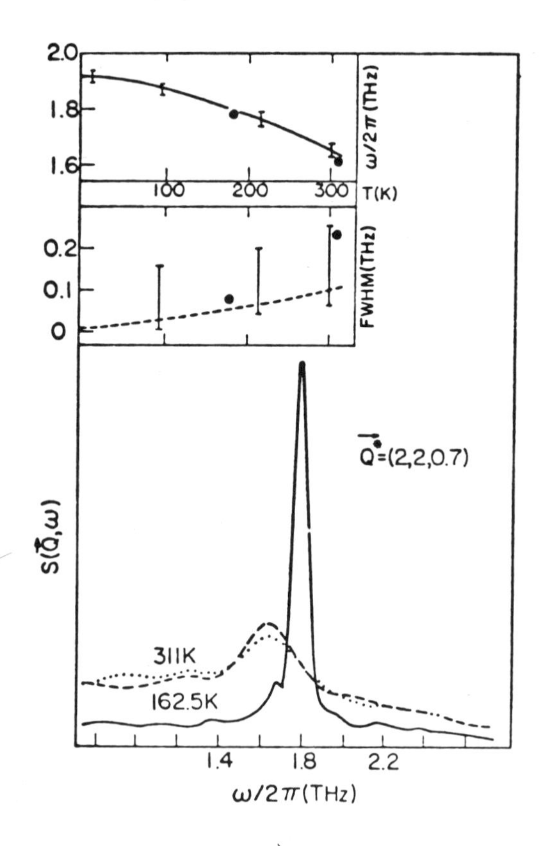

Fig. 6. MD calculation of $S(\vec{Q},\omega)$ for solid K. Peak positions and the FWHM are compared with experiment [8] in the inset.

sees that there is good agreement between the MD calculation and the data shown in figure 5. Detailed anharmonic phonon calculations have also been carried out for K using the same interatomic potential as used in the MD calculations [4]. These calculations show that the large temperature shift in this particular phonon frequency is due to the fact that both first and second order anharmonic effects give negative contributions that are rather large at high temperatures.

Figure 7 shows part of the longitudinal <100> phonon dispersion curve calculated by MD for solid Na at a temperature close to its melting point [9]. This calculation was carried out as part of a more extensive study of diffusion mechanisms [10]. We see that even at these elevated temperatures the low-$\vec{Q}$ response of the crystal is well defined. This is related to the fact that the alkali metals have rather harmonic potentials [3] which in turn are responsible for the relatively small Grüneisen parameters, small differences between C_p and C_V [11], and the relatively weak Rayleigh peak in figure 7.

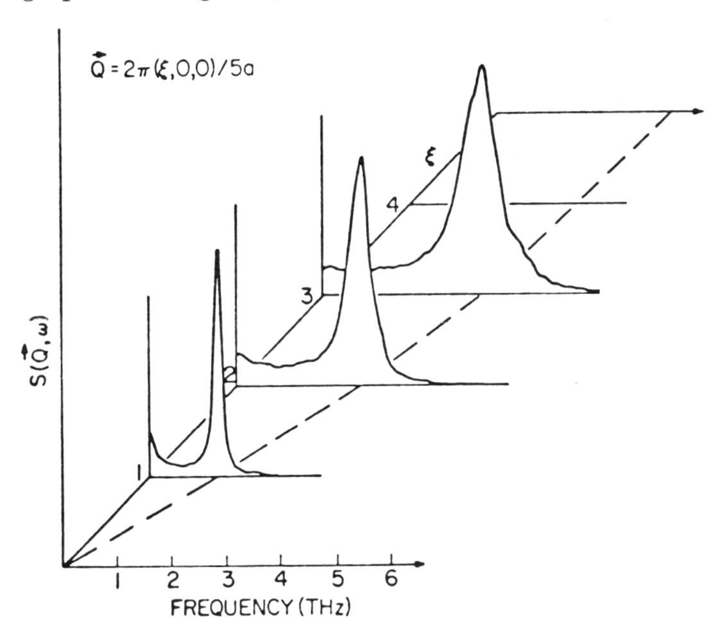

Fig. 7. A MD calculation for longitudinal phonons propagating in the <100> direction of solid Na at 345 K.

Even though the low-$\vec{Q}$ phonons are well defined, at sufficiently large values of $\vec{Q}$ and sufficiently high temperatures the one-phonon response will eventually merge into the multiphonon background. This effect is shown clearly in the MD results presented in figure 8 for longitudinal phonons propagating in the

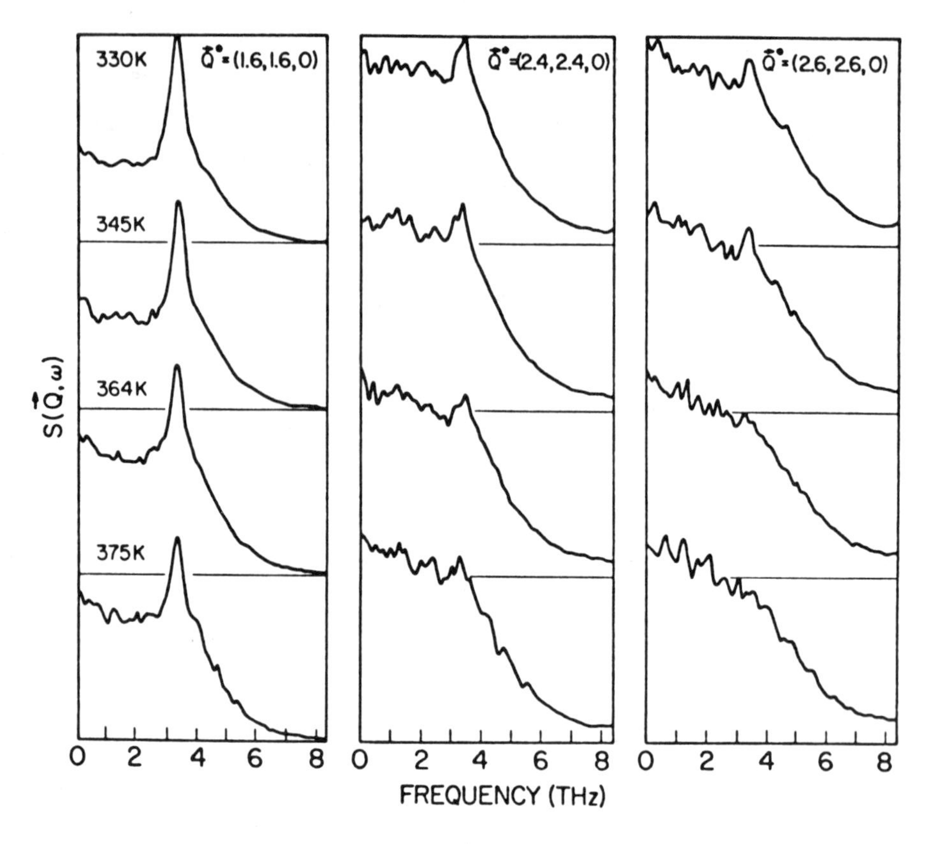

Fig. 8. A MD Calculation of the $\vec{Q}$-dependence of $S(\vec{Q},\omega)$ for Na, as a function of $\omega/2\pi$ in THz, for longitudinal phonons propagating in the <110> direction at T = 330, 345, 364 and 375 K.

<110> direction. There is a hugh difference between the phonon $\vec{Q}^* = \vec{Q}a/2\pi = (1.6, 1.6, 0)$ and $(2.4, 2.4, 0)$. These values of $\vec{Q}$ are on either side of the Bragg peak (2,2,0) and hence subject to the interference effect between the one-phonon peak and the multi-phonon background [4]. This effect causes the left hand side of the response function to be depressed for $\vec{Q}^* = (1.6, 1.6, 0)$ and enhanced for $\vec{Q}^* = (2.4, 2.4, 0)$. Finally, close to the melting point for $Q^* = (2.6, 2.6, 0)$ the one-phonon peak disappears. It will be of considerable interest to see if these predictions are borne out by experiments.

We now turn to the case of an fcc metal, solid Al. Here the interatomic potential is steeper than in the alkalis [3] and this explains in part the similarity of $S(\vec{Q},\omega)$ data for Al with that of the fcc rare gas solids [12] and the fact that the Grüneisen parameters for Al is rather large. Figure 9 shows some MD data

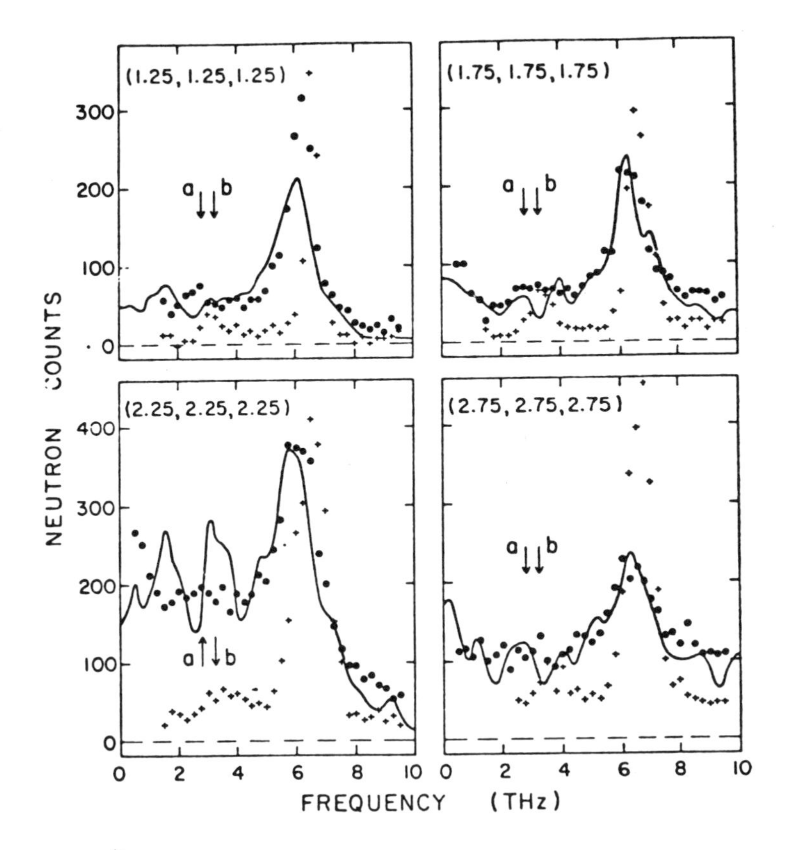

Fig. 9. $S(\vec{Q},\omega)$ data for longitudinal <111> phonons in solid Al. The lines are the MD calculations [12] for 800 K. The crosses and dots are the experimental neutron scattering data for 300 and 800 K respectively.

[12] for the <111> direction of solid Al at 800 K. Also shown in this figure are recent neutron measurements taken at Chalk River [13]. Overall there is impressive agreement between the predictions of the MD calculations and the experiments.

The MD curve for (2.25, 2.25, 2.25) deserves further comment. The two satellite peaks below the main peak at 6 THz are clearly absent in the experimental data [13]. The original MD data were therefore reanalysed and when due allowance is made for statistical noise in the $F(\vec{Q},t)$ data one can show that these peaks are spurious [13]. Additional evidence for this is contained in our MD study of solid Al with one vacant lattice site. The $S(\vec{Q},\omega)$ for the phonon (2.25, 2.25, 2.25) at 800 K is shown in figure 10. While this data also shows some noise, particularly at low frequency, the noise level is greatly reduced. Morever, the one-phonon

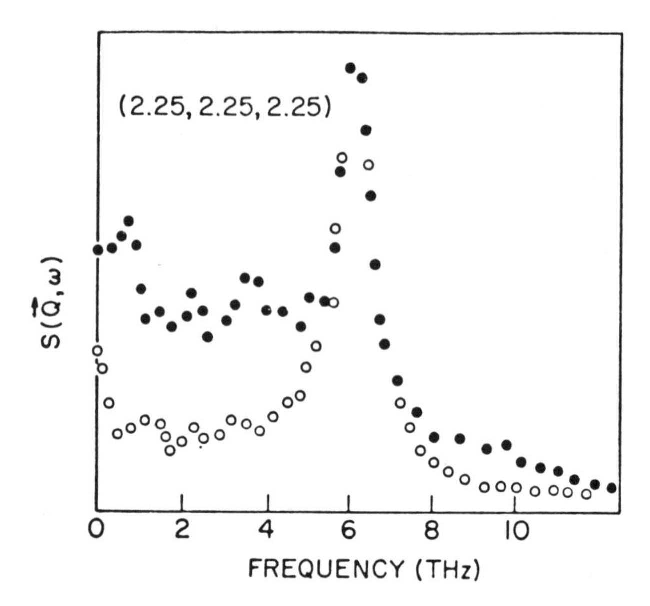

Fig. 10. MD calculation of $S(\vec{Q},\omega)$ for solid Al at 800 K. The solid dots are the full calculation and the open circles are the one-phonon approximation. The solid had one vacant lattice site.

approximation to $S(\vec{Q},\omega)$, which is also shown in figure 10, is rather smooth suggesting that it is difficult to get good statistics for the full $S(\vec{Q},\omega)$ in an highly anharmonic crystal. It should be noted that the one-phonon approximation to $S(\vec{Q},\omega)$ has a large (anharmonic) background.

Detailed comparisons between the two MD calculations for Al (not shown here) reveal that the vacant lattice site has little effect on the overall $S(\vec{Q},\omega)$ data except for a peak at $\omega = 0$ which is found in most of the low-$\vec{Q}$ data. At high-$\vec{Q}$ the Al data show the same phenomena shown in figure 8. The one-phonon peak eventually disappears into the multiphonon background when $\vec{Q}^* = (3.25, 3.25, 3.25)$ at 800 K.

The above examples show that, with care, computer simulation calculations can yield interesting results on the dynamics of metals at high temperatures. In the future, the MD technique will no doubt be applied to other, more complicated metallic systems.

5. STRUCTURE AND DYNAMICS OF AMORPHOUS SYSTEMS

We have already heard much about this topic earlier in the conference. The main point we wish to make here is that the techniques used to calculate the $S(\vec{Q})$ and $S(\vec{Q},\omega)$, as discussed in the earlier sections can be applied to amorphous systems. For example $S(\vec{Q},\omega)$ for the Lennard-Jones (12-6) system has been studied in the liquid, solid, and amorphous phases [14-16]. One can now make models of the amorphous phase rather routinely [17] and the overall structure as displayed by the g(r) and S(Q) seems to depend little on the detailed intermolecular forces that are present. However, the amorphous system of interest to experimentalists are mostly two-component systems and here much less work has been done. A calculation of $S(\vec{Q},\omega)$ in the one phonon approximation for the system $Mg_{70}Zn_{30}$, which is a two-component metallic glass, has been reported recently [18]. This calculation used realistic potentials and started from a structure appropriate to the dense random packing of hard spheres but allowing relaxation. Our own studies on the alloy system $Rb_{71}K_{29}$ have revealed the extreme importance of relaxation effects. This system (see figure 11) shows a well defined local mode associated with the K atoms. This we have shown [19] arises from a delicate balance

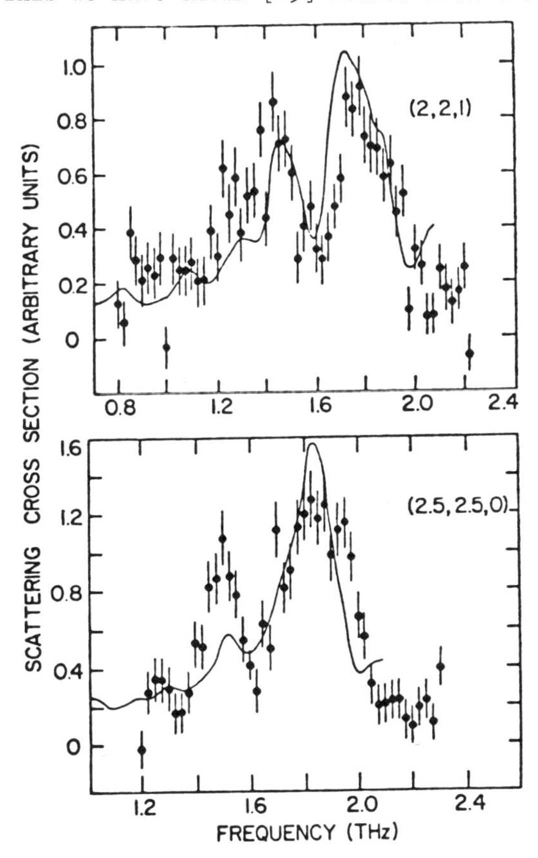

Fig. 11. Comparison of MD calculations (solid lines) and neutron scattering $S(\vec{Q},\omega)$ data for the alloy $Rb_{71}K_{29}$.

between mass changes, force constant changes, and local relaxation effects. Our MD calculations of $S(\vec{Q},\omega)$ are in excellent accord with neutron scattering data [20].

It is clear that we are only at the very beginning in this interesting area of research and that the detailed understanding of the structure and dynamics of amorphous metals (metallic glasses) will be aided immeasurably by computer simulation studies.

REFERENCES

1. J.-P. Hansen and I.R. McDonald, Theory of Simple Liquids, Academic Press, London, 1976.
2. S.S. Cohen, M.L. Klein, M.S. Duesbery, and R. Taylor, J. Phys. F:Metal Physics 6, 337, 1976.
3. L. Dagens, M. Rasolt, and R. Taylor, Phys. Rev. B11, 2726, 1975.
4. H.R. Glyde, J.-P. Hansen, and M.L. Klein, Phys. Rev. B16, 3476, 1977.
5. R.D. Murphy and M.L. Klein, Phys. Rev. A8, 2640, 1973.
6. R.B. Schierbrock and G. Fritsch, Phys. Letters 48A, 151, 1974.
7. A.J. Greenfield, J. Wellendorf, and N. Wiser, Phys. Rev. A4, 1607, 1971.
8. W.J.L. Buyers and R.A. Cowley, Phys. Rev. 180, 755, 1966.
9. G. Jacucci and M.L. Klein, Solid State Commun. 1979.
10. D. DaFano and G. Jacucci, Phys. Rev. Letters 39, 950, 1977.
11. D.L. Martin, Phys. Rev. 139, A150, 1965; 154, 571, 1967.
12. G. Jacucci and M.L. Klein, Phys. Rev. B16, 1322, 1977.
13. W.J.L. Buyers, G. Dolling, H.R. Glyde, G. Jacucci, and M.L. Klein, Phys. Rev.
14. D. Levesque, L. Verlet, and J. Kurkijarvi, Phys. Rev. A7, 1690, 1973.
15. J.-P. Hansen and M.L. Klein, Phys. Rev. B13, 878 (1976).
16. A. Rahman, M.J. Mandell, and J.P. McTague, J. Chem. Phys. 64, 1976.
17. H.R. Wendt and F. Abraham, Phys. Rev. Letters, 41, 1244, 1979.
18. L. von Heimendahl, J. Phys. F:Metal Physics 9, 161, 1979.
19. G. Jacucci, M.L. Klein, and R. Taylor, Phys. Rev. B18, 3782, 1978.
20. W.A. Kamitakahara and J.R.D. Copley, Phys. Rev. B18, 3772, 1978.

COLLECTIVE EXCITATIONS IN LIQUID METALS AND ALLOYS

G. Jacucci

Dipartimento di Matematica e Fisica,
Libera Università di Trento, Povo, Trento, Italy

I.R. McDonald

Department of Physical Chemistry, University
of Cambridge, Lensfield Road, Cambridge CB2 1EP, UK

INTRODUCTION

Computer "experiments" based on the method of molecular dynamics (MD) have in recent years provided much valuable insight into the character of collective short-wavelength excitations in classical liquids /1, 2/. It has been shown, for example, that in liquid metals /3-5/ (Na, Rb and - to a lesser extent - Al) a remnant of the hydrodynamic sound wave mode is observable at wavenumbers k which are easily accessible in neutron inelastic scattering (NIS) experiments /6/, and good agreement has been found with experimental data in the one case (Rb) where detailed comparison has been made /5,6/. Thus far, however, at least for systems of uncharged particles /7/, attention has been confined solely to the one-component case, and no attempt has been made to investigate the rich variety of phenomena which are peculiar to mixtures. Two problems, in particular, merit some attention. The first is the general question of the interplay between concentration fluctuations and fluctuations in number density, and of the role of the former in determining the NIS coherent cross section. The second concerns the decay of concentration fluctuations in the intermediate range of k between the hydrodynamic ($k \to 0$) and free particle ($k \to \infty$) regimes. The possibility of the onset of reactive rather than diffusive behaviour cannot be excluded, and standard sum rule arguments may be used to obtain an estimate of the characteristic frequency of the resulting concentration "wave" /8/.

LONGITUDINAL MODES

As a first step in the investigation of such problems we have carried out a series of MD simulations of liquid Na and K, and of the alloy Na-K at equal concentrations of the two species. We have used throughout the density-dependent effective two-body potentials of Dagens, Rasolt and Taylor /9/, which are known to give a good description of many of the static and dynamic properties of the alkali metals. In each "experiment" the equations of motion of 250 atoms (with periodic boundary conditions) were integrated numerically with a time step Δt over a total time of approximately $4 \times 10^4\ \Delta t$. Details of the thermodynamic states which were studied are given in Table 1, together with the values used for Δt. The temperatures for the pure metals are somewhat lower than the experimental melting temperatures at atmospheric pressure but we do not regard this as significant, since the structural features /10/ of the model systems are typical of those of a liquid close to its triple point.

The longitudinal collective modes of a binary mixture are conveniently discussed in terms of the three partial dynamical structure factors $S_{\alpha\beta}(k,\omega)$ ($\alpha, \beta = 1, 2$), defined /11/ as

$$S_{\alpha\beta}(k,\omega) = \frac{1}{2\pi N} \int_{-\infty}^{\infty} \exp(i\omega t) \langle \rho_\alpha(\vec{k},t)\, \rho_\beta(\vec{k},t) \rangle \, dt \qquad (1)$$

where $\rho_\alpha(\vec{k},t)$ is a Fourier component of the microscopic density of particles of species α. The main computational effort has therefore been directed at the calculation of these functions. The spectra of number density fluctuations, $S_{NN}(k,\omega)$, and of concentration fluctuations, $S_{CC}(k,\omega)$, may then be expressed as the linear combinations

$$S_{NN}(k,\omega) = cS_{11}(k,\omega) + 2\{c(1-c)\}^{\frac{1}{2}} S_{12}(k,\omega) + (1-c)S_{22}(k,\omega) \qquad (2)$$

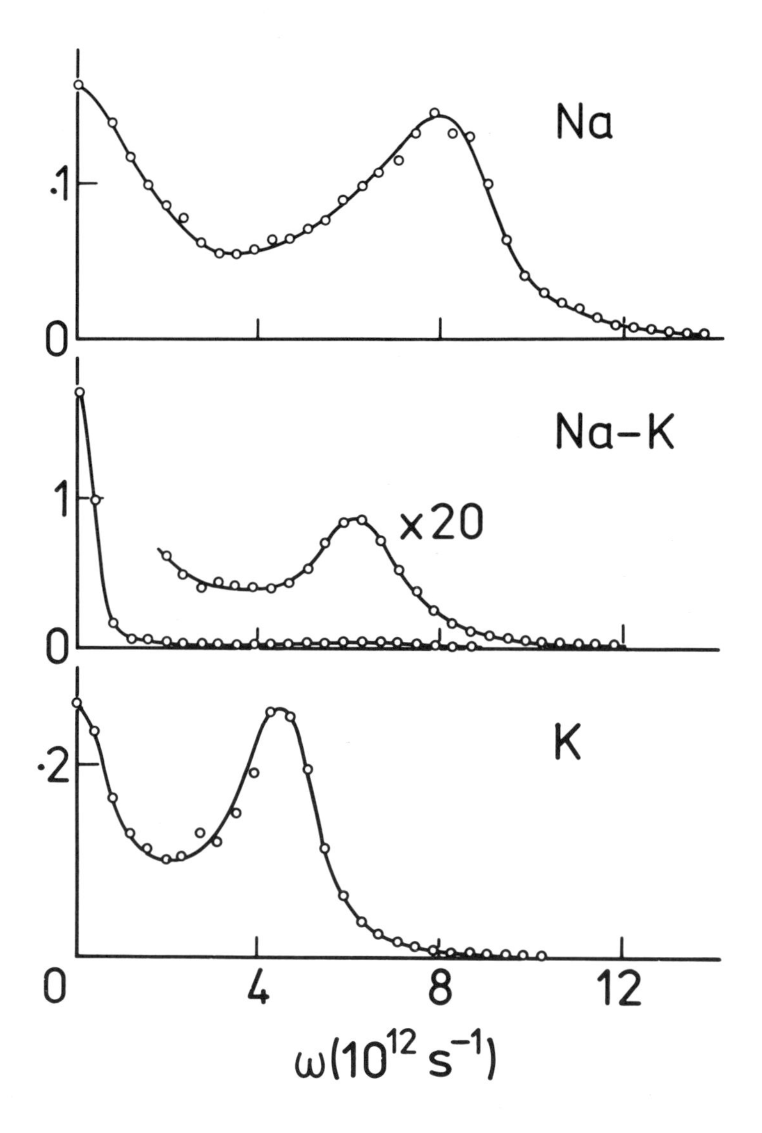

Fig. 1. The function $S_{NN}(k,\omega)$ (in units of 10^{-12}s and normalized to unit area) for the three model systems. The wavenumbers are the smallest consistent with the periodic boundary condition: $k = 0.289 Å^{-1}$ (Na), $0.258 Å^{-1}$ (Na-k) and $0.232 Å^{-1}$ (K). The high-frequency region for Na-K is also shown on a scale twenty times larger.

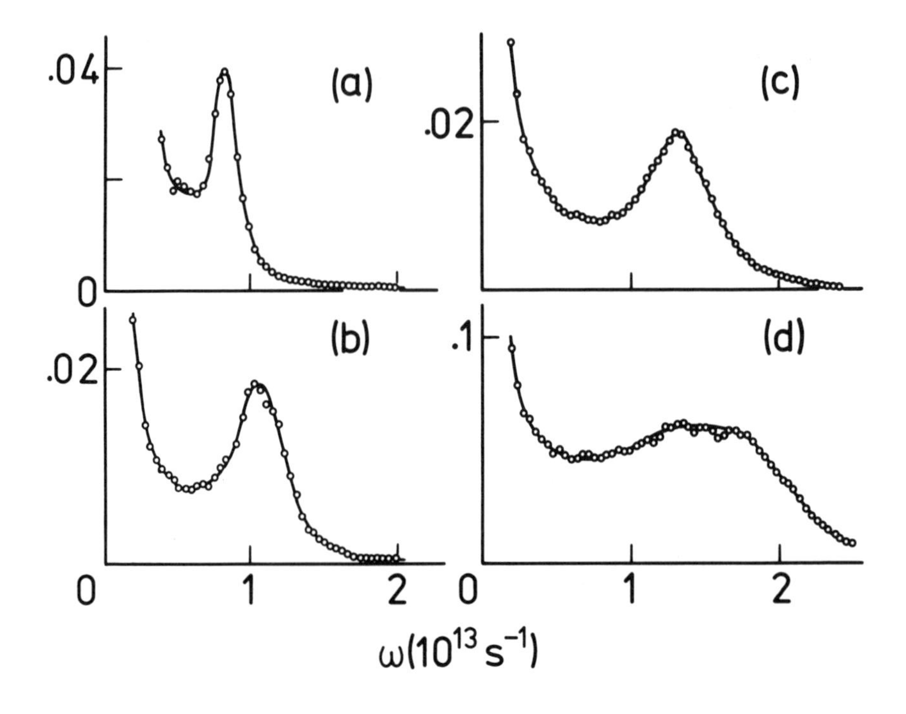

Fig. 2. The function $S_{NN}(k, \omega)$ (in units of 10^{-12} s and normalized to unit area) for Na-K at $k = 0.258$ Å^{-1} (a), 0.447 Å^{-1} (b), 0.577 Å^{-1} (c) and 1.064 Å^{-1} (d).

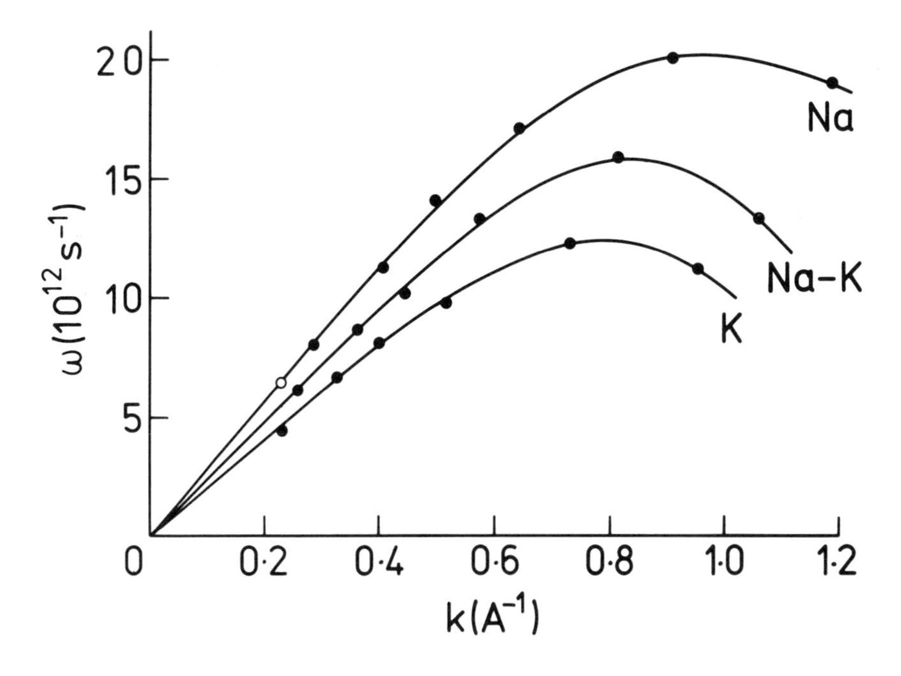

Fig. 3. Dispersion of the sound wave peak for the three model systems. Dots: present work; circle: results of Rahman /3/. The curves are drawn only to aid the eye.

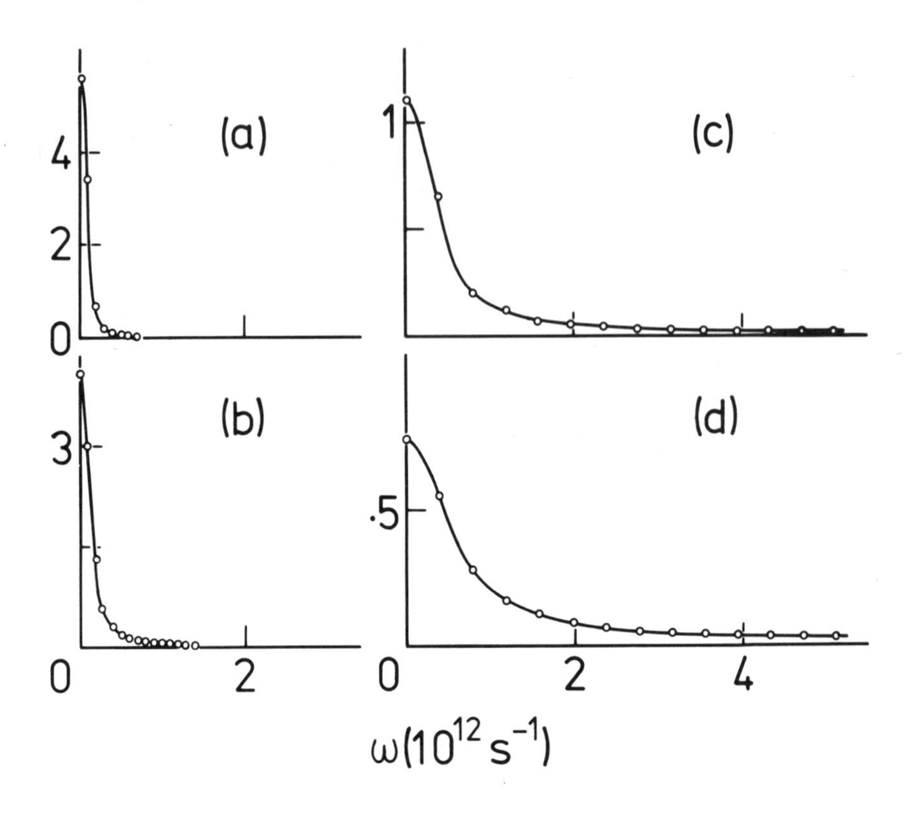

Fig. 4. The function $S_{CC}(k,\omega)$ (in units of 10^{-12} s and normalized to unit area) at k = 0.258 Å^{-1} (a), 0.577 Å^{-1} (b), 1.064 Å^{-1} (c) and 1.569 Å^{-1} (d).

$$S_{CC}(k,\omega) = (1-c)\left[(1-c)S_{11}(k,\omega) - 2\{c(1-c)\}^{\frac{1}{2}}S_{12}(k,\omega) + cS_{22}(k,\omega)\right] \quad (3)$$

where c is the number concentration of species 1.

Results on $S_{NN}(k, \omega)$ for the three systems are shown in Figure 1 for the smallest values of k compatible with the periodic boundary conditions. We see a dramatic difference in the behaviour for the alloy as compared with the two pure systems. In each case the spectrum has a hydrodynamic-like structure, with a well-resolved Brillouin side peak and a central (Rayleigh) peak. However, whereas in pure Na and K the Rayleigh and Brillouin peaks are of similar height and width, in the alloy the central line is much the more intense and narrower of the two. It follows that the density-density autocorrelation function (the Fourier transform of $S_{NN}(k, \omega)$ for the alloy has a component which decays very slowly with time.

As k increases the Brillouin peak of the alloy shifts and broadens in the manner shown in Figure 2. Its height relative to that of the central peak remains very small up to $k \simeq 1 \mathring{A}^{-1}$. At larger k the side peak broadens into a shoulder, with a rapid rise in intensity, finally disappearing at $k \simeq 1.2 \mathring{A}^{-1}$. The dispersion of the propagating mode is plotted for all three systems in Figure 3. The slopes at small k correspond to values for the speed of sound which are listed in Table 1. These are in fair agreement with experimental measurements /12/, except that they do not reproduce the marked negative departure from ideality that is observed for the alloy. The results for Na are also in good agreement with an earlier calculation of Rahman /3/ based on a potential due to Shyu et al. /13/. The figure also shows that a well-defined propagating density fluctuation persists down to wavelengths $\lambda = 2\pi/k$ which are comparable with the nearest-neighbour separation in the liquid, i.e. $\lambda = 5.0$ Å (Na), 5.7 Å (Na-K) and 6.3 Å (K). Similar behaviour is found for Rb /4/, suggesting that the relative persistence of the sound wave mode is essentially the same for all the alkali metals.

The function $S_{CC}(k, \omega)$ for the alloy is shown for several values of k in Figure 4. At long wavelengths it is known /14/ that the width of the Rayleigh component in $S_{NN}(k, \omega)$ is controlled partly by thermal conduction and partly by mutual diffusion, whereas the width of $S_{CC}(k, \omega)$ is determined almost wholly by mutual diffusion. In practice, apart from a scale factor, $S_{CC}(k,\omega)$ is almost indistinguishable from the low frequency component of $S_{NN}(k,\omega)$. (This behaviour is not apparent from comparison of Figures 1 and 4 because the spectra have been convoluted with a gaussian filter of width equal to the spacing of successive points.) We therefore have no hesitation in ascribing the dominance of the central peak in $S_{NN}(k,\omega)$ to the effects of interdiffusion. The other main feature of interest is the fact that there is no sign of any side peak which could be ascribed to an oscillation in the local concentration of the type discussed by Parrinello et al. /8/ and Abramo et al. /15/.

Two conclusions emerge concerning the dynamical properties of the simple alloy considered here. First, the spectrum of density fluctuations at small k is dominated by diffusive proccesses. When account is taken of possible instrumental broadening of the central peak, it becomes clear that it may be very difficult to detect a sound wave mode in NIS experiments on liquid alloys. Second, there is no evidence of any propagating concentration fluctuation analogous to the optic mode of molten salts and plasma /7/; the spectrum $S_{CC}(k,\omega)$ retains its diffusive character throughout the range of k that we have studied.

SHEAR WAVES

Molecular dynamics (MD) calculations /1,2/ have shown conclusively that at short wavelengths monatomic liquids can support a propagating shear wave, i.e. a collective excitation analogous to the transverse phonons found in solids. The onset of this so-called viscoelastic behaviour is linked to the fact that as the wavenumber k increases the rigidity of the liquid becomes an important factor. At long wavelengths, by contrast, the decay of transverse current fluctuations is described, in the mean, by a hydrodynamic diffusion equation /16/, the correlation time of the fluctuations being determined by the kinematic shear viscosity.

Thus far, however, attention has been focused mainly on argon-like liquids and, to our knowledge, no MD results have been reported for systems simulating liquid metals. The interest in extending the work in this direction lies in the fact that the

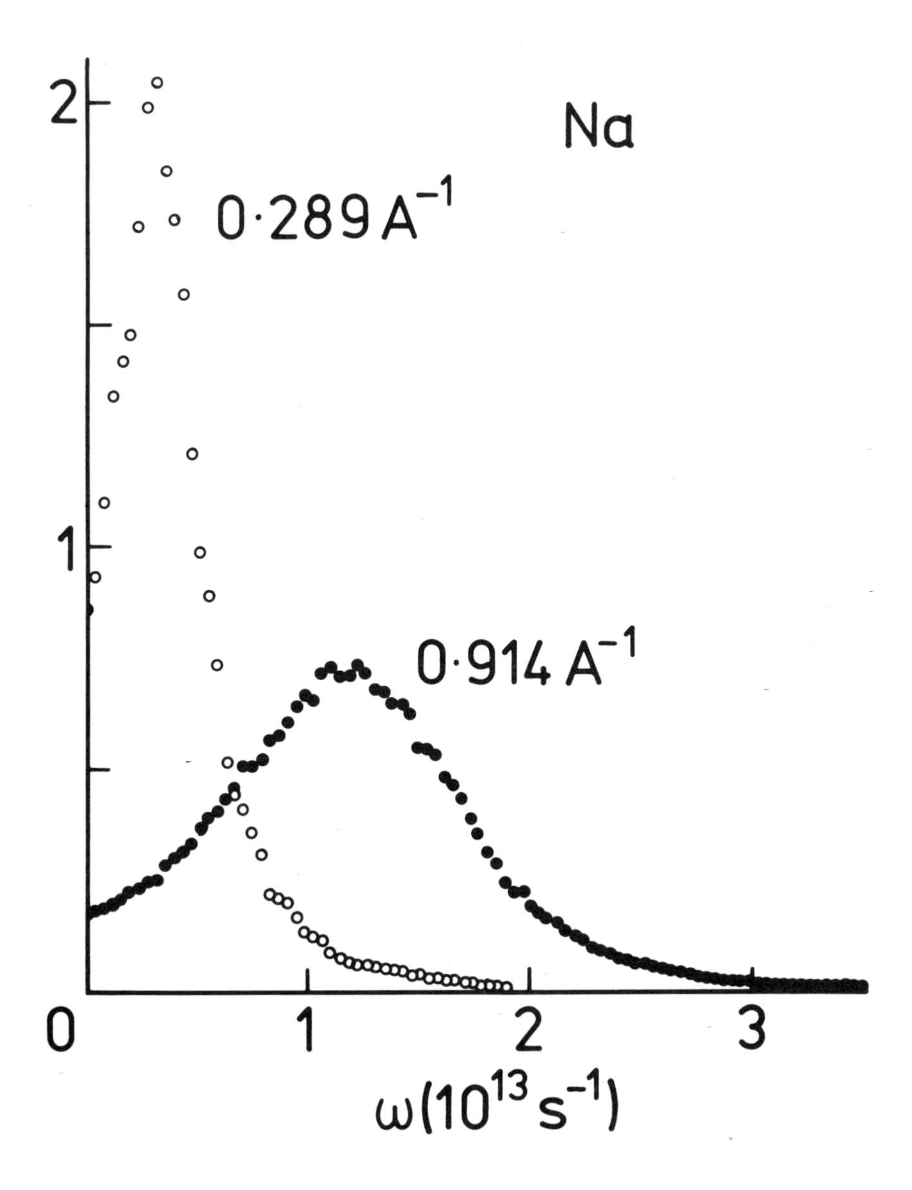

Fig. 5. The function $C_t(k,\omega)$ (in units of 10^{-13} s and normalized to unit area) for liquid Na at two values of k.

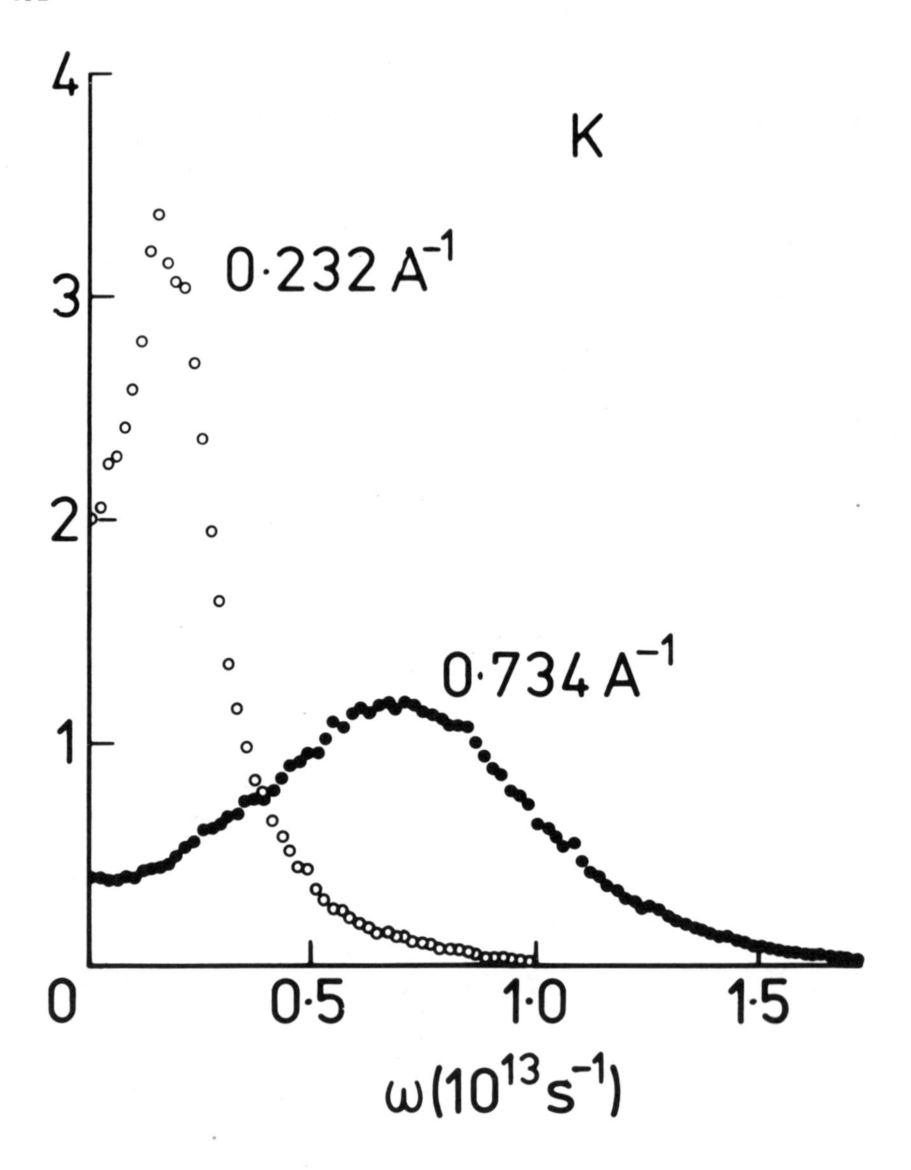

Fig. 6. The function $C_t(k, \omega)$ (in units of 10^{-13} s and normalized to unit area) for liquid K at two values of k.

longitudinal collective modes of liquid metals are in certain respects very different from those of liquids such as argon. In particular, it is known from MD studies of density fluctuations /3,4/ and from neutron inelastic scattering experiments /6/ that a remnant of the hydrodynamic sound wave mode persists in liquid alkali metals down to wavelengths which, measured relative to the spacing of neighbouring atoms, is much smaller than in argon /1,2,17/ or in neon /18/. A similarly large difference in behaviour may therefore be expected for the transverse modes.

Here we describe the results of MD calculations of the spectrum of transverse current fluctuations, which we shall denote by the symbol $C_t(k,\omega)$, for systems modelling liquid sodium and potassium. The potentials we have used are those proposed by Dagens, Rasolt and Taylor /9/, which have been shown elsewhere to account satisfactorily for many of the properties of alkali metals. Brief details of the MD "experiments" are given in Table 1, namely the densities, temperatures and values of the timestep Δt used in the numerical integrations.

The functions $C_t(k,\omega)$ introduced above is the Fourier transform of the transverse current autocorrelation function $C_t(k,t)$. The latter may be defined as

$$C_t(k,t) \;=\; (1/2N)\ \mathrm{Tr} < \underset{\sim}{k} \times \underset{\sim}{j}(\underset{\sim}{k},t)k \times \underset{\sim}{j}(-\underset{\sim}{k},0)$$

where

$$\underset{\sim}{j}(\underset{\sim}{k},t) \;=\; \sum_i \dot{\underset{\sim}{r}}_i(t)\exp\{\underset{\sim}{k}\cdot\underset{\sim}{r}_i(t)\}$$

is the microscopic particle current, $\underset{\sim}{r}_i(t)$ being the position of atom i at time t. Some representative spectra are plotted in Figures 5 (for sodium) and 6 (for potassium); the examples shown are for the smallest wavenumber compatible with the periodic boundary conditions and for a value of k some three times larger than this. The results are broadly similar to those obtained for potential models appropriate to liquid argon /1,2/. In each case there is a well-defined shear wave peak which broadens and shifts to higher frequencies as k increases; at sufficiently short wavelengths, the peak vanishes altogether, since in the limit $k \rightarrow 0$ the atoms behave as free particles and all correlations disappear. At small k, however, the peak is much sharper than that seen in liquid argon. This

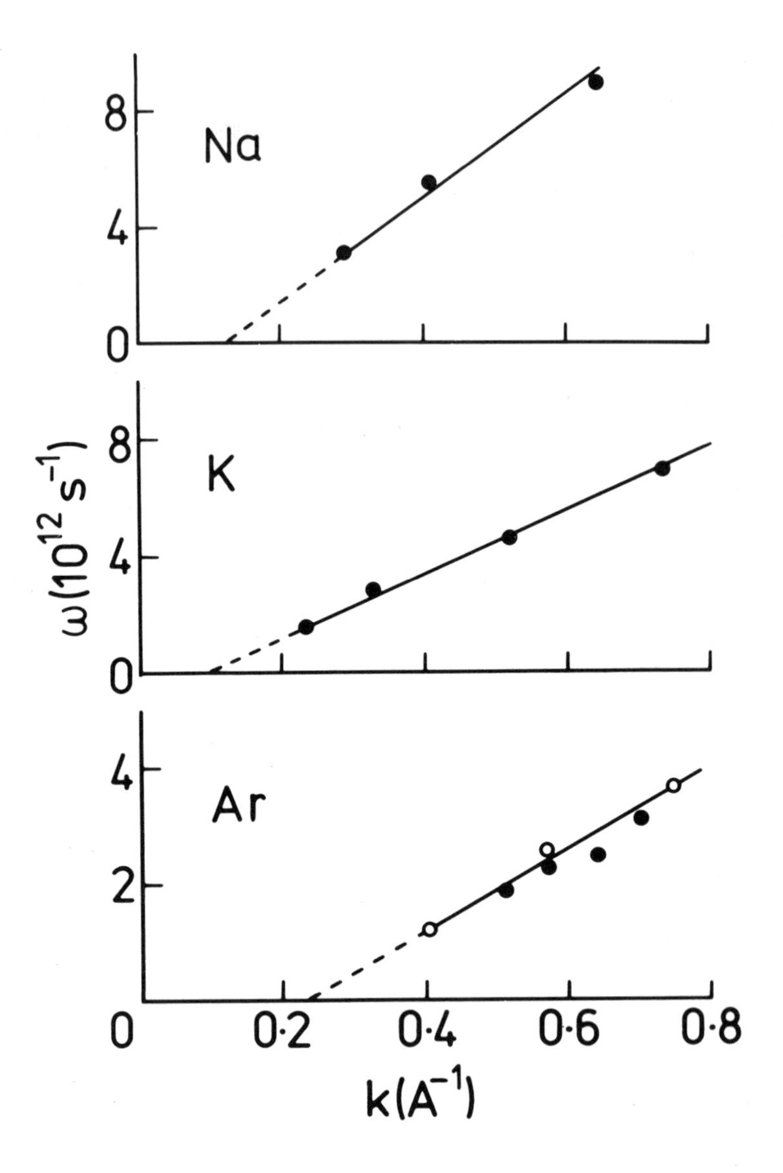

Fig. 7. Dispersion of the shear wave peak for three model systems. Results for Na and K: present work; results for Ar: from Rahman /1/ (dots), and Levesque et al. /2/ (circles).

enhancement of the elastic (solid-like) component of the spectrum is in accord with a recent calculation for liquid rubidium by Chiakwelu et al. /19/.

Dispersion curves for the propagating transverse mode are plotted for sodium, potassium and argon in Figure 7. It is clear from the figure that over a modest range of k the results may be described by dispersion relations of the form $\omega = v_t(k - k_t)$, where v_t is the speed of propagation of the shear wave and k_t is a critical wavenumber which marks the boundary of the shear wave regime; the latter quantity may be reliably estimated by backward extrapolation of the MD results. In the case of the longitudinal modes it is well-known /16/that at small k the dispersion of the sound wave peak in the dynamic structure factor is given by the hydrodynamic relation $\omega = v_l k$, where v_l is a longitudinal speed of sound. However, a well-defined sound wave peak is visible only for wavenumbers smaller than a value k_l characteristic for each liquid.

Results obtained for v_t, k_t and k_l from this and earlier work are summarized in Table 2. There is some uncertainty in the estimates quoted for the quantity k_l, since the sound wave peak does not disappear abruptly, but first broadens into a shoulder. It should be noted that, in each case, v_t is smaller than v_l, the values obtained for the ratio v_t/v_l being consistent with the measured velocities of transverse and longitudinal sound in a variety of monatomic solids.

From the discussion given above it follows that the collective modes of a given liquid may be discussed in terms of two characteristic wavelengths: $\lambda_t = 2\pi/k_t$, an upper limiting wavelength at which shear waves disappear; and $\lambda_l = 2\pi/k_l$, a lower limiting wavelength at which sound wave propagation ceases. To make sensible comparison between results for different systems it is necessary to introduce a characteristic scale of length for each liquid. This is conveniently taken to be the separation a at which the radial distribution function has its main peak, a having an obvious interpretation as the mean nearest-neighbour spacing in the liquid. Taking for a the values quoted in Table 2, we find that in the case of argon

$$\lambda_t/a \simeq 7, \quad \lambda_l/a \simeq 6 \qquad \text{(Ar)}$$

whereas for the two alkali metals

$$\lambda_t/a \simeq 14, \quad \lambda_l/a \simeq 1.4 \quad \text{(Na, K)}$$

In liquid argon, therefore, the boundaries of the shear wave and sound wave regimes are nearly coincident. From a theoretical point of view, this is a neat result, since the disappearance of the sound wave mode and the onset of shear waves may both be taken as an indication that the equations of linearized hydrodynamics are no longer of even qualitative value in predicting the spectra of fluctuations in the corresponding component of the particle current. In the liquid metals, however, λ_t is significantly larger (relative to a) than in argon, and λ_l is significantly smaller. As a consequence, there is now a wide range of wavelength, extending roughly from one (or two) to fourteen interatomic spacings, in which the liquid can support propagating collective excitations of both transverse and longitudinal character. The persistence of a shear wave mode at long wavelengths serves to confirm the view /20/ that the atomic motion in liquid metals has a much stronger collective character than in argon-like liquids.

We thank Roger Taylor for providing the potentials and Carl Moser for the hospitality of CECAM during the period in which the computations were carried out.

REFERENCES

1. A. Rahman, in Neutron Inelastic Scattering, vol.1 (IAEA, Vienna, 1968).
2. D. Levesque, L. Verlet, and J. Kürkijarvi, Phys. Rev. A 7, 1690 (1973).
3. A. Rahman, in Statistical Mechanics: New Concepts, New Problems, New Applications, ed. S.A. Rice, K.F. Freed, and J.C. Light (University of Chicago Press, Chicago, 1972).
4. A. Rahman, Phys. Rev. Lett. 32, 52 (1974); A. Rahman, Phys. Rev. A 9, 1667 (1974).
5. I. Ebbsjö, T. Kinell, and I. Waller, J. Phys. C. 11, L501 (1978).
6. J.R.D. Copley and J.M. Rowe, Phys. Rev. Lett. 32, 49 (1974); Phys. Rev. A 9, 1656 (1974).

7. Charged fluids are a special case. For MD calculations, see I.R. McDonald and J.P. Hansen, Phys. Rev. A 11, 2111 (1975); J.R.D. Copley and A. Rahman, Phys. Rev. A 13, 2276 (1976); I.R. McDonald, P. Vieillefosse, and J.P. Hansen, Phys. Rev. Lett. 39, 271 (1977); I.R. McDonald and J.P. Hansen, Phys. Rev. Lett. 41, 1379 (1978).
8. M. Parrinello, M.P. Tosi, and N.H. March, J. Phys.C 7, 2577 (1974).
9. M. Rasolt and R. Taylor, Phys. Rev. B 11, 2717 (1975); L. Dagens, M. Rasolt, and R. Taylor, Phys. Rev. B 11, 2726 (1975).
10. G. Jacucci, I.R. McDonald, and R. Taylor, J. Phys. F 8, L121 (1978).
11. N.H. March and M.P. Tosi, Atomic Dynamics in Liquids (Macmillan, London, 1976).
12. G. Abowitz and R.B. Gordon, J. Chem. Phys. 37, 125 (1962).
13. W.-M. Shyu, K.S. Singwi, and M.P. Tosi, Phys. Rev. B 3, 237 (1971).
14. A.B. Bhatia, D.E. Thornton, and N.H. March, Phys. Chem. Liq. 4, 97 (1974).
15. M.C. Abramo, M. Parrinello, M.P. Tosi, and D.E. Thornton, Phys. Lett. 43A , 483 (1973).
16. J.P. Hansen, and I.R. McDonald, Theory of Simple Liquids (Academic Press, London, 1976).
17. K. Sköld, J.M. Rowe, G. Ostrowski, and P.D. Randolph, Phys. Rev. A 6, 1107 (1972).
18. H.G. Bell, H. Moeller-Wenghoffer, A. Kollmar, R. Stockmeyer, T. Springer, and H. Stiller, Phys.Rev. A 11, 316 (1975).
19. D. Chiakwelu, T. Gaskell, and J.W. Tucker, J. Phys.C 9, 1635 (1976).
20. J.R.D. Copley, and S.W. Lovesey, Rep. Prog. Phys. 38, 461 (1975).

System	T(K)	ρ(g cm^{-3})	Δt(10^{-14}s)	c(10^5 cm s^{-1}) MD[a]	expt[b]
Na	293.5	0.925	1.0	2.8	2.53
Na-K	295.1	0.897	1.0	2.4	2.07
K	299.8	0.814	2.0	2.0	1.87

Table 1. Details of the MD calculations. c is the speed of sound: (a) present work, (b) experimental values /12/ at 373 K.

Liquid	a(Å)	v_t(10^5 cm s^{-1})	k_t(Å^{-1})	v_l(10^5 cm s^{-1})	k_l(Å^{-1})
Na[a]	3.7	1.8	0.12	2.8	1.25
K[a]	4.5	1.1	0.10	2.0	1.0
Ar[b]	3.8	0.7	0.23	1.1	0.3

Table 2. Characteristics of the collective modes. (a) present work and ref. /6/, (b) refs. /1/ and /2/.

2.2 Amorphous and glassy systems

ATOMIC SCALE STRUCTURE AND STRUCTURAL MODELS FOR AMORPHOUS METALLIC ALLOYS

G. S. Cargill III

IBM Thomas J. Watson Research Center
Yorktown Heights, N. Y. 10598, U.S.A.

ABSTRACT. A wide variety of metallic alloys have been prepared as amorphous solids by rapid quenching of liquids, sputtering, coevaporation, and other methods. Structural studies of these materials have employed x-ray, electron, or neutron scattering, and more recently EXAFS measurements. Most efforts to model atomic scale structures of amorphous alloys have been based on computer generated dense random packings. Results of scattering studies with amorphous rare earth metal-transition metal alloys agree well with relaxed, binary dense random packing models, but similar models have been much less successful when applied to amorphous (Nb, Ta)-(Si, Ge) alloys. Recent experiments indicate that chemical ordering is significantly more important for the latter materials, and that it may be more useful to describe their atomic scale structure in terms of quasi-molecular building blocks than in terms of simple dense random packing.

1. INTRODUCTION

This chapter begins with a brief review of concepts used in characterizing atomic scale structure of amorphous alloys, followed by more detailed discussion of experimental results for two types of amorphous alloys. The first consists of rare earth metals and transition metals, e.g. Gd-Co; the second, of group VA metals and group IVB metalloids, e.g. (Nb, Ta)-(Si, Ge) alloys. These two types of alloys provide a contrast between systems which seem to be well described in terms of dense random packing models and those in which local structure is more strongly affected by chemical ordering tendencies and may more usefully be described in quasi-molecular terms. Structures of amorphous metallic alloys have been discussed in several recent review articles [1-6] which are more extensive and general than the present chapter.

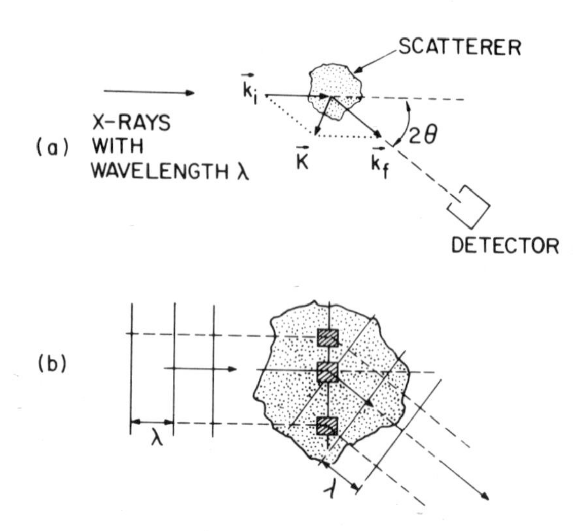

Fig. 1. (a) Essential aspects of an x-ray scattering experiment.
(b) Path length differences in plane wave scattering.

2. SCATTERING EXPERIMENTS AND DISTRIBUTION FUNCTIONS

Experimental techniques for amorphous solids are very similar to those used for liquids: x-ray, neutron, and electron scattering. Essential aspects of an x-ray scattering experiment are shown in Fig. 1. Although the present discussion is for x-ray scattering, most of the concepts are equally applicable to experiments using electrons or neutrons. A detector is used to measure the scattered intensity as a function of scattering angle 2Θ. The scattering vector $\mathbf{K}$ is defined as $\mathbf{K} = \mathbf{k}_f - \mathbf{k}_i$ and has magnitude $K = 4\pi(\sin\Theta)/\lambda$. K-dependent interference in the scattering process arises from path length differences within the scatterer, as illustrated in Fig. 1(b). The scattered amplitude for a single isolated atom is given simply by the atomic scattering factor f(K). The scattered intensity measured experimentally for a monatomic material can be divided by $f(K)^2$ to remove the K-dependence attributable only to intra-atomic interference. Normalization and division of the scattered intensity by $f(K)^2$, or by $<f(K)>^2$ for polyatomic materials, after corrections for polarization, absorption and geometrical factors, produces an interference function I(K) which oscillates about 1 at large values of K. This is illustrated schematically in Fig. 2.

Structural information contained in I(K) can best be represented in terms of a radial distribution function RDF(r), obtained by Fourier transformation of the reduced interference function $F(K)=K[I(K)-1]$. Because RDF's give only spherically averaged information on correlations in atom positions, they do not specify uniquely positions and chemical identities of atoms of an amorphous material. RDF's for crystalline and amorphous forms of a hypothetical monatomic material are shown schematically in Fig. 3, together with sketches of the corresponding atom arrangements. For the crystalline material, the RDF consists of delta functions, with weights corresponding to the numbers of neighbors within the various coordination shells.

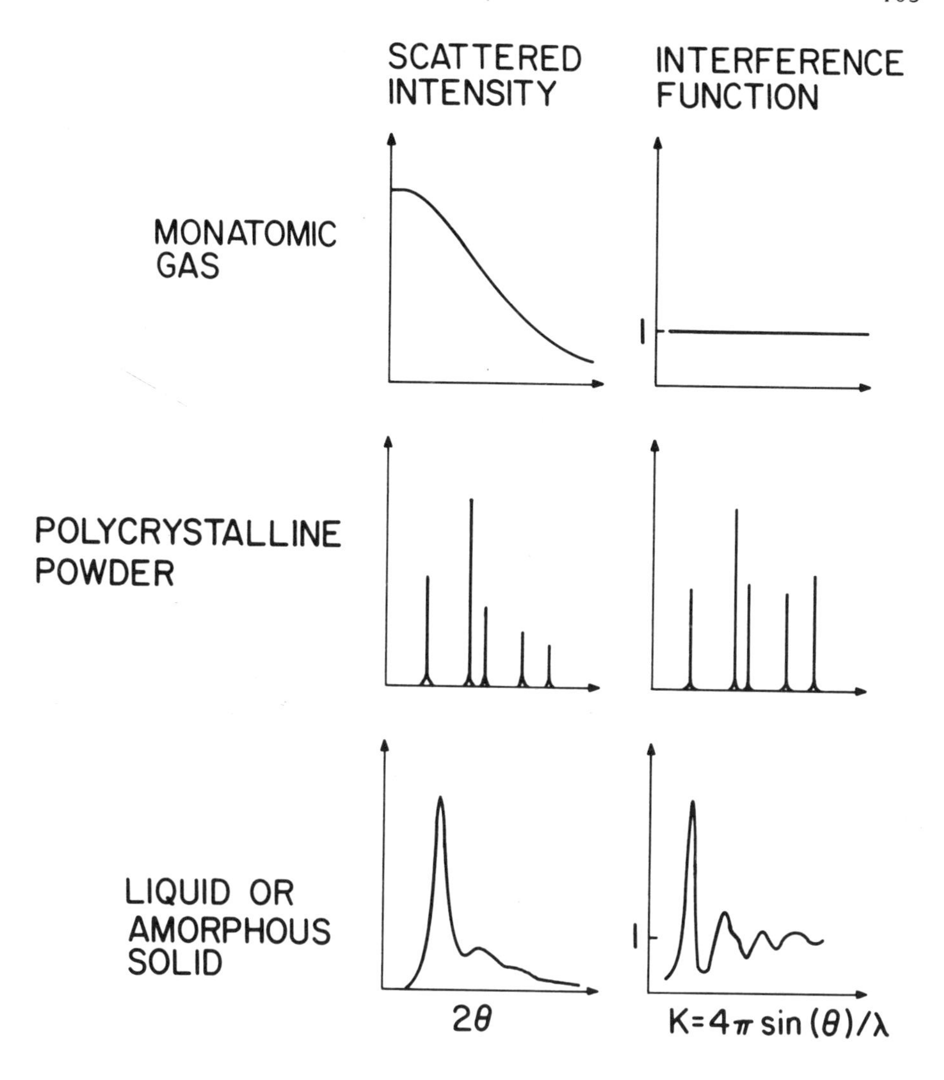

Fig. 2. Scattered intensity distributions and interference functions for three types of isotropic scatterers.

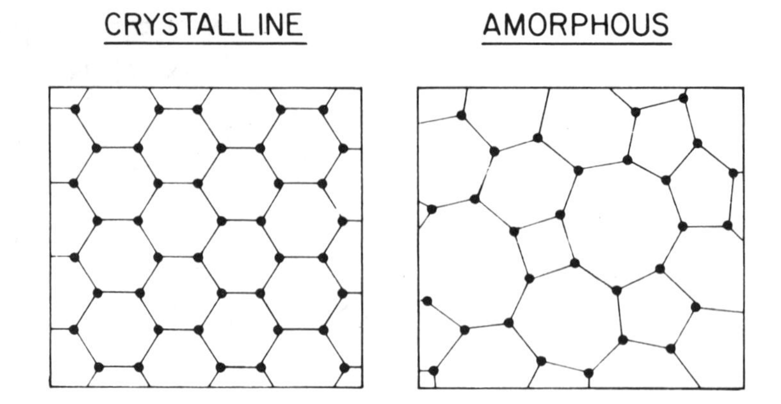

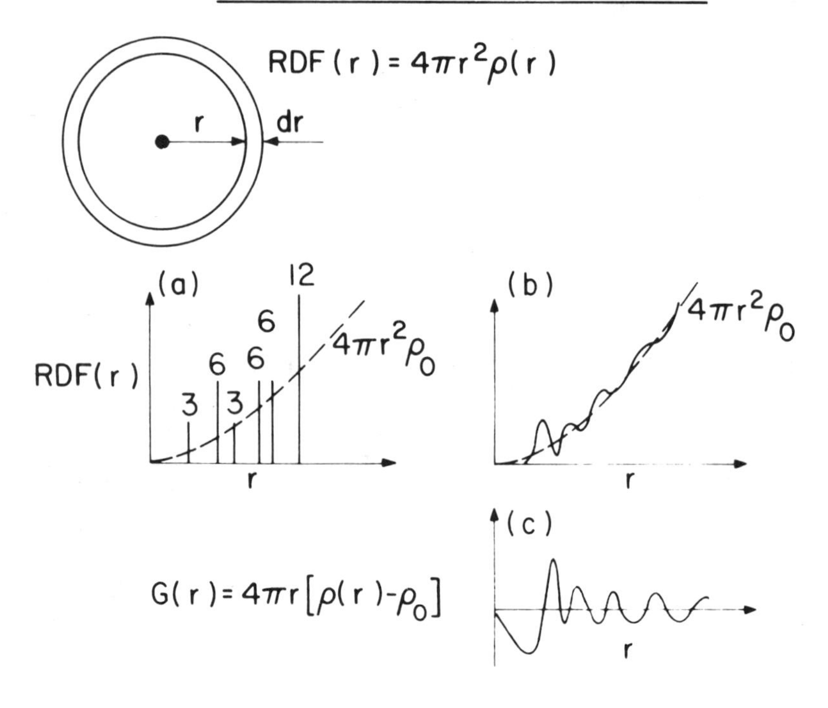

Fig. 3. Radial distribution functions for (a) crystalline and (b) amorphous materials. Also shown is the reduced distribution function (c) for the amorphous material.

For the amorphous solid the surroundings of each atom are generally different from those of other atoms, and RDF(r) is determined by averaging results obtained for spherical shells centered on each of the individual atoms. The result is a continuous function, with maxima corresponding to particularly frequently occurring atom-atom separation distances within the amorphous substance. RDF(r)=0 for r-values less than the minimum nearest neighbor distance. A reduced radial distribution function $G(r)=4\pi r[\rho(r)-\rho_o]$ is obtained from the reduced interference function F(K),

$$G(r) = (2/\pi)\int_{K_{min}}^{K_{max}} F(K) \sin (Kr)\, dK. \tag{2.1}$$

$RDF(r)=4\pi r^2\rho(r)$ can be calculated from G(r) if the density ρ_o is known. In practice the scattering pattern can be measured only over some finite range of K, and the experimentally determined distribution functions differ from true distribution functions because of termination effects in the Fourier transformation.

Radial distribution functions and their interpretation are more complicated for materials which contain more than one type of atom. An amorphous material with two types of atoms, A and B, is shown schematically in Fig. 4. Three different distribution functions $\rho_{AA}(r)$, $\rho_{BB}(r)$, and $\rho_{AB}(r)$, called partial distribution functions, are needed to describe the structure of this material. $\rho_{ij}(r)$ is obtained by considering spherical shells centered on type i atoms and by counting only type j atoms which fall within these shells. If c_A and c_B are the atom fractions for the two types of atoms, it follows that

$$\rho_{BA}(r) = (c_A/c_B)\, \rho_{AB}(r). \tag{2.2}$$

Only weighted averages of these partial distribution functions can be obtained from an individual scattering pattern,

$$\rho(r) = \sum_{i,j} W_{ij}\rho_{ij}(r)/c_j \tag{2.3}$$

with

$$W_{ij} = c_i c_j f_i f_j/\langle f\rangle^2. \tag{2.4}$$

Eq. (2.3) is strictly correct only if the ratios of scattering factors in Eq. (2.4) are independent of K, which for x-ray or electron scattering is never exactly satisfied. Eq. (2.3) is nevertheless a good approximation for most amorphous metallic alloys [2].

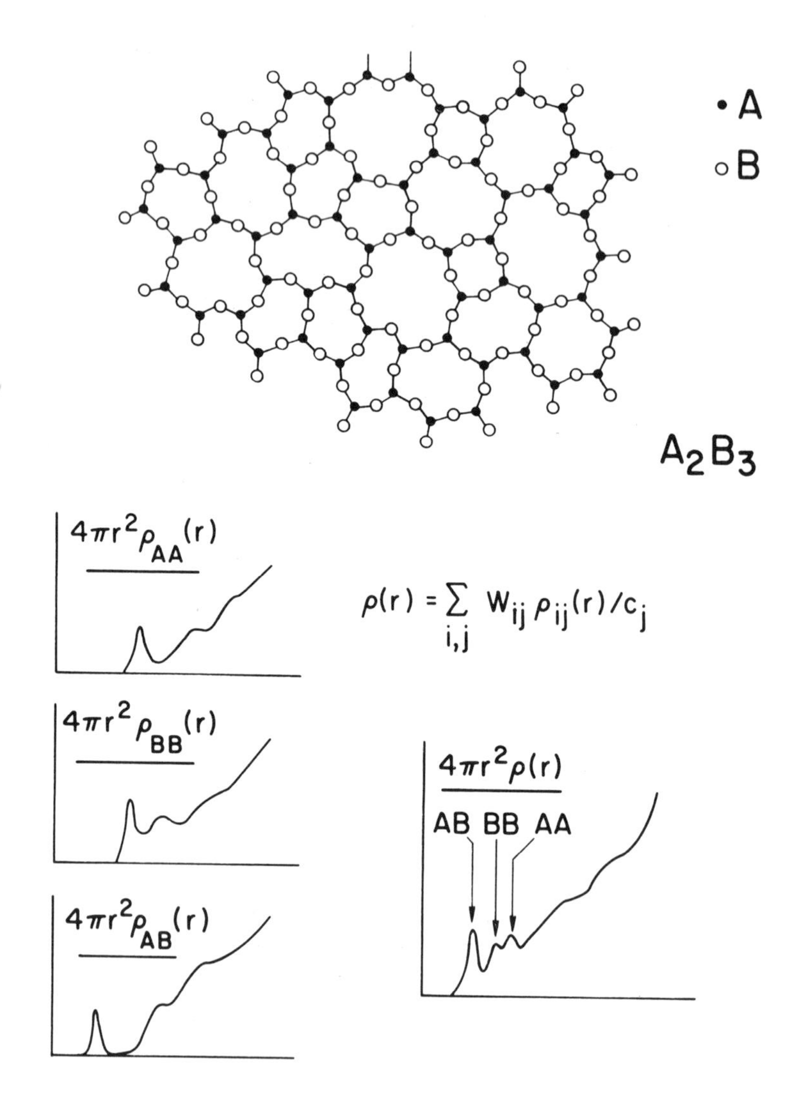

Fig. 4. Partial and composite radial distribution functions for an amorphous binary alloy.

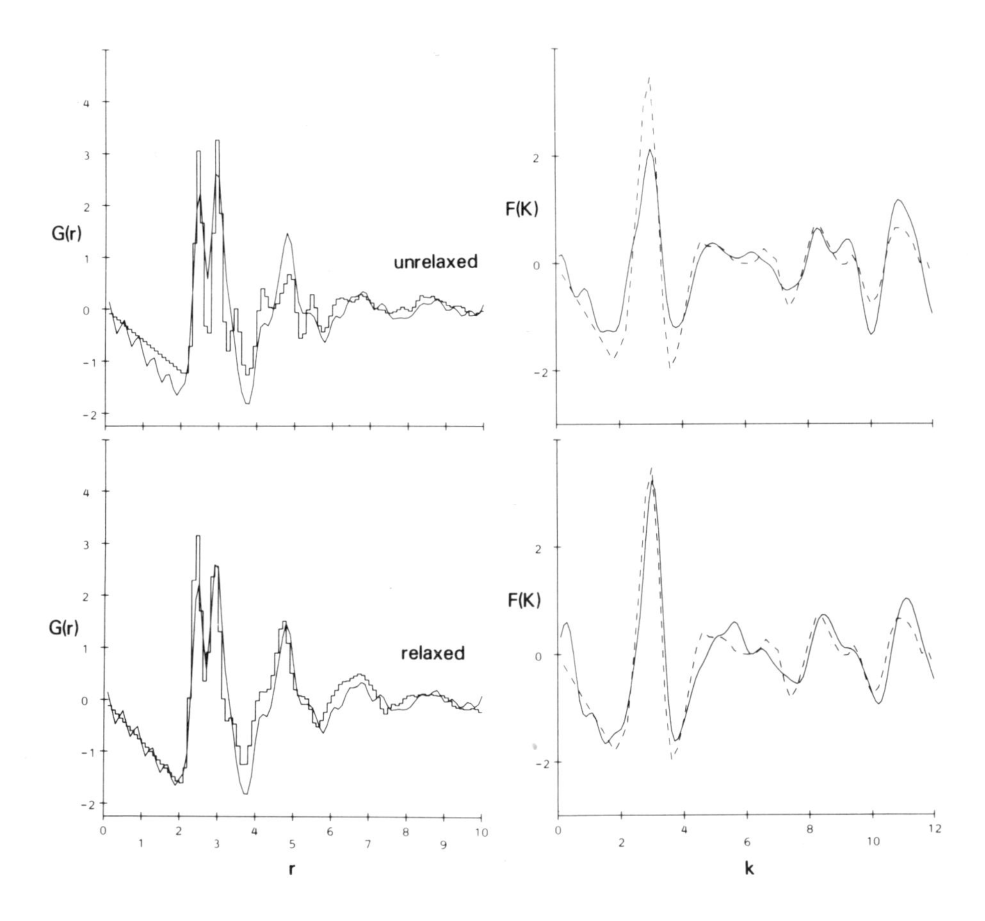

Fig. 5. Model and experimental distribution functions and interference functions for amorphous $Gd_{18}Co_{82}$. The continuous line is the experimental result for G(r), and the dashed line is the experimental result for F(K).

3. RARE EARTH METAL-TRANSITION METAL ALLOYS

3.1 Experimental Results

Amorphous films of Gd-Co and other rare earth metal-transition metal alloys (RE-TM) have been prepared over wide ranges of composition by coevaporation and by sputtering. Sputtered samples of Gd_xCo_{1-x} have been used in x-ray scattering studies [7], and in some cases contributions from the different types of nearest neighbor pairs could be resolved: TM-TM, TM-RE, and RE-RE. Reduced interference

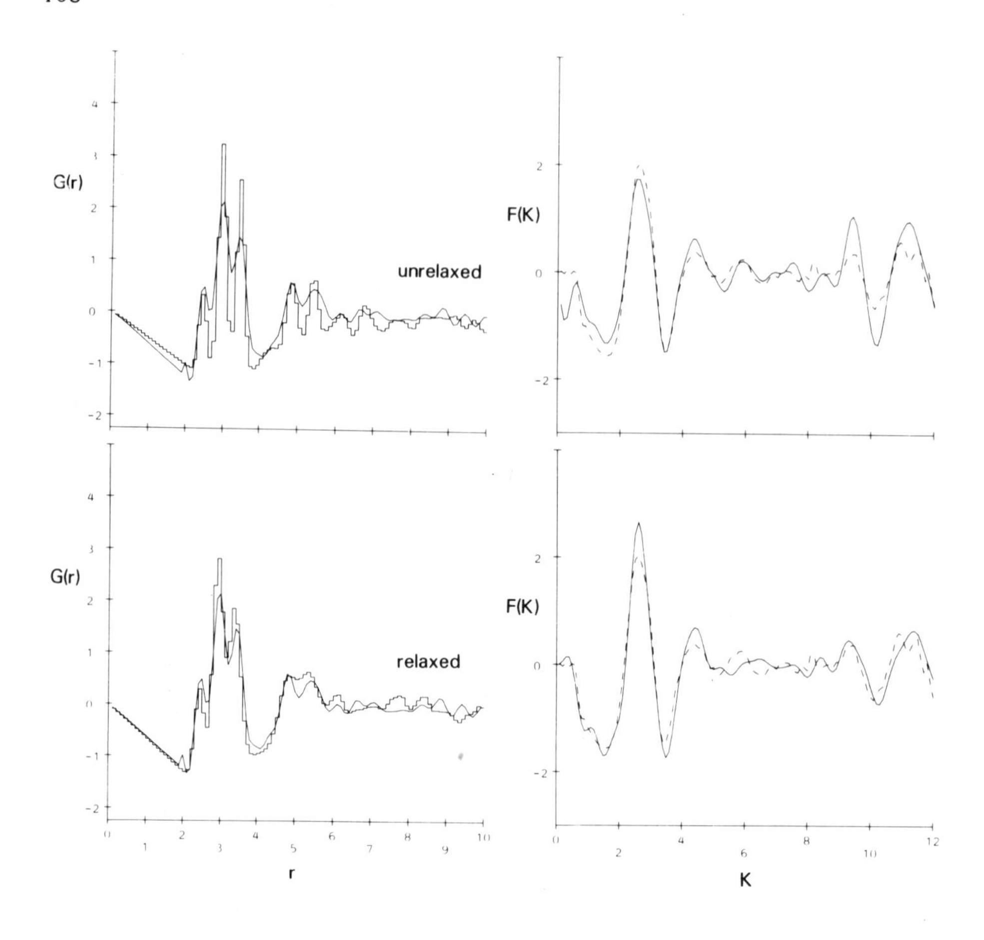

Fig. 6. As for Fig. 5, but for $Gd_{32}Co_{68}$. [Experimental data are from C. N. J. Wagner, N. Heiman, T. C. Huang, A. Onton, and W. Parrish, AIP Conf. Proc. *28*, 188 (1976).]

functions and reduced distribution functions for $Gd_{18}Co_{82}$ [2] and for $Gd_{32}Co_{68}$ [7] are shown in Fig. 5 and Fig. 6, together with results of model calculations [6]. The three nearest neighbor contributions are most clearly seen for the $Gd_{32}Co_{68}$ alloy. Experimental results for this alloy are very similar to those obtained earlier [8] for amorphous $Gd_{36}Fe_{64}$. The nearest neighbor regions of radial distribution functions for $Gd_{18}Co_{82}$ and for $Gd_{37}Fe_{64}$ are shown in Fig. 7, together with gaussian components fitted to the experimental curves by a least squares method. These components have been interpreted as arising from the three types of nearest neighbor pairs. Also shown are vertical bars, which correspond to the positions and numbers of nearest neighbor pairs in $GdFe_2$ and $GdCo_5$ crystal structures. Crystalline $GdCo_5$ has no

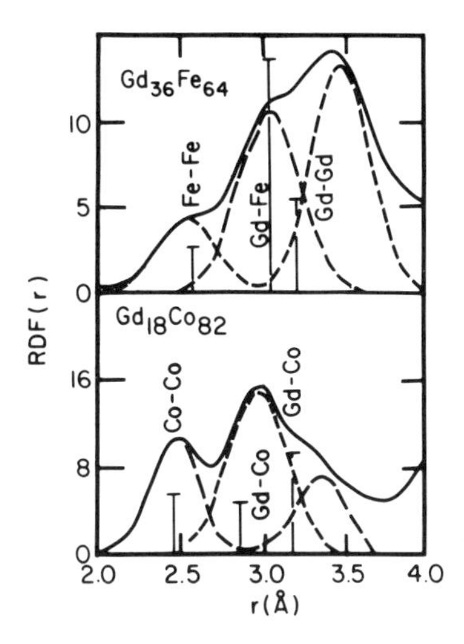

Fig. 7. Nearest neighbor regions of RDF(r) for two amorphous RE-TM alloys. Vertical bars indicate positions and coordination numbers for the crystalline compounds $GdFe_2$ and $GdCo_5$.

Gd-Gd nearest neighbors, but evidence for Gd-Gd nearest neighbors is seen for the amorphous $Gd_{18}Co_{82}$ alloy. For the amorphous $Gd_{36}Fe_{64}$ alloy, the Gd-Gd nearest neighbor distance is much larger than for crystalline $GdFe_2$, falling near that of crystalline hcp Gd.

3.2 Dense Random Packing Models for RE-TM Alloys

Dense random packing models for RE-TM amorphous alloys have been generated using a computer algorithm [9] based on one originally proposed by Bennett [10] and by Adams and Matheson [11]. The algorithm involves sequential addition of spheres at three-fold coordinated sites on the surface of a cluster. A random number generator is used to select either a large sphere, for a RE atom, or a small sphere, for a TM atom, at each addition step, with a criterion which produces models of the desired composition.

To reproduce characteristic features of the RE-TM distribution functions, it was found necessary [9] to limit the sites used for adding spheres to those for which the three spheres forming the site were in nearly "hard contact" with one another, thereby increasing the "tetrahedron perfection" [12] of the structure. Model structures of 800 spheres for RE-TM alloys were generated with a tetrahedral perfection constraint [9, 12] of 1.1, with sphere diameters scaled to 2.5Å for TM-atoms and 3.6Å for RE-atoms. No chemical constraints were introduced to either forbid or favor particular types of i-j pairs. At each sequential addition, the type of sphere selected by the random number generator was placed at the site having the prescribed tetrahedron perfection which was closest to the center of the cluster.

Partial distribution functions $4\pi r^2\rho_{ij}(r)$ were calculated from coordinates of sphere centers, using all pairs in the cluster. Because of the finite cluster size, these distribution functions became zero for distances r greater than the cluster diameter. A finite size correction [13] was applied to the partial distribution functions, and partial reduced distribution functions $G_{ij}(r)$ were calculated. Atomic scattering factors for Gd and Co were approximated by the atomic numbers Z_{Gd} and Z_{Co} in calculating the total reduced distribution function G(r). A density ρ_o was chosen for which G(r) oscillated about zero for r=6Å-10Å, and convolution with a gaussion broadening function was used to obtain nearest neighbor peak widths similar to those found experimentally. Results for these "unrelaxed" structural models are shown in Fig. 5 and 6. Reduced interference functions F(K) are also shown.

Although the structural models reproduce most of the experimentally seen features, agreement between models and experiments was improved by relaxing the model structures [14], i.e. allowing small displacements of atoms which reduce the sum of pairwise energies for the structure. Lennard-Jones potentials were used. The energy being minimized is given by

$$E = \frac{1}{2}\sum_{i \neq j}\sum \left[\left(\frac{R_{ij}}{r_{ij}}\right)^{12} - 2\left(\frac{R_{ij}}{r_{ij}}\right)^{6}\right] \tag{3.1}$$

where R_{ij} is the sum of hard sphere radii of spheres i and j, and r_{ij} is the actual distance between their centers. The interaction extended only over nearest neighbors. The relaxation was continued until reductions in energy for further displacements, as well as sizes of the displacements, became very small. Distribution functions and interference functions were calculated from coordinates for the relaxed structures and are shown in Fig. 5 and 6. Relaxation significantly improved agreement with the experimental results. Densities of the model structures ρ_o were increased by relaxation to within 5% those measured experimentally [15] for amorphous Gd-Co alloy films. This level of agreement was achieved without introduction of chemical ordering or chemical constraints in building or relaxing the model structures.

4. AMORPHOUS M_3X ALLOYS

Amorphous alloys M_3X of group VA metals (M=Nb, Ta) and group IVB metalloids (X=Si, Ge) provide an interesting contrast to the RE-TM alloys discussed above. Attempts to model amorphous M_3X alloys by simple binary dense random packing models have been unsuccessful [16]. Structural studies suggest that chemical ordering and particularly well defined M-X bonding play important roles in these materials [16,17].

4.1 X-ray Scattering Studies

Radial distribution functions for amorphous sputtered films of Nb_3Ge, Ta_3Ge, and

Ta_3Si are shown in Fig. 8. These amorphous alloys are expected to have similar structures, because Nb and Ta are chemically similar and have nearly identical metallic radii, and metalloid elements Si and Ge are likewise chemically similar, with covalent radii differeing by only 4%. However, the experimentally determined RDF's are expected to differ, because of the different weighting factors W_{ij} for ij = MM, MX, and XX in the various M_3X alloys,

$$\rho_{Nb_3Ge}(r) = 0.84\rho_{NbNb}(r) + 1.31\rho_{NbGe}(r) + 0.22\rho_{GeGe}(r) \tag{4.1}$$

$$\rho_{Ta_3Ge}(r) = 1.01\rho_{TaTa}(r) + 0.89\rho_{TaGe}(r) + 0.065\rho_{GeGe}(r) \tag{4.2}$$

$$\rho_{Ta_3Si}(r) = 1.18\rho_{TaTa}(r) + 0.45\rho_{TaSi}(r) + 0.004\rho_{SiSi}(r) \; . \tag{4.3}$$

If the alloys are assumed to be isostructural, approximate partial distribution functions for M-M and M-X pairs can be calculated from the experimental results. For example, using the results for Nb_3Ge and Ta_3Si,

$$\rho_{MX}(r) + 0.16\rho_{XX}(r) = 1.01\rho_{Nb_3Ge} - 0.72\rho_{Ta_3Si} \tag{4.4}$$

$$\rho_{MM}(r) + 0.05\rho_{XX}(r) = 1.12\rho_{Ta_3Si} - 0.39\rho_{Nb_3Ge} \; . \tag{4.5}$$

These approximate partial distribution functions, shown in Fig. 9, contain small contributions from metalloid-metalloid pairs, which can for most purposes be ignored.

The M-X nearest neighbors are less numerous and occur at a smaller separation distance than the M-M nearest neighbors. Also shown in Fig. 9 is the nearest neighbor peak which would result from a gaussian shaped distribution of M-X separations centered on 2.66Å, with 2.1 metalloid neighbors per metal atom, or equivalently 6.3 metal atoms per metalloid atom, with a full-width-at-half-maximum (FWHM) of only 0.10Å. Hayes, et al., [18] have shown that the Pd-Ge nearest neighbor distance distribution in amorphous $Pd_{80}Ge_{20}$ is also very sharp. Superimposed on the metal-metal partial function is the peak corresponding to a gaussian shaped distribution of M-M separations centered on 2.99Å, with an average of 10.7 metal neighbors per metal atom, and with a FWHM of 0.45Å. The metal-metal peak seen experimentally is much broader than that for metal-metalloid pairs and deviates more from the assumed gaussian shape.

The weighting factors W_{ij} for Ta_3Ge are intermediate between those of Nb_3Ge and Ta_3Si. Approximate partial distribution functions calculated from the Nb_3Ge and Ta_3Si experimental results were tested by using them to calculate the total

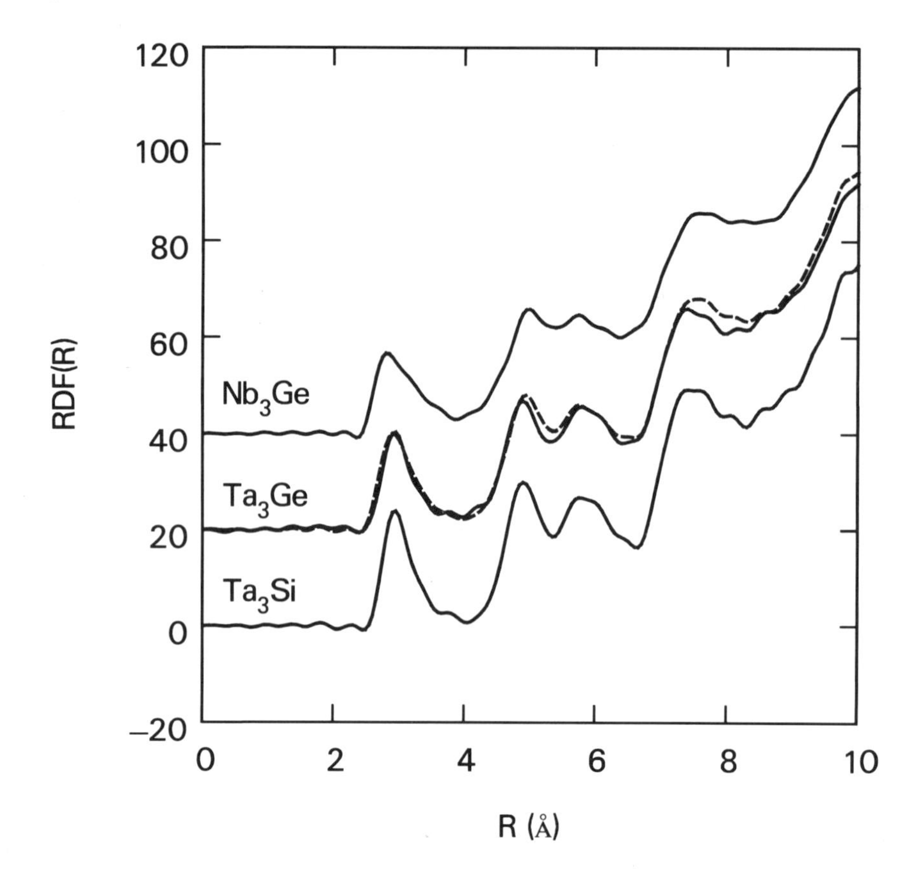

Fig. 8. Radial distribution functions for three amorphous M_3X alloys. The dashed curve is the RDF calculated for Ta_3Ge from the approximate MM and MX partial distribution functions, which were obtained from the Nb_3Ge and Ta_3Si experimental data.

distribution function for Ta_3Ge. The result, shown in Fig. 8, should contain a very small extra contribution from metalloid-metalloid pairs. The calculated and observed Ta_3Ge RDF's agree quite well. Differences between them could in principle be used to calculate $\rho_{XX}(r)$, but experimental uncertainties make this impractical.

Nearest neighbor structural parameters for the amorphous M_3X alloys are compared with those for crystalline A15 Nb_3Ge in Table I. For the amorphous alloys the M-M distance distribution shows no evidence of short intrachain distances found in the A15 crystal and occurs at a slightly larger distance than the expected metallic diameter for Nb or Ta of 2.92Å. Metal-metalloid neighbors are much closer together in the amorphous alloys than in the crystalline compound, and there are fewer of them. These measurements provide no information about metalloid-metalloid correlations in the amorphous alloys.

Table I - M_3 X Nearest Neighbor Distances r_{ij} and Coordination Numbers CN_{ij}.

	A15–Crystalline Nb_3Ge		Amorphous M_3X	
	r_{ij}[Å]	CN_{ij}	r_{ij}[Å]	CN_{ij}
MM	2.58	2	2.99	10.7
	3.16	8	(0.45 FWHM)	
MX	2.87	4	2.66	2.1
			(0.10 FWHM)	
XX	no "nearest neighbors"			

4.2 EXAFS Measurements for Nb_3Ge

Extended x-ray absorption fine structure (EXAFS) spectroscopy is proving to be useful in structural studies of amorphous alloys [19-22]. Positions and amplitudes of oscillations in the x-ray absorption coefficient within a few hundred eV of the K-shell absorption edge of an element in such alloys provide information on the nearest neighbor distances and coordination numbers involving this element, as well as on the distribution of the nearest neighbor distances. X-ray absorption spectra obtained for the amorphous Nb_3Ge alloy described in the previous section are shown in Fig. 10 as absorption μt versus x-ray photon energy $h\nu$ near the energy E_K of the Ge K-shell absorption edge. EXAFS oscillations occur on the high energy, large absorption side of the edge.

The EXAFS oscillations $\chi(h\nu)$ contained in $\mu t(h\nu)$ for $(h\nu - E_K) \sim 50eV - 1000eV$ can be intepreted in terms of interference of spherical waves emerging from the absorbing atom, representing the outgoing photoexcited K-shell electron, with spherical waves emerging from surrounding atoms, representing the backscattering of the photoelectron by these atoms. The wavevector k of the photoelectron depends on its kinetic energy, i.e. roughly on the difference between $h\nu$ and E_K. Maxima in $\chi(k)$ correspond to constructive interference between the outgoing and backscattering waves at the absorbing atom, and minima, to destructive interference. This scattering model can be used to derive the following expression for $\chi(k)$, the k-dependence of the EXAFS oscillations, [19-21]:

$$\chi(k) = \frac{m}{4\pi h^2 k} \sum_j \frac{N_j}{R_j^2} t_j(2k)\, e^{-2R_j/\lambda} \sin\left[2kR_j + 2\delta_j(k)\right] e^{-2k^2\sigma_j^2} \tag{4.6}$$

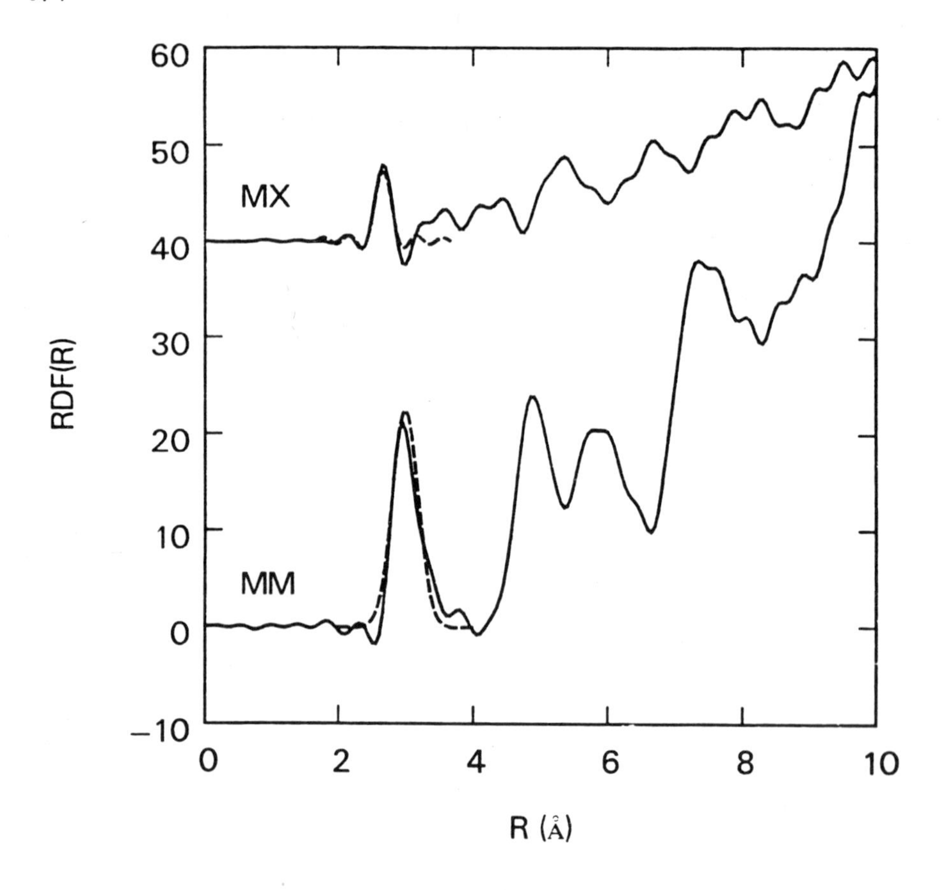

Fig. 9. Approximate partial RDF's for MX and MM pairs, calculated from experimental data for amorphous Nb_3Ge and Ta_3Si. Dashed curves are maxima calculated for M-X and M-M nearest neighbor maxima using parameters given in the text and including effects of Fourier transform termination.

where N_j is the average number of atoms in the jth coordination sphere of the absorbing atom, R_j is the average radial distance to atoms in the jth coordination sphere, $t_j(2k)$ is the back-scattering matrix element encountered by the electrons, λ is the mean free path of the electrons, the exponential containing σ_j^2 is a Debye-Waller-type term where σ_j is the rms fluctuations of distances about R_j, and $\delta_j(k)$ is a phase shift.

Eq. (4.6) can be simplified by adopting the following expression for the phase shift [21],

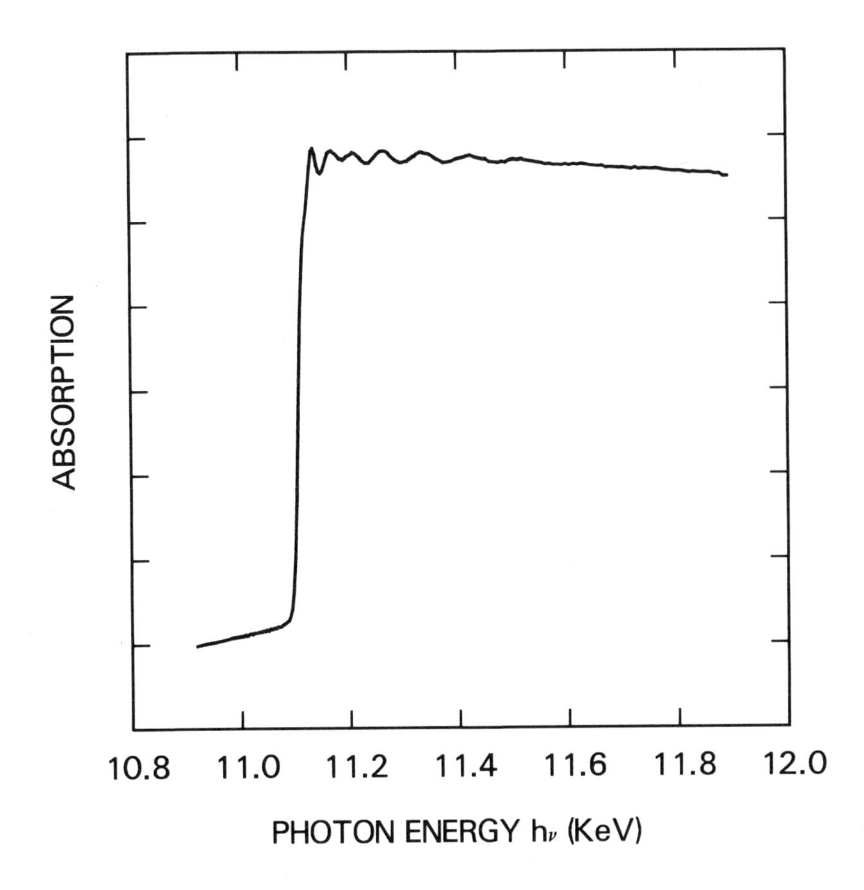

Fig. 10. EXAFS for amorphous Nb_3Ge, absorption μt versus photon energy $h\nu$.

$$\delta_j(k) = -\alpha_j k + \beta_j, \tag{4.7}$$

so that the oscillatory terms of Eq. (4.6) become $\sin[2k(R_j-\alpha_j)+2\beta_j]$. From this it is clear that each well defined coordination sphere of the absorbing atom contributes an oscillatory term of period $\pi/(R_j-\alpha_j)$, and the observed EXAFS oscillations result from a summation over coordination spheres within the range of the effect, which is determined largely by the Debye-Waller factor and electron mean free path terms. EXAFS for many amorphous materials are dominated by the nearest neighbor coordination sphere. The phase shifts $\delta_j(k)$, or α_j and β_j, complicate extraction of structural information from $\chi(k)$, but this problem can be minimized by comparing $\chi(k)$ of an amorphous alloy with $\chi(k)$ of the same material in a single phase crystalline form of know structure, since the phase shifts are thought to be characteristic of the particular atoms and independent of the surroundings for a given class of material [21].

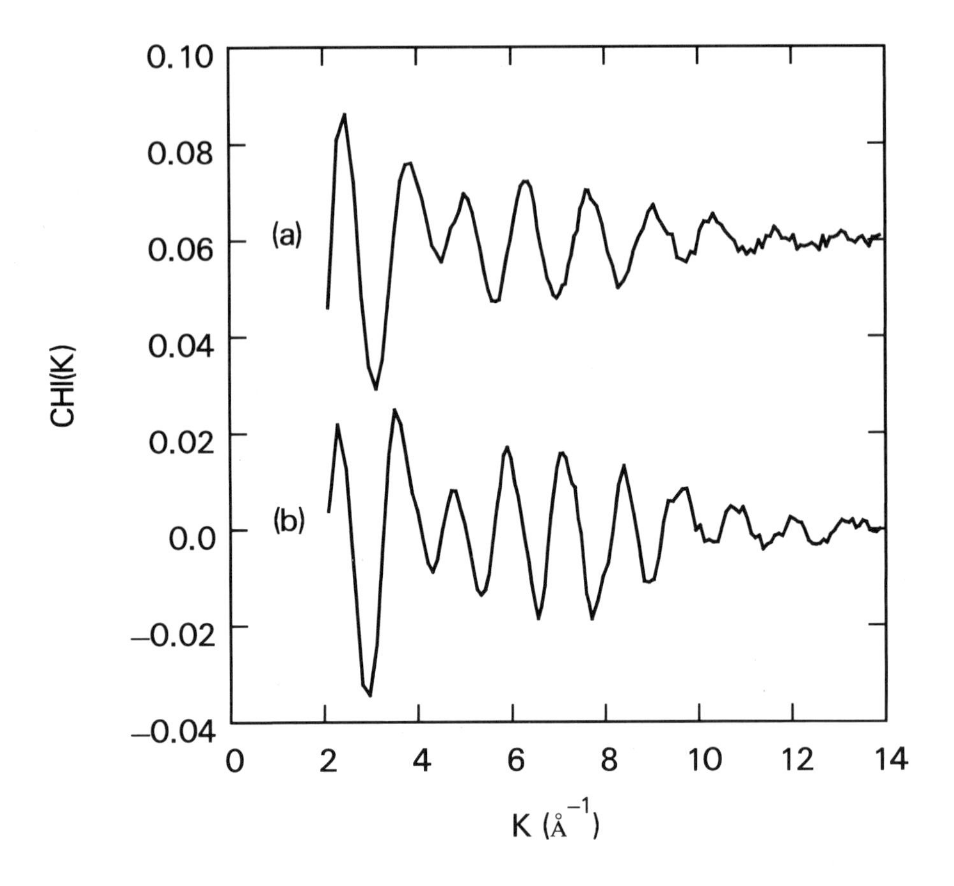

Fig. 11. Normalized EXAFS data for (a) amorphous and (b) A15 crystalline Nb_3Ge.

For EXAFS which are dominated by nearest neighbor contributions, differences in the periods of oscillations for the crystalline and amorphous forms indicate differences in nearest neighbor distances R_1 involving the absorbing element. The amplitude of the oscillations and the rate at which they decay with increasing k depend on the number of neighbors N_1 and on the width of the distribution of nearest neighbor distances σ_1.

Results for amorphous and A15 crystalline Nb_3Ge, as χ versus photoelectron wavevector k, are shown in Fig. 11. The period of oscillation for the amorphous alloy is larger than that for the crystalline material, and the rates of decay of oscillation amplitude are similar for the two materials. EXAFS oscillations for $k > 3Å^{-1}$ look like a single damped, modulated sine wave, which indicates that only one coordination shell contributes significantly in the interference process. The difference in period of oscillation for the amorphous and A15 crystalline Nb_3Ge corresponds to a difference in nearest neighbor coordination shell radius of

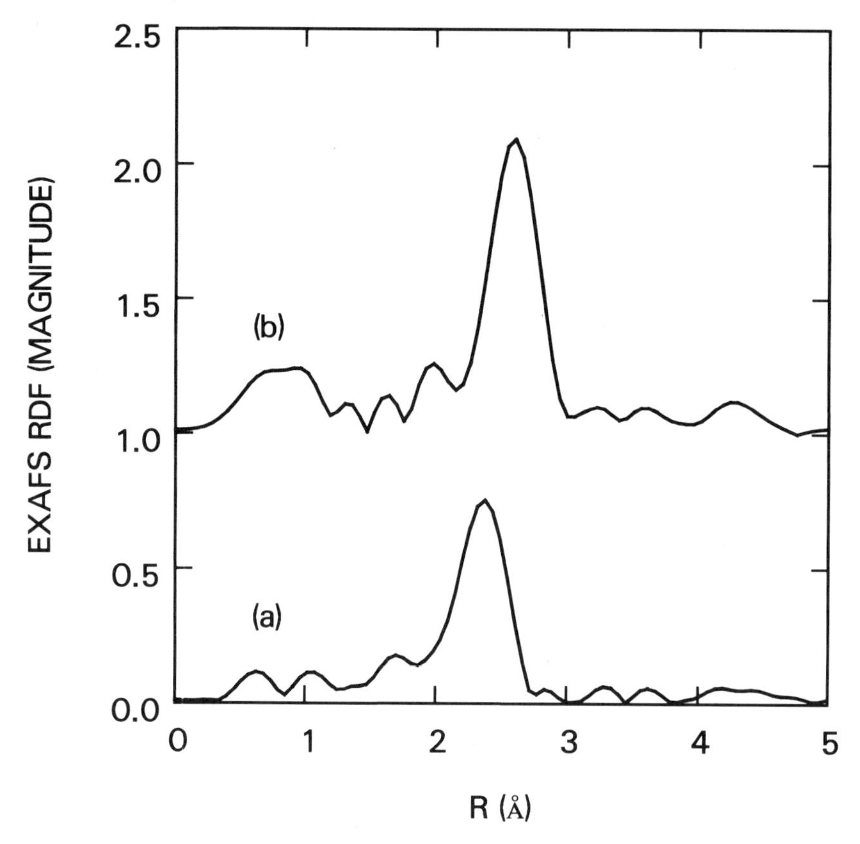

Fig. 12. EXAFS RDF's obtained by Fourier transformation of $k^2\chi(k)$ for (a) amorphous and (b) A15 crystalline Nb_3Ge.

$R_{A15} - R_{amorph} = 0.23 \pm 0.06$Å. Since both EXAFS patterns are expected to be dominated by Nb atoms surrounding the Ge absorbing atom, this yields 2.64 ± 0.06Å for the mean Ge-Nb nearest neighbor separation in the amorphous alloy, in good agreement with an earlier EXAFS result for amorphous Nb_3Ge films prepared under quite different conditions [22] and with the distance obtained from x-ray scattering measurements discussed in the previous section.

EXAFS data like those shown in Fig. 11 can also be Fourier transformed to obtain a partial distribution function for pairs containing the absorbing species [21], but effects of transform termination must be carefully evaluated, and such transforms usually contain little information about second and more distant neighbors. EXAFS distribution functions for the amorphous and A15 crystalline Nb_3Ge alloys are shown in Fig. 12. The prominent maxima centered at 2.377Å for the amorphous alloy and at 2.604Å for the crystalline one result from the oscillations in $\chi(k)$ and are shifted from the true nearest neighbor distances by the phase shift α_1.

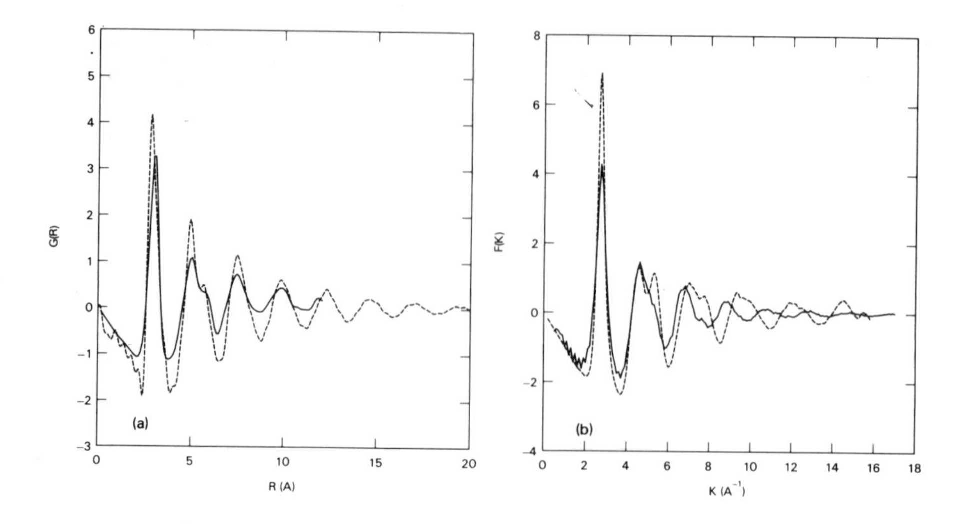

Fig. 13. Results of relaxed dense random packing model calculations for amorphous Nb_3Ge (solid curves). Dashed curves are experimental data.

4.3 Dense Random Packing Models for Amorphous Nb_3Ge

Binary dense random packing models were generated with sphere diameters scaled to sizes slightly larger than those indicated by the Nb-Nb and Nb-Ge distances given by the above described partial distribution functions [16]. For the models discussed in this section, the radii were 1.53Å and 1.22Å for spheres representing Nb and Ge atoms respectively, i.e. r_{NbNb}=3.06Å and r_{NbGe}=2.75Å, and tetrahedron perfection was fixed at 1.1. The thus generated structure containing 800 spheres was relaxed using the same procedures described above for the Gd-Co modelling. Results are shown in Fig. 13, together with experimental interference functions and distribution functions. Although qualitative agreement is achieved for G(r), the magnitude of oscillations in the model G(r) is too small. Also, the model F(K) does not reproduce the splitting seen in the second maximum of experimental F(K)'s for all of the M_3X alloys studied. Other choices of sphere sizes and of tetrahedron perfection were also tried, as well as increasing the depth of the potential function for M-X pairs and forbidding X-X nearest neighbors, but no substantial improvements were obtained in the model results.

4.4 Hydrogen Doping of Amorphous Nb_3Ge

Lanford, et al., [23] have reported that immersion of crystalline Nb_3Ge films in solutions of HF and H_2O results in incorporation of up to 23 at.% hydrogen in the

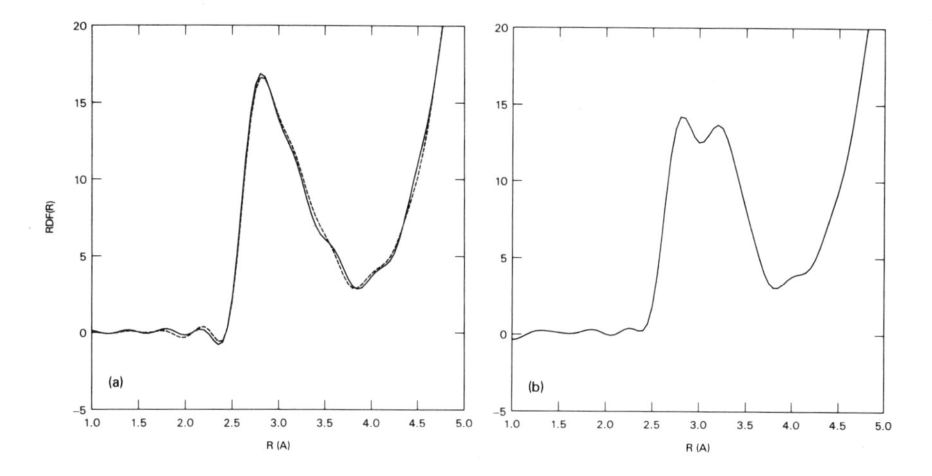

Fig. 14. First maximum in RDF(r) for amorphous Nb_3Ge, (a) two different as-deposited samples and (b) after hydrogen doping.

films, producing a 2.5% increase in lattice parameter. Internal friction measurements by Berry and Pritchet [24] have shown that similar amounts of hydrogen are incorporated in amorphous Nb_3Ge films by immersion in HF:H_2O, and that the hydrogen is expelled by annealing for 1hr at 550°C in 10^{-6} Torr vacuum. X-ray scattering measurements [16] and EXAFS measurements [25] have been used to investigate effects of hydrogen doping on atomic scale structure of amorphous Nb_3Ge. As-prepared film samples were doped with hydrogen by immersion in a 1:1 solution of HF:H_2O for about 1 hr.

Results of large angle scattering measurements on the H-doped sample are shown in Fig. 14(b) and Fig. 15. The most obvious effect of hydrogen doping is splitting of the first maximum in RDF(r) into two maxima at 2.80Å and 3.15Å, as shown in Fig. 14(b). The peak in RDF(r) at 3.15Å which is produced by hydrogen doping is probably associated with pairs of atoms having increased separations caused by incorporation of hydrogen atoms in interstitial sites of the amorphous structure.

From the RDF results alone it was not known whether the extended nearest neighbor pairs were Nb-Nb pairs, Nb-Ge pairs, or both. However, comparison of Ge-edge EXAFS measurements for as-deposited and H-doped Nb_3Ge films clearly showed the nearest neighbor environment of the Ge atoms to be unaffected by the hydrogen doping [25]. Neither the period nor the amplitude of EXAFS oscillations was changed, although the first peak in the scattering pattern was shifted to smaller angles by 1.1% for the H-doped film used in the EXAFS measurements. A shift would be clearly seen in the EXAFS oscillations if the Nb-Ge nearest neighbor separation were changed by this amount.

Attempts were made to measure EXAFS at the Nb-edge for these samples and for the A15 crystalline material, although an optimum monochromator crystal was not available for these measurements. EXAFS oscillations were seen for the A15 crystal, but were more complicated than those for the Ge-edge, because the Nb

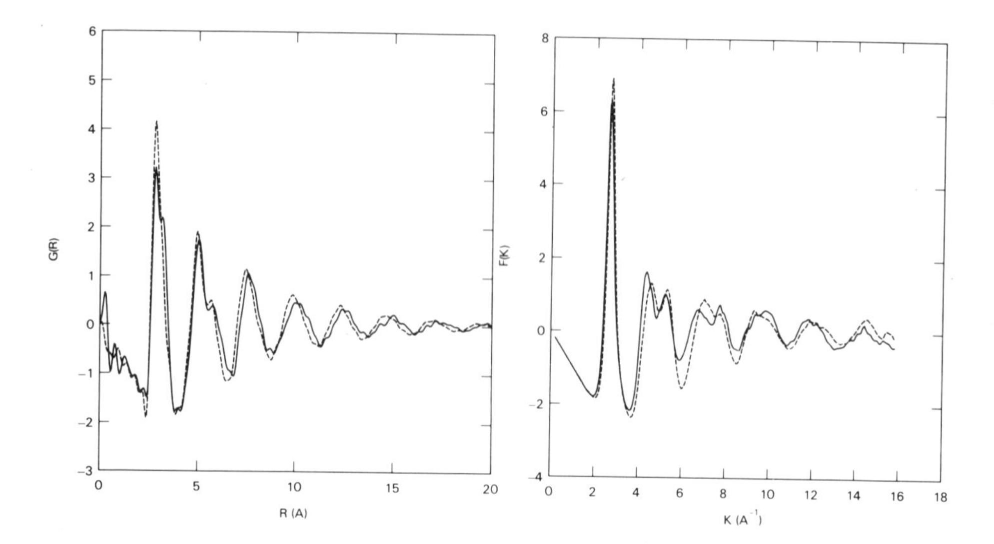

Fig. 15. Comparison of experimental results for amorphous Nb_3Ge before (dashed line) and after (solid line) hydrogen doping.

atoms have three different types of nearest neighbors: two intrachain Nb neighbors at 2.58Å, eight interchain Nb neighbors at 3.16Å, and four Ge neighbors at 2.87Å. Only very weak oscillations were seen for the two amorphous films. They were expected to be weak, because the Nb-Nb nearest neighbor distribution is much broader than the Nb-Ge distribution. Reliable evidence for differences between Nb-edge EXAFS for the as-deposited and H-doped amorphous films could not be extracted from the experimental data.

Results of the RDF measurements on amorphous M_3X alloys and on H-doped Nb_3Ge, together with EXAFS results for the Nb_3Ge alloys before and after H-doping, provide a basis for explaining the failure of simple dense random packing models for these materials. The extreme sharpness of the M-X nearest neighbor distance distribution, relative to the width of the M-M distribution, indicates that the M-X bond length is very well defined and that these M-X bonds are less tolerant to distortions than the M-M bonds. Incorporation of ~10 at.% hydrogen expands the amorphous structure, without affecting the M-X bond lengths, a further indication of the special character of these bonds. Some M-M nearest neighbor distances are expanded by as much as 5%, based on the splitting seen in RDF's for the H-doped material.

4.5 Quasi-molecular Structure for Amorphous M_3X Alloys?

Results described above support a view of amorphous Nb_3Ge in terms of quasi-molecular units consisting of a Ge atom surrounded by from six to seven strongly

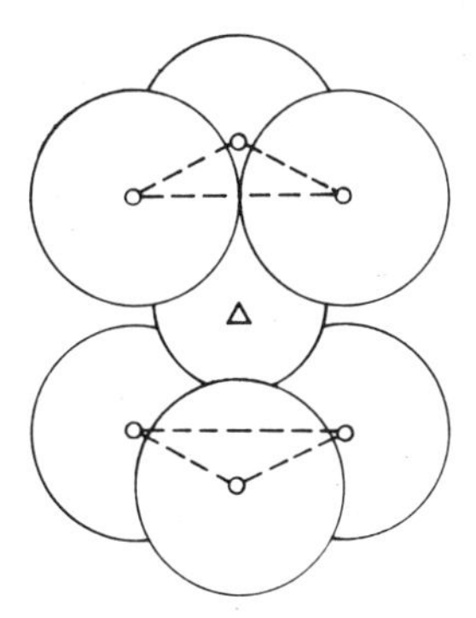

Fig. 16. M_6X structural unit proposed as a building block for the atomic scale structure of amorphous M_3X alloys like Nb_3Ge.

bound Nb atoms, perhaps in the form of an expanded octahedron, as sketched in Fig. 16. It may be useful to consider the atomic scale structure of the M_3X alloys as being built up from such units, with some sharing of Nb atoms between two or more units, as required to achieve a M_3X stoichiometry based on M_6X building blocks. Misfits in packing the units together are accomodated by the broadened M-M distance distribution. H-doping occurs by H atoms being incorporated between pairs of Nb atoms, or among larger groups of Nb atoms, which belong to different but adjacent M_6X units, thereby increasing some Nb-Nb nearest neighbor separations and producing an overall expansion of the structure without affecting the Nb-Ge nearest neighbor distance distribution.

A structural model proposed and investigated by Gaskell [26] for amorphous $Pd_{80}Si_{20}$, based on a motif of Si atoms enclosed in capped trigonal prisms of Pd atoms, with sharing of triangular faces between adjacent units, is closely related to the type of structure being proposed here for amorphous M_3X alloys like Nb_3Ge, except that the RDF measurements for the M_3X alloys indicate that the X atoms have between six and seven M neighbors, in contrast to $Pd_{80}Si_{20}$, for which neutron scattering [27] indicates that Si atoms have ~9 Pd neighbors.

5. CONCLUSIONS

Although chemical ordering seems to be an important feature of the amorphous Nb_3Ge and related alloys, there is no evidence for important chemical ordering effects in the RE-TM alloys, e.g. Gd_xCo_{1-x}, for which dense random packing models are quite successful and nearest neighbor distance distributions for the three types of pairs have similar widths. Coordination numbers for these alloys are close to those expected for complete chemical disorder [28]. Further work is needed to determine whether the structural ideas described above for amorphous Nb_3Ge have applicability to other metal-metalloid amorphous solid or liquid alloys, e.g. $Fe_{80}B_{20}$, $Ni_{75}P_{25}$, and $Pd_{80}Si_{20}$.

REFERENCES

1. B. C. Giessen and C. N. J. Wagner, in **Liquid Metals**, ed. by S. Z. Beer, Dekker, New York, 1972, p. 633.
2. G. S. Cargill III, in **Solid State Physics**, Vol. 30, ed. by H. Ehrenreich, F. Seitz, and D. Turnbull, Academic Press, New York, 1975, p. 227.
3. J. L. Finney, Nature *266*, 309 (1977).
4. C. N. J. Wagner, J. Non-Cryst. Solids *31*, 1 (1978).
5. F. Spaepen, in **Rapidly Quenched Metals III**, Vol. 2, ed. by B. Cantor, The Metals Society, London, 1978, p. 253.
6. G. S. Cargill III, in **Diffraction Studies on Non-Crystalline Substances**, ed. by I. Hargittai and W. Y. Orville-Thomas, Akademiai Kiado (Hungarian Academy of Sciences), Budapest, to be published.
7. C. N. J. Wagner, N. Heiman, T. C. Huang, A. Onton, and W. Parrish, AIP Conf. Proc. *29*, 188 (1976).
8. G. S. Cargill III, AIP Conf. Proc. *18*, 631 (1974).
9. G. S. Cargill III and S. Kirkpatrick, AIP Conf. Proc. *31*, 339 (1976).
10. C. H. Bennett, J. Appl. Phys. *44*, 2727 (1972).
11. D. J. Adams and A. J. Matheson, J. Chem. Phys. *56*, 1989 (1972).
12. T. Ichikawa, Phys. Stat. Solidi *a29*, 293 (1975).
13. See discussion in G. A. N. Connell and G. Lucovsky, J. Non-Cryst. Solids *31*, 123 (1978).
14. G. S. Cargill III, in **Thin Film Phenomena - Interfaces and Interactions**, ed. by J. E. Baglin and J. M. Poate, The Electrochemical Society, Princeton, 1978, p. 221.
15. L. J. Tao, R. J. Gambino, S. Kirkpatrick, J. J. Cuomo, and H. Lilienthal, AIP Conf. Proc. *18*, 641 (1974).
16. G. S. Cargill III and C. C. Tsuei, in **Rapidly Quenched Metals III**, Vol. 2, ed. by B Cantor, The Metals Society, London, 1978, p. 337.
17. G. S Cargill III and C. C. Tsuei, Bull. Am. Phys. Soc. *24*, 358 (1979).
18. T. M. Hayes, J. W. Allen, J. Tauc, B. C. Giessen, and J. J. Hauser, Phys. Rev. Letters *40*, 1282 (1978).
19. E. A. Stern, Phys. Rev. B *10*, 3027 (1974).
20. F. W. Lytle, D. E. Sayers, and E. A. Stern, Phys. Rev. B *11*, 4825 (1975).
21. E. A. Stern, D. E. Sayers, and F. W. Lytle, Phys. Rev. B *11*, 4836 (1975).
22. G. S. Brown, L. R. Testardi, J. H. Wernick, A. B. Hallak, and T. H. Geballe, Solid State Commun. *23*, 875 (1977).
23. W. A. Lanford, P. H. Schmidt, J. M. Rowell, J. M. Poate, R. C. Dynes, and P. D. Dernier, Appl. Phys. Letters *32*, 339 (1978).
24. B. S. Berry and W. C. Pritchet, in **Rapidly Quenched Metals III**, Vol. 2, ed. by B. Cantor, The Metals Society, London, 1978, p. 21.
25. G. S. Cargill III and C. C. Tsuei, to be published.
26. P. H. Gaskell, Nature *276*, 484 (1978).
27. K. Suzuki, T. Fukunaga, M. Misawa, and T. Masumoto, Sci. Rep. RITU *A26*, 1 (1976).
28. G. S. Cargill III and F. Spaepen, J. Non-Cryst. Solids, to be published.

CALCULATION OF THE STRUCTURE AND STABILITY OF AMORPHOUS METALLIC ALLOYS [x]

J.Hafner

Institut für Theoretische Physik der Technischen Universität Wien, A 1040 Wien, Austria

ABSTRACT. An ab-initio investigation of the structure of amorphous metallic alloys is presented. Pseudopotential methods are used to calculate the structure and the thermodynamic properties of crystalline, liquid, and amorphous phases. Cluster-relaxation and thermodynamic variational techniques are employed for the disordered phases. The interrelations between the interatomic potentials, the amorphous structure, the phase-diagram and the glass-forming ability are discussed.

1. INTRODUCTION

In a recent review-paper on the structure of amorphous metals, Hoare [1] has established a list of "some of the requirements that any good structural model of a glass might reasonably be expected to fulfil". In a first group of sine qua non conditions he lists (I have taken some freedom to reformulate and rearrange his conditions):

- (1a) Reproduction of the main essentials of the random geometric conditions for the interpretation of diffraction data.
- (1b) Explanation of the static mechanical stability of the amorphous phase at T=O K and of the nature of the free energy barrier to nucleation to the crystalline state.
- (1c) Explanation of the glass-forming ability and of the effect of composition on the stability of binary systems.
- (1d) Reasonable consistency with the gross properties, in particular with the density of the amorphous phase.

[x] This work has been supported in part by the "Fonds zur Förderung der wissenschaftlichen Forschung in Österreich."

In a second group he mentions some additional properties which he considers merely as "desirable". Among other conditions he lists

(2a) The explanation of the glass-transition and its associated thermodynamics.

(2b) Reproduction of the low-temperature vibrational spectra, if possible together with the low-temperature thermodynamic, thermal conductivity and acoustic properties.

(2c) The calculated structure should be a reasonable starting point for the computation of the major electronic (including electronic transport) and magnetic properties.

(2d) Some support for theories of strength, plasticity, self-diffusion, annealing etc.

If we think what the current structural models actually can explain, we find that even the first part of the list is too exacting, not to speak of the second part. Nonetheless, I think that even the sine qua non conditions are incomplete, since they ignore two very important correlations: (i) the well-established correlation between glass-formation and the formation of eutectic phase-diagrams and (ii) the recently demonstrated correlation between glass-formation and the formation of certain types of crystalline compounds.

The correlation between the glassy and a particularly stable liquid state has served as a starting point to the earliest attempts of model building. After the very considerable refinements introduced in the last years [2] , the packing models allow a quite accurate reproduction of the diffraction data,but they bring no or only very little progress in the other items of our list. The reason is that hard-packing models contain no and soft-packing models (using Lennard-Jones or Morse-type potentials) only very little alloy-chemistry. Evidently the most urgent need is for realistic models for the interatomic interactions. Here the correlation between the amorphous and the crystalline structures is an important help in the task of constructing and testing interatomic potentials. It is clear however, that the theory of the structures of crystalline intermetallic compounds is not so much further advanced than the theory of the amorphous structure. Only in a few special cases where some progress has recently been achieved, a quantum-mechanical calculation of the bonding forces is possible.

Polk and Giessen [3] were the first to point out a remarkable coincidence between glass-formation and the formation of certain classes of stoichiometric crystalline compounds. If we pursue this interrelation somewhat further, we find that the metallic glasses may be roughly classified in three main groups: the first contains the now classical transition metal - metalloid (T-M) glasses, examples are Pd-Si, Fe-B etc. In the extreme case we can include the Rb-O and Cs-O glasses [4]. Typical crystal structures belonging to this group are the Fe_3C-(cementite), Fe_2P-, and MnP-types. A common feature of these lattices are their metalloid-site coordinations by transition metals only, in the form of slightly distorted trigonal prisms [5]. Again, the "anti-cluster" structures of the Rb and Cs suboxides with their octahedral coordination of the O-atoms by alkali atoms [6] fit losely into this scheme.

The simple-metals glasses (S-S, e.g. Mg-Zn, Ca-Mg, Ca-Al), the simple metal - transition metal (S-T, e.g. Ti-Be), the simple metal - rare earth (S-R, e.g. La-Ga), and the transition metal - rare earth glasses (T-R, e.g. Gd-Co, Tb-Fe) form a second group. Their phase-diagrams are characterized by the formation of highly stable AB_2 Laves-phases or closely related Frank-Kasper phases ($CaCu_5$-, Th_2Mn_{17}-, $BaCd_{11}$-, $NaZn_{13}$-types [7]) at majority concentrations of the smaller B-atoms. Glasses are formed at majority concentrations of the larger A-atoms (though the glass-forming region may by extended to include the Laves-phase by high-speed quenching techniques, e.g. for T-R glasses). In this region there are usually no or only relatively unstable crystalline compounds. Unifying features of the stable structures are the principle of tetrahedral close-packing and the coordination in form of certain types of polyhedra. The smaller majority atoms have coordination number (CN) 12, icosahedral for the Laves-phases, not or partially not icosahedral for the other structures [8]. The CN12 polyhedra are linked by large interpenetrating coordination polyhedra (CN16 Friauf-polyhedra for the Laves-phases, and even larger polyhedra - CN18, CN20, CN22 - for the latter structures).

The glasses formed by transition metals only (T-T, e.g. Nb-Ni, Ta-Ir) constitute a third group characterized by complex tetrahedrally close-packed Frank-Kasper structures such as the μ- and σ-phases. Icosahedral coordination (CN12) around the smaller atoms alternates with CN14, CN15 (Frank-Kasper polyhedra) around the larger atoms. Contrary to the second group the glass-forming region overlaps with the often rather broad homogeneity range of the crystalline compounds. While the crystalline phases of the former group are strictly ordered and characterized by a size-ratio $R_A/R_B \geq 1.15$, the μ- and σ-phases show a tendency towards substitutional disorder with a size ratio $R_A/R_B \leq 1.15$.

2. STRUCTURAL MODELS

Two different paths have been followed to arrive at plausible model structures for amorphous alloys. The first takes the Bernal dense-random-packing of hard spheres (DRPHS) [9] as a starting point and different techniques are used to improve the agreement of the calculated diffraction patterns with experiment. The principle underlying the second approach is the design of large non-crystalline clusters based on local motifs with icosahedral and five-fold symmetries.

2.1 Dense-random-packing models

Cargill [2]has compared the radial distribution function (RDF) of a $Ni_{76}P_{24}$ glass with that obtained obtained by Finney [10] (laboratory-built single-sphere DRPHS) or Benett [11] (computer-generated DRPHS) and pointed out that there is an overall correspondence, though agreement on peak shapes and intensities is lacking. The use of a single-sphere DRPHS model for composite glasses is usually justified by the Polk hypothesis [12] which assumes that the smaller and softer metalloid atoms are located in the large cavities (Bernal holes) of the DRPHS network. From an analysis of the number and sizes of these holes one might reasonably well predict the observed concentration dependence for the stability of T-M glasses.

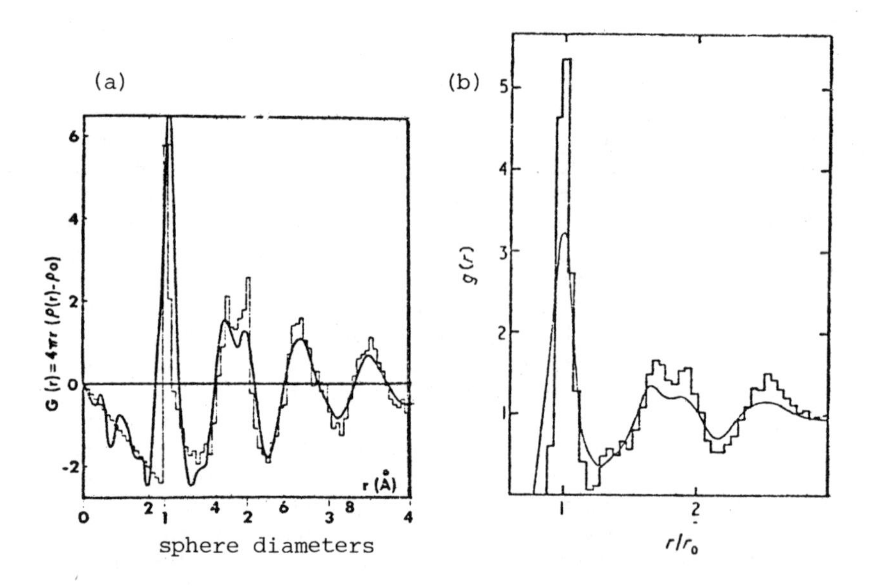

Fig. 1. (a) Comparison of the reduced radial distribution function for Finney's unrelaxed DRPHS structure and for amorphous $Ni_{76}P_{24}$ (after Cargill [2]). (b) Comparison of the pair distribution function for Heimendahl's relaxed DRPHS model and for amorphous $Ni_{76}P_{24}$ (after von Heimendahl [13]).

Evidently, different types of refinements might be imagined for the DRPHS model. Von Heimendahl [13] and Barker et al. [14] were the first to relax the Bernal structure under soft pair-interactions of a Lennard-Jones type. The fitting of the second peak in the pair distribution functions is greatly improved. The second possibility to improve the DRPHS model is to force a certain degree of "tetrahedron perfection" on nearest neighbour sites. This technique has been applied by Cargill et al. [15] to T-R glasses. Some justification for this step is provided by the comparison with the tetrahedrally close-packed Laves-phases found in these systems. Requiring a high degree of tetrahedron perfection causes the structure to be quite porous, and a relaxation using soft pair-potentials is necessary to arrive at an acceptable density of the glass. The step from the single-sphere DRPHS to a binary model with two sizes of spheres is perhaps the most obvious improvement. The first such model has been reported by Sadoc et al. [16] for T-M glasses, binary DRPHS structures for T-R glasses have been generated by Cargill and Kirkpatrick [15]. They were able to associate the structure in the first peak of the RDF with well defined T-T, T-R, and R-R distances. More recent work of Cargill on Gd-Co glasses combines all three types of refinements and produces very accurate X-ray diffraction patterns [17].

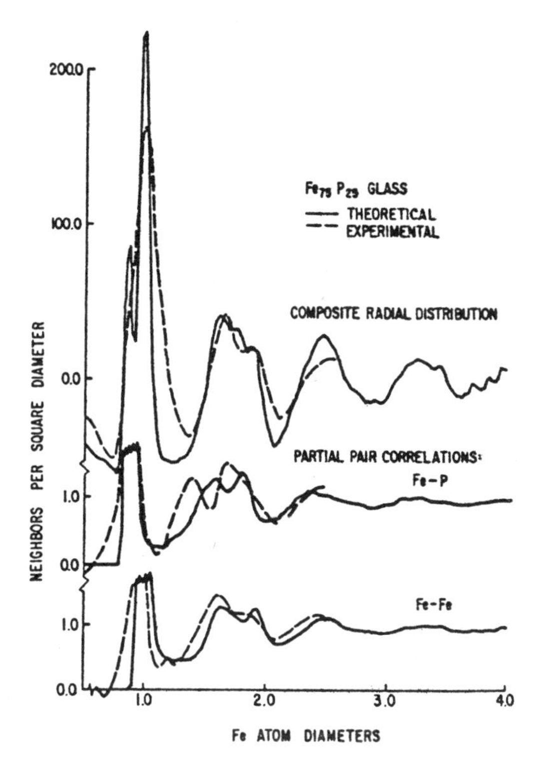

Fig. 2. Comparison of calculated and measured radial distribution function and partial pair distribution functions for $Fe_{75}P_{25}$ (after Boudreaux [18]).

The very recent work of Boudreaux [18] reports relaxed binary DRPHS structures for T-M glasses, with explicit suppression of M-M neighbours, showing excellent agreement with experiment for the composite (X-ray weighted) RDF, but the agreement for the partial PDF's is much less good, especially for the T-M correlations. It is difficult to decide wether this is due to the over-simplified LJ-potentials or to the data reduction procedure used by Waseda et al. [19] to extract their "experimental" partial PDF from a set of global interference functions. Quite generally, there is an urgent need for reliable partial scattering functions for more sensitive tests of the theoretical models.

All the calculations cited above refer to finite clusters, typically of some thousand atoms. In a cluster of that size more than thirty percent of the atoms are located at the surface. Their near-neighbour statistic is quite different from atoms in the bulk and and the un-balancing of the interatomic forces will introduce considerable distortions in the cluster. For this reason, surface atoms are usually ignored in establishing the statistics of the cluster. An other way out of the dilemma is the use of periodic boundary conditions [20] as usually done in Monte-Carlo or molecular dynamics calculations of disordered systems. It is often argued that the periodic replication of the cluster introduces some elements of lattice symmetry in the results and exerts a perturbing influence even for distances well below the repeat-unit. However, I think that as long as the interaction range between the atoms is small compared to the cluster dimensions, the advantages of using periodic boundary conditions, namely: (i) avoidance of systematic density variations from the center to the surface of the cluster and (ii) all atoms are equally representative for the structure and may be used to calculate the PDF far outweight their possible shortcomings.

Finally, I would like to mention another useful aspect of hard-sphere systems: they allow to formulate certain approximate euations of state [21] which can serve as a starting point for the calculation of the thermodynamic properties of realistic systems. In the last decade, much progress in the theory of liquid metals and alloys has been achieved along theses lines [22,23](cf. also Prof. Evans lecture). I shall show that this technique is also useful for investigating the thermodynamics of glass-formation. Weeks [24] has proposed that the analytical expressions for the partial structure factors (calculated in the Percus-Yevick or mean-spherical approximations) should be a reasonable first approximation to the structure of metallic glasses.

2.2 Designed structures

A number of workers [25-27] have advanced an alternative model for the structure of amorphous metals based on icosahedral packing. The structure is envisioned to be built up by 13-atom or larger icosahedral units (e.g. the 55-atom Mackay icosahedron, Fig.3), which

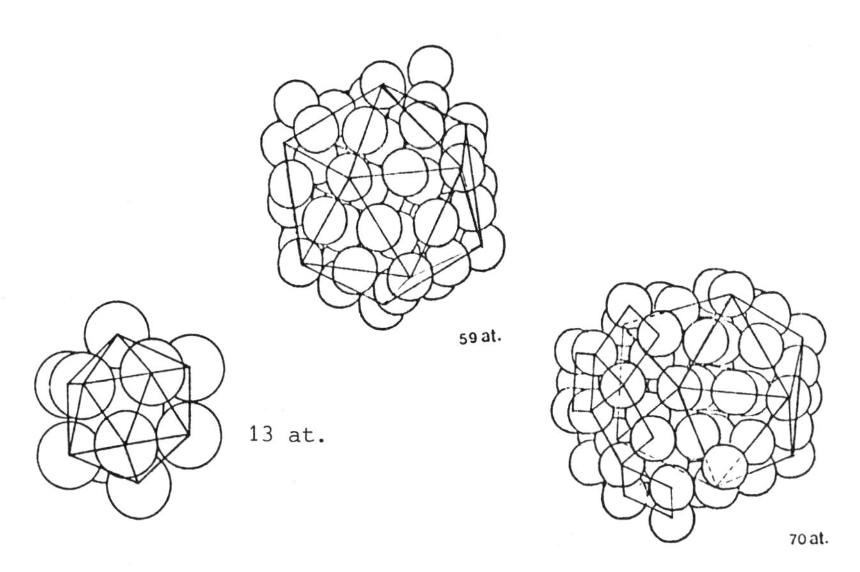

Fig. 3. The 13-atom icosahedron and structures produced by slow MD cooling of liquid LJ-droplets. Results of Farges et al. [26] as reported by Hoare [1].

will sometimes interlock. The basic ordered units are surrounded by less ordered material. This construction is motivated by three facts: (i) For atoms interacting via an attractive pair-potential it has been established theoretically and experimentally that the equilibrium structure of small microclusters is not identical to the bulk structure, but is rather based on the 13-atom icosahedron. (ii) Structures produced by slow molecular dynamic (MD) cooling of liquid LJ-droplets show icosahedral packing [26,27]. (iii) Larger units constructed by interpenetrating smaller icosahedral units have the desired property of self-limiting growth [1]. Another argument comes from the frequent occurence of the icosahedron in the crystalline structures characterizing the glass-forming binary metal systems. Though Briant and Burton [27] have shown that the interference functions of assemblys of the smaller of these microclusters possess the features characteristic for metallic glasses, it is still far from established that they really exist in amorphous bulk phases. An investigation of one of the more realistic packing models in terms of the local coordination appears to be desirable. Designed structures for T-M glasses based on trigonal-prismatic units have recently been proposed by Gaskell [5].

3. INTERATOMIC FORCES AND CRYSTAL STRUCTURES

In my opinion, the most urgent need for a further refinement of the present structural models is for realistic interatomic interactions.

Pseudopotential theory has enormously contributed to the advance in the microscopic theory of the binding, the crystal structure, the phase transitions, the vibrational and the electronic properties of simple metals [28]. A realistic picture has emerged of a potential energy function in which the ions move as the sum of a volume-dependent energy and a sum over interatomic pair potentials. Progress for alloys has been much slower. The reason is that the pseudopotential is not an atomic property which is independent of the atom's surrounding, but rather a collective property which describes the scattering of electrons by an ion in a given effective medium (the average potential of the other ions and electrons in the crystal). If the surrounding of the atoms changes, it's pseudopotential changes, too. I have recently put forward an optimized pseudopotential theory for binary alloys which is based on the expansion of the conduction elctron states in orthogonalized plane waves. Different electronic effects contribute to change the pseudopotential on alloying: (i) the conduction electrons have to be orthogonalized to two different sets of core states. This will alter the strength of the repulsive part of the pseudopotential. (ii) The orthogonalization is equivalent to push electrons out of the core region: the ion "digs an orthogonalization-hole". If the ion A in an A-B alloy sees an enhanced electron-density compared to the pure metal, it must dig a bigger orthogonalization hole while for the B-ion a smaller one is sufficient. This is equivalent to a charge transfer in the direction that is expected from the electronegativity difference (for a discussion of the interrelation between orthogonalization and electronegativity, cf. Hafner and Sommer [30]). (iii) The electrical neutrality of the pseudoatom requires the screening charge to compensate for the change in the orthogonalization charges. In order to maintain the electroneutrality even on a local level, the additional screening charge of the electropositive ion (the one with the larger orthogonalization hole) is concentrated in the core region. This is accompanied by an inward-shift of the Friedel-oscillations in the screening charge (Fig.4). Around the electronegative ion, the effect is just the inverse. The whole story corresponds to Pauling's early discussion of charge transfer effects in intermetallic compounds[31].

The interatomic potentials follow the re-distribution of the screening charge. The change in the short-range part is equivalent to a compression of the more electropositive ion and an expansion of the more electronegative ion. This is an effect which is well known in the metallurgical literature as the "chemical compression". The effects discussed here exist in all alloys, but they are certainly more important in alloys with large differences in electronegativity and electron density. In two recent papers [29] I have demonstrated how such effects contribute to stabilize the tetrahedrally close-packed structures of stoichiometric alkali-alkali compounds (Na_2K, Na_2Cs, K_2Cs - hexagonal Laves-phases, K_7Cs_6 - hexagonal stacking variant of a μ-phase). The effects are even more prominent in heterovalent alloys such as the series $CaLi_2$-$CaMg_2$-$CaAl_2$

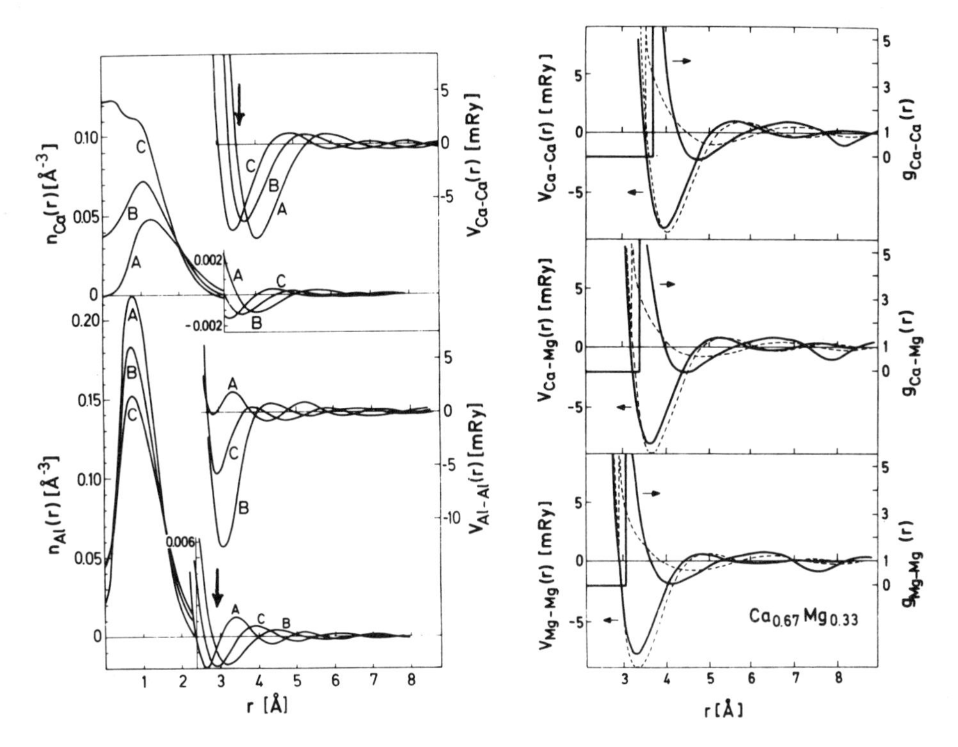

Fig. 4(left). Pseudoatom charge density n(r) and pair potential V(r) for Ca and Al ions in the pure metals (curve A) and in alloys of the composition $Ca_{0.67}Al_{0.33}$ (curve B) and $Ca_{0.33}Al_{0.67}$ (curve C), demonstrating the "chemical compression" of the Ca-ion in the alloy. The arrows indicate the nearest-neighbour distance in the Laves-phase Ca Al_2 [32].

Fig. 5(right). Pair potentials V(r) and pair distribution functions g(r) for a $Ca_{0.67}Mg_{0.33}$ alloy in the amorphous (bold lines, T=25°C) and in the liquid (broken lines, T=900°C) states.

Laves phases. Topologically close-packed structures are usually thought to be determined by the size-ratio of the constituent atoms. For the Laves-phases, the ideal size ratio is R_A/R_B=1.225. For our examples the size-ratios deduced from their Goldschmidt-radii (or equivalently from their pair-potentials) do not scale with their observed stabilty: $CaAl_2$ with it's highly non-ideal size-ratio (1.4) is an extremely stable compound with a large negative enthalpy of formation, $CaLi_2$ with a nearly ideal size-ratio (1.26) is relatively unstable and decomposes peritectically at low temperature. If we compute the size-ratios in terms of the effective pair potentials in the alloy, they agree with the observed stabilities, for the $CaMg_2$ and $CaAl_2$ phases. the interatomic distances fit exactly into the minima of the pair potentials (Fig.4), i.e. the calculated potential energy function allows for a formation of nearly strain-free A-A, A-B, and B-B bonds. It is worthwhile to go one step further: the free parameters of the Laves-phase structure (i.e. those not determined by space group symmetry) may be calculated by minimizing the energy in the configuration spce of the atomic positions. This calculation shows that the electronic effects discussed above are vital in stabilizing the tetrahedral networks of both the A- and B-sunlattices. The structural energy differences are also correctly predicted by our theory: in accordance with experiment, $CaAl_2$ is shown to be cubic ($MgCu_2$-type), while $MgZn_2$, $CaMg_2$, and $CaLi_2$ are hexagonal ($MgZn_2$-type). Since the nearest-neighbour configuration is identical for all Laves-phase variants, this shows that the potentials are realistic even for larger distances. Without presenting any details, I would like to mention that a full calculation of the thermochemical quantities of the crystalline compounds shows very good agreement with experiment, I will come back to this point later.

Using the Gibbs-Bogoljubov thermodynamic variational method [22,23] our potentials allow a calculation of the structure (via a kind of dynamic hard-sphere simulation) and the thermodynamics of liquid alloys. Again we find that the lowering of the total electronic energy in very stable liquid alloys arises from an optimal embedding of the atoms into the minima of the potential energy function. This is visualized by the coincidence of the maxima in the partial PDF's with the minima in the corresponding pair potentials (cf. Fig.5).

4. THE STRUCTURE AND STABILITY OF SIMPLE-METAL GLASSES

Using the pair potentials described in the previous section, von Heimendahl [20,32] has performed a cluster-relaxation calculation for the $Mg_{0.70}Zn_{0.30}$ glass. The ingredients of his computation are as follows: the relaxation was started from regular rhombic dodecahedron containing 800 atoms which was cut out of the center of Finneys DRPHS model [10]. Periodic boundary conditions were applied. The particular choice of the geometry was motivated by the fact that it approximates a sphere most closely. The relaxation was essentially equivalent to the steepest-descent algorithm (in the configuration space of all 2400 coordinates) described by Hoare [1]: the total force on each atoms is given by the summ of the pair forces. The displacement necessary to bring each atom in a force-free position (while the remaining atoms are kept fixed at their sites) is calculated in a linear approximation. In each relaxation run the displacement vectors $\delta\vec{r}_1$ of all atoms are calculated and stored. Then new atomic positions $\vec{r}_1'$ are calculated by adding a certain fraction f to $\vec{r}_1$: $\vec{r}_1'=\vec{r}_1+f.\delta\vec{r}_1$. Starting from the new positions the procedure is

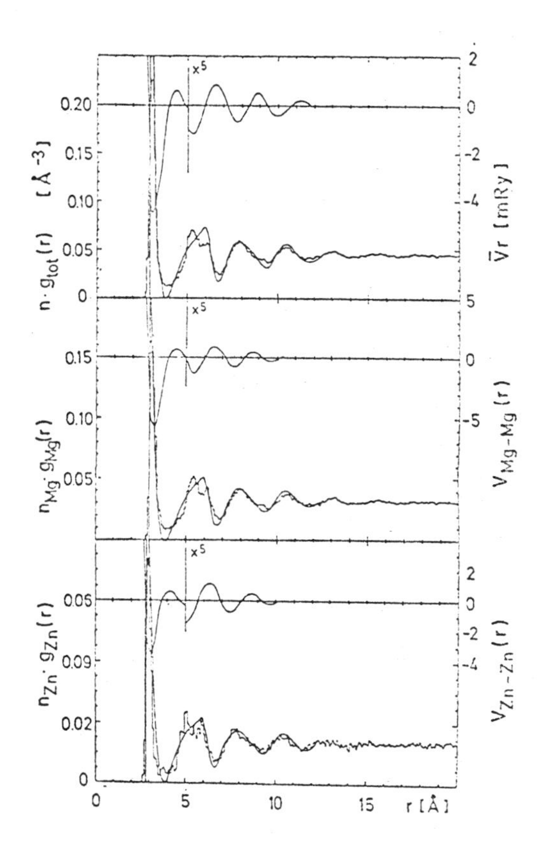

Fig. 6. Total and partial pair distribution functions g(r) for the $Mg_{0.7}Zn_{0.3}$ glass. The histogram shows the result of the cluster-relaxation calculation, the line the result of the thermodynamic variational technique (T=25°C). The average and the partial pair potentials are also shown [32].

repeated until the last set of $\delta\vec{r}_1$ is negligible. In Fig.6 the total and two partial PDF's are shown together with the pair potentials. The position of the first peaks in the PDF's agrees very well with the first minimum in the potentials. This one-to-one correspondence extends to second and third-nearest neighbours. This will be an important element in the discussion of the stability of the glassy state.

The splitting of the second peak in the partial PDF's is very distinct, it is somewhat smeared out in the total PDF. The second peaks in both the total and the partial structure factors are split in the same way as for the PDF's [20], which is surprising since it is not a simple consequence of the Fourier transform. Both features

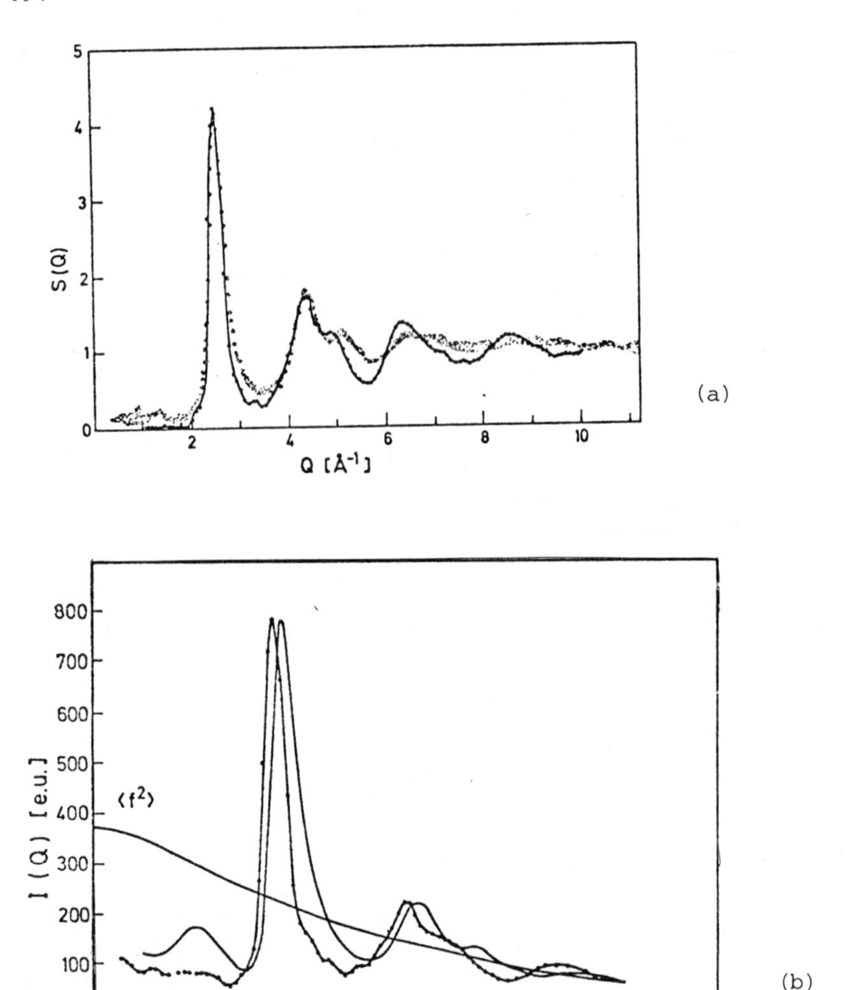

Fig. 7. (a) Interference function S(Q) for neutron scattering in a $Mg_{0.7}Zn_{0.3}$ glass. Solid line theory, the shaded area represents the spread of the experimental points. (b) X-ray scattering intensity. Dots: theory, continous line: experiment. The experimental results have been kindly communicated by Prof.Rudin prior to publication.

suggest a relatively high degree of topological (but not substitutional) short-range order.

The structure factors may be used to calculate the coherent

elastic scattering intensities for neutrons and X-rays (using the Fourier-transforms of the calculated electronic densities). In comparing the theoretical results with the experiments of Rudin [33], we have to keep in mind that the neutron scattering length of Mg and Zn is approximately equal, while the X-ray form factor of Zn is much larger than that of Mg, ≈85% of the scattering comes from the Zn atoms. Hence we can conclude that the Mg-potential seems to be very accurate, while the Zn-potential yields Zn-Zn distances which are approximately 0.12 Å larger than found in the experiment.

Before we discuss some elements of the local structure, it is interesting to reveal some implications of the calculated structure on the stability of the glass. The correlation between the PDF's and the potentials suggests that the principle underlying the chemical bonding is the same in all stable structures: the formation of as many strain-free bonds as possible. It would be interesting to pursue this correlation to other concentrations and in other binary systems. On the basis of cluster-relaxation calculations, this would be quite expensive. A simple and cheaper procedure is to use the Gibbs-Bogoljubov variational technique - and it offers the additional advantage of furnishing thermodynamic information on the supercooled liquid state. For a supercooled (≈glassy) Mg-Zn alloy the free-energy minimization procedure yields PDF's and partial structure factors in very good agreement with the cluster-relaxation (Fig.6). This allows to study the relationship PDF-potential in other alloys. For the glass-forming systems Mg-Zn, Ca-Mg, and Ca-Zn potentials and PDF's are very well "in phase" for the whole range of non-negligible interatomic forces and for all concentrations, for the Ca-Al system this is true only on the Ca-rich side. The known assymptotic behaviour of the oscillations of pair potentials ($V_{ij} \to \cos(2k_F r)/r^3$) and PDF's ($g_{ij}-1 \to \sin(Q^p_{ij} r)/r$, where Q^p_{ij} is the wave vector of the first peak in the partial structure factor S_{ij}) suggests an interpretation in terms of the Nagel-Tauc rule [35] which says $Q^p_{ij} \simeq Q^p \simeq 2k_F$ [36]. However, due to the size-difference of the components, the three Q^p_{ij} are different and the three PDF's canot oscillate in phase in the assymptotic region, while the screening charge density is the same for all three V_{ij}'s which do oscillate in phase. Thus there can be non matching between the pair potentials and the PDF's in the assymptotic region. Anyway, this is much more important for short and intermediate distances where the potentials (and especially that of the minority component) deviates quite substantially from it's assymptotic form. The existence of this type of "constructive interference" is seen to be a more complicated function of the usual alloy-chemical factors: (i) size-ratio, (ii) strong chemical bonding (charge transfer and screening), and (iii) valence-electron concentration. The empirical validity of these correlations has been established by Giessen and co-workers [34].

Without anticipating too much of Prof.Beck's lecture on the

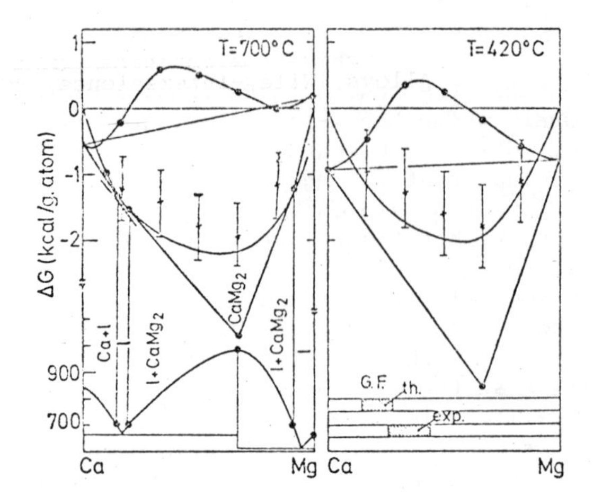

Fig. 8. Free enthalpy of formation ΔG (relative to the pure liquid metals) of Ca-Mg alloys at T=700°C and T=420°C (i.e. close to the recrystallization temperature of the glass): o solid solution, x liquid (resp. supercooled liquid) solution, o Laves phase. The error bars represent the estimated uncertainty, the full line a tentative interpolation. The common-tangent construction for the phase diagram and the glass-forming range is indicated (cf. ref.37).

stability of amorphous metals, I would like to close this section by combining all the results on the thermodynamic properties of the crystalline, liquid, and supercooled liquid (≃amorphous) phases in the form of temperature-dependent ΔG-diagrams (Fig.8). The limits of stability of the different phases are determined by the usual common tangent construction. In this way we derive a fully realistic phase diagram. The picture demonstrates very clearly in which way the eutectic composition is determined by the relative magnitudes of the competing liquid and crystalline phases. Below the eutectic temperature, there is a limited concentration range in which ΔG for the supercooled liquid is very close to the straight line representing the stable ($Ca+CaMg_2$) two-phase mixture. If there is no crystalline phase with a lower energy (and indeed it can be shown that possible crystal structures have a higher energy [37]) and the system is constrained against phase-separation, glass-formation will occur with great ease.

5. LOCAL COORDINATION

The number and type of near-neighbours coordinating the atoms of a given type is an interesting guide to the short-range order. For all the simple-metal glasses investigated here, the total number of nearest neighbours for both types of atoms has nearly constant values of ≃12, the partial coordination numbers do not show any preferred coordinations. Similar results for T-T glasses have been obtained experimentally by Chen and Waseda [38]. This type of coordination has been shown to be an important element in the stabilization of the amorphous phase relative to possible crystalline phases of $CuAl_2$-,Mg_2Cu - or related structures types via an energetically more favourable coordination of the smaller minority atoms [37].

This type of coordination contrasts the behaviour found in T-M glasses, where apparently very strong T-M interactions prohibit the M-atoms from being nearest neighbours ([18], cf. also Cargill in these proceedings). The results of Boudreaux demonstrate that the M-T coordination is very similar in the crystalline and in the amorphous alloys. This suggests some guide-lines for further model-building studies: (i) to study the interrelation between the relaxed DRPHS-models and the designed icosahedral structures for metal-metal glasses, and (ii) to introduce some element of "atomic-scale crystallinity" in the modelling of T-M glasses.

6. CONCLUSIONS

The reader will agree, I hope, that the microscopic theory for the structure and stabilty of simple-metal glasses meets the four *sine qua non* conditions for a good structural model. Beyond satisfying these basic requirements, the theory is a good starting point to study other properties: some results on vibrational properties are reported in these proceedings, the first results on elctronic properties (electrical transport, optical reflectivity) are quite promising.

ACKNOWLEDGEMENT

Many of the results presented here are due to the fruitful cooperation of Dr. L. von Heimendahl.

REFERENCES

1. M.R.Hoare, J.Non-cryst.Sol. 31,157(1978).
2. G.S.Cargill, Solid State Physics 30,227(1975).
3. D.E.Polk and B.C.Giessen, in Metallic Glasses, p.1, American Society for Metals, Metals Park, Ohio, 1978.
4. W.Bauhofer and A.Simon, Phys.Rev.Lett. 40,1730(1978).
5. P.H.Gaskell, Nature 276,484(1979).
6. A.Simon, Z.Anorg.Allgem.Chem. 395,301(1973); A.Simon, H.J.Deiseroth, E.Westerbeck, and B.Hillenköter, Z.Anorg.Allgem.Chem. 423, 203(1976).

7. F.C.Frank and J.S.Kasper, Acta.Cryst. 11,184(1958); 12,483(1959).
8. W.B.Pearson, The Crystal Chemistry and Physics of Metals and Alloys, Wiley-Interscience, New York, 1972.
9. J.D.Bernal, Proc.Roy.Soc. A280,299(1964).
10. J.L.Finney, Proc.Roy.Soc. A319,479(1970).
11. C.H.Bennett, J.Appl.Phys. 43,2727(1972).
12. D.E.Polk, Scr.Metallurg. 4,117(1970).
13. L.von Heimendahl, J.Phys. F5,L141(1975).
14. J.A.Barker, J.L.Finney, and M.R.Hoare, Nature 257,120(1975).
15. G.S.Cargill and S.Kirkpatrick, AIP Conf.Proc. 31,339(1976).
16. J.F.Sadoc, J.Dixmier, and A.Guinier, J.Non-cryst.Sol. 12,46 (1973).
17. G.S.Cargill, in Thin Film Phenomena - Interfaces and Interactions, p.22, J.E.Baglin and J.M.Poate, eds., The Electrochemical Society, Princeton, 1978.
18. D.S.Boudreaux, Phys.Rev. B18,4039(1978).
19. Y.Waseda, H.Okezaki, and T.Matsumoto, Proc. of the Int.Conf. on the Structure of Non-cryst. Materials, Cambridge 1976, p.202, Taylor and Francis, London, 1977.
20. L. von Heimendahl, J.Phys. F9,161(1979).
21. G.A.Mansoori, N.F.Carnahan, K.E.Starling, and T.W.Leland, J.Chem.Phys. 54,1523(1971). This version of the hard-sphere equation-of-state has been used in the calculations.
22. N.W.Ashcroft and D.Stroud, Solid State Physics 33,1(1978).
23. J.Hafner, Phys.Rev. A16,351(1977).
24. J.D.Weeks, Philos.Mag. 35,1345(1977).
25. M.R.Hoare, Ann.N.Y.Acad.Sci. 279,186(1976).
26. J.Farges, B.Raoult, and G.Torchet, J.Chem.Phys. 59,3454(1973).
27. C.L.Briant and J.J.Burton, phys.stat.sol. (b) 85,393(1978).
28. V.Heine and D.Weaire, Solid State Physics 24,247(1971).
29. J.Hafner, J.Phys. F6,1243(1976); Phys.Rev. B15,617(1977); Phys. Rev. B19,5094(1979).
30. J.Hafner and F.Sommer, CALPHAD 1,325(1977).
31. L.Pauling, The Nature of the Chemical Bond, 3rd ed. sec. 11-15., Cornell University Press, Ithaca, 1960.
32. J.Hafner and L. von Heimendahl, Phys.Rev.Lett. 42,386(1979).
33. H.Rudin, private communication.
34. B.C.Giessen, J.Hong, L.Kabacoff, D.E.Polk, R.Raman and R.St. Amand, Proc. 3rd Int.Conf. on Rapidly Quenched Metals, Brighton 1978; R.St.Amand and B.C.Giessen, Scr.Metallurg. 12,1021, (1978).
35. S.R.Nagel and J.Tauc, Phys.Rev.Lett. 35,380(1975).
36. H.Beck and R.Oberle, Proc.3rd Int.Conf. on Rapidly Quenched Metals, Brighton 1978.
37. J.Hafner, to be published.
38. H.S.Chen and Y.Waseda, phys.stat.sol. (a) 51,593(1979).

THE DYNAMICAL PROPERTIES OF METALLIC GLASSES

J.Hafner

Institut für Theoretische Physik der Technischen
Universität Wien, A 1040 Wien, Austria

ABSTRACT. Different theoretical attempts to investigate the dynamical properties of metalic glasses - the equation-of-motion method, molecular dynamics and continued-fraction techniques - are reviewed.

1. INTRODUCTION

There has been considerable progress in recent years in understanding the vibrational properties of amorphous semiconductors and oxide glasses [1]. Much of this has been stimulated by the wealth of experimental information offered by optical spectroscopy. Similarly exciting experimental results are still lacking for the amorphous metals and the advance in the theory of the dynamical properties of metallic glasses has been comparatively slow.

In metallic glasses (as in all amorphous materials) there is no conservation of crystal momentum to help us in the theoretical treatment of vibrational properties. Even the simplest model with a harmonic potential is quite difficult to handle. The origin of the difficulty lies in the disordered atomic arrangements. There are three aspects in this disorder. First there is a <u>topological disorder</u> which is associated with the absence of a large scale self-repeating lattice. Second, there is a <u>quantitative disorder</u> in the strength of the interatomic interactions. This is an aspect which is typical for amorphous metals. In amorphous semiconductors, bond-distances vary by at most 2-3%, bond-angles by 6-10%, so the neglect of the variation of the interaction is a good first approximation. For metallic glasses on the other side there is ample evidence for a variation of the nearest-neighbour (n.n.) distances of up to 15%. Metallic interatomic potentials are known to vary rapidly around the n.n. distance, so quantitative disorder will be a very important aspect. Furthermore, metallic interactions are long-ranged and quantitative disorder will be increasingly important for higher than n.n. interactions. Finally all metallic glasses are composite systems, hence we have substitutional disorder which adds to the quantitative disorder. In my lecture on the static structure and stability of metallic glasses I have emphasized the necessity of using realistic interatomic forces in any discussion of the interrelation between these forces and the equilibrium structure. It is clear that this will be even more important for vibrational properties.

In the first part of this lecture I shall discuss different possible methods for obtaining the vibrational spectra of amorphous metallic alloys: (i) the equation-of-motion method, (ii) the molecular dynamics approach, and (iii) two techniques based on continued-fraction expansions, namely the Mori-technique and the recursion method. In the second part recent theoretical results for the dynamical structure factor of the $Mg_{0.7}Zn_{0.3}$ glass will be presented.

2. THE EQUATION-OF-MOTION METHOD

The equation-of-motion method [2,3] is a very efficient technique to calculate the properties of systems whose Hamiltonian is quadratic in the dynamical variables. Such a Hamiltonian may be transformed to normal coordinates, e.g. the displacements $\vec{u}_l(t)$ of the atoms of a system of N particles may be written as

$$\vec{u}_l(t) = m_l^{-1/2} \sum_i \vec{e}_{il} (A_i \cos(\omega_i t) + B_i \sin(\omega_i t)/\omega_i) \tag{1}$$

m_l denotes the mass of the atom in the equilibrium position $\vec{r}_l$. The normal modes are characterized by their frequencies ω_i and their polarisation vectors $\vec{e}_{il}$, who satisfy the usual orthonormality and closure relations. The amplitudes A_i and B_i are determined by the initial conditions $\vec{u}_l(0)$ and $\dot{\vec{u}}_l(0)$ via

$$A_i = \sum_l \vec{e}_{il} \vec{u}_l(0) m_l^{-1/2} \quad ; \quad B_i = \sum_l \vec{e}_{il} \dot{\vec{u}}_l(0) m_l^{-1/2} \tag{2}$$

Eqs. (1) and (2) are valid for ordered as well as disordered systems. ω_i and $\vec{e}_{il}$ contain all the dynamical information on the system nd can in principle be calculated by diagonalizing the Hamiltonian. For ordered systems this is a very efficient procedure, since the lattice periodicity allows to reduce the problem of diagonalizing the 3Nx3N force constant matrix to the repeated diagonalization of the 3rx3r (r is the number of particles in the unit cell) dynamical matrix. For disordered systems this is impossible and we would have to diagonalize the full 3Nx3N matrix. The direct diagonalization is of course very inefficient. If the range of the interatomic forces is small compared to the dimensions of the N-particle cluster, the matrix is of band-diagonal form. In a direct diagonalization, this advantage may be exploited only for one-dimensional systems. The equation-of-motion methods starts from the Newtonian equation of motion

$$\ddot{\vec{u}}_l(t) = m_l^{-1}\,\vec{F}_l(t) \tag{3}$$

For the atoms in a solid, the force $\vec{F}_l$ on the l-th particle is given by

$$F_{l\alpha}(t) = \sum_{m,\beta} \Phi_{\alpha\beta}(l,m)\; u_{m\beta}(t) \tag{4}$$

where Φ is the force-constant matrix. It is clear that only the non-zero elements of Φ contribute to the equation of motion, thus the sparseness of the matrix is fully exploited. The integration of (3) is done numerically after an appropriate choice of the initial conditions $\vec{u}_l(0)$ and $\dot{u}_l(0)$. In the simplest form this is done by the difference equation

$$\vec{u}_l(t+\Delta) = \vec{u}_l(t-\Delta) - 2\vec{u}_l(t) + \Delta^2 m_l^{-1}\vec{F}_l(t) + O(\Delta^4) \tag{5}$$

with the time step Δ. Equ. (5) may be substituted by more efficient integration formulae, such as Runge-Kutta or predictor-corrector formulae [3]. Once the $\vec{u}_l(t)$ are known, interesting quantities such as the density of states or the dynamical structure factor may be computed.

2.1 The dynamical structure factor

For ordered as well as disordered systems the one-phonon scattering cross-section for neutrons is given in the harmonic approximation by

$$\begin{aligned}
\frac{d^2\sigma_1}{d\omega d\Omega} &= \frac{k_f}{2\pi\hbar k_i} \sum_{ll'} a_l a_{l'} e^{-i\vec{k}(\vec{r}_l-\vec{r}_{l'})} \int_{-\infty}^{\infty} dt\; e^{-i\omega t} \langle \vec{k}.\vec{u}_l(t)\,\vec{k}.\vec{u}_{l'}(0)\rangle \\
&= \hbar k_f/k_i \sum_{ll'} a_l a_{l'} (m_l m_{l'})^{-1/2}\, e^{-i\vec{k}(\vec{r}_l-\vec{r}_{l'})} \\
&\quad \times \sum_i \vec{k}.\vec{e}_{il'} \frac{1}{2\omega} [n(\omega)\delta(\omega+\omega_i) + (n(\omega)+1)\delta(\omega-\omega_i)]\vec{k}.\vec{e}_{il} \\
&= \hbar k_f/k_i\; (n(\omega)+\Theta(\omega))\; S_1(k,\omega)/\,\omega\ .
\end{aligned} \tag{6}$$

Here $\hbar\vec{k}_i$ and $\hbar\vec{k}_f$ are the initial and final momenta of the neutron, $\hbar\vec{k}$ and $\hbar\omega$ are the momentum and energy transfer, a_l is the product of the scattering length and the Debye-Waller factor of the l-th atom and $n(\omega)$ is the Bose-Einstein distribution function [4]. The last equality defines the one-phonon dynamical structure factor $S_1(\vec{k},\omega)$. In the equation of motion technique, one proceeds as follows: the $\vec{u}_l(t)$ are calculated twice, using two different sets of initial conditions:

$$\text{(i)}\ \vec{u}_1(0) = a_1\vec{k}\cos(\vec{k}.\vec{r}_1)/m_1 \qquad \text{(ii)}\ \vec{u}_1(0) = a_1\vec{k}\sin(\vec{k}.\vec{r}_1)/m_1$$
$$\dot{\vec{u}}_1(0) = 0 \qquad\qquad \dot{\vec{u}}_1(0) = 0 \tag{7}$$

From the two sets of $u_1(t)$ the sums $F_{cos}(\vec{k},t)$ and $F_{sin}(\vec{k},t)$

$$\left.\begin{array}{l} F_{cos}(\vec{k},t) = \\ F_{sin}(\vec{k},t) = \end{array}\right\} \sum_1 m_1\ \vec{u}_1(t).\vec{u}_1(0) \tag{8}$$

are calculated. The dynamical structure factor is just the cosine-Fourier transform of $(F_{cos}+F_{sin})$

$$S_1(\vec{k},\omega) = \pi^{-1}\int_0^\infty \cos(\omega t)\,(F_{cos}(\vec{k},t)+F_{sin}(\vec{k},t))\,dt \tag{9}$$

This can be seen by calculating $\vec{u}_1(t)$ from (1) to (3), using the initial conditions (7). The sums in (8) are simplified using the orthonormality and closure relations for the $\vec{e}_{il}$ and yield

$$F_{cos}(\vec{k},t) = \sum_{1,1'} \frac{a_1 a_{1'}}{(m_1 m_{1'})} \cos(\vec{k}.\vec{r}_1)\cos(\vec{k}.\vec{r}_{1'}) \sum_i \vec{k}.\vec{e}_{i1}\vec{k}.\vec{e}_{i1'}\cos(\omega_i t) \tag{10}$$

and similarly for F_{sin}. Performing the Fourier transform (9) we get

$$S_1(\vec{k},\omega) = \sum_{1,1'} \frac{a_1 a_{1'}}{(m_1 m_{1'})} \cos(\vec{k}(r_1 - r_{1'})) \sum_i \vec{k}.\vec{e}_{i1}\ \vec{k}.\vec{e}_{i1'}$$
$$\times\ 0.5[\delta(\omega+\omega_i)+\delta(\omega-\omega_i)] \tag{11}$$

Comparing (11) with (6) completes the proof. In an actual calculation the integration (9) has to be truncated at a finite T and a damping function f(t/T) has to be introduced in (9) to eliminate unwanted oscillations. For any technical detail, see von Heimendahl's article [3].

2.2 The density of states

The density of states projected onto a given initial state $\vec{u}_1(0)$ of the system is given by the following integral representation [5]

$$g(\omega) = \pi^{-1}\int_0^\infty \sum_1 \vec{u}_1(0).\vec{u}_1(t)\ \cos(\omega t)\ dt \tag{12}$$

Eqs. (12) and (3) are again evaluated numerically, with due caution in the cut-off procedure. The total density of states is obtained statistically, using initial conditions with random phases.

2.3 Model calculations

In the first application of the method to model systems represen-

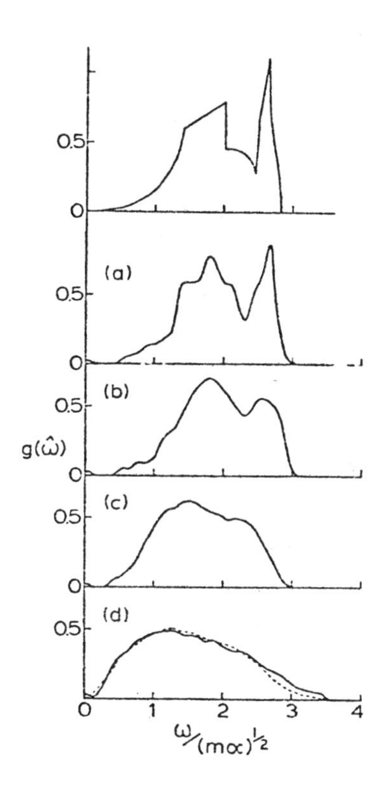

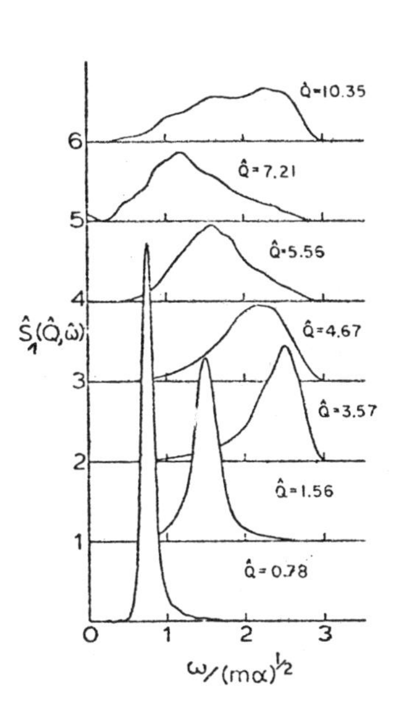

Fig. 1. (left) Vibrational density of states for (o) fcc-crystal, (a) 480-atom fcc cluster, (b)-(d) 500-atom amorphous model with (b) constant force constants, (c) a Morse-potential, and (d) a Lennard-Jones potential. (Right) Dynamical structure factor $S_1(Q,\omega)$ for the amorphous model, calculated using a Morse potential. After Rehr and Alben [5].

ting amorphous metals, von Heimendahl and Thorpe [6] considered the effect of topological disorder alone. Their calculated density of states turned out to be very similar to that of the crystalline close-packed metals, with separated peaks for transverse- and longitudinal-type modes. The effect of quantitative disorder was first investigated by Rehr and Alben [5]. Their calculation was based on a static structure factor determined by Rahman et al. [7] using molecular dynamics and different models for the interatomic forces, only nearest-neighbour interactions being considered. They concluded that quantitative disorder almost completely destroys the structure in the vibrational density of states. The effect depends on the amount of quantitative disorder. It is stronger for a Lennard-Jones potential than for a Morse potential (Fig.1), because the force constant derived from the former potentialshows a larger variation over the width of the first peak in the radial distribution function.

On the other side, the dynamical structure factor $S_1(Q,\omega)$ does not depend dramatically on the amount of quantitative disorder. A sharp longitudinal acoustic mode is seen for small Q. As Q increases, there is a minimum in the energy in the vicinity of the first peak in the static structure factor. This shows that some effects of long range order are preserved in the amorphous structure. At these Q-values, the structure can statically absorb momentum, allowing low-energy vibrations to contribute to the scattering [5] .

3. MOLECULAR DYNAMICS

Rahman et al. [7] have studied the low-temperature properties of an amorphous system of 500 Lennard-Jones particles using molecular dynamics techniques. Their calculated static structure factor agrees well with the soft-packing models. They calculated the density of states by directly diagonalizing the dynamical matrix. This problem was tractable only for a subsystem of 200 particles, embedded in the original cluster of 500. This was done at different moments in time and in different regions of the 500-particle box. Their calculated density of states agress well with the equation-of-motion result of Rehr and Alben (cf. the dotted line in Fig.1).

4. CONTINUED-FRACTION METHODS

The dynamical structure factor $S(\vec{k},\omega)$ for coherent scattering is given by the Fourier-transform of a density-correlation function

$$S(\vec{k},\omega)=\frac{1}{2\pi N}\sum_{l,l'} e^{-i\vec{k}(\vec{r}_l-\vec{r}_{l'})}\int_{-\infty}^{\infty} e^{-i\omega t}\langle e^{-i\vec{k}\vec{u}_l(0)}e^{i\vec{k}\vec{u}_{l'}(t)}\rangle dt \quad (13)$$

(this definition is slightly different from that used by v.Heimendahl, cf. equ.(6)). Such correlation functions for dynamical variables are very efficiently calculated using Mori's [8] technique. A relaxation shape function $F(\vec{k},\omega)$ is defined by

$$F(\vec{k},\omega)=\frac{1}{2\pi}\int_{-\infty}^{\infty}\frac{\sum_{ll'} e^{-i\vec{k}(\vec{r}_l-\vec{r}_{l'})}\langle e^{-i\vec{k}\vec{u}_l(0)}e^{i\vec{k}\vec{u}_{l'}(t)}\rangle}{\sum_{ll'} e^{-i\vec{k}(\vec{r}_l-\vec{r}_{l'})}\langle e^{-i\vec{k}\vec{u}_l(0)}e^{i\vec{k}\vec{u}_{l'}(0)}\rangle}\, e^{-i\omega t}dt \quad (14)$$

so that for $\vec{k}\neq 0$ we have

$$S(\vec{k},\omega)= S(\vec{k})\, F(\vec{k},\omega) \quad (15)$$

where $S(\vec{k})$ is the static structure factor. Moris treatment leads to a continuedfraction expansion for the Laplace-transform of the relaxation shape function

$$F(\vec{k},\omega)= \pi^{-1}\,\mathrm{Re}\,\tilde{F}(\vec{k},i\omega) \quad (16)$$

$$\tilde{F}(\vec{k},s)=\cfrac{1}{s+\delta_1\cfrac{}{s+\delta_2\cfrac{}{s+\delta_3\cfrac{}{s+\ldots\ldots\ldots\ldots}}}} \qquad (17)$$

where $\tilde{F}$ denotes the Laplace-transformed of F. Using (15) this leads to the following expansion for the dynamical structure factor (Beck and Tomanek [9], Tucker [10])

$$S(\vec{k},\omega)=\pi^{-1}\, S(\vec{k})\ \mathrm{Re}\cfrac{1}{s+\delta_1\cfrac{}{s+\delta_2\cfrac{}{s+\ldots\ldots}}} \qquad (18)$$

with $s=i\omega$. The coefficients δ_i are related to the frequency moments $\langle\omega^n\rangle$ of $S(\vec{k},\omega)$

$$\langle\omega^n\rangle=\int_{-\infty}^{\infty}\omega^n\, S(\vec{k},\omega)\, d\omega \qquad (19)$$

For the first few terms, this relation is given by

$$\begin{aligned}\delta_1&=\langle\omega^2\rangle/S(\vec{k})\\ \delta_2&=\langle\omega^4\rangle/\langle\omega^2\rangle-\langle\omega^2\rangle/S(\vec{k})\\ \delta_3&=(\langle\omega^6\rangle-\langle\omega^4\rangle^2/\langle\omega^2\rangle)/(\langle\omega^4\rangle-\langle\omega^2\rangle^2/S(\vec{k})).\end{aligned} \qquad (20)$$

To proceed further, the continued fraction has to be terminated at some stage. This is a rather critical step and we refer to the specialized literature ([9] and references cited therein).

The recursion method of Haydock, Heine and Kelly [11] is similar to the Mori technique. It allows the calculation of the density of states without explicitly calculating the normal mode frequencies. It starts from the following expression for the local density of states in terms of the matrix elements of the resolvent operator

$$g(\omega^2)=-\pi^{-1}\ \mathrm{Im}\ G_{oo}(\omega^2+i\varepsilon) \qquad (21)$$

$$G=(\omega^2 I-\Phi)^{-1} \qquad (22)$$

where Φ is the force-constant matrix. A continued-fraction expansion for G_{oo} is obtained by setting up a new basis in which Φ is tridiagonal. In this basis, it is possible to invert $(\omega^2 I-\Phi)$ analytically to obtain the continued-fraction expansion. As yet the method has not been applied to amorphous metals, but recently Meeks [12] used the recursion method to discuss the interrelation between the vibrational spectra and the topological structure of tetrahedrally bonded amorphous semiconductors.

4. THE DYNAMICS OF AMORPHOUS $Mg_{0.7}Zn_{0.3}$

Starting from the pair potentials and the equilibrium structure described in my first lecture von Heimendahl [13] has calculated the dynamical structure factor $S_1(\vec{k},\omega)$ for the $Mg_{0.7}Zn_{0.3}$ glass. Full account was taken of the topological, quantitative, and substitutional disorder, i.e. all force constants between any atom and all its neighbours within the interaction range of r=6.68 Å (the interaction range corresponds to the position of the minimum in the radial distribution function after the split second peak, it contained in the average 57 neighbours) were calculated at their relaxed positions. The results for both longitudinal and transverse excitations are shown in Fig.2, the $\vec{k}$-vectors are in the [100]-direction. Spectra with $\vec{k}$-vector along [110] or [111] interpolate smoothly between the results shown. The results are qualitatively similar to the spectra of Rehr and Alben (Fig.1). The k-dependence of the peaks is

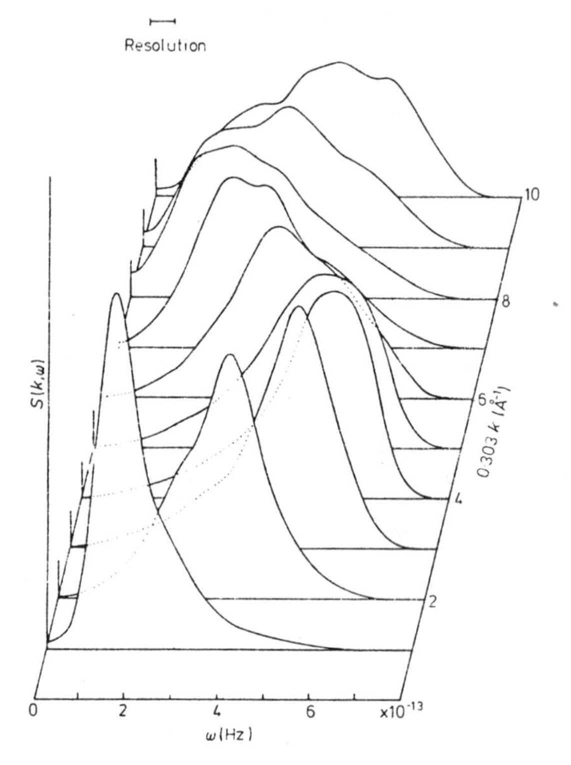

Fig. 2(a). The dynamical structure factor for longitudinal excitations. After von Heimendahl [13].

plotted in Fig.3. These "dispersion relations" show a clear maximum for longitudinal excitations at $k \simeq 1.21$ Å^{-1}, i.e. at approximately half the value of the wavevector of the first peak in the static

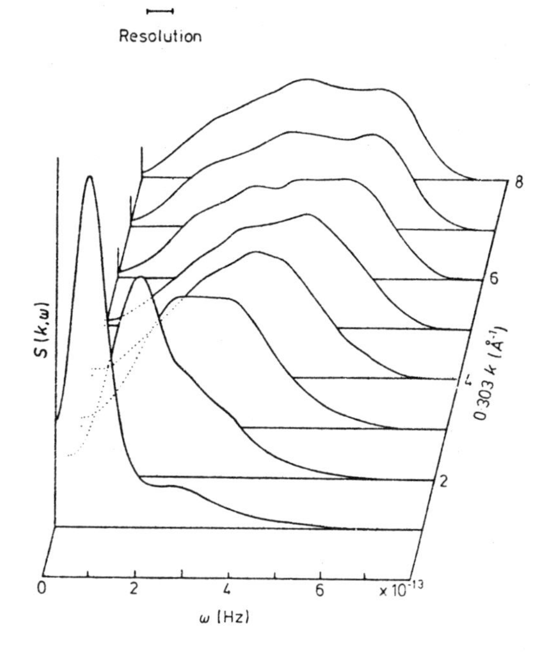

Fig. 2(b). The dynamical structure factor for transverse excitations. After von Heimendahl [13].

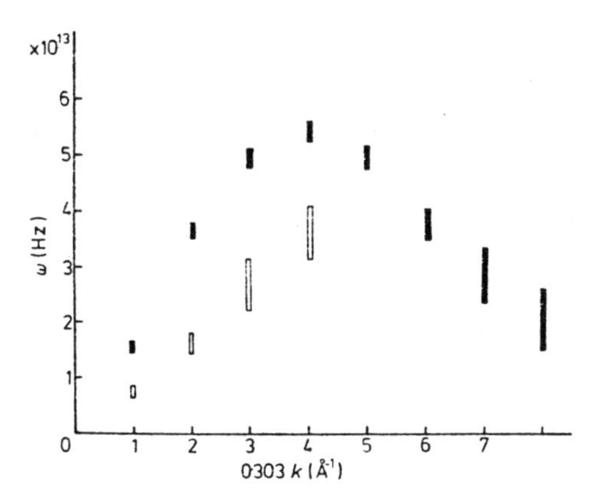

Fig. 3. The "dispersion relations" for the Mg-Zn glass [13].

structure factor (Q_p=2.54 Å^{-1}). The subsequent minimum is not really identifiable, because the peaks in the structure factor become too broad. For transverse waves the vibrational energy is well defined only for very small momenta, but apparently there is no peak in the dispersion relations for transverse excitations. The results of the inelastic neutron-scattering experiments is yet unknown. At present the only experimental information comes from the Brillouin scattering experiments of Grimsditch and Güntherodt [14] who measured a surface sound-velocity of $v_{surf}=2.13\times10^5$ cm/sec. If we assume a Poisson ratio of $\sigma \simeq 0.30$ (this is approximately equal to the σ of pure Mg and Zn metals), we can deduce longitudinal and transverse sound velocities of $v_L=4.3\times10^5$ cm/sec, $v_T=2.3\times10^5$ cm/sec. From von Heimendahl's dispersion relations one might estimate $v_L=5.1\times10^5$ cm/sec and $v_T=2.5\times10^5$ cm/sec.

Beck and Tomanek [9] used the Mori-technique to calculate the dynamical structure factor, using some simplifying assumptions: (i) simpler pair-potentials based on empty-core pseudopotentials, (ii) analytical hard-core structure factors in the Percus-Yevick approximation. Using this information they are able to calculate the fourth moment $\langle\omega^4\rangle$, the calculation of the sixth moment already necessitates further assumptions on the unknown three-particle correlation functions. Despite of these limitations, they are able to reproduce qualitatively von Heimendahl's results. They calculated explicity partial structure factors $S_{\alpha\beta}(\vec{k},\omega)$ in terms of the partial moments $\langle\omega^n_{\alpha\beta}\rangle$. They try to define "dispersion relations" by transforming the matrix of the fourth moment $\langle\omega^4_{\alpha\beta}\rangle$ to principal axis form. In a vague analogy to the crystal it is possible to distinguish "optical" and "acoustic" modes. For small k this assignement is justified by the form of the eigenvectors. This offers a possible explanation for the structure found in von Heimendahl's $S(\vec{k},\omega)$.

The phonon-dispersion relations for crystalline Mg have been calculated by Hafner and Eschrig [15], for Zn by Hafner [16], for the Laves-phase compound $MgZn_2$ by Eschrig and co-workers [17], using the same type of potentials. The comparison of these results with those for the glass shows that the maximum frequencies vary roughly linearily with concentration, independently of the structure. Thus, there is no phonon-softening in the amorphous phase for larger k-vectors. Anomalously low vibrational frequencies have been predicted by Eschrig et al. for some TO-modes in $MgZn_2$. They are a particularity of the structure and of the mass-ratio of this compound. No related features are found in the glass. The comparison of the elastic part of the dispersion relationsdoes not reveal any softening of the long-wave part of the spectrum of the metallic glass.

REFERENCES

1. M.H.Brodsky and M.Cardona, J.Non-cryst.Sol. 31,81(1978).
2. R.Alben, L.von Heimendahl, P.Galison, and M.L.Long, J.Phys. C8,L468(1975).
3. L.von Heimendahl, phys.stat.sol. (b) 86,549(1978) and further references cited therein.
4. A.A.Maradudin, E.W.Montroll, G.H.Weiss, and I.P.Ipatova, Theory of Lattice Dynamics in the Harmonic Approximation, 2nd ed.,

Academic Press, New York, 1971.
5. J.J.Rehr and R.Alben, Phys.Rev. B16,2400(1977).
6. L.von Heimendahl and M.F.Thorpe, J.Phys. F5,L87(1975).
7. A.Rahman, M.J.Mandell, and J.P.McTague, J.Chem.Phys. 64,1564 (1976).
8. H.Mori. Prog.Theor.Phys. 34,399(1965).
9. H.Beck and D.Tomanek, unpublished.
10. J.W.Tucker, Sol.State Comm. 18,43(1976).
11. R.Haydock, V.Heine, and M.J.Kelly, J.Phys.C5,2845(1972); ibid. C8,2591(1975).
12. P.E.Meek, Philos.Mag. 33,897(1976).
13. L.von Heimendahl, J.Phys. F9,161(1979).
14. M.Grimsditch and G.Güntherodt, private communication.
15. J.Hafner and H.Eschrig, phys.stat.sol. (b) 72,179(1975).
16. J.Hafner, unpublished.
17. H.Eschrig, K.Feldmann, K.Hennig and L.Weiss, in: Neutron Inelastic Scattering, Grenoble 1972, p.157, IAEA, Vienna, 1972.

THE ELECTRICAL CONDUCTIVITY OF STRONGLY DISORDERED SYSTEMS

W. Götze

Physik-Department, Technische Universität München and Max-Planck Institut für Physik München, West Germany

The well established theory for the electrical conductivity of metals, based on Boltzmann's kinetic equation, requires the assumption that the electron scattering mechanism is sufficiently weak. This is particularly evident for Ziman's (1) derivation of the current relaxation rate $1/\tau$ in liquid metals, since he worked out a leading order approximation using the electron pseudopotential as a small parameter. In liquid and in amorpheous metals the electron mean free path ℓ, estimated from the observed resistivity, is of the order of the electron wave length λ which is of the same order as the range L_o of the potential fluctuations. It is obvious, however, that the very concept of a mean free path loses any meaning if ℓ approaches λ or L_o (2). So there appears the necessity of inventing a theory for the electron mobility taking into account the new phenomena to be expected in systems with strong disorder.

Indeed, it is well known that for sufficiently strong electron random potential coupling the system does not exhibit a metallic conductivity at all, but rather it behaves like a semi-conductor (3). According to Mott (4) the conductor insulator transition in Anderson's sense occurs, if ℓ, calculated in the usual fashion, comes close to the Ioffe-Regel limiting value λ or L_o. Hence, a theory for the conductivity in strongly disordered systems should describe Anderson's transition. It should be stressed that the transition under consideration occurs for a system of noninteracting electrons.

Coulomb interactions and phonon effects are important but the essence of the phenomenon is one for a single electron moving in a random potential.

Recently an approximation scheme for the relevant correlation functions of a zero temperature gas of non-interacting electrons moving in a strong three-dimensional random potential has been proposed (5). The central concept of this theory is the current relaxation kernel $M(\omega)$, a causal function of frequency ω yielding the dynamical conductivity via

$$\sigma(\omega) = \frac{i}{\omega + M(\omega)} \left(\frac{ne^2}{m}\right). \qquad (1)$$

A leading order correlation approximation was used to express $M(\omega)$ in terms of Ziman's coupling function $\omega(q) = s(q)|U(q)|^2$ and the electron correlation function $\phi(q,\omega)$ for density fluctuations of wave number q:

$$M(\omega) = \int d^3q\ q^2\omega(q)\phi(q,\omega). \qquad (2)$$

If one replaces $\phi(q,\omega)$ by the correlation function $\phi^o(q,\omega)$ of the free electron gas, equation (2) yields $M(\omega) = i/\tau$ with $1/\tau$ identical with Ziman's (1) expression; equation (1) then is Drude's formula. Equation(2) introduces a feed-back or self-consistency problem. To calculate $M(\omega)$, i.e. the conductivity, one needs the density correlations. The motion of density fluctuations is closely connected with current propagations which are influenced by the conductivity. So one needs $M(\omega)$ in order to calculate $\phi(q,\omega)$. An approximate nonlinear connection between $\phi(q,\omega)$ and $M(\omega)$ has been obtained by truncating the exact Mori equation for the electron phase space density response:

$$\phi(q,\omega) = \frac{\phi^o(q,\omega+M(\omega))}{1+M(\omega)\phi^o(q,\omega+M(\omega))/\rho_F} \qquad (3)$$

If one replaces $M(\omega)$ by its weak coupling result $1/\tau$, equation (3) is identical with the kinetic equation result for $\phi(q,\omega)$ with the collision integral replaced by a simple rate factor. The current decay rate, given by the absorptive part $M''(\omega)$ of the kernel $M(\omega)$ is non-zero, since the single current excitation couples to other modes of the system. The simplest modes, occuring as current decay channels, are pair excitations consisting of density fluctuations with spectral density $\phi''(q,\omega)$ and static potential fluctuations. The single mode excitations have to be calculated self-consistently with the pair excitations mentioned. The self-consistency mechanism yields a frequency dependence and thus a reactive part $M'(\omega)$ of the kernel $M(\omega)$. It acts like a potential with a tendency to trap the electron. The metal insulator transition is characterized by $M'(\omega)$ being as important as $M''(\omega)$ and both diverge for zero frequency. In the insulator phase $M'(\omega)$ dominates in comparison to $M''(\omega)$. It has been shown (5) that near the Anderson transition point the equations (2,3) can be solved in terms of elementary functions.

In the present lecture the complete solution of equations (2,3) is presented for the simplified model $\omega(q)\propto\lambda$, $q<2\pi/L_o$; it is obtained by carrying out asymptotic expansions in three regimes. First, power series expansion of $M(\omega)$ in terms of λ yields in first order the kinetic equation result and in second order the leading correction to the classical expression for the mobility. Second, a power series expansion of $M(\omega)$ in terms of $(1/\lambda)$ yields a molecular field approximation for Anderson's Fermi glass. Upon decreasing disorder the polarizability of the insulator increases and exhibity a typical Curie Weiss instability. Third, in the transition regime equations (2,3) can be reduced asymptotically to a cubic algebraic equation for $M(\omega)$. The analytic discussion is backed up by presenting the correct solution of equations (2,3), obtained numerically. As main result a representative set of $\sigma(\omega)$ versus ω curves for various degrees of disorder is presented and discussed. The lecture is based on a paper submitted for publication elsewhere (6).

References

(1) J.M. Ziman, Phil.Mag. 6, 1013 (1961)
(2) A.F.Ioffe and A.R.Regel, Progr.in Semicond. 4,237 (1960)
(3) P.W. Anderson, Phys.Rev. 109, 1492 (1958)
(4) N.F. Mott, Phil.Mag. 13, 989 (1966)
(5) W.Götze, Sol.St.Comm. 27, 1393 (1978);J.Phys.C 12, 1279 (1979)
(6) W.Götze, preprint, submitted to Phil.Mag., Sept. 1979

THEORY FOR THE ELECTRONIC STRUCTURE OF DISORDERED TRANSITION-METALS

K.H. Bennemann

Institute for Theoretical Physics, FU Berlin
1000 Berlin 33

I. Introduction

II. Theory for Electronic Density of States

III. Applications

I. INTRODUCTION

Due to strong electron-electron interactions and electron-lattice coupling transition-metals have particular interesting properties. For many purposes transition-metal alloys are of particular interest. A quantity characterizing well the electronic structure of the transition-metal is the electronic density of states,

$$N(\varepsilon)$$

respectively, the local electronic density of states at atomic site i,

$$N_i(\varepsilon).$$

For many metals this electronic density of states has been determined by photoemission experiments. In Fig. (1) we show a typical $N(\varepsilon)$ for transition metals.

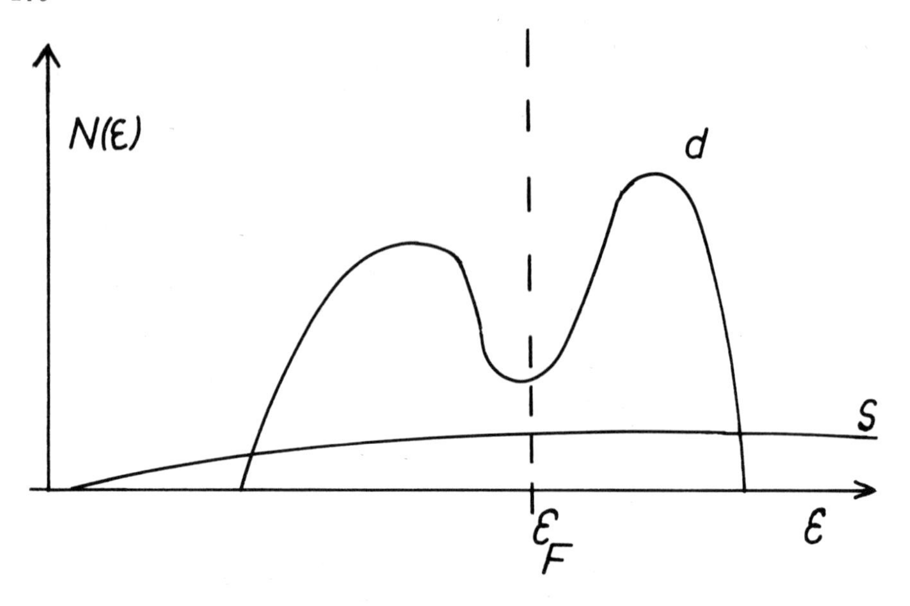

Fig. 1: Illustration of the electronic density of states $N(\varepsilon)$ for transition metals. Typically for f.c.c. - the 2 peak structure is more pronounced than for b.c.c. metals.

Neglecting fine structure, $N(\varepsilon)$ consists of two peaks resulting from the covalent splitting of the quasi atomic like electronic energy levels into bonding and antibonding states. In a simple treatment this covalent splitting is proportional to the hopping integral t_{ij} describing the electronic transitions between the atomic site i and j. Fine structure in $N(\varepsilon)$ may result partly from the fact that t_{ij} is different for the 5d-orbitals e_g and t_{2g} for cubic systems and from t_{ij} referring to next nearest neighbors and overnext nearest neighbors. The width of the d-band is approximately determined by the coordination number and the s-d hybridization.

The electronic density of states determines many properties of the transition-metal like the electronic energy ($E \sim \int d\varepsilon N(\varepsilon)\varepsilon + ..$), cohesion, stability of the system, magnetic properties (spin-susceptibility, magnetic moments,...) and electronic charge transfers and thus metal-insulator transitions and catalytic properties at surfaces, for example. Due to the quasi local character of the d-electrons the electronic structure of transition-metals in particular its alloys has been determined by using a tight-binding type Hamiltonian. In the following we present theoretical

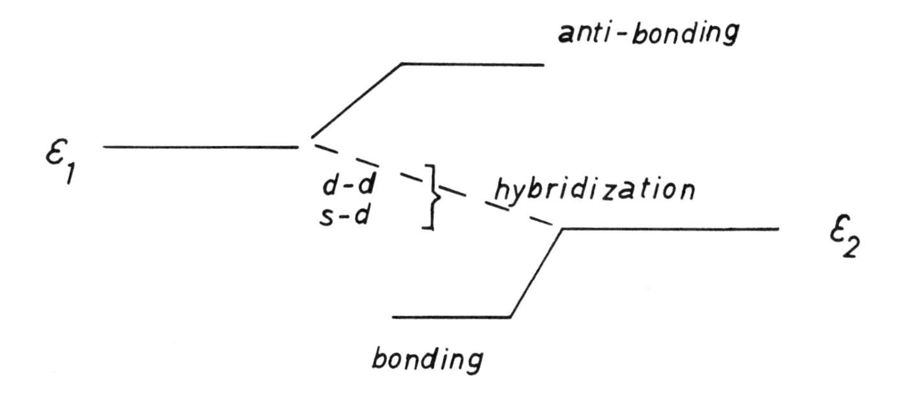

Fig. 2: Illustration of covalent splitting of electronic states 1, 2 due to hybridization

methods for determining the electronic density of states which permit also to determine local properties of transition-metals. It is shown how the basic theory can be used for determining electron-charge transfer leading to metal-insulator transitions (s. CsAu$\rightarrow$Cs$^{(+)}$Au$^{(-)}$) for understanding the electronic structure of amorphous transition metal alloys and magnetic properties of transition metals.

II. THEORY FOR ELECTRONIC STRUCTURE OF TRANSITION-METAL (ALLOYS)-DETERMINATION OF N(ε).

In the following we present a theory for determining the electronic properties of transition-metal alloys. For simplicity we discuss first the coherent-potential approximation as developed by Soven[1] and Ehrenreich[2] and others which neglect s-d hybridization and disorder in the hopping integral and a dependence of the hopping integral on the character of the d-orbitals (e_g, t_{2g}).

(a) C.P.A. theory

The d-electrons of a transition-metal alloy $A_{1-x}B_x$ are treated by using the simplified tight-binding Hamiltonian

$$H = \sum_{i,\sigma} \varepsilon_{i\sigma}\, n_{i\sigma} + \sum_{i,j} t_{ij}\, c^{\dagger}_{i\sigma} c_{j\sigma} + \text{h.c.} \qquad (II.1)$$

Here,

$$\varepsilon_{i\sigma} = \varepsilon_i + U_i n_{i-\sigma} \tag{II.2}$$

As illustrated in Fig. 3, $\varepsilon_{i\sigma}$ describes the energy for electrons with spin σ at atomic site i and the hopping integral t_{ij} describes the electronic transitions between atomic sites i and j. U_i denotes the intra-atomic Coulomb interaction at site i.

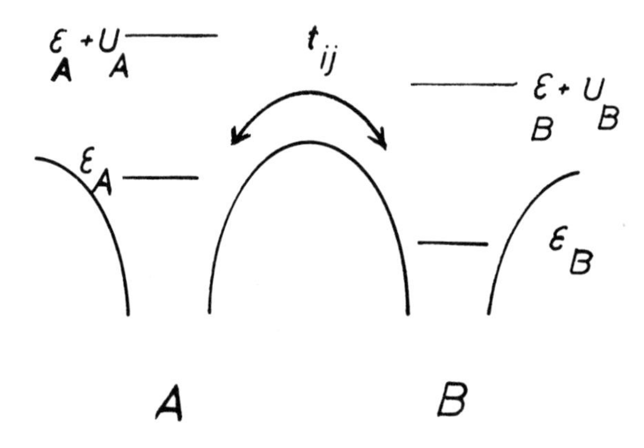

Fig. 3: Illustration of the physical significance of the parameters ε_i, U_i, t_{ij} used in Eq. (II,1)

We split the Hamiltonian H into

$$H = H_0 + V , \qquad V = \sum_i V_i , \tag{II.3}$$

where V denotes the perturbing potential and V_i the perturbing potential due to the atom i. The electronic properties are determined by the electron Green's function

$$\mathcal{G}(z) = \mathrm{Tr}\left(\frac{1}{H-z}\right)$$

defined by the operator eq.

$$(z-H)\mathcal{G} = 1$$

Defining $G = \mathcal{G}(V=o)$ by $(z-H_o)G = 1$, it follows straightforwardly

$$\mathcal{G}(z) = G(z) + GV\mathcal{G}$$

Iteration yields

$$\mathcal{G} = G + GtG + GtGtG + \ldots\ldots, \qquad (II.4)$$

where the t-matrix

$$t = V + VGt \qquad (II.5)$$

describes multiple scattering by the potential V. To obtain rapide convergence of the perturbation series we write

$$H = H_o + \sum_{i,\sigma} (\varepsilon_{i\sigma} - \Sigma_\sigma) n_{i\sigma} , \qquad (II.6)$$

where it follows from Eq. (II.1) that

$$H_o = \sum_{i,\sigma} \Sigma_\sigma(z) n_{i\sigma} + \sum_{i,j,\sigma} c^\dagger_{i\sigma} c_{j\sigma} \quad . \qquad (II.7)$$

Note, H_o corresponds to the Hamiltonian of a pure metal with energy levels $\Sigma_\sigma(z)$ at each atomic site rather than $\varepsilon_{A\sigma}$ or $\varepsilon_{B\sigma}$. The perturbing potential

$$V_{i\sigma} = (\varepsilon_{i\sigma} - \Sigma_\sigma)$$

which scatters electrons with spin σ at site i represents potential fluctuations in the transition metal alloy with respect to the average energy $\Sigma_\sigma(z)$. It follows from Eq. (II.7) that in Bloch-representation

$$G_\sigma(k,z) = \langle k|(z-H_o)^{-1}|k\rangle = (z-\Sigma_\sigma(z)-\varepsilon_k)^{-1} \qquad (II.8)$$

with

$$\varepsilon_k = \sum_j t_{ij}\, e^{-i\mathbf{k}\cdot\mathbf{r}_{ij}} \quad .$$

Thus, in Wannier representation one obtains

$$G_{ij\sigma}(z) = \frac{1}{N}\sum_{k} G_{\sigma}(k,z)e^{i\underset{\sim}{k}\cdot\underset{\sim}{r}_{ij}}$$

and in particular

$$G_{\sigma}(k,z) = \frac{1}{N}\sum_{k} \frac{1}{z-\Sigma_{\sigma}-\varepsilon_{k}} = F_{\sigma}(z-\Sigma_{\sigma}) \tag{II.9}$$

Obiously,

$$N(z) = \frac{1}{N}\sum_{k} \delta(z-\varepsilon_{k}) \tag{II.10}$$

is the density of states per atom (N = total number of atoms) for electrons with energy z. Using Eq. (II.9) we obtain for $\langle \mathcal{G} \rangle = G$

$$N_{\sigma}(z) = -\frac{1}{\pi N}\,\mathrm{Im}\,F_{\sigma}(z-\Sigma_{\sigma}+i\delta) \tag{II.11}$$

for the average alloy density of states of electrons with spin σ. To determine the electron density of states via Eq. (II.11) the average electron self-energy $\Sigma_{\sigma}(z)$ needs to be calculated. Clearly, averaging $\mathcal{G}$ over all alloy configurations, which gives $\langle \mathcal{G}_{\sigma} \rangle$, one gets

$$\langle \mathcal{G}_{\sigma} \rangle = G_{\sigma} \quad ,$$

which implies

$$\langle t_{i\sigma} \rangle = 0 \tag{II.12}$$

for the t-matrix averaged over all alloy configurations. Eq. (II.12) is called the coherent-potential-approximation. One finds from Eq. (II.12) for an alloy A_xB_{1-x} that

$$\langle t_{i\sigma} \rangle = x t_{A\sigma} + (1-x)\, t_{B\sigma} = 0 \quad . \tag{II.13}$$

Since

$$t_{i\sigma} = \frac{V_{i\sigma}}{1-V_{i\sigma}F_\sigma} \quad , \qquad \text{(II.14}$$

as follows from solving Eq. (II.4), Eq. (II.13) can be rewritten as

$$x \frac{\varepsilon_{A\sigma} - \Sigma_\sigma}{1-(\varepsilon_{A\sigma}-\Sigma_\sigma)F_\sigma} + (1-x) \frac{\varepsilon_{B\sigma} - \Sigma_\sigma}{1-(\varepsilon_{B\sigma}-\Sigma_\sigma)F_\sigma} = 0 \quad .$$

From this one obtains easily

$$\Sigma_\sigma(z) = \varepsilon_\sigma - (\varepsilon_{A\sigma}-\Sigma_\sigma)F\ (z-\Sigma_\sigma)(\varepsilon_{B\sigma}-\Sigma_\sigma) \quad , \qquad \text{(II.15)}$$

with

$$\varepsilon_\sigma \equiv x\varepsilon_{A\sigma} + (1-x)\,\varepsilon_{B\sigma} \,.$$

This Eq. determines the electron self-energy $\Sigma_\sigma(z)$. To discuss limiting cases it is useful to introduce the parameter

$$\delta_\sigma = \frac{\varepsilon_{A\sigma} - \varepsilon_{B\sigma}}{W} \quad ,$$

which measures the difference of the d-levels on A and B sites. W denotes half of the average band width. Then,

$$\varepsilon_\sigma - \varepsilon_{B\sigma} = x\delta_\sigma \quad , \quad \varepsilon_\sigma - \varepsilon_{A\sigma} = -(1-x)\delta_\sigma$$

$$\Sigma_\sigma - \varepsilon_{B\sigma} = x\delta_\sigma + (\varepsilon_{A\sigma} - \Sigma_\sigma)\ F_\sigma(\Sigma_\sigma - \varepsilon_{B\sigma}) \quad .$$

The Eq. for $\Sigma_\sigma(z)$ can now be solved easily in the following limiting cases.

(1) Dilute alloy limit: Then, $x \ll 1$ and one may expand $\Sigma(z)$ linearly in x. Since.

$$\varepsilon_{A\sigma} - \Sigma_\sigma \simeq \varepsilon_{A\sigma} - \varepsilon_\sigma \simeq \delta_\sigma$$

and $\Sigma_\sigma - \varepsilon_{B\sigma} = x\delta_\sigma/(1-\delta_\sigma F_\sigma)$,

one obtains

$$\Sigma_\sigma(z) = \varepsilon_\sigma + x\delta_\sigma^2 \frac{F_\sigma}{1-\delta_\sigma F_\sigma} \quad . \tag{II.16}$$

(2) Virtual crystal approximation: Then, $\delta_\sigma \ll 1$ and consequently weak alloy effects are expected. Therefore, one puts on the right hand side of Eq. (II,15) $\Sigma_\sigma \simeq \varepsilon_\sigma$ and obtains

$$\Sigma_\sigma(z) = \varepsilon_\sigma + x(1-x)\delta_\sigma^2 F(z-\Sigma_\sigma) \quad . \tag{II.17}$$

The first term corresponds to the rigid band approximation. The second term corresponds to the result of weak coupling multiple scattering theory by Edwards, Beeby et al. Eq. (II,17) should be applicable in particular to isoelectronic transition-metal alloys.
(3) Split-band limit, $\delta \gg 1$: Note, $\delta \gg 1$ may result from $W \to 0$, narrow bands (atomic limit) or from large separation of ε_A, ε_B. This is the case, for example, for 3d-5d alloys (NiAg, etc), or for NiAl. Then,

$$F(z-\Sigma) = \sum_k \frac{1}{\varepsilon - \varepsilon_k - \Sigma} - \frac{1}{z-\Sigma} \quad .$$

Consequently,

$$\Sigma_\sigma(z) = \varepsilon_\sigma + x(1-x) \frac{\delta_\sigma^2}{z-\Sigma_\sigma} \quad , \tag{II.18}$$

since $\varepsilon_{A\sigma} - \Sigma_\sigma \simeq \varepsilon_{A\sigma} - \varepsilon_\sigma = (1-x)\delta_\sigma + \ldots$.

In summary, the electron density of states $N(\varepsilon)$ for transition metal alloys can be calculated from Eqs. (II,11), (II,9) and (II,15).

For many properties the local electron density of states $N_i(\varepsilon)$ at an atomic site i is of importance and which is given by

$$N_{i\sigma}(\varepsilon) = -\frac{1}{\pi} \operatorname{Im} G_{ii}(\varepsilon) \quad , \tag{II.19}$$

with

$$G_{ii\sigma} = \langle \mathcal{G}_{ii,\sigma} \rangle_i \tag{II.20}$$

Here, $\langle \mathcal{G}_\sigma \rangle_i$ denotes averaging over all atomic sites except site i. Thus,

$$G_{ii\sigma} = F_\sigma + F_\sigma\, t_{i\sigma}\, F_\sigma = \frac{F_\sigma}{1-V_{i\sigma}F_\sigma} \quad . \tag{II.21}$$

Note, the average number of electrons with spin σ per atom is given by

$$n_\sigma = \int dz\ f(z)\ N_\sigma(z) = xn_{A\sigma} + (1-x)n_{B\sigma} \quad , \tag{II.22}$$

where

$$n_{i\sigma} = -\frac{1}{\pi}\int dz\ f(z)\ \mathrm{Im}\ G_{ii,\sigma}(z)$$

gives the number of electrons on site i with spin σ.

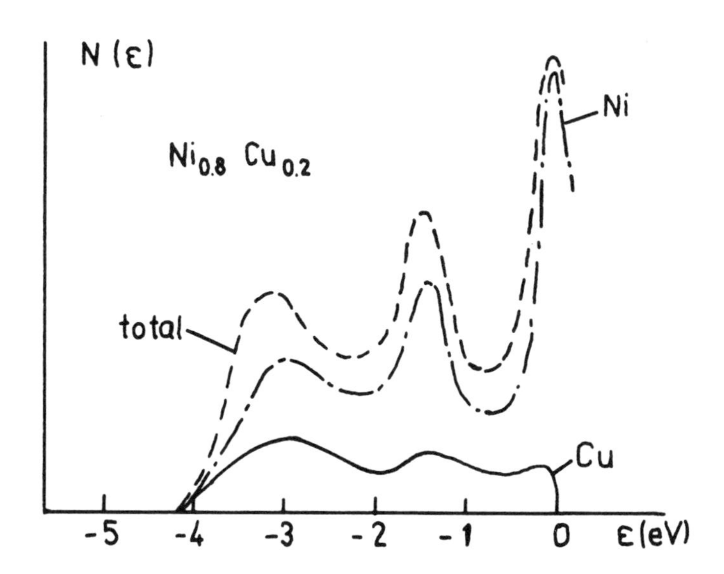

Fig.4: Illustration of $N(\varepsilon,x)$ and $N_i(\varepsilon)$ for an alloy A_xB_{1-x}. We put $\varepsilon_F = 0$.

(b) Continued-Fraction type-Theory

A very useful alternative way to derive CPA, which also permits to take into account easily local atomic environment effects (s. magnetic-moments in $Ni_{1-x}Cu_x$, surface problems, etc.), is the continued-fraction method. Using in Eq. (II,1) the hopping integral t_{ij} as the expansion parameter, then the electron Green's function G is given by

$$G_\sigma = G^0_\sigma + G^0_\sigma t_\sigma G_\sigma \quad ,$$

with

$$G^0_\sigma = \frac{1}{z-\varepsilon_{i\sigma}} \quad ,$$

Then, solving $G_{ii} = G^0_{ii} + G^0_{ii} t_{ij} G_{ji}$

by iteration one obtains[3]

$$G_{ii\sigma}(z) = \frac{1}{z-\varepsilon_{i\sigma}-\Delta_{i\sigma}} \tag{II.23}$$

with

$$\Delta_{i\sigma} = \sum_j \frac{t^2_{ij}}{z-\varepsilon_{j\sigma}} + \sum_{jl} \frac{t_{ij}t_{jl}t_{li}}{(z-\varepsilon_{j\sigma})(z-\varepsilon_{l\sigma})} \tag{II.24}$$

Here, the electron self-energy $\Delta_{i\sigma}$ results from all electron hopping paths which start at i and return to i and avoid the site i in-between. We replace now in $\Delta_{i\sigma}$ all $\varepsilon_{j\sigma}, \varepsilon_{l\sigma}$, etc. by the mean energy $\Sigma_\sigma(z)$, which corresponds to molecular field approximation. Then, $\Delta_{i\sigma}$ is the same as in a pure metal with electron energy $\varepsilon_{i\sigma} \equiv \Sigma_\sigma$. Thus,

$$\langle G_{ii\sigma}\rangle = \frac{1}{z-\Sigma_\sigma-\Delta_{i\sigma}} = \sum_k \frac{1}{z-\Sigma_\sigma-\varepsilon_k} \quad ,$$

which gives the relation

$$\Delta_{i\sigma}(z) = z-\Sigma_\sigma - F_\sigma^{-1} \quad . \tag{II.25}$$

The mean-field energy Σ_σ is determined by using the self-consistency eq.

$$F_\sigma = xG_{AA\sigma} + (1-x)\, G_{BB\sigma} \quad , \tag{II.26}$$

with

$$G_{ii\sigma}^{-1} = z - \varepsilon_{i\sigma} - \Delta_{i\sigma} \quad .$$

It can be easily shown that Eqs. (II.25) and (II.26) give the same expression for Σ_σ as previously derived.

Note,

$$G_{AA\sigma} = F_\sigma/(1-(\varepsilon_{A\sigma}-\Sigma_\sigma)F_\sigma) \quad .$$

To include local environment effects we proceed as follows[3]. The series expansion for $\Delta_{i\sigma}$ is renormalized as[4]

$$\Delta_{i\sigma} = \sum_j \frac{t_{ij}^2}{z-\varepsilon_{j\sigma}-\Delta_{j\sigma}^i} \quad , \tag{II.27}$$

where

$$\Delta_{j\sigma}^i = \sum_l \frac{t_{jl}\, t_{lj}}{z-\varepsilon_{l\sigma}-\Delta_{l\sigma}^{ij}} + \dots \tag{II.28}$$

includes all contributions from electron hopping paths from j-> j and avoiding in-between i and j. (We replace now in $\Delta_{j\sigma}^i$ all energies $\varepsilon_{l\sigma}$ etc. by Σ_σ. Hence, $\Delta_{j\sigma}^i$ is the same as in a pure metal with $\varepsilon_{i\sigma}=\Sigma_\sigma$). Note, $\Delta_{l\sigma}^{ij}$ is similarly expressed as $\Delta_{j\sigma}^i$ and results from all electron hopping l-> l and avoiding in-between sites i,j,l. Replacing in $\Delta_{i\sigma}$ the energy $\varepsilon_{j\sigma}$ by Σ_σ one obtains approximately

$$\Delta_{i\sigma} \simeq n_1\, t^2/(z-\Sigma_\sigma-\Delta_{j\sigma}^i) \quad ,$$

where n_1 denotes the number of nearest neighbors to atom i and $t_{ij} \equiv t$.

$t_{ij} \equiv t$. Thus,

$$\Delta^{i}_{j\sigma} \simeq \varepsilon_{\sigma} - \Sigma_{\sigma} - \frac{n_1 t^2}{\Delta_{i\sigma}} \quad . \tag{II.29}$$

Again, Σ_{σ} is determined self-consistently by Eq. (II,26) with $G_{ii\sigma} = (z - \varepsilon_{i\sigma} - \Delta_{i\sigma})^{-1}$ and $\Delta_{i\sigma}$ given by Eq. (II,27) using for $\Delta^{i}_{j\sigma}$ the approximate expression given by Eq. (II,29). Thus, we take explicitly into account the immediate atomic configuration surrounding the site i. This treatment corresponds to the Bethe-Peierls approximation. Note, the approximation

$$\Delta^{i}_{j\sigma} \simeq \frac{n_1 - 1}{n_1} \Delta_{i\sigma}$$

closing the continued fraction expansion (II,27), (II,28), is often useful for obtaining results to approximate order. This approximation makes Eqs. (II,23), (II,27) etc. also useful for pure transition-metals[5], liquids, amorphous metals etc. For a pure metal $G_{ii\sigma}$ can be written as

$$G_{ii\sigma}(z) = \cfrac{1}{z - \cfrac{n_1 t^2}{z - \cfrac{(n_1-1)t^2}{z - \cfrac{(n_1-1)t^2}{z - \ddots}}}}$$

which can be solved as[5]

$$G_{ii\sigma}(z) = \frac{2(n_1-1)}{(n_1-2)z + n_1 (z^2 - 4(n_1-1)t^2)^{1/2}} \quad .$$

Thus, one obtains a band width $W = 4\sqrt{n_1 - 1}\, t$ which due to neglecting some hopping paths is smaller than $W = 2 n_1 t$ as obtained within tight-binding approximation.

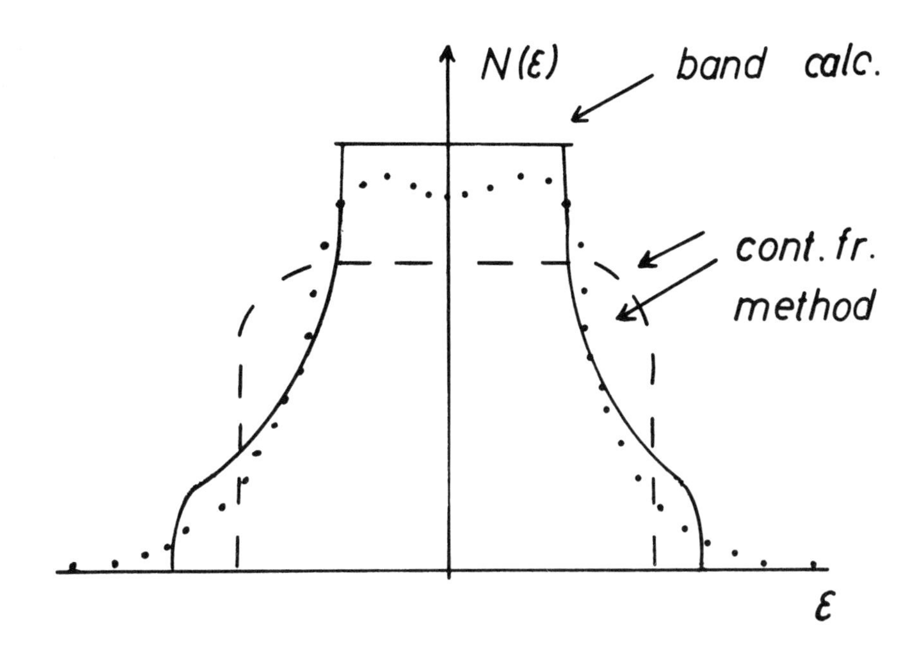

Fig. 5: Density of states N(ε) as determined by continued fraction method for t_{ij} = const and s.c. symmetry[5].

(c) Alloy theory including s-d hybridization

For many problems and properties the s-d hybridization is significant. For example, to treat electron charge transfer self-consistently one has to take into account the coupling between the s- and d-electrons. It is now relatively easy to extend the previous theory. We use[6]

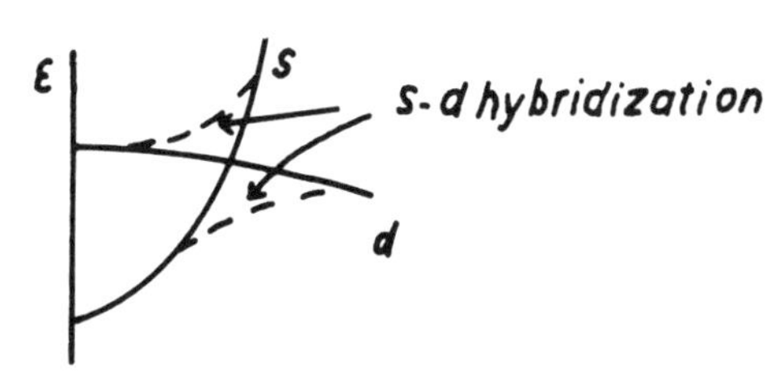

Fig. 6: Illustration of s-d hybridization.

$$H = \sum \varepsilon_{is\sigma} n_{s\sigma} + \sum \varepsilon_{id\sigma} n_{d\sigma} +$$

$$\sum_{ij} t^{s}_{ij} c^{+}_{is} c_{js} + \sum_{ij} t_{ij} d^{+}_{i\sigma} d_{j\sigma} + \text{h.c.} +$$

$$\sum \gamma_i c^{+}_{i\sigma} d_{j\sigma} + \text{h.c.} \quad , \qquad \text{(II.30)}$$

where $\varepsilon_s(k) = \sum t^{s}_{oj} e^{i\underset{\sim}{k}\cdot\underset{\sim}{r}_{oj}}$ and where s and d refer to s- and d-electrons and γ_i denotes the hybridization constant at site i. Rewriting H in matrix-form with

$$\varepsilon_i = \begin{pmatrix} \varepsilon_{is\sigma} & \gamma_i \\ \gamma_i & \varepsilon_{id} \end{pmatrix} \quad ,$$

it becomes immediately obvious that G is again given by the matrix equation

$\mathcal{G} = G + GtG + \ldots$

with

$t = V + VGt$

and the matrix scattering potential

$$V_{i\sigma} = \varepsilon_{i\sigma} - \Sigma_\sigma \quad .$$

Here, the electron self-energy matrix Σ_σ is given by

$$\Sigma_\sigma = \begin{pmatrix} \Sigma_{ss} & \Sigma_{sd} \\ \Sigma_{ds} & \Sigma_{dd} \end{pmatrix} \quad .$$

From $\langle t_{i\sigma} \rangle = 0$, one obtains again

$$\Sigma_\sigma = \varepsilon_\sigma - (\varepsilon_{A\sigma} - \Sigma_\sigma)\, F_\sigma\, (\varepsilon_{B\sigma} - \Sigma_\sigma) \quad ,$$

with

$$\varepsilon_\sigma = x\, \varepsilon_{A\sigma} + (1-x)\, \varepsilon_{B\sigma}$$

and

$$F_\sigma = G_{ii\sigma} = \frac{1}{N_k} \Sigma G(k,z) \quad .$$

Here,

$$G_\sigma(k,z) = \begin{pmatrix} z-\varepsilon_s(k)-\Sigma_{ss\sigma}, & -\Sigma_{sd} \\ -\Sigma_{ds}, & z-\varepsilon_{d\sigma}(k)-\Sigma_{dd\sigma} \end{pmatrix}^{-1} \quad .$$

We may use $t^d_{ij}=\alpha t^s_{ij}$. From these expressions which follow directly from the one-band theory by analogy one obtains using

$$GG^{-1} = 1 \;, \quad FF^{-1} = 1 \quad ,$$

the formulae

$$F_{ss} = \Big\{ (z-\Sigma_{dd}-\alpha\varepsilon_1)\, F^0_s\, (\varepsilon_1) - (z-\Sigma_{dd}-\alpha\varepsilon_2)\, F^0_s\, (\varepsilon_2) \; \frac{1}{\alpha(\varepsilon_2-\varepsilon_1)} \quad ,$$

$$F_{sd} = \Sigma_{sd}\, (F^0_s(\varepsilon_1)-F^0_s(\varepsilon_2))\, \frac{1}{\alpha(\varepsilon_2-\varepsilon_1)}$$

and

$$F_{dd} = \Big\{ (z-\Sigma_{ss}-\varepsilon_1)\, F^0_s\, (\varepsilon_1) - (z-\Sigma_{ss}-\varepsilon_2)\, F^0_s\, (\varepsilon_2) \Big\} \frac{1}{\alpha(\varepsilon_2-\varepsilon_1)} \quad .$$

Here,

$$\varepsilon_{1,2} = \frac{1}{2} \left\{ z - \Sigma_{ss} + \frac{z-\Sigma_{dd}}{\alpha} \pm \left(\left[z - \Sigma_{ss} - \frac{z-\Sigma_{dd}}{\alpha} \right]^2 + \frac{4\Sigma_{dd}^2}{\alpha} \right)^{1/2} \right\} .$$

The local Green's function is given by

$$G_{ii\sigma} = G_\sigma / (1 - (\varepsilon_{i\sigma} - \Sigma_\sigma) G_\sigma) .$$

Electron charge transfer between atoms A and B in an alloy A_xB_{1-x} is given by

$$\Delta n_{i\sigma}^{s,d} = \int_{-\alpha}^{\varepsilon_F(x)} d\varepsilon \, N_{i\sigma}^{s,d}(\varepsilon) - n_i^{s,d} ,$$

where $n_i^{s,d}$ refers to the number of s-, d-electrons at site i in the pure metal. Note, charge conservation requires

$$x(\Delta n_a^s + \Delta n_A^d) + (1-x)(\Delta n_B^s + \Delta n_B^d) = 0$$

Furthermore, due to the Coulomb interaction between s- and d-electrons, one uses[6,7]

$$\varepsilon_i^d = \varepsilon_{id}^o + U_{dd}^i n_{id} + U_{ds}^i n_{is}$$

and

$$\varepsilon_i^s = \varepsilon_{is}^o + U_{ss}^i \, n_{is} + U_{sd}^i \, n_{id} \quad .$$

Note, as a result of these Coulomb interaction the electron energy levels $\varepsilon_{i\sigma}$ will shift upon electron charge-transfer.

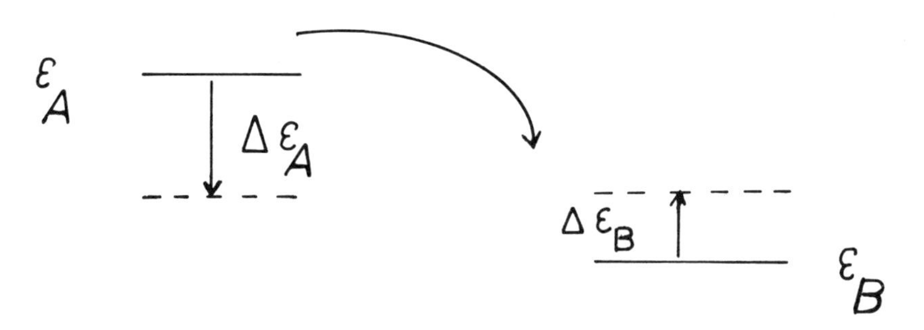

Fig. 7: Illustration of energy shifts $\Delta\varepsilon_i$ due to electron charge transfer from atom A$\rightarrow$B.

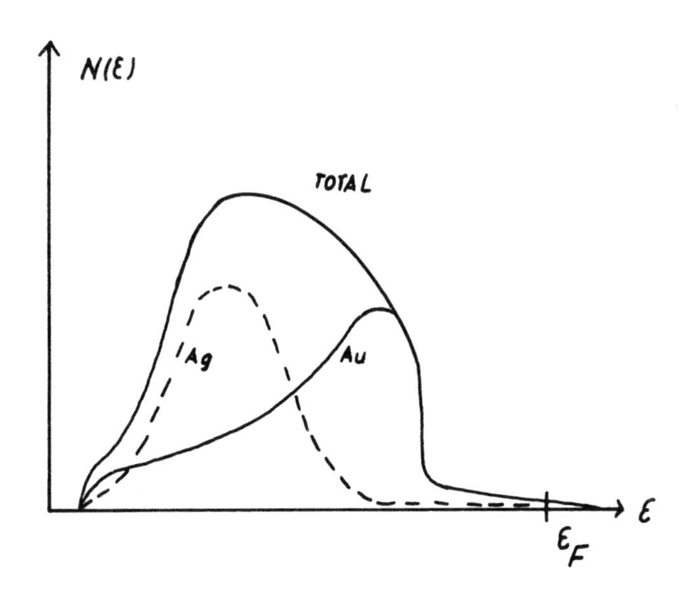

Fig. 8: Schematic illustration of N(ε) and $N_i(\varepsilon)$ for Au_xAg_{1-x}[6] .

III. APPLICATIONS

In the following we outline briefly how the theories described in Sec. II can be applied to interesting problems.

(1) Spin-Susceptibility in Transition-Metal Alloys

The spin-susceptibility χ in transition-metals is strongly determined by the d-electron density of states, $N(\varepsilon \simeq \varepsilon_F)$. Knight-shift N.M.R. etc. determine the local spin-susceptibility χ_i which depends sensitively on the local density of states $N_i(\varepsilon \simeq \varepsilon_F)$.

Since, the average alloy spin-susceptibility is given by

$$\chi = \lim_{H \to o} \mu_B \frac{n_\sigma(H) - n_{-\sigma}(H)}{H} , \qquad \text{(III.1)}$$

where H denotes the external magnetic field, one obtains

$$\chi = x \chi_A(x) + (1-x) \chi_B(x) . \qquad \text{(III.2)}$$

Here, $\chi_i = \lim_{H \to o} \mu_B \frac{n_{i\sigma} - n_{i-\sigma}}{H}$ denotes the local spin-susceptibility at site i. Since apparently

$$n_{i\sigma}(H) = n_{i\sigma}(o) + \frac{1}{2} \mu_B \sigma \chi_i H ,$$

one may find χ_i by expanding $n_{i\sigma}(H)$ to lowest order in H. Since

$$\partial n_{i\sigma} = (\partial n_{i\sigma}/\partial H) = -\partial n_{i-\sigma} \quad \text{and} \quad \chi_i \propto \partial n_{i\sigma} ,$$

we have to calculate $\partial n_{i\sigma} = \partial n_{i\sigma}(\varepsilon_{A\sigma}(H), \varepsilon_{B\sigma}(H))$.
Using

$$\partial n_{i\sigma} = \frac{\partial n_{i\sigma}}{\partial \varepsilon_{A\sigma}}\bigg|_{H=o} \partial \varepsilon_{A\sigma} + \frac{\partial n_{i\sigma}}{\partial \varepsilon_{B\sigma}}\bigg|_{H=o} \partial \varepsilon_{B\sigma}$$

and

$$\varepsilon_{i\sigma}(H) = \varepsilon_i + U_i n_{i-\sigma} - \mu_B \sigma \cdot H \quad ,$$

which yields

$$\partial \varepsilon_{i\sigma} = U_i \, \partial n_{i-\sigma} - \mu_B \sigma \quad ,$$

one obtains

$$\partial n_{A\sigma} = f_{AA} \, (-\mu_B \sigma + U_A \partial n_{A-\sigma}) +$$

$$f_{AB} \, (-\mu_B \sigma + U_B \partial n_{B-\sigma}) \quad .$$

Here,

$$f_{ij} = \frac{\partial n_{i\sigma}}{\partial \varepsilon_{j\sigma}} \quad .$$

Note, the eq. for $\partial n_{B\sigma}$ results from the expression for $\partial n_{A\sigma}$ by the replacements $A \rightleftarrows B$. Then, solving the coupled eqs. for $\partial n_{A\sigma}$ and $\partial n_{B\sigma}$ using $\partial n_{i\sigma} = -\partial n_{i-\sigma}$ one finds[8]

$$\chi_A(x,T) = \frac{2\mu_B^2}{D} \left\{ (1 + U_B \frac{\partial n_B}{\partial \varepsilon_B}) \frac{\partial n_A}{\partial \varepsilon_F} - U_B \frac{\partial n_A}{\partial \varepsilon_B} \frac{\partial n_B}{\partial \varepsilon_F} \right\} \quad , \qquad \text{(III.3)}$$

where

$$D \equiv (1 + U_A \frac{\partial n_A}{\partial \varepsilon_A})(1 + U_B \frac{\partial n_B}{\partial \varepsilon_B}) - U_A U_B \frac{\partial n_A}{\partial \varepsilon_B} \frac{\partial n_B}{\partial \varepsilon_A} \quad . \qquad \text{(III.4)}$$

The expression for χ_B is obtained from χ_A by using $A \rightleftarrows B$. Note, for simplicity we assumed that the 5d-states are degenerate. Therefore, χ is the spin-susceptibility per atom for a single one of the five d-bonds. It follows from the general expressions above that in the weak scattering limit ($\delta \ll 1$) one obtains

$$\chi = 2\mu_B^2 \frac{N(o,x)}{1-U_{eff}N(o,x)} \quad , \tag{III.5}$$

where $N(o,x)$ refers to the average alloy density of states as $\varepsilon = \varepsilon_F$ and where

$$U_{eff}(x) = x\, U_A + (1-x)\, U_B - (U_A - U_{eff}) \frac{\chi_o}{1-U_{eff}\chi_o} (U_B - U_{eff}) \quad . \tag{III.6}$$

Here,

$$\chi_o \equiv \mathrm{Im} \int \frac{1}{\pi}\, d\varepsilon\, f(\varepsilon)\, F^2\, (\varepsilon - \Sigma) \quad .$$

Furthermore, for isoelectronic dilute alloys ($x \ll 1$) one finds

$$\chi(x) = \chi_B\, (x=o) + \Delta\chi \tag{III.7}$$

where

$$\chi_B = 2\mu_B^2\, N_B\, (o,x=o)/(1-U_B N_B(o,x=o)) \tag{III.8}$$

and

$$\Delta\chi = (U_A - U_B) \frac{\chi_B^2(x=o)/2\mu_B^2}{1-(U_A-U_B)\chi_o}\, x + \ldots . \tag{III.9}$$

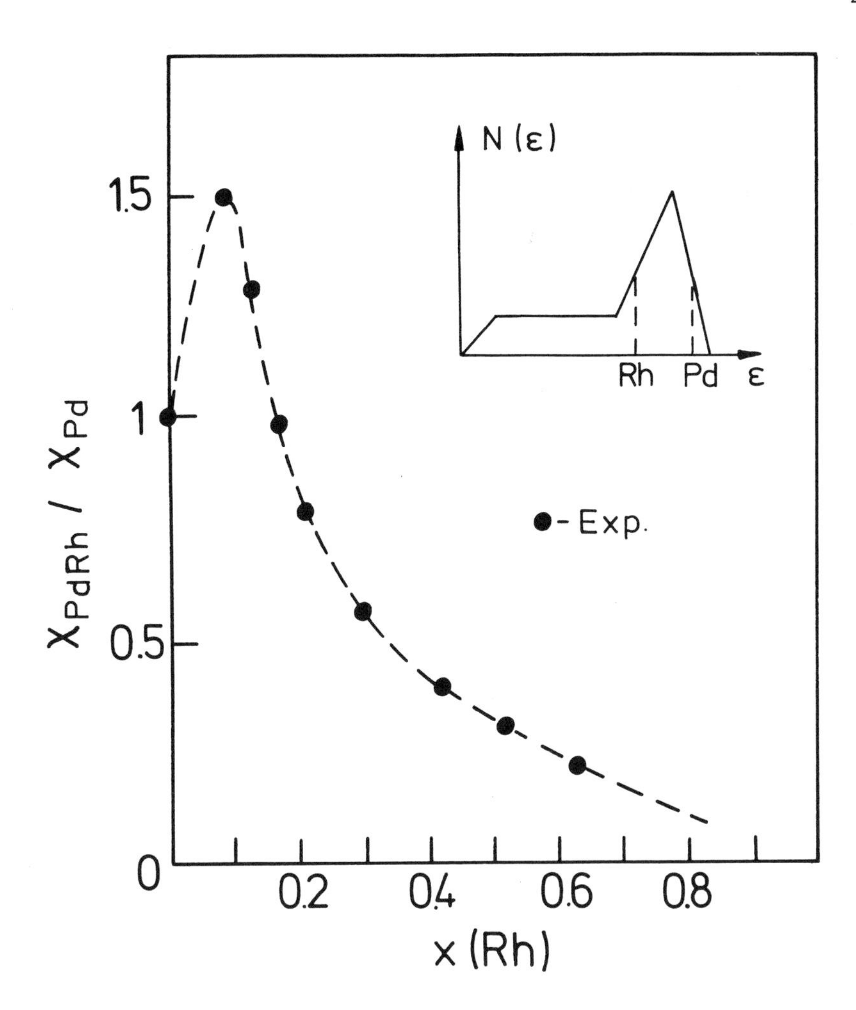

Fig. 9: Spin susceptibility χ in Rh_xPd_{1-x}.

Of course, the results for χ_i can also be derived by using the continued fraction-technique. Using instead of C.P.A. the Bethe-Peierls approximation it is possible to include local atomic environment in determining χ_i and local spin-polarization,

$$\mu_i \propto (n_{i\sigma} - n_{i-\sigma}) \quad .$$

(2) Metal-Insulator Transition in Alloys ($Cs_{1-x}Au_x$)

Here, we show briefly how the continued fraction method discussed in Sec. II can be used to describe the metal-insulator transition observed in liquid $Cs_{1-x}Au_x$. One expects intuitively that this transition results from electron charge transfer from $Cs \rightarrow Au$ ($CsAu \rightarrow Cs^{(+)}Au^{(-)}$) changing the originally metallic bonds to ionic bonds.

We use the Hamiltonian[9]

$$H = \sum_i \varepsilon_i n_i + \sum_{ij} t_{ij} c_i^+ c_j \quad , \qquad \text{(III.10)}$$

with

$$\varepsilon_i = \varepsilon_i^0 + U_0 \Delta n_i + 2 \sum_j U_{ij} \Delta n_j \quad . \qquad \text{(III.11)}$$

Here, $U_0 = U_A = U_B$ denotes the intra-atomic Coulomb interaction and U_{ij} the interatomic Coulomb interaction. The electron charge transfers are denoted by Δn_i. Charge neutrality gives

$$x \Delta n_A + (1-x) \Delta n_B = 0 \quad .$$

$N_i(\varepsilon)$, $N(\varepsilon)$ and Δn_i are determined from

$$G_{ii}(z) = \frac{1}{(z - \varepsilon_i - \Delta_i)} \quad .$$

Here, using the results derived in Sec. II we obtain for the electron self-energy Δ_i, the approximate expression

$$\Delta_A = P_{AA} t_{AA}^2 G_{AA} + P_{AB} t_{AB}^2 G_{BB} \quad .$$

In this expression P_{AA} denotes the probability to find the atom A at distance r from the Atom A. Similarly, P_{AB} is defined. t_{ij} denote the hopping integral of n.n. neighbors. Note, since the length of n.n. neighbor A^+A^+-bonds is larger than the A^+B^--bond length t_{AA} and t_{AB} are different.

We use

$$P_{AA} = n_1 (x + [1-x] \sigma_1) \quad ,$$

where σ_1 denotes the first shell shortrange order parameter. Note, it is important to take into account shortrange order, since physically one expects that as a result of electron charge transfer the positively charged up atom (Cs) prefers negatively charged atoms as n.n. neighbors. Fig. 10 shows results obtained by using the above formulae.

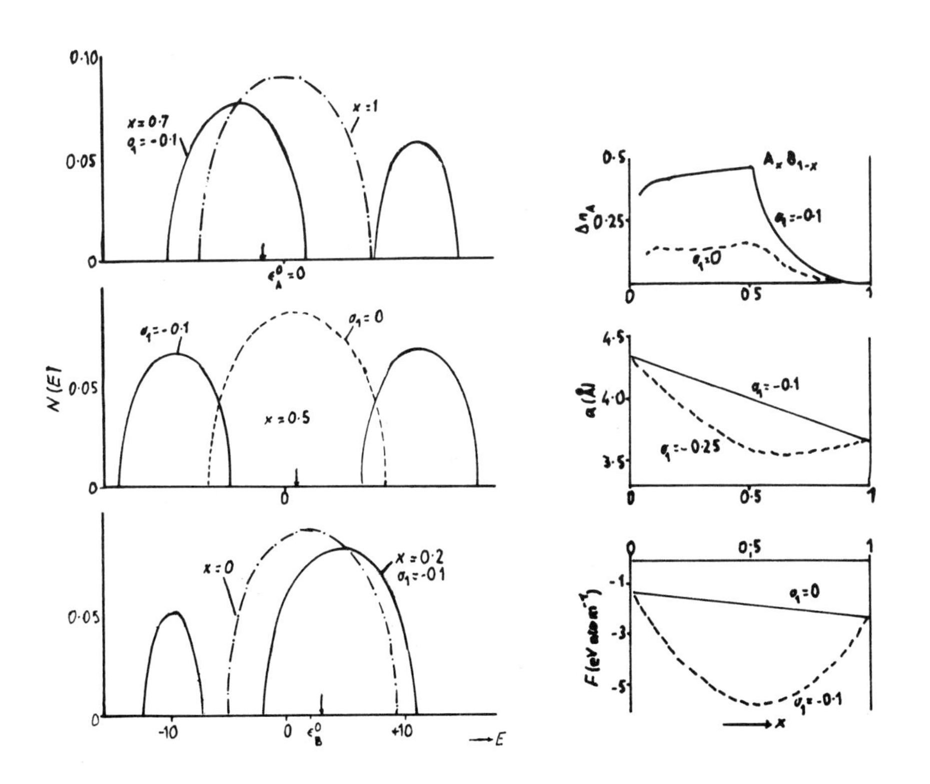

Fig. 10: Results for Δn_i and $N(\varepsilon, x)$ in $Cs_{1-x}Au_x$.

(3) Electron density of states in Liquids

As already mentioned the continued fraction method derived in Sec. II can be used also to treat electrons in liquids[10]. Then,

$$G_{ii}(z) = \frac{1}{z-\varepsilon_i-\Sigma} \quad , \tag{III.12}$$

where now due to the positional atomic disorder in the liquid the electron self-energy is given by

$$\Sigma(z) = \int \frac{d^3k}{(2\pi)^3} t_k t_k \, G\,(k,z) \quad . \tag{III.13}$$

Here,

$$t_k = \int d^3r \; t(r) \; g(r) \; e^{i\underline{k}\cdot\underline{r}} \quad . \tag{III.14}$$

$g(r)$ denotes the atomic correlation function. Note, for a random liquid with $g(r) = 1$ one recovers the results of Sec. II. The electron Green's function is given by

$$G(k,z) = \frac{1}{z-\varepsilon_0-nt_k-\Sigma} \tag{III.15}$$

Here, n is the atomic density and ε_0 the "atomic" energy level of the electrons. As a result of the positional disorder the electron density of states gets "washed out" relative to the crystalline case. The expressions given above apply also to amorphous metals.

In the case of liquid or amorphous alloys we have to treat in a combined way positional and "chemical" disorder. We write then

$$H = H_0 + \sum_i (\varepsilon_i-\varepsilon_0)n_i \quad , \tag{III.16}$$

where

$$H_o = \sum_i \varepsilon_o n_i + \sum_{ij} t_{ij} c_i^+ c_j$$

Note, H_O describes "liquid" atomic disorder and the 2. term in Eq. (III,16) the "chemical" disorder which approximately can be assumed to be independent of the liquid disorder. Thus, we have

$$\mathcal{G} = G + GtG + \ldots.$$

where $\mathcal{G}$ and G denote the Green's functions which have been averaged over all liquid configurations. Averaging then also over all alloy configurations, $\mathcal{G} \rightarrow \langle \mathcal{G} \rangle$ and using C.P.A., $\langle t_i \rangle = o$, one obtains

$$G_{ii\sigma}(z) = \sum_k \frac{1}{z - \varepsilon_\sigma^o - \Sigma_\sigma} \quad . \tag{III.17}$$

Here, Σ_σ is given by Eq. (III,13) and the average self-energy ε_σ^o due to scattering by the "chemical" disorder is given by the C.P.A. equation (II,15) using for F $(z - \varepsilon_\sigma^o)$ the expression

$$F_\sigma (z - \varepsilon_\sigma^o) = \int \frac{d\omega}{2\pi} \frac{N_\sigma^o(\omega)}{\omega - (z - \varepsilon_\sigma^o)} \quad ,$$

which is just the spectral representation of Eq. (III,17). The electron density of states N_σ^o refers to electrons with energy ε_o and hopping integral $t_{ij} = t$ in a liquid. Assuming that one can successively average over liquid- and chemical disorder this is the expected result.
Again, the average G and N(z) may be obtained from

$$G = x\, G_A + (1-x)\, G_B \quad .$$

The Fig. indicates what changes one might expect for $N(\varepsilon)$ in the liquid state.

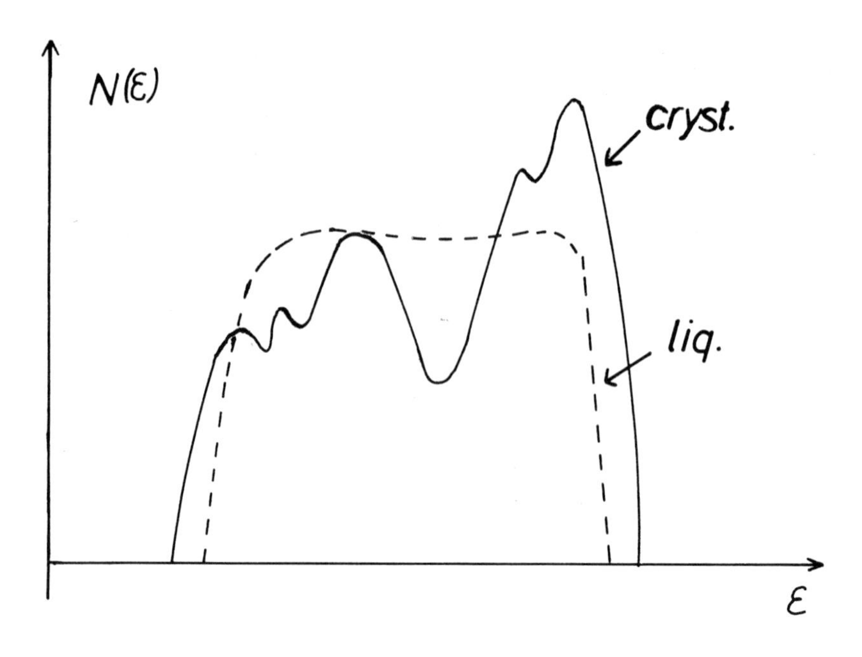

(4) Electronic density of states of amorphous Transition-Metal Alloys

The electronic structure of transition metal alloys and how it determines the alloy cohesion and how it relates to the possibility to form an amorphous alloy is of particular interest. Since only a few transition-metal alloys form crystalline solutions over a wide composition range, further studies are needed for clarifying the above questions and to understand more generally how alloy electronic density of states, $N(\varepsilon)$, can be interpreted in terms of a rigid band model or split band model, local density of states, $N_i(\varepsilon)$, and how the band width of the alloy components changes with alloy compositions.

Recently, it has become possible, to form various glassy alloys[11]. This can be used to study how the electronic structure of the alloy results from the one of the pure components. To help the understanding of glass formation it is of particular interest to study systems which components are from different ends of the periodic table (CuZr, PdZr,...) and to compare their electronic structure with the one observed for alloys with components which are neighbors in the periodic table and which do not form glasses (CuNi, AgPd,...).

For pure transition metals (d-band metals) $N(\varepsilon)$ has an overall structure which is generally understood as resulting from the covalent splitting of the e_g, t_{2g} states. Upon alloying additional "splitting" of d-states is expected to result from hybridization of the d-states of the two components. This additional splitting will be stronger the more unlike the atoms are and most clearly seen if the component $N_i(\varepsilon)$ do not completely overlapp. Decomposing the total $N(\varepsilon)$ into Lorentzian-like $N(\varepsilon)$ with center of gravities ε_i, then approximately ($\varphi_A, \varphi_B \rightarrow \varphi = a\,\varphi_A + b\,\varphi_B$) hybridization via the average hopping integral, t, leads to

$$\varepsilon_{1,2} = \frac{1}{2}(\varepsilon_A+\varepsilon_B) \pm \frac{1}{2}\sqrt{(\varepsilon_A-\varepsilon_B)^2+4t^2} \quad ,$$

or

$$\varepsilon_{1,2} = \begin{pmatrix} \varepsilon_A \\ \varepsilon_B \end{pmatrix} \pm \frac{t^2}{\varepsilon_B-\varepsilon_A}$$

if ε_1 is close to ε_A and ε_2 close to ε_B. Thus, as indicated in the Fig.11 one expects for states with same energy that they are split apart by 2t and that in the alloy a "narrowing" results not only from $W_i \sim \sqrt{n_i t_i}$ (W_i = band width of component i with hopping integral t_i and n_i nearest neighbors of same kind) but also strongly from the distortion of $N(\varepsilon)$ due to the "repulsion" of d-states of unlike atoms[12]. It follows directly from the above simple formula that this distortion depends only weakly on composition and increases when the energy difference of the different d-states decreases. Thus, states around ε_2, s. Fig. 11, are expected to be more repelled than states around ε_1. Consequently, an apparent "narrowing" of the d-bond results due to this non-uniform distortion of $N_i(\varepsilon)$. Since these shifts and distortions of $N_i(\varepsilon)$ will lead to changes in $N(\varepsilon_F = o)$, one expects changes in the screening of the ionic potentials by the s,d valence electrons and thus shifts in core states ($\varepsilon_{3d}, \ldots$) and of their density of states symmetry[13]. Furthermore, shifts in $N_i(o)$ cause changes in the magnetic properties ($\chi_i, \mu_i, \ldots$), chemisorption, etc.

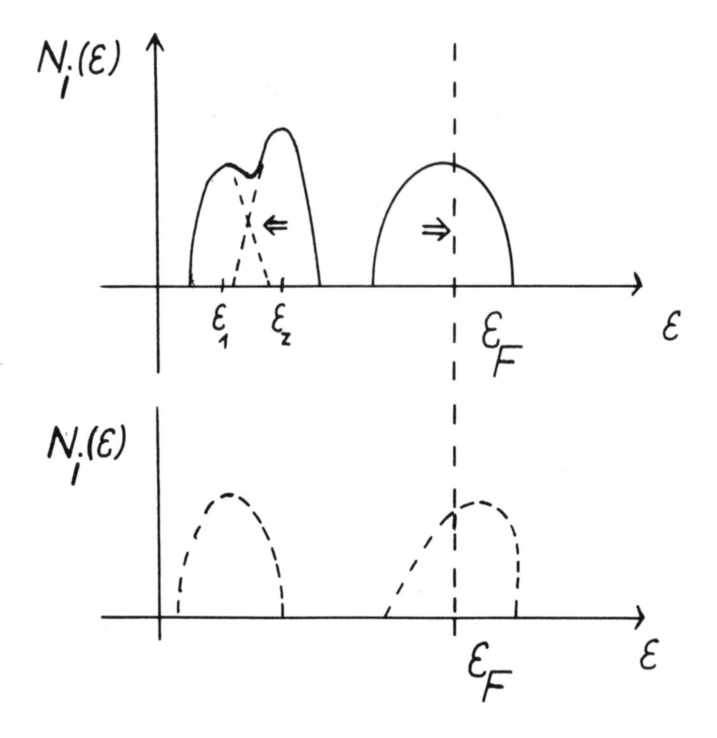

Fig. 11: "Repulsion" of d-states

In Fig. 12 experimental results[11] for the electron density of states of amorphous Cu_xZr_{1-x} are shown. Upon alloying Cu with Zr one notes that the Cu d-states with lowest binding energy "disappear" respectively, get pushed towards higher binding energies and the opposite seems to happen with the Zr d-states. This repulsion is not strongly dependent on composition.

$$\varepsilon_F^{Zr}\ (\text{alloy}) \approx \varepsilon_F^{Zr} \quad \text{and} \quad \varepsilon_{B-d}^{Cu}\ (x) \approx \varepsilon_B^{Cu}$$

In Fig. 13 experimental results[11] for $N(\varepsilon, x)$ of amorphous Zr_xPd_{1-x} alloys are shown. Note, upon alloying the Pd-states seem to get pushed away from ε_F to higher binding energies. $N(\varepsilon)$ for $- 2\ \text{ev} \leq \varepsilon \leq \varepsilon_F$ is nearly independent of x. Again, $\varepsilon_{B-d}^{Pd}(x) \approx \varepsilon_{B-d}^{Pd}$ and $\varepsilon_F^{Zr}\ (x)\ \varepsilon_F^{Zr}$.

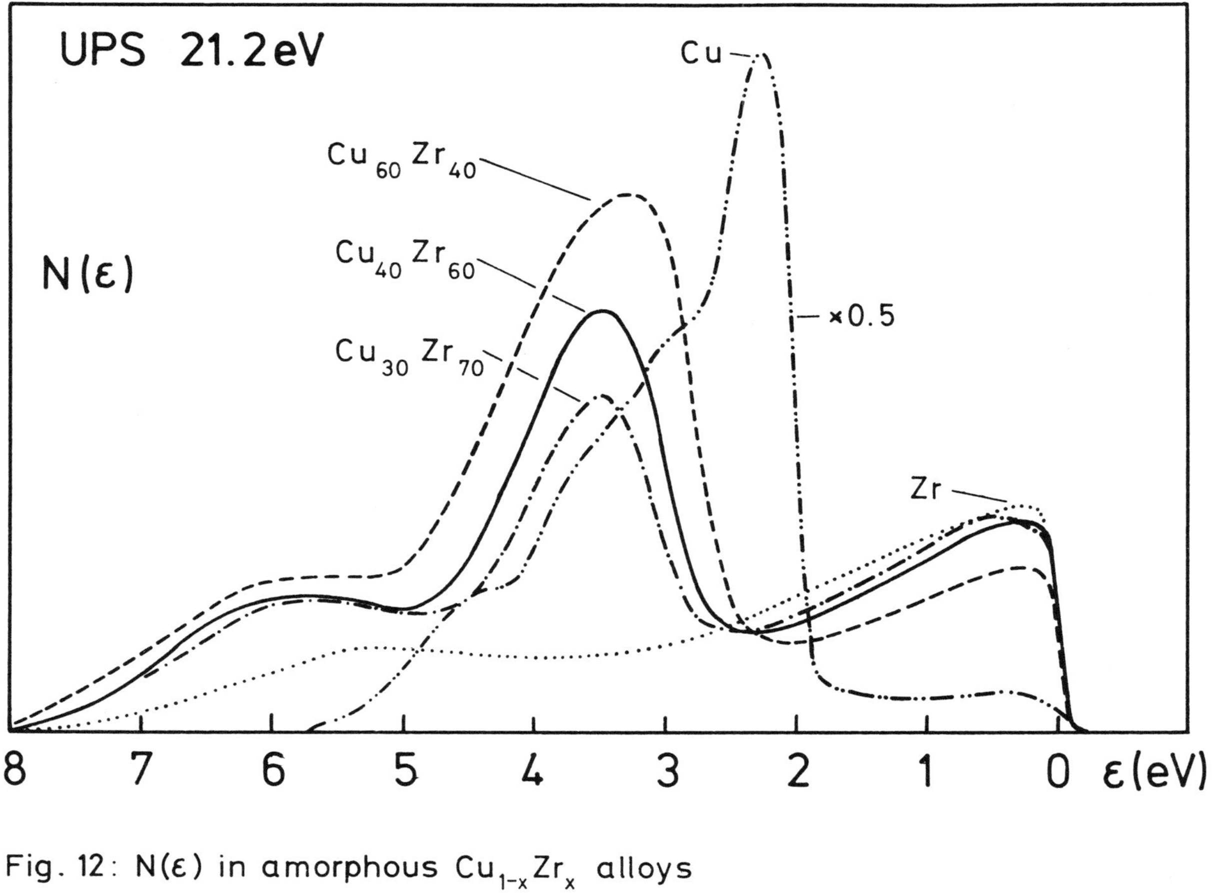

Fig. 12: N(ε) in amorphous $Cu_{1-x}Zr_x$ alloys

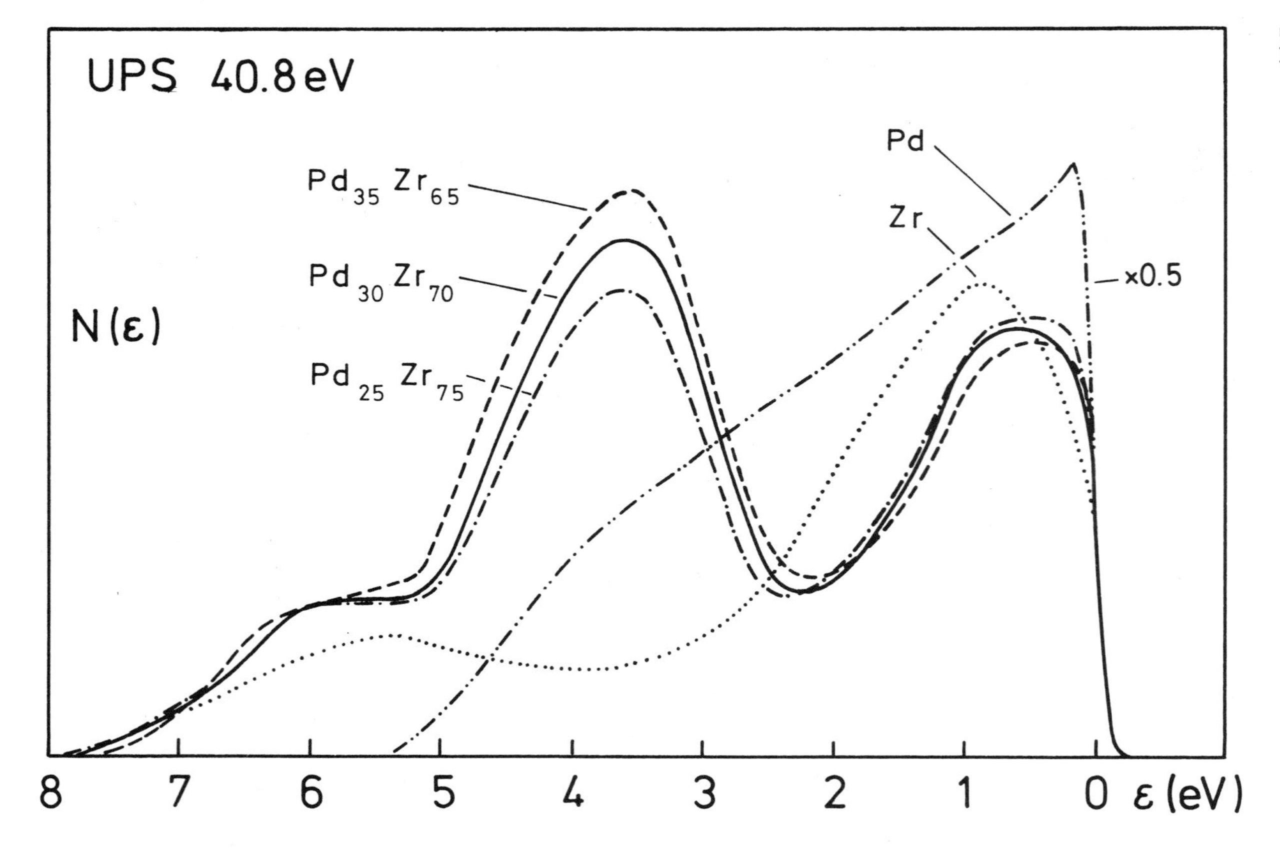

Fig. 13: N(ε) in amorphous $Pd_{1-x}Zr_x$ alloys

In the following we attempt to analyze these results in terms of changes in the Fermi energy F and d-state "repulsion" caused by hybridization of d-states located at different atoms. We use for the analysis the parameters

	t (eV)	ε_d^1	ε_d^2	ε_d(eV)	ε_F(eV)
Cu	~0.9	-3.8	-2.5	- 3.4	4.4
Zr	~1.8	-1.7	~2	0.2	3.9
Pd	1.35	~-2.5	~-0.4	- 2.15	4.8

which are taken from band calculations. Then the following band distortion results.

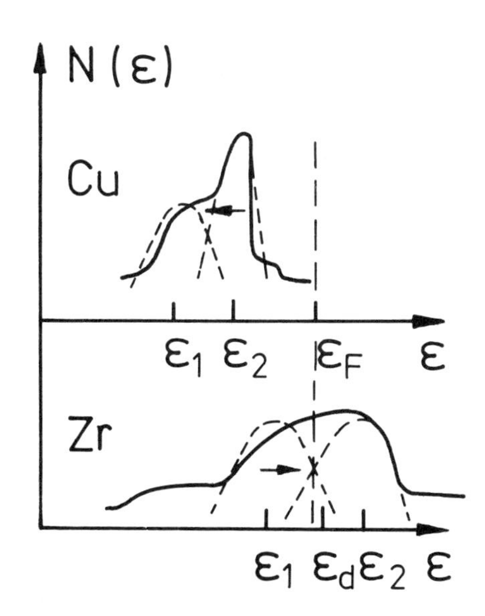

Fig. 14: Illustration of N(ε) for Cu_xZr_{1-x}. Note, energy scales are related by putting

$$\varepsilon_F(x) = x\,\varepsilon_F^{Cu} + (1-x)\,\varepsilon_F^{Zr}$$

Clearly, states around ε_2^{Cu} are more repelled than those around $\varepsilon_1{}^{Cu}$. We use t=1.35 eV. Then, interaction between $\varepsilon \approx \varepsilon_2{}^{Cu}$ states and $\varepsilon \approx \varepsilon_1{}^{Zr}$ gives a shift

$$\Delta \varepsilon_2^{Cu} \simeq 0.7 \text{ eV}$$

If the shifted states also interact with Zr-states around ε_2^{Zr} then approximately a total shift

$$\Delta \varepsilon_2^{Cu} \simeq 1.1 \text{ eV}$$

results. To this a shift $\Delta\varepsilon \lesssim 0.3$ eV due to $\Delta\varepsilon_F$ should be added ($x \approx 0.4$). This simple analysis predicts

$$\Delta \varepsilon_B^{Cu} \approx 0.3 \text{ eV} \quad .$$

Thus, the resultant physical picture for N(ε) alloy is that Cu d-states around $\varepsilon_2{}^{Cu}$ are pushed to higher binding energies into the peak of $N_{Cu}(\varepsilon)$ around $\varepsilon_1{}^{Cu}$ and the change in the apparent Cu d-band width results strongly from this d-state repulsion. The energy shifts of the Zr d-states are in "opposite" direction and should give rise to N (ε) shape distortions above ε_F which should be observable in optical experiments. Of course, the shifted states result from hybridization and should only approximately be interpreted as Cu or Zr shifted states.

In the case of amorphous Zr_xPd_{1-x} the shape of N(ε,x) can be understood as follows. We use t = 1.57 eV

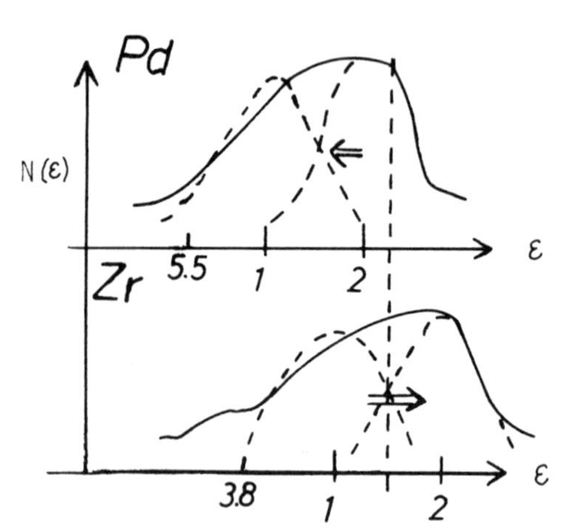

Fig. 15: N(ε) for Zr_xPd_{1-x}.

Since states around ε_2^{Pd} and d-states of Zr around ε_1^{Zr} overlapp one finds

$$\Delta\varepsilon_2^{Pd} \simeq 1.5 \text{ eV}$$

and a shift of states around ε_1^{Zr} in opposite direction. To this a (interaction between states around $\varepsilon_2{}^{Pd}$ and $\varepsilon_d{}^{Zr}$ gives $\Delta\varepsilon_2{}^{Pd} \approx 1.1$ eV) shift $\Delta\varepsilon \approx 0.6$ eV should be added ($x \approx 0.7$) due to $\Delta\varepsilon_F$. Furthermore, $\Delta\varepsilon_B{}^{Pd} \approx 0.3$ eV may be estimated.

Thus, we have the following physical picture for $N(\varepsilon)_{alloy}$. Pd-states around $\varepsilon_2{}^{Pd}$ are pushed away from ε_F by about 1.5 eV, more than Pd-states at higher binding energies. This contributes strongly to the apparent Pd d-band width ($\delta W \sim \sqrt{0.7\ 12}\cdot 1.35$). Zr-band distortion should be most apparent above ε_F. Since states at $-2 \leq \varepsilon \leq 0$ consists of Pd and Zr states $N(\varepsilon)_{alloy}$ is fairly independent of x in that energy range.

Some interesting consequences of $N(\varepsilon)$ distortions should be observed:

Core level shifts: Since the "valence" d-states contribute to the screening of the ionic potentials, a change in $N_i(\varepsilon)$, (i = A,B) better the local density of states of A and B metal in the alloy, will cause shifts in core levels. In particular, using

$$V(r) = -\frac{Ze^2}{4\pi r}\, e^{-\lambda r} \quad ,$$

$$\lambda = \sqrt{4\pi e^2 N(o)} \quad ,$$

and writing

$$N_i^{alloy} = N_i + \delta N_i$$

one expects from a Taylor expansion in terms of $\delta N_i(o)$ a shift

$$\Delta\varepsilon_{3d}^i \sim \delta N_i(o)$$

due to a change in screening.

Assymmetry of $N(\varepsilon)$ of core levels: Furthermore, since assymetry of core level $N(\varepsilon)$ is proportional to $N_i(o)$, again via Taylor expansion one expects

$$\delta(\tfrac{a}{b}) \sim \delta N_i(o)$$

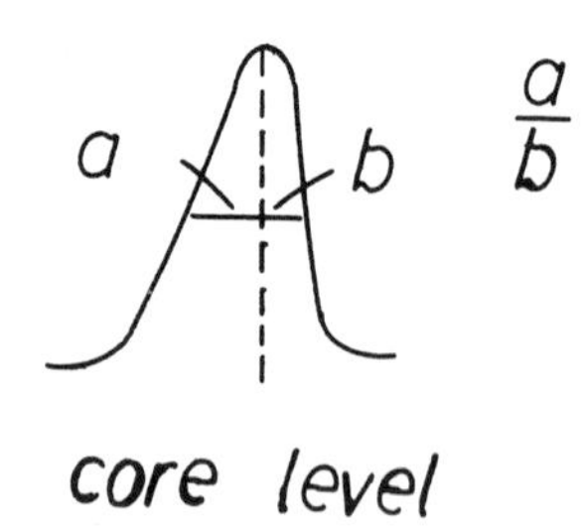

Fig. 16: Illustration of core-level assymmetry.

Spin-susceptibility: Since

$$\chi_i = \frac{\chi_i^o}{1-U_i\chi_i^o}\ , \quad \chi_i^o \sim N_i(o)$$

one expects changes in χ_i due to change in $N_i(o)$.

Chemisorption: Due to changes in $N_i(o)$, $N_i(\varepsilon)$, chemisorption capacity should change. For example, in view of the results for PdZr one expects that PdZr due to Zr oxidizes more rapidly than Pd.

Experimental results for core level shifts and assymmetry give for Cu_xZr_{1-x} alloys for $3d^{Zr}_{5/2}$ a level shift $\Delta\varepsilon \simeq 0.1$ eV and the assymmetry

$$\delta(\tfrac{a}{b}) \simeq 0.1$$

Consequently, $N_{Zr}(o)$ in the alloy must have changed only slightly to a lower value. In the case of Pd_xZr_{1-x} one observes for

$$3d^{Zr}_{5/2}$$

a level shift $\Delta\varepsilon \simeq 0.2$ eV and the assymmetry $\delta(\frac{a}{b}) \simeq 0$.

The level shift of $3d^{Pd}_{5/2}$ is $\Delta\varepsilon \simeq 1.5$ eV and the assymmetry

$$\delta(\tfrac{a}{b}) \simeq 0.8 \text{ eV}.$$

Thus, if distortion is interpreted as pushing down of Pd d-states and pushing up of Zr d-states, one has lower $N_{Pd}(o)$ and bigger (slightly) $N_{Zr}(o)$ in PdZr alloys. Consequently, in view of the experimental results one expects large core level shift and assymmetry change for Pd in PdZr due to the strong distortion of the local Pd density of states. Furthermore, then $\chi_{alloy}(Pd)$

$$(\chi_{alloy} = x\,\chi_{Pd} + (1-x)\,\chi_{Zr})\ .$$

should be much smaller than χ_{Pd}. Possibly Fe impurities in PdZr alloys should not induce "giant" moments. These results show that the observed distortions of $N(\varepsilon)$ can be also interpreted in terms of local density of states.

Comparing the distortions in $N_i(\varepsilon)$ observed for PdZr, CuZr with those observed for CuNi, AgPd, which are fairly small, one may speculate that these distortions play a significant role for the formation of glasses. Clearly, in comparing crystalline alloys and amorphous alloys one should take into account that $t \sim e^{-qr}$ (due to volume changes) and Z_i change. Clearly, the (effects observed for CuZr, PdZr) distortions of $N(\varepsilon)$ due to alloying should be small for like atoms (same ε_d, t).

In summary, we have shown how to interprete the electronic structure of amorphous transition-metal alloys. Our results are consistent with core level shifts, changes of their assymmetry. Further experiments are suggested. The studies described should be useful for understanding the d-electron contribution to the stability of glasses.

(5) Itinerant Ferromagnetism-Alloy treatment

For many years itinerant ferromagnetism in Ni, Co, Fe etc. has been studied. Yet no completely satisfactory theory describing properly local and non-local behaviour of magnetism in transition metals has been presented. Only very recently the first promising calculations of the Curie-temperature T_c, of short-range spin correlations above T_c, of magnetic moments and Zeeman-splitting above T_c have been published[13]. In the following we outline briefly by using alloy-theory how an itinerant magnetic theory with local magnetic moments and Zeeman splitting in $N(\varepsilon)$ above T_c can be developed.[13,14]. For this alloy theory is relevant since the Hamiltonian

$$H = \sum_{i,\sigma} \varepsilon_{i\sigma} n_{i\sigma} + \sum_{i,j} t_{ij} d_i^+ d_j + \ldots, \qquad \text{(III.18)}$$

with

$$\varepsilon_{i\sigma} = \varepsilon_i^0 + U_i n_{i-\sigma} \quad ,$$

can be viewed as a fictitious alloy. Note, an electron with spin σ sees an alloy $A_{n-\sigma} B_{1-n-\sigma}$ where A denotes atoms occupied by one electron and B denotes atoms with no electrons. Clearly, the concentration of atoms A is $n_{-\sigma}$, which is the average number of electrons with spin $(-\sigma)$ per atom. Similarly, electrons with spin$(-\sigma)$ see a fictitious alloy $A_{n_\sigma} B_{1-n_\sigma}$. Rewriting

H as $H = H_o + V$, where

$$H_o \equiv \sum_{i,\sigma} \Sigma_\sigma n_i + \Sigma t_{ij} d_i^+ d_j$$ and

$$V \equiv \sum_{i,\sigma} (\varepsilon_{i\sigma} - \Sigma_\sigma) n_{i\sigma} \quad ,$$ one may apply C.P.A. and

determine $\Sigma_\sigma(\varepsilon)$ from $\langle H \rangle = \langle H_0 \rangle$. To include important "short range" order spin-correlations (above T_C) we may use the continued-fraction-local environment theory. Then,

$$G_{ii\sigma}(\varepsilon) = (\varepsilon - \varepsilon_{i\sigma} - \Delta_{i\sigma})^{-1} \qquad \text{(III.19)}$$

and the average Green's function is

$$G_\sigma(\varepsilon) = n_{-\sigma} G_{AA\sigma}(\varepsilon) + (1 - n_{-\sigma}) G_{BB\sigma}(\varepsilon) \quad , \qquad \text{(III.20)}$$

Here, the spin-dependent self-energy $\Delta_{i\sigma}$ is determined according to C.P.A. theory as previously outlined. To include local spin configuration environment we use in determining $\Delta_{i\sigma}$ that

$$P_{AA}^\sigma = n_1 \; (n_{-\sigma} + [1 - n_{-\sigma}] \, \sigma_1) \quad .$$

Here, P_{AA}^σ is the probability for finding in the nearest-neighbor shell with n_1 atoms around the atom A an atom A also with spin σ. σ_1, σ_2, ... denote short-range spin order parameters referring to the nearest-, overnext nearest neighbor shell, etc. These are defined as usually in alloy theory. Determining now the free energy $F = U - TS$, where the entropy S is given by the usual gas mixing entropy expression or, what is far better, by the Kichuchi pair-entropy expression, then the short-range spin order parameters σ_1, σ_2 etc. are calculated from

$$\partial F / \partial \sigma_i = 0 + \text{CORR.} \qquad \text{(III.21)}$$

Here, the corrections arise from Lagrange multipliers due to electron spin conservation. Now, we determine the average magnetization

$$M \propto (n_{\uparrow} - n_{\downarrow}) \quad ,$$

where the average number of electrons with spin σ

$$n_{\sigma} = \int d\varepsilon \, N_{\sigma}(\varepsilon) \, f(\varepsilon)$$

is determined from $G_{\sigma}(\varepsilon)$. The local spin-polarization and magnetic moments are determined by

$$\mu_i(T) \propto (n_{i\uparrow} - n_{i\downarrow}) \quad ,$$

where $$n_{i\sigma} = \int d\varepsilon \, N_{i\sigma}(\varepsilon) \, f(\varepsilon) \quad .$$

Then, calculating T_C from $M(T)=0$ and the disappearance of μ_i at T_{μ} from $\mu_i(T) = 0$ one finds in general (depending on U/W,...) the result $T_C < T_{\mu}$ and that Zeeman-splitting "survives" in $N_i(\varepsilon)$ above T_C. Further details of this theory can be obtained in a straightforward way. A most interesting check of the theory would be its application to FeCr alloys and similar alloys exhibiting phase-diagrams involving ferromagnetism and anti-ferromagnetism.

REFERENCES

1. P. Soven, Phys. Rev. 156, 809
   and Phys. Rev. 178, 1136.
2. B. Velický et al., Phys. Rev. 175, 747.
3. F. Brouers et al., Phys. Rev. B 7, 4370.
4. K. Watson, Phys. Rev. 105, 1388.
5. R. Haydock et al., J. Phys. C 5, 2845.
6. C.D. Gelatt et al., Phys. Rev. B 10, 398.
7. L. Hodges et al., Phys. Rev. B 5, 3953.
8. K. Levin et al., Phys. Rev. Lett. 27, 589,
   Phys. Rev. B 6, 1865.
   J. Kamamori et al., J. Phys. Soc. Jap. 31, 382.
9. A. ten-Bosch et al., J. Phys. C 11, 2959.
10. B. Movaghar et al., J. Phys. F 4, 687.
11. P. Oelhafen et al., to be published.

12. J.F. Janak et al., Phys. Rev. B 12, 1257.
13. J. Hubbard, Phys. Rev. B 19, 2626,
    H. Hasegawa, J. Phys. Soc. Jap. 46, 1504.
14. K. Levin and K.H. Bennemann, Phys. Rev. B 5, 3770.

SUPERCONDUCTIVITY IN AMORPHOUS TRANSITION-METALS

K.H. Bennemann

Institute for Theoretical Physics, FU Berlin
1000 Berlin 33

ABSTRACT. A superconductivity in amorphous transition-metals is discussed. A simple theory is presented providing a physical understanding of how atomic disorder affects superconductivity.

I. INTRODUCTION

Superconductivity in transition-metals and alloys thereof is of particular interest due to continuing interest in high superconducting transition-temperatures T_c and large critical magnetic fields H_c. By studying amorphous transition-metals one hopes to learn how important atomic structure, crystal symmetry etc. is for various superconductivity properties (surface superconductivity, type II superconductivity - vortex structure, etc.)

In the following we describe first the experimental situation. In Fig. 1 we show how T_c varies as a function of valence in crystalline and vapour quenched amorphous transition-metals and their alloys[1]. Note, that in contrast to the behaviour of simple metals with $T_c \geq T_c^0$ one observes $T_c \gtrless T_c^0$, where T_c^0 refers to the crystalline metals. Furthermore, note that in the amorphous metals the valence dependence of T_c is rather different than what it is in the crystalline metals (Matthias-rules). In tables 1 and 2 further

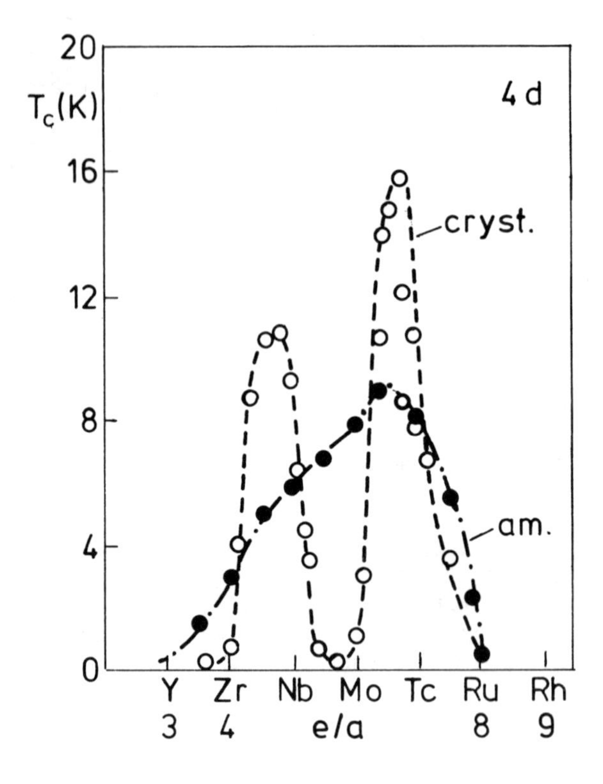

Fig. 1 The superconducting transition-temperature T_C as a function of valence in amorphous transition-metals.

results for T_C in amorphous A-15 compounds[2] and liquid quenched transition-metal alloys[3] are shown.

Amorphous A - 15

	$T_c(k)$	$T_c^o(k)$
Nb_3Ge	3.9	21.8
Nb_3Si	3.8	18
Nb_3Sn	~3	18
V_3Si	?	17.1

Table 1: Observed values for the superconducting transition temperature T_c in amorphous A-15 compounds

The results for T_c of amorphous transition-metal alloys containing Ni, Pd, etc. atoms are particularly interesting since they show also how magnetism changes upon alloying and disordering.

Amorphous alloys

	$T_c(k)$	$T_c^o(k)$
La(β)	?	6.1
$La_{.7}Cu_{.3}$	3.5	
$La_{.78}Ni_{.22}$	3	
Zr	~3	0.7
$Zr_{.65}Pd_{.35}$	3.5	
$Zr_{.7}Be_{.3}$	2.8	
Pd	?	

Table 2: T_c in liquid quenched amorphous transition-metal alloys.

In view of the increase of T_c in Zr upon adding Pd one might speculate as indicated in the Table 2 about superconductivity in "amorphous" Pd.

II. THEORY

In the following a simplified theory for the superconducting transition-temperature in highly disordered d-electron superconductors is presented. Our main purpose is to understand how atomic disorder affects the most important electronic parameters and phonons which determine superconductivity.

First, we study the effect of atomic disorder on the electron-lattice coupling characterized by the coupling constant[4,5]

$$\lambda = N(o) \frac{J^2}{M\langle\omega^2\rangle} \qquad \text{(II.1)}$$

where N(o) denotes the electronic density of states at the Fermi-energy $\mathcal{E}_F$, M the ionic mass per unit volume, J^2 the sum over all electron-lattice matrix-element products of the form[5]

$$\langle i | \nabla V_l | m \rangle \langle m | \nabla V_{l'} | j \rangle \qquad \text{(II.2)}$$

resulting from interatomic transitions $i \rightarrow m$, $m \rightarrow j$ due to the vibrating atoms l and l' with potentials V_l and $V_{l'}$. The phonon spectrum is characterized by[4,5]

$$\langle \omega^2 \rangle^{-1} = \int d\omega \frac{\alpha^2 F}{\omega} \Big| \int d\omega\, \omega \alpha^2 F$$

Atomic disorder will change J^2, $\langle\omega^2\rangle$ and N(o). We expect that in amorphous metals the density of states for phonons $D(\omega)$ and for electrons $N(\mathcal{E})$ will approximately change as illustrated in Figs. 2 and 3.

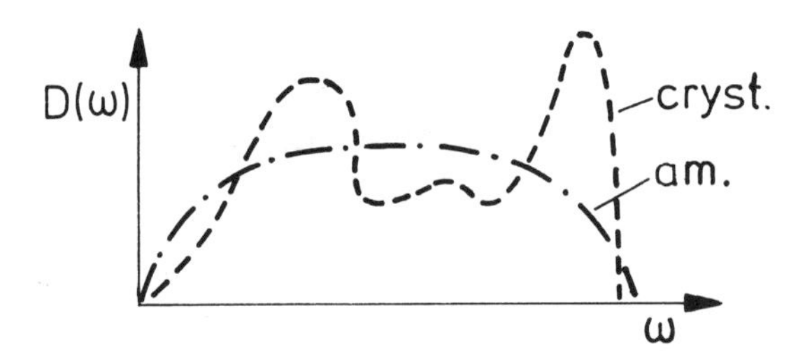

Fig. 2 Illustration of change of phonon density of states $D(\omega)$ due to atomic disorder

Note, one should expect that transverse-phonons are affected differently than longitudinal phonons. This would be significant to take into account if the electron-lattice coupling constant is rather different for coupling to transverse or longitudinal models.

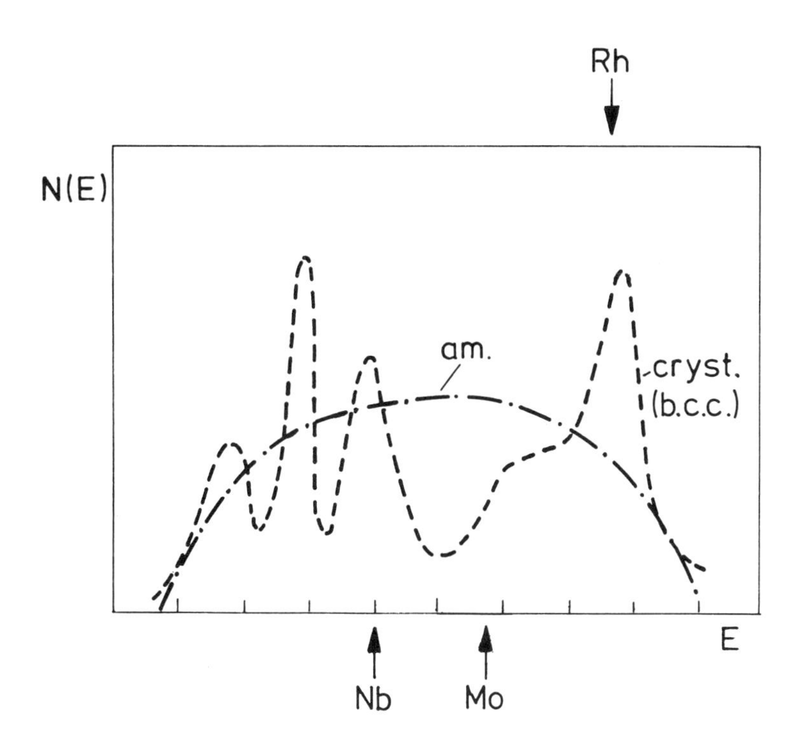

Fig. 3 Illustration of change of electron-density of states $N(\varepsilon)$ due to atomic disorder

How reasonable the simplifying approximation for $N(\varepsilon)$ in amorphous transition-metals is could be checked by NMR-specific-heat experiments, etc. In Table 3 resultant values for $N(o)/N_o(o)$ are shown.

	$N(o)/N_o(o)$
La(β)	~1
Zr	1.2
Nb	0.9
Mo	~2
Tc	~1
Ru	1.3
Rh	~0.5

Table 3: The ratio $(N(o)/N_o(o))$ for the electronic density of states obtained by approximating $N(\varepsilon)$ as shown in Fig. 3.

The quantity $M\langle\omega^2\rangle$ in amorphous metals is determined by using

$$M\langle\omega^2\rangle > \alpha T_m \quad , \tag{II.3}$$

where T_m denotes the melting temperature. In Fig. 4 results are shown for $M\langle\omega^2\rangle$ as a function of valence. Note, that $(M\langle\omega^2\rangle)^{-1}$

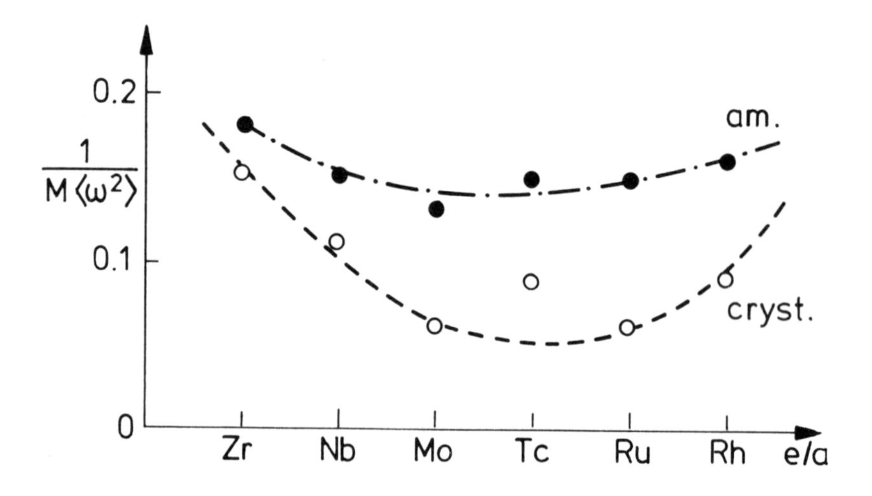

Fig. 4 Valence dependence of $(M\langle\omega^2\rangle)^{-1}$.

varies much smoother with valence in amorphous metals than in crystalline metals. This, of course, is expected if $D(\omega)$ in amorphous metals is approximately given as shown in Fig. 2.

The important electronic parameter J^2 is determined by noting that[5]

$$J^2 \simeq f \frac{\eta(\Omega)}{K^2} \quad , \qquad \text{(II.4)}$$

where k is the dielectric screening function and f a factor which does not vary much with volume Ω and valence. The dependence of η on valence is shown in Fig. 5 for crystalline transition-metals. η is calculated[5] from T_c^0. Since one finds in general that due to atomic disorder the atomic volume changes as

$$\Omega_0 \rightarrow \Omega = \Omega_0 + \Delta\Omega$$

with

$$\Delta\Omega \propto \Delta\Omega_m \quad ,$$

one should use for amorphous metals

$$\eta(\Omega) = \eta_0 (\Omega_0) + \Delta\eta.$$

Note, $\Delta\Omega_m$ denotes the volume change observed upon melting and

$$\Delta\eta = \frac{\partial\eta}{\partial\Omega} \Delta\Omega \quad . \qquad \text{(II.5)}$$

$\partial\eta / \partial\Omega$ is determined as in the case of pressure dependence of T_c^0 [6,7]. Combining now these results the change of the electron-phonon coupling constant λ due to atomic disorder can be calculated. Furthermore, by using the Mc.Millan formula[4] for T_c

$$T_c \simeq \frac{\langle\omega^2\rangle^{1/2}}{1.2} \exp\left[- \frac{1+\lambda+\mu_{spin}}{0.96\lambda-(1+0.6\lambda)(\mu^*+\mu_{spin})}\right]$$

which should be accurate enough to determine the relative change of T_c and its valence dependence, one can calculate also T_c in amorphous metals. Results obtained this way are shown in Table 4.

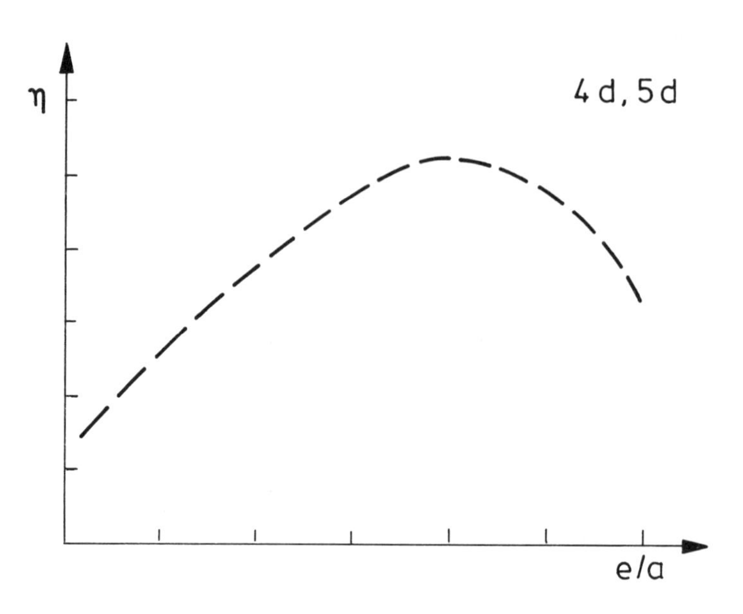

Fig. 5 The valence-dependence of η in transition-metals.

4d	λ_o	λ	calc. T_c(k)	exp. T_c(k)
Zr	~0.4	~0.5	1.5	3.5
Nb	0.9	~0.8	7.3	~6
Mo	0.4	~0.9	8.6	8.5
Tc	~0.7	~0.9	9	8
Ru	0.4	~0.6	~2	$\lesssim$2
Rh	-	~0.7	~4	?

Table 4: Numerical results for λ and T_c using $\mu \approx 0.1$ and $\mu_{spin} \simeq 0$.

For determining T_c in amorphous A-15 compounds we assume that structural change-destruction of quasi 1d-transition-metal atomic chains is dominant. Consequently, amorphous A_3B should have similar electronic structures as the corresponding $A_{0.75}B_{0.25}$ alloy. Also, using $M\langle\omega^2\rangle \propto \Theta_D$ (Debye-temperature) or $M\langle\omega^2\rangle \propto T_m$, one may estimate $M\langle\omega^2\rangle$ for an amorphous A-15 compound from $M\langle\omega^2\rangle$ for the corresponding alloy. For this alloy we use

$$N(o) = x\, N_A(o) + (1-x)\, N_B(o)$$

Also, due to

$$\eta \propto \left| \sum_{ij} \{ (\ldots) \langle i | \nabla V_i | j \rangle + (\ldots) \langle i | \nabla V_j | i \rangle \} \right|^2$$

η is expected to vary between x and x^2. Thus, we may estimate[5] $T_c(x)$ for $A_{0.75}B_{0.25}$ from λ, etc. if known for A and B metal. Thus, we obtain results shown in Table 5.

	$T_c^o(k)$	$T_c^{exp.}$	$T_c^{calc.}$	calc. $N(o)/N_o$	exp. N/N_o
Nb_3Ge	21.8	3.9	~4	~0.4	~0.5
Nb_3Sn	18	~3	~4	~0.5	~0.3
Nb_3Si		3.8	~4		
V_3Si	17.1		~6	~0.4	
Zr_3Rh		4.5+	~2.2		
V_3Au	3		~6	~0.7	
$Mo_{.68}Si_{.32}$		6.7	~5.6		

+$T_c(Zr)$ = 3.5k

Table 5: Results for $N(o)/N_o(o)$ and T_c in disordered A-15 systems.

For determining T_c in amorphous alloys containing Ni, Pd, etc. we have to estimate[5,8]

$$\mu_{spin}\ (N(o)) \quad ,$$

which measures the strength of electron-local spin interaction. Note, μ_{spin} tends to suppress superconductivity. One expects[8] that μ_{spin} (N(o)) varies as shown in Fig. 6. Note, due to

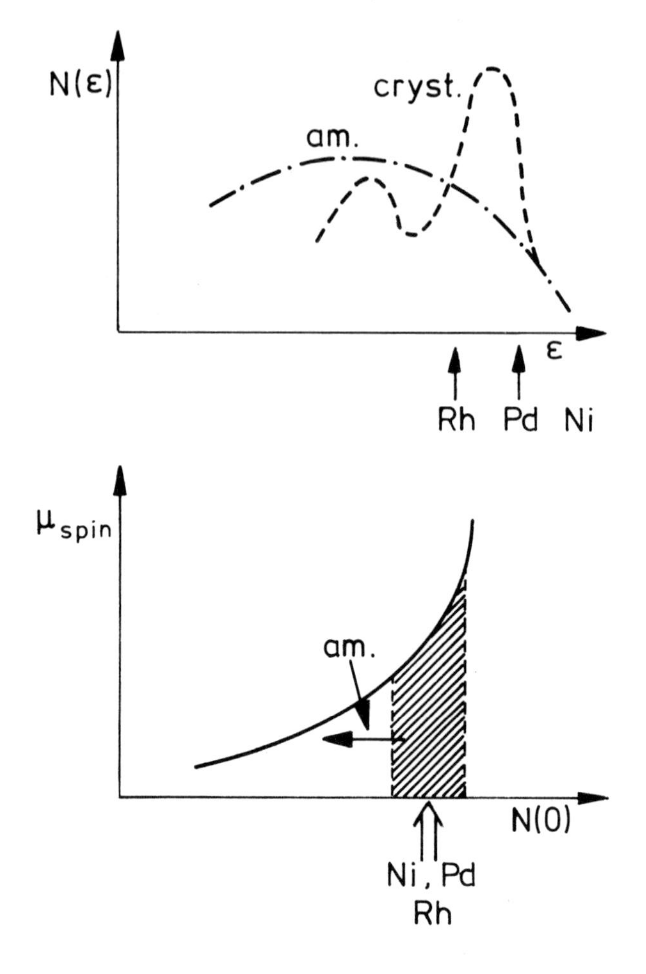

Fig. 6 Illustration of the dependence of μ_{spin} on N(o).

$N_o(o) \rightarrow N(o)$ upon disorder, large changes $\mu^o_{spin} \rightarrow \mu_{spin}$ may result. Thus, we estimate for Pd, Rh, Ni that $\mu_{spin} \simeq 0$. Thus, one finds for T_c the results shown in Table 6. Note, we neglect contributions to λ due to Cu, Ni, etc. as suggested by $T_c(x)$ shown in Fig. 7 and also plausible if Ni d-states as the Cu d-states are filled. In the case of $Zr_{.65}Pd_{0.35}$ we assume

	calc. T_c(k)	exp. T_c(k)
La	~5	--
$La_{.7}Cu_{.3}$	~4	3.5
$La_{.79}Ni_{.22}$	~4	3
Zr	1.5	~3
$Zr_{.65}Pd_{.35}$	~4	3.5
$Zr_{.7}Be_{.3}$	~2	2.8

Table 6 Results for T_c in amorphous alloys.

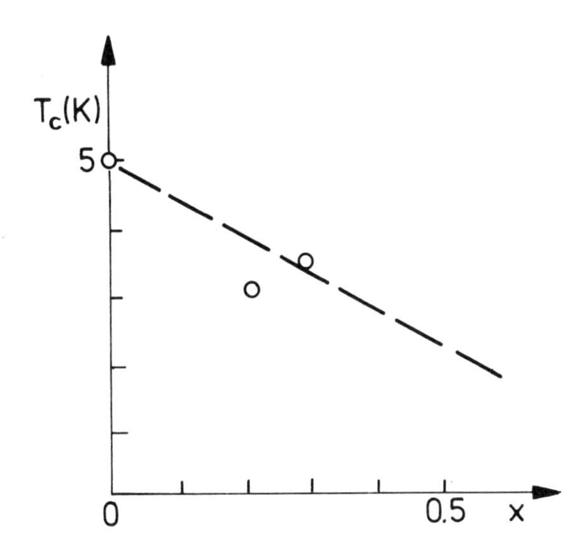

Fig. 7: Experimental results for $T_c(x)$

Pd d-states at ε_F, $\mu_{spin} = 0$ and $\lambda_{Zr} \simeq \lambda_{Pd}$. ($T_c(Zr) \sim 3k$). In the case of $Zr_{.7}Be_{.3}$ we neglected Be contributions to A. Note, $T_c(ZrPd) > T_c(Zr)$.

III. DISCUSSION

To understand the valence dependence of λ and T_c in amorphous transition-metals, note that in view of Figs. 3 and 4 the valence dependence of λ results essentially from η,

$$\lambda \simeq \lambda(\eta) \quad ,$$

and consequently,

$$T_c(\frac{e}{a}) \simeq T_c(\eta(\frac{e}{a})) \quad .$$

This explains the breakdown of Matthias-rules in amorphous 4d- and 5d-metals, since η varies as shown in Fig. 5. We find in general that ΔT_c is strongly determined by $\delta\eta$, $\delta(M\langle\omega^2\rangle)$, $\delta N(o)$ and $\delta\mu_{spin}$.

The change ΔT_c observed for A-15 compounds can be understood be viewing amorphous A_3B as corresponding to the alloy $A_{0.75}B_{0.25}$. Clearly, if $\lambda_B \simeq 0$, then $T_c(A_3B) < T_c(A)$ where both T_c refer to the amorphous state. Our simple physical picture for amorphous A-15 systems gives for the Debye-temperature

$$\theta_D^{am}(A_3B) \approx \theta_D(A)$$

Indeed, for Nb_3Ge one observes $\Theta_D^{am} \simeq 220$ k $\simeq \Theta_{DNb}$. From our results and comparison with experiment we conclude that in amorphous Nb_3Ge, V_3Au, etc. superconductivity arises essentially from the Nb, V, etc. d-electrons. In the case of Zr_3Rh both Zr and Rh d-electrons contribute to superconductivity. Due to the decrease of N(o) at the Rh-sites $\mu_{spin}(Rh) \simeq 0$ in amorphous Zr_3Rh.

The results obtained for amorphous transition-metals containing Cu, Ni, Pd, etc. suggest that for all d-electrons is typically larger than λ for s-electrons (s. T_c of LaCu-alloys, etc.) and that $\mu_{spin} \simeq 0$ for amorphous alloys with Pd, Rh, Ni, etc. which support our assumption for $\delta N(o)$. If $\mu_{spin} \simeq 0$ in amorphous Pd, then we expect a relatively large T_c ($T_c \approx 7$k) for amorphous Pd. Note, recently T_c was observed[9] to be $\sim$ 3k in presumably partially disordered Pd.

Note, we have not studied explicitly changes in the coordination number and that in amorphous metals electron scattering processes are less restricted due to

$$S_{cryst}(\underset{\sim}{q}) \rightarrow S_{am}(\underset{\sim}{q})$$

where $\mathcal{S}(q)$ denotes the structure factor. We may assume

$$S_{am}(\underset{\sim}{q}) \approx S_{liq}(\underset{\sim}{q}) \quad .$$

(S_{liq} refers to the liquid state). T_c depends on S(q) via J(S(q)).

Note,
$$J = \sum \int d^3r \, \varphi_{k}{}^{*}, \, (\varepsilon_{q,r} \cdot \nabla V_i) \, \varphi_k{}^{ik \cdot r_i}$$

and thus
$$J^2 \sim \int \ldots \, S(q) \, |V_q - q'|^2 \; .$$

It would be interesting to demonstrate the dependence of J^2 on $\mathcal{S}(q)$ by systematic numerical analysis.

REFERENCES

1. M.M. Collver and Hammond, Phys. Rev. Lett. 30, 92.
2. C.C. Tsuei et al., Phys. Rev. Lett. 41, 664;
   C.C. Tsuei, Phys. Rev. B 18, 6385.
3. W.L. Johnson, Proc. 24. Conf. Magnetism
   (Cleveland - Ohio, Nov. 1978).
4. W.L. McMillan, Phys. Rev. 167, 331.
5. K.H. Bennemann and J.W. Garland, Superconductivity in
   d- and f- Band Metals, AIP Conf. Proc. No. 4, 103
   (ed. by D. Douglass, Im. Inst. of Physics, New York, 1972).
6. J.W. Garland and K.H. Bennemann, Superconductivity in d-
   and f-Band Metals, AIP Conf. Proc. No. 4, 255 (ed. by
   D. Douglass, Am. Inst. of Physics, New York, 1972).
7. S. Barisic, Phys. Lett. 34A, 188.
8. G. Gladstone et al., Superconductivity Vol. 2, 665
   (R.D. Parks edt.), New York 1969, Marcel Dekker.
9. B. Striethker, Phys. Rev. Lett, 42, 1769.

TWO LEVEL SYSTEMS IN GLASSES

H. Beck

Institut de Physique de l'Université,
CH - 2000 Neuchâtel, Switzerland

The dynamics of atoms in a solid is most suitably discussed on the basis of interatomic potentials. At $T = 0$ any configuration - crystalline or disordered - is stable if there is no force acting on any particle. The potential $W(\vec{r})$ in which a given particle moves when all the remaining are kept at their equilibrium position has a (local) minimum at the rest place of the former. This is the key for building computer models of metallic glasses (MG) : random hard sphere structures are "relaxed" under the influence of realistic pair potentials, until all forces vanish [1].

In an amorphous solid such $W(\vec{r})$ are obviously very complicated and anisotropic in detail. Let us consider a one-dimensional system with nearest neighbor Lennard-Jones potentials $V(r)$ (with minimum at $r = r_0$ and inflection point at r_2). $W(r)$ for some particle is the sum $W(r) = V(r + d) + V(d - r)$ of the contributions of its two neighbors, located at $\pm d$. $W(r)$ has a single minimum at $r = 0$ if $d < r_2$, but two minima at $\pm r_1$ for $d > r_2$. Thus, if the two neighbors are further apart than $2r_2$, there are two equivalent equilibrium positions for the intermediate particle, whereas $r = 0$ is now an unstable maximum of W. A chain of such particles can be built by fixing the first (and later the last) by an external force, corresponding to a given pressure. Fig. 1 shows the possible distances d_ℓ and d_r between an atom and its left and right neighbors. When the distance d between the first and second particle is less than r_0 the third has to follow again after d_1 (regular arrangement), but for $d_1 > r_0$ there are two possible distances for the third, a_1 and a_2. In order to have stability, an a_2 distance has to be surrounded by two a_1 distances. Fig. 2 shows a segment of a possible disordered ("amorphous") chain. Given the length of such a chain, its ground state is highly degenerate : a particle sitting in one minimum of a double well (DW) can equally sit in the opposite well, and whole groups of particles can be shifted such that the long distances a_2 occur at new places. We emphasize that in one dimension a stable disordered array is only possible by creating such "bistable" configurations : the degeneracy is intimately linked with the amorphous phase.

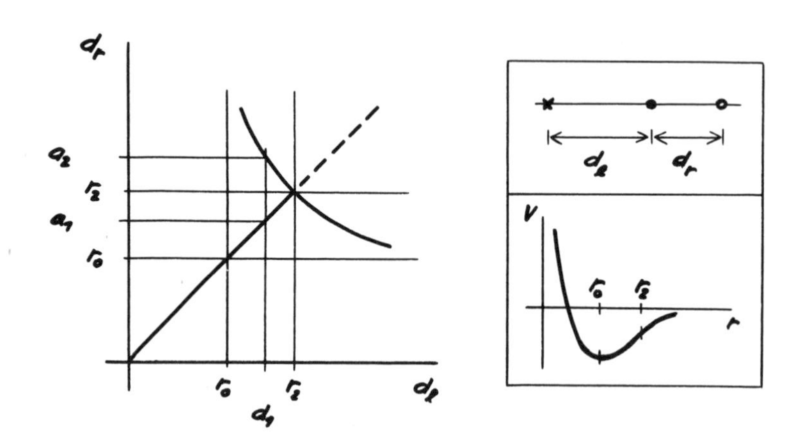

Fig. 1 : Possible equilibrium distances d_r between a particle (•) and its right neighbor (o), when d_ℓ, the distance from the left neighbor (x) is given (see upper inset) for a 1-d chain interacting by nearest neighbor Lennard-Jones potentials V (lower inset). The dashed line represents unstable configurations. For the remaining quantities see text.

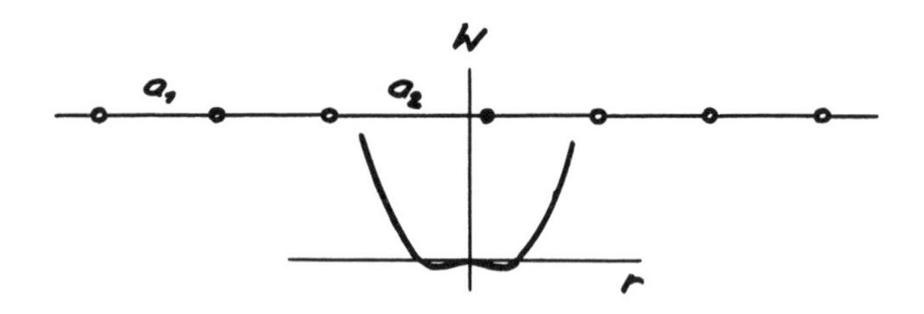

Fig. 2 : Stable configuration of a disordered chain as described in the text and in Fig. 1. Underneath the fourth particle (o) its local double well potential W(r) with two minima at $\pm r_1$ is shown.

The three-dimensional case is more complicated. Again, $W(\vec{r})$ can have more than one minimum, but their depths can be different. A DW can be created by two neighbors in roughly opposite directions, each further away than r_2, but also by two such neighbors which are "too close", creating a repulsive ridge the central particle tends to avoid. It is difficult to assess how many particles will move in "multiminimum" potentials, but since the pair correlation function $g(R)$ admits a broad spectrum of neighbor distances such situations should definitely exist.

The fact that in any "normal" amorphous structure some particles or groups of particles can take two or more almost degenerate equilibrium positions has interesting consequences. As a model situation consider an entity of mass M moving in a DW potential

$$W(r) = A_0 + A_1 r + A_2 r + A_3 r^3 + A_4 r^4 + \dots \; ; \; A_2 < 0 . \tag{1}$$

(r can be something more general than the coordinate of <u>one</u> particle). Keeping the surrounding atoms - and thus the coefficients A_i - fixed at their mean values we can evaluate the energy levels E_α of $W(r)$. At low T the levels below the threshold (of height W_o) between the wells are important. They appear in pairs separated by energies $h\omega_o$ of vibration within a (virtually harmonic) well, whereas the splitting <u>within</u> a pair is determined by the asymmetry of W and the tunnel energy $\Delta_o \approx \hbar\omega_o \exp(-2r_1\sqrt{2MW_o}/\hbar)$. If these DW-systems are sufficiently distant their direct interaction is weak, but they will interact with electrons and lattice vibrations ("phonons"), which influence W. To lowest order the coefficients A_i can be written as

$$A_i = A_i^{(0)} + \sum_q f_i(q) Q_q + \sum_q g_i(q) \rho_q \tag{2}$$

where Q_q and ρ_q represent Fourier components of ionic displacements and electronic density, respectively. Most of the interesting effects of this coupling will be at low T where only the two lowest tunnelling levels with splitting E_o have to be considered. Such a "Two level system" (TLS) is formally described as a spin $\vec{\sigma}$ with $S = \frac{1}{2}$ and Hamiltonian [2-4] :

$$H_{TLS} = E_o \sigma_z + \sum_q (f_\perp \sigma_x + f_{//} \sigma_z) Q + \sum_q (g_\perp \sigma_x + g_{//} \sigma_z) \rho_q . \tag{3}$$

Let us discuss the physical consequences of (3) in three categories :

(i) <u>Thermodynamics</u> : The specific heat of a TLS without coupling is given by

$$C = \frac{k_B}{4} (\beta E_o)^2 \, \mathrm{Sech}^2 (\frac{\beta E_o}{2}) \; ; \qquad \beta = (k_B T)^{-1} \; . \tag{4}$$

Here an important fact has to be considered : due to the randomness of the glass there will be DW's with a whole variety of coefficients A_i. The sum over the contributions (4) from all TLS can be replaced by an integral over a distribution $N(E_o)$ of splittings. If $N(E)$ is essentially constant between zero and $E \approx k_B T$ this yields

$$C_{total} \approx \frac{k_B^2}{4} N(0) T \int dx \, x^2 \, \mathrm{Sech}^2 (\frac{x}{2}) \; . \tag{5}$$

The linear T-dependence of C was first explained in this tunnelling model by Anderson, Halperin and Varma [5]. Obviously the averaging procedure is important in order to arrive at (5), as in many other contexts [4].

(ii) Interaction TLS-phonons : The formal theory was developed by interpreting the various terms of (3) as magnetic fields and solving the analogous of Bloch's equations for the spin dynamics [2,3]. Without going into details we present some results :

(a) The interaction introduces "spin flips" and leads to finite lifetimes T_1 of the levels [2,3] :

$$T_1^{-1} = \sum_\lambda \frac{M_\lambda}{c_\lambda^{\,5}} E_o^{\,3} \, \mathrm{Ctg} \, (\frac{\beta E_o}{2}) \; ; \qquad M_\lambda \; \alpha \; f_i^2 \; . \tag{6}$$

The sum runs over phonon polarizations λ with sound velocities c_λ.

(b) Phonons with energies $\hbar\omega \approx E_o$ undergo "resonant absorption" leading to a spin flip. The corresponding phonon mean free path is

$$\ell_{res}^{-1}(\lambda) = B_\lambda \omega \, \mathrm{Tgh} \, (\frac{\beta\omega}{2}) \; ; \qquad B_\lambda \; \alpha \; f_i^2 \; . \tag{7}$$

This scattering process dominates the thermal conductivity k at low T. In a Debye approximation $k = k_o T^2$ [5]. Sound waves ($\omega << k_B T$) will be attenuated by

$$\alpha \quad (\lambda) = B_\lambda \, \omega^2/(2k_B T) \; . \tag{8}$$

In practice this increase (!) of sound absorption as T^{-1} is only observed at low ultrasonic power I, since a strong pulse "saturates" the TLS : once they are all in the upper state there is no more resonant absorption. Non-linear solutions of Bloch's equations [2] yield $\alpha = \alpha_o (1 + I/I_c)^{\,1/2}$, where I_c = saturation intensity. Saturation is a very selective test of the TLS-model : any anharmonic phonon type mechanism with oscillator level sequences would not explain it. The same resonant interaction introduces shifts in sound velocities. For $\omega << k_B T$:

$$c_\lambda(T) - c_\lambda(T_0) = \pi B_\lambda \, \ell n(T/T_0) \, . \tag{9}$$

Phonon echoes [6,7] also demonstrate the validity of the spin 1/2 model (3).
(c) Away from resonance there are relaxation processes : the elastic strain shifts the energy levels and causes readjustement of the occupation numbers. The feedback to the sound wave reduces the velocity and causes absorption [2] :

$$\delta c_{rel}(\lambda) = - G_\lambda \, \mathrm{Sech}^2(\frac{\beta E_0}{2})(1 + \omega^2 T_1^2)^{-2} \quad ; \; G_\lambda \propto f_{//}^2 \, , \tag{10}$$

$$\ell_{rel}^{-1}(\lambda) = \frac{2G_\lambda}{c_\lambda^2} \, \mathrm{Sech}^2(\frac{\beta E_0}{2}) \omega^2 T_1 (1 + \omega^2 T_1^2)^{-1} \tag{11}$$

These results should again be integrated over $N(E_0)$, yielding, e.g., $\ell_{rel}^{-1} \propto T^3$ for low T. Expressions like (10), (11) hold quite generally for systems which relax towards new equilibrium configurations under the strain field of sound waves. The same type of DW, representing a TLS at low T (relaxing by tunnelling) can relax at higher T by classical, thermally induced jumps across the barrier [2,3]. Depending on the corresponding T , one or the other process will be more important for a given DW [3].

Before leaving this subject we briefly discuss the experimental situation in metallic glasses :
(a) A term aT in the specific heat is seen in superconducting Pd_3Zr_7 [8]. The coefficient a is about 2% of the electronic term above T_c. Measurements or glassy Pd Cu Si [9] in the normal state also indicate an additional linear term on top of the electronic part.
(b) Thermal conductivity behaves like T^m with $m \approx 2$ at low T in PdZr [8], TiBe [10] and Pb-Cu-films [11], all superconducting, as well as in normal PdSi, NiP, FeP [10] after subtraction of the electronic part using resistivity ρ and the Wiedemann-Franz law.
(c) The typical logarithmic increase (9) of the sound velocity was found in NiP [12], PdSi [13] and CoP [14]. Some experiments also show a decrease at higher T, presumably due to relaxational processes, although the almost linear T-dependence seems not well understood [14].
(d) An intensity dependent ultrasonic absorption was found in Pd Si Cu [4,15].
(e) Ultrasound experiments at higher T [16] on $Cu_{50}Zr_{50}$, $Co_{35}Y_{65}$,

$Co_{35}Dy_{65}$ etc. reveal relaxational peaks in the internal friction which may interpreted as the effect of thermally activated jumping of DW-particles. The relaxation times follow an Arrhenius law with activation energies of about 0.5 eV.

Besides demonstrating the success of the TLS model such experiments, to some extent, allow for a determination of coupling constants and average numbers of TLS in some energy domain. For example, N(0) entering (5) is of the order $10^{33}/(\text{erg cm}^3)$ [2,8]. The effective TLS-phonon coupling in (7) - (11) seems to be about an order of magnitude smaller in metallic than in dielectric glasses [12,14].

Fig. 3 shows a very qualitative summary of $\alpha(T)$ and $c(T)$ for (insulating) glasses based on data in Fig. 3,7,14,15 of Ref. [2].

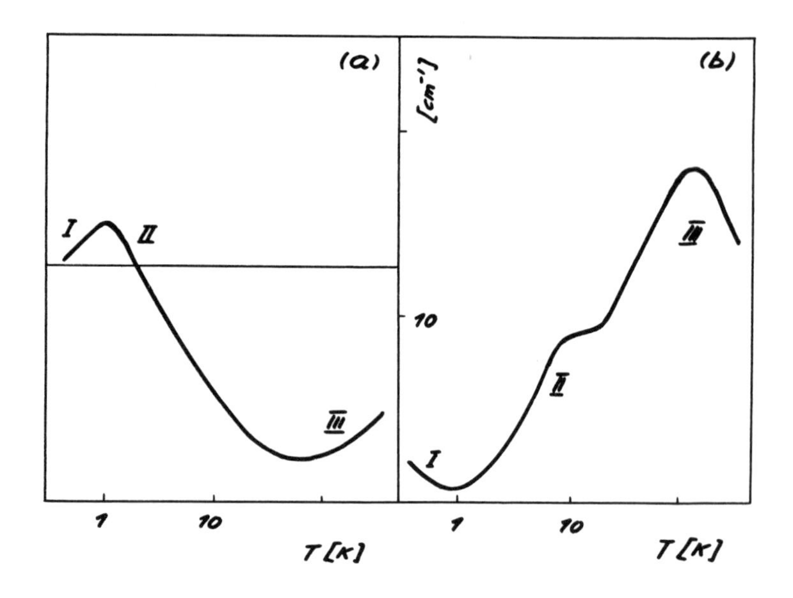

Fig. 3 : (a) Relative variation $\Delta c/c$ of the sound velocity (for true scales see [2]), (b) ultrasonic inverse meanfree path for a typical glass. Roughly three regimes can be distinguished : I : TLS resonance (see eq.'s (8), (9)); II : TLS relaxation (see eq.'s (10), (11)); III : classical thermally activated DW relaxation with characteristic peaks in ℓ^{-1} at $\omega T_1 \approx 1$.

(iii) Interaction TLS-electrons : This specific interaction in MG's adds a decay rate

$$T_{1_{el}}^{-1} = R\ D(\varepsilon_F)E_o\ Ctg(\frac{\beta E_o}{2})\ , \qquad R \ \alpha\ g_{\perp}^2 \qquad (12)$$

to (6). $D(\varepsilon_F)$ is the electronic density of states at the Fermi energy. For low T (12) is more important than (6) owing to the lower power of E_o. Thus the saturation intensity I_c of Pd Si Cu [4] is much higher and T_1 much shorter (4 orders of magnitude !) than in fused silica. Therefore the range $\omega T_1 < 1$ is accessible in acoustic experiments, which also leads to different behaviors of c (deviations from ℓnT in the resonance regime) and α (contributions $\alpha\ \ell nT$) [4,17]. Black and Fulde [18] recently suggested that T_1 should increase drastically in a metallic glass below the transition temperature T_c to the superconducting state. Experiments performed on PdZr [19] do show a strong decrease of the sound absorption, which occurs, however, about 1 K below T_c. Thus the coupling between TLS and conduction electrons seems not yet well understood.

Last, but not least, this coupling influences the conduction electrons. A recent explanation of the Kondo-like increase of ρ at low T in some MG [20] seems, however, to be based on a model which is not appropriate for almost free conduction electrons [21]. Instead, a treatment based on (3) leads to very small second order changes in ρ [22]. Divergent terms in higher order (similar, but not the same as in the Kondo problem) [21] even point to a reduction of the effect on ρ at lower T, where the divergent Fermion response makes perturbation theory break down. This problem certainly needs further clarification.

This work was supported by the Swiss National Science Foundation.

REFERENCES

1. L. von Heimendall, J. Phys. F9, 161 (1979).
2. S. Hunklinger, W. Arnold, "Physical Acoustics", ed. R.N. Thurston, W.P. Mason (Acad. Press, N.Y.), 1976, vol.12, p. 155.
3. J. Jäckle, L. Piché, W. Arnold, S. Hunklinger, J. Non-Cryst. Sol. 20, 365 (1976).
4. B. Golding, J.E. Graebner, A.B. Kane, J.L. Black, Phys. Rev. Lett. 41, 1487 (1978).
5. P.W. Anderson, B.I. Halperin, C. Varma, Phil. Mag. 25, 1 (1972).

6. B. Bolding, J.E. Graebner, Phys. Rev. Lett. 37, 852 (1976).
7. J.L. Black, B.I. Halperin, Phys. Rev. B16, 2879 (1977).
8. J.E. Graebner, B. Golding, R.J. Schutz, F.S.L. Hsu, H.S. Chen, Phys. Rev. Lett. 39, 1480 (1977).
9. J.C. Lasjaunias, A. Ravex, D. Thoulouze, J. Phys. F9, 803 (1979).
10. J.R. Matey, A.C. Anderson, Phys. Rev. B16, 3406 (1977).
11. H. v. Löhneysen, F. Steglich, Phys. Rev. Lett. 39, 1205 (1977).
12. G. Bellessa, P. Doussineau, A. Levelut, J. de Physique - Lettres 38, L-65 (1977).
13. G. Bellessa, O. Bethoux, Phys. Lett. 62A, 125 (1977).
14. G. Bellessa, J. Phys. C10, L-285 (1977).
15. P. Doussineau, P. Legros, A. Levelut, A. Robin, J. de Physique - Lettres 39, L-265 (1978).
16. H.-U. Künzi, article in this volume, and references therein.
17. G. Bellessa, Phys. Rev. Lett. 40, 1456 (1978).
18. J.L. Black, P. Fulde, Phys. Rev. Lett. 43, 453 (1979).
19. G. Weiss, W. Arnold, K. Dransfeld, H.J. Güntherodt, to be published.
20. R. Cochrane, R. Harris, J. Ström-Olson, M. Zuckermann, Phys. Rev. Lett. 35, 676 (1975).
21. J.L. Black, B.L. Gyorffy, Phys. Rev. Lett. 41, 1595 (1979).
22. J.L. Black, B.L. Gyorffy, J. Jäckle, to appear.

STABILITY OF METALLIC GLASSES

H. Beck

Institut de Physique de l'Université,
CH - 2000 Neuchâtel, Switzerland

Up to now metallic glass (MG) alloys have only been produced for limited concentrations, such as [1] : $Au-Si_{18-40}$, $Pd-Si_{15-22}$, $Fe-B_{9-26}$, $Ni-Si_{4-18}-B_{10-26}$, $Nb-Ni_{25-60}$, $Zr-Co_{20-90}$ [2] etc. Quantitative predictions of this range for a given substance, though highly desirable, are very difficult. This is not surprising : first principles determinations of crystal structures are already quite delicate, but for a metastable MG, in addition to comparing the free energy with that of a competing crystalline phase one should also estimate energy barriers between the two. Moreover, the amorphous phase is not just one point in phase space, like, e.g. a fcc crystal.

The problem can be discussed from two points of view. One aspect is "glass forming ability" : the possibility of obtaining a MG at a given quench rate. Here the discussion [1] involves arguments about nucleation and growth of crystallites in the undercooled liquid : one has to pass the region between melting (T_m) and glass (T_g) temperatures rapidly enough, in order to preserve the liquid structure. The key quantity is the viscosity $\eta(T)$ in this domain. A second, though related, aspect is "thermal stability". The latter is usually characterized by T_g or T_c (crystallization temperature), giving the upper temperature limit of the range of existence of the glassy state.

There are a few empirical criteria : for example most MG are formed in the vicinity of a relatively deep entectic. Evidently this shows that the liquid disorder - favored by entropy at high T - can be stabilized down to relatively low T. On the other hand a high T_g/T_m - ratio is favorable for glass forming. Quantitative representations of, say, the influence of changing some constituent of the alloy on the stability by plotting T_g or T_c should, however, be interpreted with care, since a low T_g (hinting at low stability) can be accompanied by low T_m (high "stability of disorder") and thus high T_g/T_m.

More microscopically the relative size of the various ions is an important stability factor. Random arrangements of ions can be "locked in" by an appropriate concentration of small spheres filling the "holes" between larger ones [3]. This may be a reason for the absence of elemental MG's : random packings of a single species lack this stabilizing factor.

For crystalline simple metals the pseudopotential concept is a fruitful approach to calculating energies and structures [4]. In second order perturbation theory the structure dependent part of the energy U of a metal is determined by an effective ionic pair potential $\phi(R)$. The latter is volume dependent and should be used for discussing possible structures at given volume only : $\phi(R)$ = "rearrangement potential" [4]. Metallic pair potentials have a characteristic property : they show long range oscillations, decaying like $\cos(2k_F R + \delta)/R^{-m}$ with $m \approx 3$ or 4. Their wave length $\lambda_F = 2\pi/2k_F$ is determined by the number of conduction electrons. This peculiar form of ϕ can have interesting consequences for crystal structures : Heine and Weaire [4] show quantitatively that, as an example, Ga has a complicated crystal structure, because a "normal" close packed structure (fcc etc.), favored at that density by the large structure independent part of U, would place atoms near the first maximum of $\phi(R)$. Thus the crystal structures of Ga, Hg, In, as well as the non-ideal c/a ratio of Zn and Cd are a consequence of the interplay between the volume dependent parts of U and the oscillations of the "rearrangement potential" [4]. Evidently the same facts are at the base of the "Hume-Rothery rules" [5].

Similar arguments can be invoked for discussing the above mentioned stability aspects of MG's. The liquid alloy, which is to be quenched, is characterized by partial pair potentials $\phi_{\alpha\beta}$ and pair correlation functions $g_{\alpha\beta}$ (α,β denoting species). The latter are often hard-sphere-like (Fig. 1). The distances between consecutive maxima, i.e. between neighbor shells, are approximately equal and given by $\lambda_p \approx 2\pi/k_p$, where k_p is the wave vector of the first peak of the corresponding structure factor. The liquid structure is mainly determined by density and ionic radii, the long range part of $\phi_{\alpha\beta}$ being less important (see below). A MG is essentially a frozen liquid. The possibility of its formation and its stability are therefore determined by the degree of compatibility between the random liquid structure and the rearrangement potential, the details of which will now be important, since T is low.

In a strongly simplified description quenching means arresting instantaneously the liquid motion and allowing only for small amplitude vibrations of the particles around their mean position. Thermodynamically the quenched configuration is at best metastable, and its lifetime with respect to structural transformations should be estimated by examining energy barriers and driving forces in phase space. Keeping the volume constant variations of the internal energy U due to displacements $\delta\vec{R}(\ell)$ of the ions are given by

$$\delta U = \tfrac{1}{2}\,\delta \sum_{\ell,\ell'}{}' \phi(\vec{R}(\ell)-\vec{R}(\ell')) = \tfrac{1}{2}\int d^3R\,\phi_{\alpha\beta}(R)\,\delta g_{\alpha\beta}(R) \qquad (1)$$

where the sum over all pairs has been replaced - in the sense of a configurational average - by an integral. It is immediately clear that negative δU's, pointing the a structural instability, are best avoided if, in the quenched structure, ions are arranged favorably with respect to $\phi_{\alpha\beta}(R)$; i.e. when most successive neighbors of a given particle are placed near a minimum of $\phi_{\alpha\beta}$. This conditions has two aspects :

(i) Nearest neighbors should predominantly sit in the first minimum. This is where the effect of ionic radii, which are determined by the size of the closed ionic shells, by pseudopotential and electron density, comes in.

(ii) Since $\phi_{\alpha\beta}$ is long ranged more distant neighbors also matter. They can predominantly sit in the minima of the potential if the wave lengths λ_F and λ_p are equal (such a "favorable" situation is shown in Fig. 1); i.e. if

$$2k_F \approx k_p. \qquad (2)$$

This criterion was put forward by Nagel and Tauc [6] upon observing that many MG's show negative temperature derivatives of the electrical resistivity ρ, which is predicted by Ziman's theory if (2) holds. On the other hand it is unclear, whether (2) is equivalent to the Fermi energy lying near a minimum of the density of states [6]. We emphasize that our stability arguments leading to (2) do not invoke electronic densities of states.

Since (2) is a very simple criterion, involving only the "basic electronic and structural lengths", its predictive power is limited. First, one should obviously consider partial structure factors (each with its own k_p) and partial $\phi_{\alpha\beta}$. Moreover in the case of transition or rare earth metals the pseudopotential approach is not useful. On the other hand the concept of a Fermi sphere of conduction electrons and condition (2) for negative $d\rho/dT$ has proven very valuable also in that case [7]. Criterion (2) is qualitatively useful for alloys of partners with different valence, like $Au_x - Si_{1-x}$, $Fe_x - B_{1-x}$ etc., in which $2k_F$ is shifted strongly by varying x. Here (2) predicts MG's in the vicinity of $x \approx 0.7$ or 0.8.

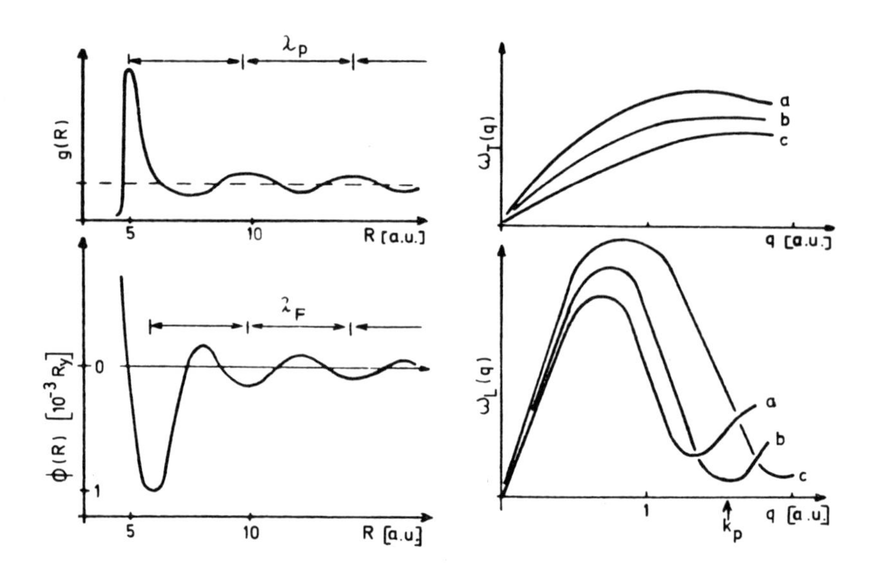

Fig. 1 : Left side: a typical pair correlation function g(R) and a pair potential $\phi(R)$ for a "matching"situation. Right hand side : longitudinal (ω_L) and transverse (ω_T) "average phonon frequencies", defined by (4). Curve b would be for a Lennard-Jones type potential, whereas for a metallic potential the curves correspond to $2k_F < k_p$ (a), $2k_F \approx k_p$ (b), $2k_F > k_p$ (c).

In order to corroborate our claim of the importance of the ratio between k_p and $2k_F$ for structural problems we may note a further consequence for liquid polyvalent metals like Ga, In, Sn, Ge, Sb etc. They show a shoulder on the high q slope of the main peak of S(q). It occurs at $q = k_s$, a few percent less than the

Fermi sphere diameter $2k_F$, evaluated in the free electron approximation. We have recently calculated $S(q)$ for such systems [8] on the basis of a pair potential $\phi(R)$, as in Fig. 1, by using well known perturbation expansions [9] for potentials with attractive tails. We indeed found that the predominant Fourier components of ϕ near $2k_F$ forces $S(q)$ to develop a shoulder at k_s somewhat below $2k_F$. This is another manifestation of the influence of electron structure on ionic structure.

Of course it would be highly desirable to test the mechanical stability of randomly packed ions interacting by metallic pair potentials in a more quantitative way. The ideal tool would certainly be computer investigations of the energy surfaces in phase space. The first analytic calculation of the phase diagram for Mg - Zn and Ca - Mg by Hafner and Heimendahl [10] which predicts, on the basis of ab initio pseudopotentials, the existence of intermetallic compounds in one concentration range and the possibility of forming a MG in another, indeed shows that in both phases the interatomic distances are such that they tend to coincide with the potential minima. This "matching" of potential and structure seems thus to be a valuable, general stability criterion which is more detailed than the somewhat crude form (2).

Analytic investigations of the consequences of "matching" and "non-matching" configurations generally do not yield too much insight because the almost inevitable "configurational average" does not say much about the energetics of individual ions or clusters thereof. We have calculated two interesting quantities for various $2k_F/k_p$ ratios and compared them with the result for a non-oscillating, Lennard-Jones-like (LJ) potential :

(i) The "on-site potential"

$$W(\vec{r}) \equiv \sum_{\ell \neq n} \phi(\vec{r} - \vec{R}(\ell)) \tag{3}$$

for particle n, produced by all others, has a parabolic shape around $r = 0$, when the neighbors are arranged on spherical shells with probability $g(R)$. Compared to the "normal" LJ case, $W(r)$ is shallower if $2k_F \neq k_p$ (non-matching), but deeper and narrower if $2k_F \approx k_p$ (matching).

(ii) Stability in the sense of (1) can be tested by displacing each particle by $\delta\vec{R}(\ell) = \vec{A}(q)\exp(i\vec{q}\cdot\vec{R}_\ell)$ and calculating the averaged second derivative

$$\left\langle \frac{\partial^2 U}{\partial A_i(q)\,\partial A_i(-q)} \right\rangle = \int a_r^3\, g(r)\, \nabla_i^2\phi(r)(1 - \cos\vec{q}\cdot\vec{r}) \equiv \langle \omega_i^2(q) \rangle . \tag{4}$$

In a crystal this would yield the dynamical matrix and thus the phonon frequencies, in the glass these "average frequencies" contribute to the fourth moment of the dynamical structure factor [11]. Transverse "frequencies" show a monotonic behavior, but the longitudinal branch has a "roton-like" minimum (Fig. 1), which was confirmed by computer simulations of Mg - Zn [12]. The LJ-system produces the minimum at $q \approx k_p$, corresponding to the first reciprocal lattice vector in a crystal. The small value of (4) at k_p arises, because in the most important r-range ($r \approx R_{NN}$, nearest neighbor distance) $\cos k_p R_{NN} \approx 1$. Thus the minimum of $\omega_L(q)$ is entirely due to the structure of the glass. Displacements with wave length $\lambda = 2\pi/k_p$ cost little energy because $\lambda \approx R_{NN}$ fits, at least on average, into the disordered structure. Turning now to the oscillatory potential the "matching case" also produces the minimum at k_p (curve b). In the non-matching situations, however, the minimum is shifted from k_p towards $q = 2k_F$. It is now the potential that dictates the wave length λ_F of the low energy displacements ! Again a detailed analysis would require the normal modes and their frequencies for such a system in order to detect possible dynamical instabilities. The fact that now the average frequency for a wave length $\lambda_F \neq R_{NN}$ is lowest, may point to the existence of unstable modes with typical wave lengths which do not fit into the random structure. Comparing curves (b) and (c) with one of the LJ-type (a), that represents some kind of a "bare spectrum" [without the influence of the long range oscillations] suggests the analogy with a Kohn anomaly in crystalline metals. There, as it was dramatically demonstrated for metallic hydrogen [13], electron-ion interaction can lower the phonon frequencies at $|\vec{q} + \vec{Q}| \approx 2k_F$ (for some reciprocal lattice vector $\vec{Q}$) and lead to instabilities.

Experimentally, stability has been investigated by examining the influence of varying the concentration of some constituents on T_g, T_c and other properties. As mentioned above, this may or may not be too relevant. Some such experiments can be understood, at least qualitatively, on the basis of our stability consideration :

(i) Giessen et al [14] examined $(CuZr)_{95}\,TM_5$ glasses and found that those transition metals (TM) yielding high Young's moduli also lead to high T_c and T_g, demonstrating the link between purely mechanical and thermodynamical stability.

(ii) Donald and Davies [15] found the highest T_c for $Ni_{1-x}(Si - B)_x$ at $x \approx 0.32$, which roughly corresponds to 1.8 electron/ion, where $2k_F$ is expected to be at k_p. Similarly Buschow and Beekmans [2] found a decreasing $T_c(x)$ for $Zr_{1-x}\,Co_x$ for large x-values where simple thermodynamic considerations would point to a further increase of stability.

(iii) Hillmann and Hilzinger [16] measured T_c and $d\rho/dT$ of Fe - Ni - alloys. They found that T_c increased when $d\rho/dT$ (which was positive in their case) decreased : increasing stability, when the system comes closer to (2), which predicts negative $d\rho/dT$.

(iv) Leitz and Buckel [17] studied thin films of $Au_{1-x}Sb_x$. They seem to be amorphous for $0.2 \lesssim x \lesssim 0.6$. In this concentration range the distances between the maxima of g(R) decrease with increasing x, in the same way as the distances between consecutive minima of V(R) are supposed to shrink as a consequence of increasing $2k_F$. In this film electron-ion interaction appears to be strong enough to force the ions into a structure with complies with the oscillations of the pair potential.

This work has been supported by the Swiss National Science Foundation.

REFERENCES

1. H.A. Davies, Third Intern. Conf. on Rapidly Quenched Metals, Brighton, (publ. : Metals Society, No 198), 1978, p. 1.
2. K.H.J. Buschow, N.M. Beekmans, Phys. Rev. B19, 3843 (1979).
3. D.E. Polk, Acta Metall. 20, 485 (1972).
4. V. Heine, D. Weaire, Sol. State Phys. 24, 249 (1970).
5. D. Stroud, N.W. Ashcroft, J. Phys. F1, 113 (1971).
6. S.R. Nagel, J. Tauc, Phys. Rev. Lett. 35, 380 (1975).
7. H. Beck, same volume.
8. R. Oberle, H. Beck, to be published.
9. H.C. Andersen, D. Chandler, J.D. Weeks, Adv. Chem. Phys. 34, 105 (1976).
10. J. Hafner, L. von Heimendahl, Phys. Rev. Lett. 42, 386 (1979).
11. D. Tomanek, H. Beck, to be published.
12. L. von Heimendahl, J. Phys. F9, 161 (1979).
13. H. Beck, O. Straus, Helv. Phys. Acta 48, 655 (1975).
14. B.C. Giessen, J. Hong, L. Kabacoff, D.E. Polk, R. Raman, R. St.Amand, p. 249 of Ref. 1.
15. I.W. Donald, H.A. Davies, p. 273 of Ref. 1.
16. H. Hillmann, H.R. Hilzinger, p. 371 of Ref. 1.
17. H. Leitz, W. Buckel, Z. Physik B, to appear.

LIQUIDS AND GLASSES

R. M. J. Cotterill

Department of Structural Properties of Materials
The Technical University of Denmark
Building 307, DK-2800 Lyngby, Denmark

1. INTRODUCTION

The object of this communication is to explore the wider implications of one currently favored theory of the melting transition and to extend the general approach that it advocates to the glass transition and the liquid-vapor transition. The theory in question attributes melting to the sudden proliferation of dislocations, this happening at a temperature at which the free energies of a perfect crystal and one saturated with dislocations are equal. Several predictions of such dislocation-mediated melting have been published [1, 2, 3, 4] and corroboration has been provided both by computer simulation [5, 6] and experiment [7, 8]. Efforts have already been made [4, 9] to generalize the dislocation theory so as to include the glass transition, and the new aspect of the present work is the emphasis given to dislocations as dynamic and perhaps transient excitations rather than static and persistent defects. The other novel feature to be presented here concerns the application of the dislocation concept in a liquid right up to the liquid-vapor transition. It is to this question that we in fact first turn our attention.

2. THE LIQUID-VAPOR TRANSITION

The great break-through in the understanding of the liquid-vapor transition in general, and the existence of a critical point in particular, was of course provided by the classical work of Van der Waals in 1873. It not only explained the shapes of the isotherms above and below the critical temperature, but also provided direct evidence of molecular size and intermolecular forces. Until now, it has not proved possible to match this success by working from first principles, using known interatomic potentials. We attempt here to construct such an ab initio theory.

We make the minimal assumption that the interatomic potential, V(r), is repulsive at short range, has a minimum at r_o, and that beyond r_o the attraction goes monotonically and asymptotically to zero at large distance. We also use, as a first approximation, a naive model of a vapor in which the atoms are uniformly spaced in a hexagonal close-packed arrangement: a so-called lattice gas. The modification resulting from the extension to a more realistic model will be discussed later. This idealized gas is depicted (schematically in two dimensions) in Figure 1, and we enquire as to the relative potential energies associated with various positions in the immediate environment of a typical atom.

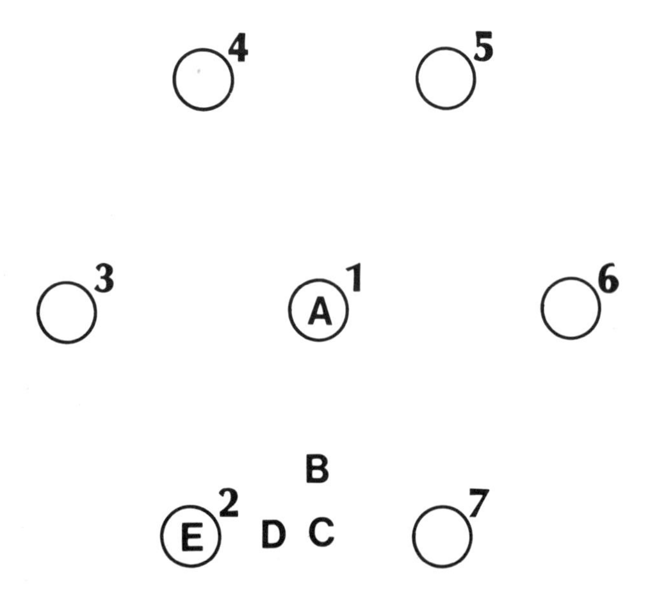

Figure 1. Two-dimensional lattice gas configuration used in the ab initio calculation of liquid-vapor critical point properties.

This latter is labelled "1" in the figure, and its nearest-neighbor shell consists of the six atoms 2-7. When lying symmetrically with respect to these neighbors, atom 1 is located at position A, and we wish to know the potential energy of atom 1 when it is located both there and also at B, C, and D. (If it were located at E, the energy would of course be infinite, because atom 3 is there already.) We now note that the relative energies at these various positions will depend on the distance scale relative to r_o, and will thus depend on the degree of compression of our lattice gas. Let the distance from A to E, the nearest-neighbor distance, be d_o. At high compression, with d_o comparable or even less than r_o, position A will correspond to a local minimum and will have $E_A < E_B < E_C < E_D$, with a saddle point at C, by symmetry. At lower compression, with d_o somewhat greater than r_o, we will have $E_B < E_A$, E_C (and E_A, $E_C < E_D$). Further decompression will take us through a situation in which $E_C < E_B < E_A$ (and $E_C < E_D$), and when $d_o = 2\ r_o$, position C will actually correspond to minimum. Finally, at still lower compression we will have $E_D < E_C < E_B < E_A$.

Just what influence these various situations have on the mobility of atom 1, within its environment, will depend on its kinetic energy, E_{KIN}, that is on the temperature, T. This is of course given by, on average, $3\ k_BT/2$, where k_B is the Boltzmann constant. For the compressed situation in which $E_A < E_B < E_C$, atom 1 will be able to pass the saddle point at C, and thus escape from its "cage", if E_{KIN} is sufficiently large. Upon further compression, however, a point must be reached at which the saddle point becomes too high, and atom 1 will then no longer be able to move freely. We note that the single-particle motion that is the hallmark of a gas will no longer be possible, and the motion must go over to the many-body regime. We return to this later. Now consider lower degrees of compression for which $E_B < E_A$, E_C and $E_A < E_C$. If E_{KIN} is sufficient to mount the barrier at C, the smaller barrier at A will be without consequence. If the temperature falls too low, it will be barrier C which dictates the onset of many-body motion.

Now consider the interesting situation in which $E_B < E_A$, E_C but $E_A = E_C$. If the temperature falls too low, the barriers at A and C will be equally effective, and at slightly lower compression it will actually be barrier A which restricts the motion of atom 1. This situation is quite different from those discussed earlier because the symmetrical configuration is now no longer accessible. A rearrangement of the assembly must occur, and the lattice gas must contract. We believe that the situation in which $E_A = E_C$ corresponds to the critical volume, and that when $E_{KIN} = E_A = E_C$, the corresponding temperature is a first approximation to the critical temperature.

This conjecture must be put to a numerical test, and we take data for argon, for which V(r) is known to the desired degree of accuracy. The Lennard-Jones function is a good approximation to V(r) for this element, and we have

$$V(r) = \varepsilon \left\{ (r_o/r)^{12} - 2\ (r_o/r)^6 \right\} \tag{1}$$

where $r_o = 3.87 \times 10^{-10}$ m and $\varepsilon/k_B = 119.3$ K, as given by Pollack [10]. We recall that one must divide by a factor of two when computing potential energy so as to avoid counting each bond twice. Now the critical volume of argon is 1.248×10^{-22} cm^3 per atom, which corresponds to $d_o = 5.609 \times 10^{-10}$ m. If the cage windows are imagined as adopting an average shape of equilateral triangles (in three dimensions of course), with side length equal to d_o, the escaping central atom (i. e. atom 1) will have to pass out through the triangle's centroid (corresponding to position C in our schematic two-dimensional picture in Figure 1). This centroid lies at a distance $d_o/3^{\frac{1}{2}}$ from each apex, so its distance to the nearest-lying atoms will be 3.238×10^{-10} m. Since there are three such nearest-lying atoms, the required potential energy is (again dividing by k_B) $E_C = 3 \times 216.2$ K $= 648.6$ K. When atom 1 is at the symmetry position, it is surrounded by twelve nearest neighbors all lying at distance d_o. This gives an energy $E_A = 12 \times 47.5$ K $= 570.0$ K. Thus we see that the critical volume does indeed correspond to a situation in which E_A and E_C are approximately equal. To find out how effective these barriers are to the motion of atom 1 we must find the depth of the lowest point on the potential energy surface. There must be a point equidistant from the three atoms which constitute a cage window and lying at r_o from them. This point will correspond to zero energy, since our energies have all been calculated relative to the bottom of the well of V(r). This would indicate a critical temperature of about 2×600 K/3 = 400 K, if the mounting of the barrier is achieved solely by the motion of atom 1, while the cage atoms remain frozen in their original places. The actual critical temperature of argon is 150.9 K [10], so we have produced an over-estimate.

The nature of the desired refinement to the above approach is fairly obvious, but clearly difficult. The instantaneous positions of the atoms is not as regular as in the model. The pair distribution function could be used for a better calculation of E_A, but it would be of no assistance for E_B, E_C, etc. Then again there are the dynamics of the situation; the cage window atoms obviously relax away as they feel atom 1 approaching. Finally, it is highly unlikely that the lowest energy position really will be as low as has been calculated since the formation of an instantaneous equilateral triangle is an improbable event.

Within the obvious limitations of the model it does seem that one can account for the occurrence of a critical point using only the gross features of V(r), and we use the findings of this section to propose a new definition of the critical temperature: *The critical temperature is the lowest temperature at which the transition from single-particle to many-particle motion in a fluid occurs continuously, the critical point being that point on the critical isotherm at which the transition is located.* We now go on to discuss the nature of the collective motions in a liquid.

3. THE LIQUID STATE

We have seen that for any temperature a certain calculable degree of compression is sufficient to inhibit single-particle motion in a fluid. We now enquire as to the nature of the many-particle motions that will then nevertheless be possible, and our discussion will be necessarily qualitative. Consider first the motion of the moving atom when it is still just able to escape, unaided, from its cage. One can use the Stokes equation (which strictly applies only to far more macroscopic systems) to emphasize the salient features of the situation. A sphere of radius a, moving in a fluid of viscosity η requires a force f to keep it moving at a velocity v, where

$$f = 6\pi\, a\, v\, \eta \tag{2}$$

Now in the present context the η can be thought of as being related to the height of the barrier E_C, while f arises from the thermal motion of atom 1. If E_C is too high, the viscosity will be large enough to make f ineffective and the escape velocity cannot be achieved. But this applies to our single atom (atom 1) with radius a. If we increase the effective radius by making the atoms move cooperatively, the situation might be more conducive to motion if f could be made to thereby rise more rapidly than the radius. We now wish to suggest that this is indeed what will happen, at least up to a certain degree of cooperativity. The point is that if a clump of atoms moves cooperatively, the effective f will increase as the volume of the clump whereas the effective value of a will increase only as the clump diameter. But there is a natural limit to such cooperativity that is set by the velocity of sound in the fluid, v_s. If the time constant for passing the barrier C is τ, atoms more distant from that barrier than a distance $v_s \tau$ cannot contribute to f. A typical value of τ would be 10^{-12} m, while v_s will be approximately 10^3 m s^{-1}. Thus the radius of the clump will be roughly 10^{-9} m, and it will contain some tens of atoms. (This agrees well with a recent estimate of this quantity published by Weisskopf [11]).

We now suggest that the cooperative motion of such a clump of atoms will proceed by mutual slipping at instantaneous surfaces of weakness. It need not occur simultaneously at all points around the clump circumference, and probably takes place by consecutive jumps. Such consecutive motion is described by the general dislocation of the Volterra type (which refers to a continuum and requires no reference lattice). It has in fact recently been observed in a simulated two-dimensional liquid [12], and dislocationlike structures have been seen in a simulated three-dimensional liquid [13].

Viscous flow in a liquid is a process of shear, and this is here envisaged as occurring by dislocation motion. The shear rate

$\dot{\gamma}$ is related to the shear stress σ through the viscosity η by

$$\dot{\gamma} = \sigma/\eta \tag{3}$$

but it is also related to the dislocation density ρ and dislocation velocity v_d through

$$\dot{\gamma} = b\,\rho\,v_d \tag{4}$$

where b is the Burgers vector. Eliminating $\dot{\gamma}$ we have

$$\eta = \frac{\sigma}{b\,\rho\,v_d} \tag{5}$$

from which one can obtain η if the relationship between σ and v_d is known. The recent simulation result [13] gave a minimum ρ of 10^{18} m^{-2}, and one can assume $b = 3 \times 10^{-10}$ m. For (crystalline) LiF Johnston and Gilman [14] found that $v_d = 10^2$ m s^{-1} when $\sigma = 9.8 \times 10^7$ N m^{-2}. If this value can be applied to a liquid, we find, from Equation (5), $\eta = 3.27 \times 10^{-3}$ N m^{-2} s. This is about twice as high as the value for mercury at 0° C, for instance, and the agreement with experiment can be deemed satisfactory. Dislocations in liquids must have smaller widths than their crystalline counterparts, so the stress required per unit velocity would be higher. This would increase the calculated value of viscosity.

4. THE MELTING AND GLASS TRANSITIONS

Evidence for a dislocation mechanism of melting has already been cited, and we now seek merely to bring this fact into general framework of the present approach; to relate dislocation generation to the features of the interatomic potential. It is found that the dislocations produced during the melting of an FCC crystal are of the shear type, and that they are formed on {111} planes. There are two potentially important contributions to the energy barrier: the close approach of atoms during the shear, and the magnitude of the "on-site" potential. Before the shear, an atom on the moving plane lies equidistant from three atoms in the plane below, and just d_o from them. The height of the tetrahedron thus formed is $0.8165\, d_o$, and during slipping the closest approach of the moving atom to two of the base atoms is $0.9574\, d_o$, the slip direction being <112>. For argon at the triple point $d_o = 3.865 \times 10^{-10}$ m, so the nearest approach distance is 3.700×10^{-10} m. This corresponds to an energy, calculated as earlier, of 11 K. The triple point of argon lies at 83.8 K [10] so the kinetic energy of an average atom is 125.7 K, again dividing by k_B. This first factor obviously cannot be determining whether dislocations are being formed.

To appreciate the second factor consider the middle atom of three successive atoms in any row. The potential it experiences due to its immediate neighbors, the "on-site" potential, monotonically increases with positive or negative displacement so long as d_o is roughly the same as r_o. But V(r) has a point of inflection at r*, where, for the Lennard-Jones function, r* is about 15% greater than r_o. It is easy to show that the on-site potential bifurcates if the nearest neighbors are more distant than r*, and it has been argued [15] that this favors the generation of solitons. The qualitative picture is that as the temperature increases, a point is reached at which the harmonic generation of phonons gives way for the anharmonic production of solitons. These have the wave function ψ given by

$$\psi = \psi_o \tan^{-1}\left[\exp \pm \left\{(x - vt)/(1 - v^2)^{\frac{1}{2}}\right\}\right]$$

where v is the soliton velocity, relative to a limiting value of unity. This bears a striking resemblance to the envelope of the Frenkel-Kontorova solution for the one-dimensional dislocation, namely that the displacement u is given by

$$\tan\left\{(\pi/4) + (\pi\ u/2b)\right\} = \exp - (\pi n/\ell)$$

where b is the Burgers vector, n is an integer, and ℓ is a measure of the dislocation width. It is interesting to note that the 15% appearing above is in excellent agreement with the idea underlying Lindemann's melting rule [16].

It thus appears that anharmonicity will force a sort of "bunching up" along the atomic rows, above a certain temperature, and that these, propagating through the structure as solitary waves, will be essentially dislocations. But it is important to stress that the latter are a natural consequence of the motions of the atoms rather than the crystal structure. In three dimensions this is envisaged as producing the shear dislocation loops. The question remains: what density of these defects will be produced? Since the transition must occur at constant free energy, the answer must be that as many will be produced as gives equality of the change in enthalpy and the change in entropy multiplied by temperature. Although this is clearly a formal requirement, it might be valid to adopt a different approach. We earlier invoked the idea of a fundamental "coherence length", τv_s, over which the atoms can function collectively. This might be the natural spacing between the dislocations generated by the anharmonicity. Indeed, until so many dislocations are generated that the spacing is down to this distance, the crystal might be able to maintain its integrity, with the individual loops shrinking back to nothing without having influenced the structure's fluidity. This idea is rather reminiscent of the concept of percolation. Since the dislocation generation propagates at the velocity of sound, the situation is adiabatic and must conserve free energy. This would indicate that the entropy will simply increase until $T\Delta S$ equals ΔE, the latter being simply the latent heat (which is determined by the dislocation density and energy).

It is interesting to note that the above picture admits of a sharp melting point even though dislocations might appear spontaneously and transiently at lower temperatures. The point is that the smaller the thermally generated dislocation loops are, the greater is the force of attraction of points at opposite ends of any diameter, and it is easy to show that this force easily exceeds the Peierls-Nabarro force. The inevitable fate of such a loop, therefore, is reversion back to zero diameter unless it is disturbed by an adjacent loop that is sufficiently close. This latter condition will obtain only at the saturation situation, when the inter-loop distance equals the loop diameter. As noted above, the coherence distance, τv_s, is probably a good measure of the dislocation density at saturation. At a temperature below the melting point, the influence of a transient loop is not just small but identically zero because the effect of the dislocation segment at one extreme of any diameter will be exactly counteracted by the dislocation segment lying at the other extreme.

It is intriguing to note that the dislocation loop generation concept offers a microscopic explanation of Maxwell's approach to the dynamics of a liquid. This regarded a liquid as being capable of sustaining a certain amount of shearing stress for a short time, after which it breaks down, and the shear recommences. According to the picture presented here, the duration during which a shearing stress can be supported would of course be τ.

Turning, finally, to the glassy state, we attempt to place this too in the dislocation framework. The possibility that the amorphous state can adequately be described in terms of dislocations [9, 18] or both disclinations and dislocations [19] has already been raised in the literature. It is still not clear whether this is the appropriate description or whether there is an ideal non-crystalline state that is fundamentally different from the crystal. It is in fact possible that dislocations might act as an agency for disrupting the crystal, but that when their density becomes sufficiently high, a fundamental change occurs in the assembly so as to produce a configuration not describable in terms of dislocations [20]. It is to be emphasized that this does not mean that transient dislocations would then play no role in the fluidity of a glass. On the contrary, this suggests a simple idea for the magnitude of T_g, the glass transition temperature. We must first digress and invoke the observation by Tallon et al. [21] that the shear moduli of alkali halides do not fall to zero at the melting point, as proposed by Born [22], but that a zero value is attained on extrapolating to the dilation corresponding to the liquid at T_m. This underlines the fact that the activation energy for dislocation generation is sensitive to changes in volume, since the shear modulus, μ, appears in the expression for dislocation energy, E_d. The point is that a glass always occupies a greater volume than the corresponding crystal, so the critical dislocation concentration will be achieved at a lower temperature than in the crystal. *The glass temperature, T_g, can be looked upon as a false melting temperature which lies below T_m because the glass occupies a larger volume than the crystal.*

The physical picture is that below T_g insufficient dislocation loops will be present at any one time in order to reach our percolation threshold, and the glass will not flow. At T_g the transient loops will interact, rather than shrink back to nothing, and flow will commence. The flow will increase with temperature simply because there is more thermal activation both to generate loops and to make them move. The question remains as to whether one can, on the basis of this model, explain the crystallization of some glasses, particularly the glassy metals, whereas other glasses, such as those based on silicates, seem to be trapped in the glassy state. Traditional nucleation theory produces the result that a large difference in the free energies of the liquid and crystalline states gives a small critical nucleus diameter, and vice versa. Dislocation-dislocation interactions must inevitably produce some mutual annihilations, which would be equivalent to forming a small volume of perfect crystal, i.e. a potential nucleus. If this volume was smaller than the critical nucleus, nothing will have been achieved by the event, and the tiny piece of crystal will soon be re-dislocated. Conversely, if the volume exceeds that of the critical nucleus, spontaneous crystallization might be triggered.

5. CONCLUSION

Working from first principles, it is possible to derive a first approximation to the liquid-vapor critical parameters, using only the characteristics of the interatomic potential. The transition from single-body to many-body motion in a fluid is believed to occur even above the critical temperature, and it might be an example of a third or fourth order phase transition [23]. The many-body motion in a fluid is believed to be such that groups of a few tens of atoms, acting within a certain coherence distance, move cooperatively by a sliding process describable by dislocations. This predicts a viscosity of the correct order of magnitude. The melting point corresponds to a sudden proliferation of dislocations, and it is reached when the concentration of thermally generated transient dislocation loops attains a threshold value. The latter is believed to be related to the above-mentioned coherence length and resembles a percolation limit. The anharmonicity of the interatomic potential is ultimately responsible for the generation of dislocations, which, being dynamic excitations, have the character of solitons. The glass temperature is seen as a sort of false melting temperature, occurring at a lower temperature because the glass occupies a larger volume than the corresponding crystal.

References

[1] S. Mizushima, J. Phys. Soc. Japan 15, 70 (1960).
[2] A. Ookawa, J. Phys. Soc. Japan 15, 2191 (1960).
[3] D. Kuhlmann-Wilsdorf, Phys. Rev. 140, 1599 (1965).
[4] S.F. Edwards, Polymer 17, 933 (1976).
[5] R.M.J. Cotterill and L.B. Pedersen, Solid State Comm. 10, 439 (1972).
[6] R.M.J. Cotterill, W. Damgaard Kristensen, and E.J. Jensen, Phil. Mag. 30, 245 (1974).
[7] R.M.J. Cotterill and J. Klæstrup Kristensen, Phil. Mag. 36, 453 (1977).
[8] R.K. Crawford (to be published).
[9] R.M.J. Cotterill, E.J. Jensen, W. Damgaard Kristensen, R. Paetsch, and P.O. Esbjørn, J. de Physique 36, C2-35 (1975).
[10] G.L. Pollack, Rev. Mod. Phys. 36, 748 (1964).
[11] V.F. Weisskopf, Trans. New York Acad. Sci. 38, 202 (1977).
[12] R.M.J. Cotterill, Physica Scripta 20, 109 (1979).
[13] R.M.J. Cotterill, Phys. Rev. Letters 42, 1541 (1979).
[14] W.G. Johnston and J.J. Gilman, J. Appl. Phys. 30, 129 (1959).
[15] R.M.J. Cotterill, Physica Scripta 18, 37 (1978).
[16] F.A. Lindemann, Z. Phys. 11, 609 (1910).
[17] R.M.J. Cotterill, Nature 273, 371 (1978).
[18] J.C.M. Li, Frontiers in Materials Science, ed. by L.E. Murr and C. Stein (Marcel Dekker, 1976) p. 527.
[19] R.C. Morris, J. Appl. Phys. 50, 3250 (1979).
[20] R.M.J. Cotterill, Physics Letters 60A, 61 (1977).
[21] J.L. Tallon, W.H. Robinson, and S.I. Smedley, Nature 266, 337 (1977).
[22] M. Born, J. Chem. Phys. 7, 591 (1939).
[23] A.B. Pippard, The Elements of Classical Thermodynamics (Cambridge University Press, 1964) p. 136.

3. EXPERIMENTAL RESULTS ON DISORDERED MATERIALS

3.1 Liquid metals and alloys

Structure and Dynamics of Liquid Metals by Neutron Scattering

W. Gläser

Physik-Department, Technische Universität München

ABSTRACT. The experimental approach to study the structure and dynamics of liquid metals by neutron scattering will be outlined. The progress made in recent years in describing measured structure factors with realistic potentials will be briefly discussed. Also an outline of the obtainable information from temperature and density dependent data will be given.

Similarly more recent experimental data on the dynamic structure of liquid metals will be discussed. It is concluded that especially for dynamical investigations experimental techniques are in general on the way to reach sufficient accuracy to allow more than qualitative conclusions on interesting details of liquid dynamics.

I. INTRODUCTION

The problem of an appropriate description of the structure and related properties of liquids arise from their intermediate situation between ideal gases and solids. Whereas crystalline solids having long range order are completely characterized by their symmetry properties liquids having no such periodicity can be characterized only by distribution or correlation functions. The term "structure" as used for liquids means something different than for solids.

The basic function introduced to describe the structure of a liquid is the pair distribution function g(r). In simple atomic liquids like liquid metals only positional configurations are of interest. If we denote the position of an atom i in a liquid of density ρ by $\underline{r}_i$ and an element of position space by $d\underline{r}_i$, then the pair correlation function $g(\underline{r}_1,\underline{r}_2)$ is defined so that $\rho g(\underline{r}_1,\underline{r}_2)d\underline{r}_2$ is the average number of atoms in $d\underline{r}_2$, when an other atom is at $\underline{r}_1$. Because liquids are isotropic, $g(\underline{r}_1,\underline{r}_2)$ depends only on the magnitude of $\underline{r}_2-\underline{r}_1$ ($|\underline{r}_2-\underline{r}_1| = r$) and consequently $g(\underline{r}_1,\underline{r}_2) = g(r)$.

If the motion of atoms has to be taken into account a proper description uses the space-time correlation function $G(\underline{r},t)$. $G(\underline{r},t)$ is a generalization of $g(r)$ and has the following meaning: Given an atom at the origin at time 0, $G(\underline{r},t)$ represents the probability to find also an atom at position $\underline{r}$ and time t. Individual and collective motion of atoms in liquids can be described in this way.

Microscopic theories have to construct appropriate models for these correlation functions. Experimental techniques aim for an accurate determination of these or related quantities.

The technique of thermal neutron scattering has proved to be a very useful tool in the study of structural and dynamic properties of simple liquids. After a short outline of this technique and its possibilities we will discuss some more recent results.

II. STATIC AND DYNAMIC STRUCTURE FACTOR

The scheme of a scattering experiment is illustrated in Fig.1

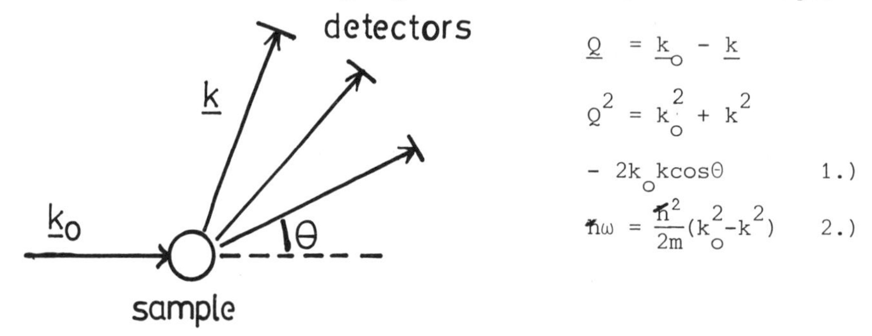

$$\underline{Q} = \underline{k}_o - \underline{k}$$

$$Q^2 = k_o^2 + k^2 - 2k_o k\cos\Theta \qquad 1.)$$

$$\hbar\omega = \frac{\hbar^2}{2m}(k_o^2 - k^2) \qquad 2.)$$

$\underline{k}_o$, $\underline{k}$ are wave vectors of incident and scattered neutrons; $\hbar\underline{Q}$, $\hbar\omega$ momentum and energy transfer in the scattering process and m = mass of the neutron. Monochromatic neutrons impinge on the sample. In the most general experiment the scattered intensity is measured not only as a function of scattering angle Θ but also of the change in wave vector magnitude or energy of the neutron. The ability of energy analysis is a unique feature of the neutron probe and presently not possible with X-rays.

Formally the scattering can be described by a double differential cross-section $d^2\sigma/d\Omega d\omega$ which represents the fraction of incident neutrons that is scattered into a solid angle $d\Omega$ and an energy interval $d\hbar\omega$. For an ensemble of nuclei in a liquid this cross-section is evaluated readily by standard techniques, that is, using the Born approximation and replacing the potential of the scatterer by the Fermi pseudopotential. The result for a scatterer with N particles is per atom /1/:

$$\frac{d^2\sigma}{d\Omega d\omega} = \frac{1}{2\pi N}\frac{k}{k_o}\iint e^{i(\underline{Q}\,\underline{r} - \omega t)}\, d\underline{r}\, dt \left\langle \sum_{ij} \int_V b_i \delta\left[\underline{r} + \underline{r}_i(0) - \underline{r}'\right] \times b_j \delta\left[\underline{r}' - \underline{r}_j(t)\right] d\underline{r}' \right\rangle \quad 3.)$$

where b_i is the bound scattering length of the nucleus i and $\langle --- \rangle$ denotes a thermal average. The expression in brackets is closely related to the space time correlation function $G(\underline{r},t)$ introduced above and formally defined as:

$$G(\underline{r},t) = \frac{1}{N}\left\langle \sum_{ij} \int_V d\underline{r}'\ \delta\left[\underline{r} + \underline{r}_i(0) - \underline{r}'\right] \delta\left[\underline{r}' - \underline{r}_j(t)\right]\right\rangle \quad 4.)$$

$d^2\sigma/d\Omega d\omega$ can be expressed in terms of $G(\underline{r},t)$ if the scattering lengths are taken out. The b_i vary from isotope to isotope and depend on the relative orientation of neutron and nuclear spins. The averaging of the b_i over all these states may be expressed in terms of coherent and incoherent scattering lengths:

$$\overline{b_i b_j} = b^2_{coh} + b^2_{incoh}\ \delta_{ij} \quad 5.)$$

With this notation $d^2\sigma/d\Omega d\omega$ for a sample of one atomic species can be written

$$\frac{d^2\sigma}{d\Omega d\omega} = \frac{1}{2\pi}\frac{k}{k_o}\iint e^{i(\underline{Q}\,\underline{r} - \omega t)}\left[b^2_{coh} G(\underline{r},t) + b^2_{incoh} G_s(\underline{r},t)\right] d\underline{r}\, dt \quad 6.)$$

where

$$G_s(\underline{r},t) = \frac{1}{N}\left\langle \sum_i \int_V d\underline{r}'\,\delta\left[\underline{r}+\underline{r}_i(0)-\underline{r}'\right]\ \delta\left[\underline{r}'-\underline{r}_i(t)\right]\right\rangle$$

is the socalled self correlation function. The difference $G(\underline{r},t) - G_s(\underline{r},t) = G_d(\underline{r},t)$ is called the distinct correlation function. Finally the space-time Fourier transformations of $G(\underline{r},t)$ and $G_s(\underline{r},t)$ are called scattering laws and denoted by $S(\underline{Q},\omega)$ and $S_s(\underline{Q},\omega)$ respectively. With this notation we have

$$\frac{d^2\sigma}{d\Omega d\omega} = \frac{k}{k_o}\left(b^2_{coh}\ S(\underline{Q},\omega) + b^2_{incoh}\ S_s(\underline{Q},\omega)\right) \quad 7.)$$

this formalism can be easily generalized for the case of samples containing different atomic species. Instead of elaborating on this we should mention some general properties of the introduced functions. From equ.4.) a simple connection between the static pair correlation function $g(r)$ and $G(\underline{r},t)$ can be derived.

$$G(\underline{r},0) = \delta(r) + \rho\, g(r) \qquad 8.)$$

Other important general properties are given by the mth moments of the scattering laws:

coherent

$$<\omega^m> = \int_{-\infty}^{+\infty} \omega^m S(Q,\omega)\, d\omega$$

$$<\omega^0> = S(Q)$$

$$<\omega^2> = \frac{k_B T}{M} Q^2 \qquad 9.)$$

$$<\omega^4> = 3\left(\frac{k_B T}{M}\right)^2 Q^4 + \frac{k_B T}{M^2} Q^2 \rho \int_V g(r) \frac{\partial^2 u(r)}{\partial x^2} \left[1 - \cos(Qx)\right] d\underline{r}$$

incoherent

$$<\omega^m>^s = \int_{-\infty}^{+\infty} \omega^m S_s(Q,\omega)\, d\omega$$

$$<\omega^0>^s = 1$$

$$<\omega^2>^s = \frac{k_B T}{M} Q^2$$

$$<\omega^4>^s = 3\left(\frac{k_B T}{M}\right)^2 Q^4 + \frac{k_B T}{M^2} Q^2 \rho \int_V g(r) \frac{\partial^2 u(r)}{\partial x^2}\, d\underline{r}$$

where M is the atomic mass.

We are particularly interested in the zeroth moment of $S(Q,\omega)$ which is called the static structure factor $S(Q)$:

$$\int_{-\infty}^{+\infty} S(Q,\omega)\, d\omega = \frac{1}{2\pi} \iiint e^{i(\underline{Q}\underline{r} - \omega t)} \left[G(\underline{r},t) - \rho\right] d\underline{r}\, dt\, d\omega$$

$$= \iint \delta(t)\, e^{i\underline{Q}\underline{r}} \left[G(\underline{r},t) - \rho\right] d\underline{r}\, dt$$

$$= \int_V e^{i\underline{Q}\underline{r}} \left[G(\underline{r},0) - \rho\right] d\underline{r}$$

$$S(Q) = 1 + \rho \int_V e^{i\underline{Q}\underline{r}} \left[g(r) - 1\right] d\underline{r} \qquad 10.)$$

S(Q) contains the complete structural information on a liquid. In an X-ray experiment energy changes cannot be resolved the ω-integration is performed automatically.

If in a neutron scattering experiment the energy analysis is omitted and the scattered intensity measured only as a function of the angle Θ or momentum transfer Q then to a first approximation S(Q) is also obtained. But it should be noted that this procedure does not give a proper integration over ω unless the neutron energy is high compared to $\hbar\omega$.

Nevertheless experiments on the structure of liquids are normally performed in this way and the experimental data are corrected by using the higher moments, the socalled Placzek corrections to come closer to equ.1o.). One of the approximations very often adapted is /2/:

$$\left.\frac{d\sigma}{d\Omega}\right|_{exp} = Nb^2_{coh}\left(S(Q) + \frac{\bar{K}m}{3E_o M} - \frac{\hbar^2 Q^2}{2ME_o}\,\frac{1}{D(k)}\right) \qquad 11.)$$

where E_o is the energy of incident neutrons, $\bar{K}$ the mean kinetic energy of atoms of mass M and D(k) the detector efficiency.

A typical S(Q) for a liquid is illustrated in Fig. 2a

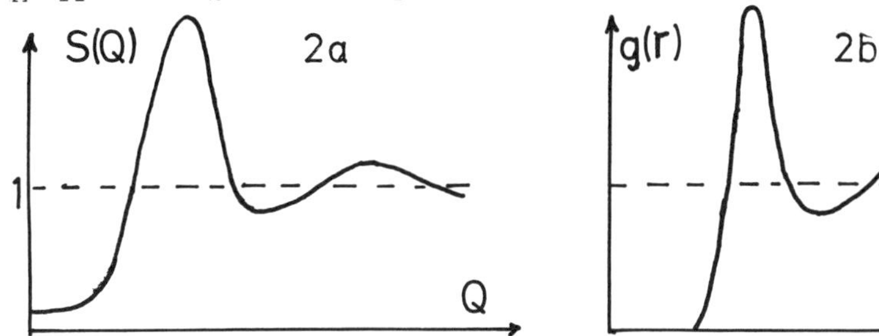

If S(Q) is measured for a sufficient large Q-range then it is obvious from equ.1o.) that g(r) can be determined by Fourier transformation

$$g(r) = 1 + \frac{1}{8\pi^3\rho}\int\left[S(Q) - 1\right] e^{i\underline{Q}\underline{r}}\, d\underline{Q} \qquad 12.)$$

In this way S(Q) and g(r) have been determined for many liquid metals near their melting point. Less data have been collected at higher temperatures or as a function of temperature.

III. RELATION BETWEEN g(r) AND THE PAIR POTENTIAL u(r)

Experimental determined S(Q) or g(r) are not only of interest for evaluating integral and transport properties of liquids but offer also a possibility to study the microscopic forces in a

liquid because the details of $g(r)$ are governed by the interactions between particles. The aim of a microscopic theory should be a satisfactory prediction of $g(r)$ from a known pair potential $u(r)$ or vice versa to extract a pair potential from a measured $g(r)$.
The formal relation between $g(r)$ and $u(r)$ is well established in statistical mechanics and usually represented in the form of the Bogoliubov-Born-Green-Kirkwood-Yvon (BBGKY)-hierarchy. Informative for our purposes is the second equation of this hierarchy, which can be written /1/:

$$\frac{\partial}{\partial r}\left[\ln g(r) + \frac{u(r)}{k_B T}\right] = -\frac{\rho}{k_B T g(r)}\int \frac{\partial u(s)}{\partial \underline{r}_1} g_3(\underline{r}_1,\underline{r}_2,\underline{r}_3)\, d\underline{r}_3 \qquad 13.)$$

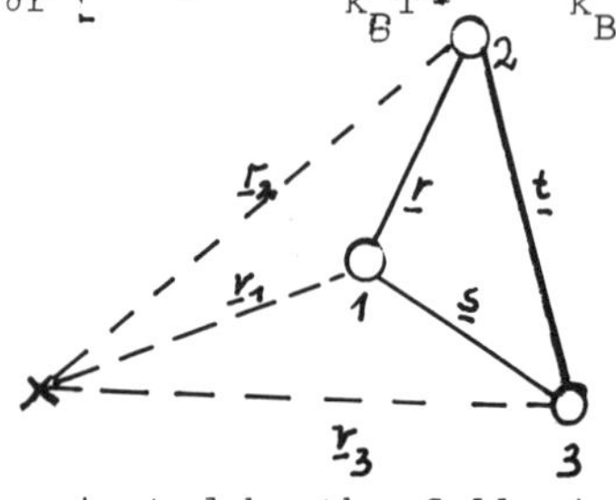

The problem is the unknown triplet correlation function $g_3(\underline{r}_1,\underline{r}_2,\underline{r}_3)$ which is a measure for the probability to find an atom at $\underline{r}_3$ if two other atoms are at $\underline{r}_1$ and $\underline{r}_2$.

By using the Kirkwood superposition approximation $g_3(\underline{r}_1,\underline{r}_2,\underline{r}_3) = g(r)\, g(s)\, g(t)$, equ. 13.) can be approximated by the following closed relation between $g(r)$ and $u(r)$, the Yvon-Born-Green (YBG)-equation:

$$\ln g(r) + \frac{u(r)}{k_B T} = -\rho\int E(|\underline{r}' - \underline{r}|)\left[g(r') - 1\right] d\underline{r}' \qquad 14.)$$

with $E(r) = \frac{1}{k_B T}\int_\infty^r g(x)\, \frac{\partial u(x)}{\partial x}\, dx$

Another approximation is the hypernetted chain (HCN) equation which follows formally from equ. 14.) by substituting $E(r)$ by the direct correlation function $c(r)$:

$$\ln g(r) + \frac{u(r)}{k_B T} = \rho\int c(|\underline{r} - \underline{r}'|)\left[g(\underline{r}') - 1\right] d\underline{r}' \qquad 15.)$$

$c(r)$ is defined through the Ornstein-Zernike relation /3/:

$$h(r) = c(r) + \rho\int c(|\underline{r} - \underline{r}'|)\, h(\underline{r}')\, d\underline{r}'; \quad h(r) = g(r)-1 \qquad 16.)$$

In fact equ. 15.) is obtained from 13.) by carrying out a cluster expansion for g_3 and omitting difficult terms. In a similar manner, by neglecting even more terms in the cluster expansion the Percus Yevick (PY)-equation is obtained /4/:

$$\ln g(r) + \frac{u(r)}{k_B T} = \ln\left[g(r) - c(r)\right] \qquad 17.)$$

Equ. 14.), 15.) and 17.) have been extensively compared with experimental data and their short-comings discussed. They are accurate for low densities. Because the PY-equation has an analytical solution for a liquid of hard spheres it is very often used to describe measured structure factors.
Indeed it has been shown by Ashcroft and Leckner /5/ that the hard sphere model can give a reasonable good overall description of the structure factors of simple liquid metals near the triple point which indicates that the repulsive part of the potential dominates the interaction. The agreement is good at large Q-values as one expects but it diminishes at lower Q where the effect of the attractive part of the potential may be most pronounced.

Realistic potentials may be treated by perturbation methods. Here the idea is that the properties of a real liquid can be described by those of a simpler reference liquid if they are corrected by a perturbing potential. Several techniques have been used for this purpose. We mention the method of Weeks. Chandler and Anderson (WCA) /6/ in which the real potential u(r) is divided in a repulsive and an attractive part as illustrated in Fig. 3

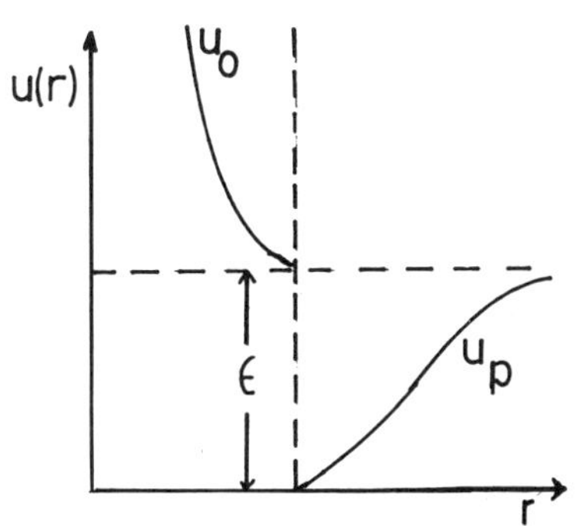

$$u(r) = u_o(r) + u_p(r) \qquad 18.)$$

$$u_o(r) = u(r)+\varepsilon\ , \quad u_p(r) = -\varepsilon \quad \text{for } r < r_o$$

$$u_o(r) = 0\ , \quad u_p(r) = u(r) \quad \text{for } r > r_o$$

where r_o is the position of the minimum of u(r).
a) The reference system u_o is first approximated by a hard sphere potential and $S_{hs}(Q)$ determined for this system, e.g. with the PY-equation;

b) the perturbation $u_p(r)$ is taken into account in random phase approximation;

c) and finally the softness of the repulse part of the potential is taken into account.

Although this method yielded very good results for Lennard-Jones systems it seems less convincing for liquid metals.
Recently a different approach based on a dynamic response theory by Singwi et al. (STLS), and called self-consistent-field method /7/ has been applied to evaluate S(Q) for liquid metals. The starting point is the relation:

$$S(Q) = \frac{1}{1-\rho c(Q)} \qquad 19.)$$

In the response theory c(Q) is the Fourier transformation of an effective field $-\psi(r)/k_B T$ which is defined by

$$\frac{\partial \Psi(r)}{\partial r} = g(r)\frac{\partial u(r)}{\partial r} \qquad 2o.)$$

Although equ. 19.) and 2o.) can be solved in principle by iteration this is not practicable for systems with strong repulsive interactions. Therefore similar as with perturbation methods a trial ansatz is used for $r < r_o$ and optimized by minimizing the free energy of the system. An example of the use of the last two methods is given in Fig.4

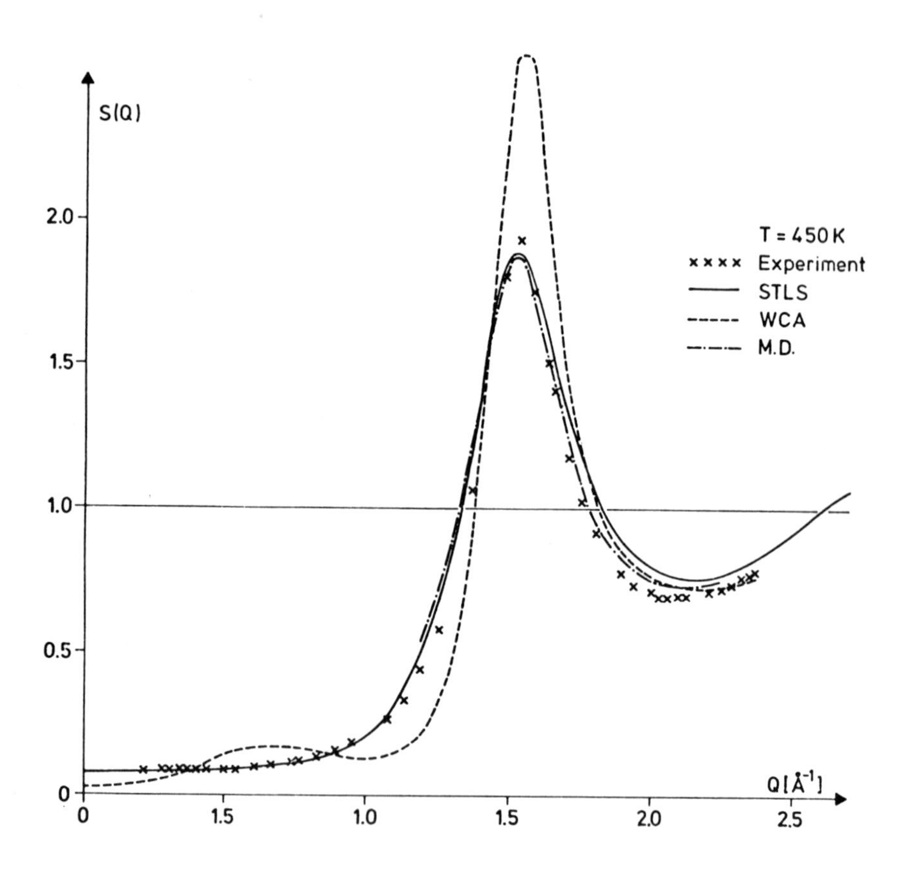

In Fig.4 the measured S(Q) of rubidium at 45oK is compared with calculations using the WCA and STLS methods. The potential used was a local pseudo potential due to Shyu and Gaspari /8 /. The same potential has also been taken for the molecular dynamics calculation. STLS-calculations have also been applied to describe the temperature dependence of S(Q) of liquid rubidium. A few results are illustrated in Fig.5

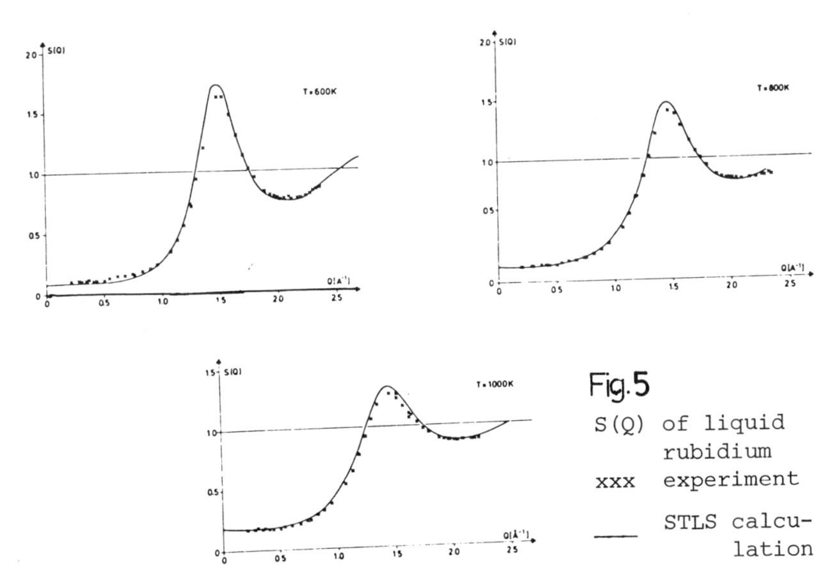

Fig.5

S(Q) of liquid rubidium

xxx experiment

—— STLS calculation

It is worthwhile to mention that in these calculations no free parameters are used. The deepness of the screened potential increases by about 6 % in the covered temperature range as one expects from the change in electronic density. The measurements on the temperature dependence of S(Q) have been extended to higher T (2oooK) and preliminary results reported / 9 /.

It is interesting that at higher T (above 12OOK) the used pseudopotential fails to reproduce the low Q-region of S(Q). This effect, although explainable by an additional deepening and broadening of the potential valley, is not understood.
For illustration Fig. 6 shows that a fit with a temperature dependent parameter hard sphere modell fails to describe this effect.

IV. DENSITY DEPENDENCE OF THE STRUCTURE FACTOR

As we have seen above the scattering of neutrons from liquids

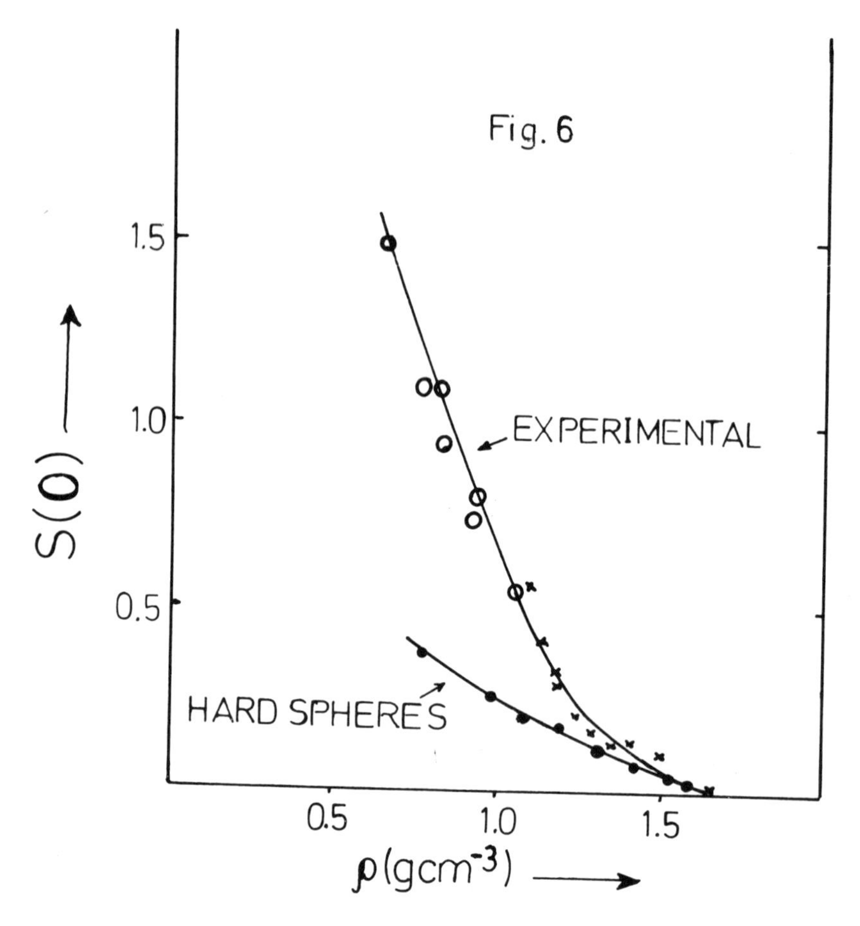

yields information only on the pair correlation functions. The only known way to go beyond this and to try to get information on higher order correlations is to change thermodynamic parameters and measure eventually corresponding derivatives of $S(Q)$ or $S(Q,\omega)$. The temperature derivative of $g(r)$ is rather complex because it involves not only the potential but also the three and four particle correlation functions /11 /. Therefore this would certainly not be the basis for an experimental approach to study higher correlations. More transparent and useful from an experimental point of view are the pressure derivatives.

The isothermal pressure derivative of the pair correlation function $g(r)$ has been derived by Yvon /12/ to be:

$$k_B T \frac{\partial \rho^2 g(r)}{\partial p} \Big|_T = 2\rho g(r) + \rho^2 \int \left[g_3(\underline{r},\underline{s}) - g(r) \right] d\,\underline{s} \qquad 21.)$$

The pressure derivative of the structure factor is essentially the Fourier transformation of 21.) and given by

$$\frac{\partial S(Q)}{\partial p}\Big|_T = \int \frac{\partial g(r)}{\partial p}\Big|_T e^{i\underline{Qr}} d\underline{r} + \frac{S(0)(S(Q)-1)}{\rho k_B T} \qquad 22.)$$

Although no direct measurement of g_3 is possible in this way, equation 21.) or 22.) may be useful for testing proposed models for the triplet correlation function if the pressure derivative of S(Q) can be measured accurately enough.

Because these are difficult experiments only a few examples have been reported in literature so far /13/. The results on liquid rubidium near the triple point lead to the conclusion that the Kirkwood superposition approximation is not a good approximation in this case.

V. EXPERIMENTS ON LIQUID DYNAMICS

In order to get information on the motions of atoms in liquids with the neutron technique the scattered neutrons have to be analysed due to the experienced momentum and energy change. In this way the double diffential scattering cross-section (equ.7.)) can be measured. If the scattering is predominantly coherent the scattering law $S(Q,\omega)$ can be determined, if it predominantly incoherent the self part $S_s(Q,\omega)$ results.

A technique which can be used for this purpose is triple axis spectrometry well established for studying excitations like phonons in solids but feasible only if a limited range of Q and ω is of interest. If the interest is in a significant range of these variables the time-of-flight technique is better suited. A modern instrument for the latter approach is the TOF-Spectrometer IN4 at the high flux reactor in Grenoble.

Up to now only a few inelastic scattering experiments on liquid metals have been performed. This is not only due to the fact that these are rather elaborate and time-consuming experiments but also because double differential scattering cross-sections of liquids are in general rather featureless smooth curves. It is necessary to achieve sufficient accuracy if significant information should be extracted from the data.

With the availability of intense neutron sources and the elaboration of the necessary evaluation procedures including the important corrections for multiple scattering /14/, the accuracy of the experiments has been greatly improved during the last few years.

1. Inelastic incoherent scattering

Among the metallic elements only vanadium scatters predominantly

incoherent but in some other cases the change in isotope composition can be used, to separate coherent and incoherent scattering.
The last technique has been used recently /15/ to study the self part of the scattering functions $S_s(Q,\omega)$ of nickel. Fig. 7 illustrates by one of the results on nickel that $S_s(Q,\omega)$ is a rather featureless function.

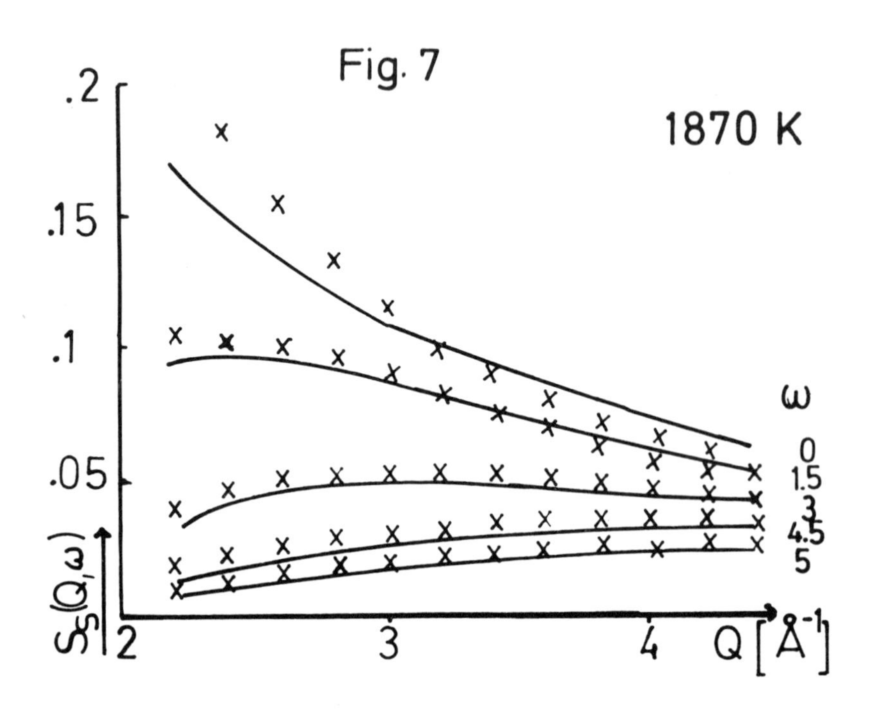

An appropriate description of single particle motion in a liquid can be given by the velocity autocorrelation function or its Fourier transform the spectral density:

$$z(t) = \langle v_x(0)v_x(t)\rangle \; ; \; p(\omega) = \frac{1}{2\pi}\int_{-\infty}^{+\infty}\langle v_x(0)v_x(t)\rangle e^{-i\omega t}dt \qquad 23.)$$

Because in the case of solids $p(\omega)$ is proportional to the frequency spectrum of phonons, in analogy $p(\omega)$ of a liquid can be interpreted as frequency spectrum of excitations. Experimentally $p(\omega)$ can be determined from $S_s(Q,\omega)$ by an extrapolation procedure /1/:

$$p(\omega) = \omega^2 \lim_{Q\to 0} \frac{S_s(Q,\omega)}{Q^2} \quad \text{with } p(0) = \frac{D}{\pi} \qquad 24.)$$

The result of this extrapolation procedure for the nickel data is shown in Fig.8. Comparison with solid data can give some information on the change of forces. Particular interest arose recently in the small ω region of p(ω) from which information on the long-time -behaviour of z(t) can be deduced.

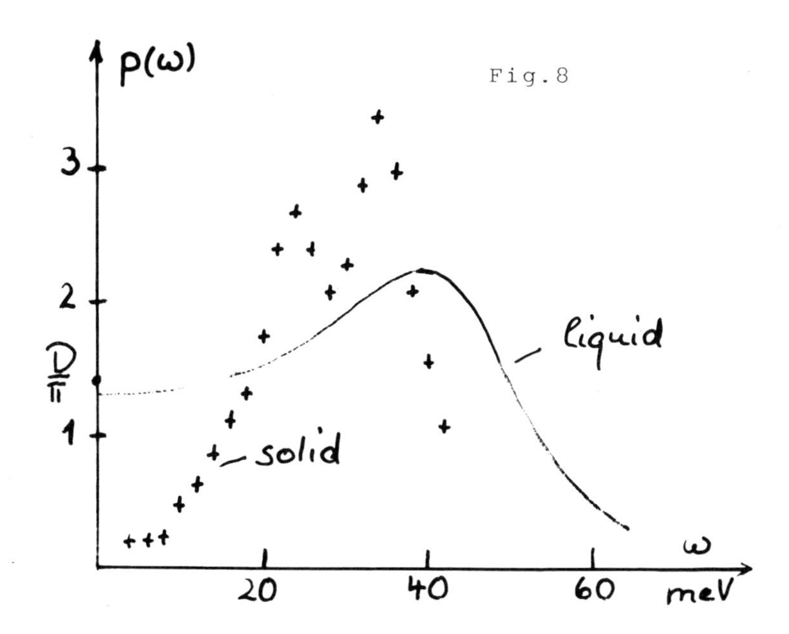

It was noticed first in molecular dynamics calculations that z(t) has a long tail, decaying at long times as $t^{-3/2}$. This effect should show in the Fourier transform p(ω) at small ω .

The $t^{-3/2}$ tail would imply that the leading term in an expansion of p(ω) should be proportional to $\omega^{1/2}$:

$$p(\omega) = \frac{D}{\pi} + a_1\omega^{1/2} + a_2\omega + a_3\omega^{3/2} \ldots \qquad 25.)$$

In the hydrodynamic limit the coefficient a_1 can be evaluated /16/ to be

$$a_1 = -\frac{\sqrt{2}}{12\pi^2}\frac{k_B T}{\rho M}\left[D + \frac{\eta}{\rho M}\right]^{-3/2} \qquad 26.)$$

where D is the diffusion constant and η the viscosity.

Clearly equation 25.) predicts a cusp with infinite slope of $p(\omega)$ at $\omega = 0$.
There have been attempts to see this effect also by reanalyzing earlier data /17/. But up to now it has not been seen, simply because the ω-resolution normaly used for liquid studies is to poor and has to be improved considerably.

2. Inelastic Coherent Scattering

a) Analysis of higher moments

Similar to the experience for solids more detailed information can in principle be deduced from the coherent scattering function $S(Q,\omega)$. As we have seen in section II, the dynamic behaviour of a system can quite generally be expressed by the higher frequency moments of $S(Q,\omega)$. The first higher moment containing non trivial information on particle interaction - assuming that $S(Q)$ is allready known - is the fourth moment. $\langle\omega^4(Q)\rangle$ can be written:

$$\langle\omega^4(Q)\rangle = 3\langle\omega^2(Q)\rangle^2 + \langle\omega^2(Q)\rangle \frac{\rho}{M}\int\left[1 - \cos(Qx)\right] g(r)\frac{\partial^2 u(r)}{\partial x^2}\, d\underline{r} \qquad 27.)$$

Rahman and others /18/ recently proposed an inversion procedure of equ. 27.) for extracting the pair potential. For this purpose the integral in 27.) may conveniently be rewritten in the form:

$$\frac{\rho}{M}\int\left[1 - \cos(Qx)\right] g(r) \frac{\partial^2 u(r)}{\partial x^2}\, d\underline{r} = P(Q) - P(0) \qquad 28.)$$

where after carrying out the angular integrations, $P(Q)$ takes the form

$$P(Q) = -\frac{4\pi\rho}{M}\int_0^\infty r^2 g(r)\, j_o(QR)\, \frac{d^2u(r)}{dr^2}\, dr - K(Q) \qquad 29.)$$

with

$$K(Q) = -\frac{8\pi\rho}{Q^2 M}\int_0^\infty g(r)\left[j_o(Qr) - \cos Qr\right]\left[\frac{d^2u(r)}{dr^2} - \frac{1}{r}\frac{du(r)}{dr}\right] dr$$

According to 27.) and 28.) $P(Q)$ can be expressed by measured quantities:

$$P(Q) - P(0) = \langle\omega^4(Q)\rangle / \langle\omega^2(Q)\rangle - 3\langle\omega^2(Q)\rangle$$

And if $K(Q)$ in equ. 29.) is neglected in a first approximation, the inversion yields

$$\frac{d^2u(r)}{dr^2} = -\frac{M}{8\pi^3\rho g(r)}\int P(Q)\, j_o(Qr)\, dQ \qquad 30.)$$

The result of equ. 3o.) may be used in equ.29.) in an iterative way until selfconsistency between d^2u/dr^2 and du/dr is reached. Sometimes the following approximation for P(Q):

$$P(Q) = \frac{3\omega_E^2}{(Qr_o)^3}\left[1 - (Qr_o)^2 \sin/Qr_o) - 2Qr_o \cos(Qr_o) + 2\sin(Qr_o)\right] \quad 31)$$

with

$$\omega_E^2 = \frac{4\pi\rho}{3M} \int_0^\infty r^2 g(r) \frac{d^2u(r)}{dr^2}\, dr$$

is used to describe experimental results.

In any case $<\omega^4(Q)>$ must be known fairly accurately. Its experimental determination is not an easy task.

One of the very few experimental determinations of $<\omega^4(Q)>$ has been performed by Suck /19/ for a restricted Q-range in liquid rubidium. His results are shown in Fig.9.

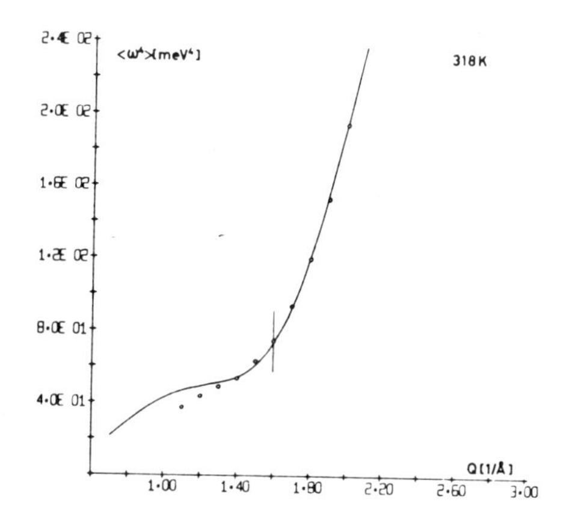

ooo experimental data

___ calculated according to equ. 31.)

The deviations at small Q may be due to incomplete experimental data.

Clearly experimental determinations of the fourth moment need further improvement before significant conclusions can be drawn from such data. Although it would be appealing to go further and to analyse the sixth frequency moment which would reveal even more details of the dynamics because of obvious experimental difficulties no reliable results have been reported up to now.

b) Structure of $S(Q,\omega)$, collective excitations

Since the first coherent inelastic neutron scattering experiments on liquids there was particular interest in detecting indications of propagating collective modes. Although interpretations of the earlier results on liquid metals were somewhat controversial it is now generally accepted that the behaviour of high frequency collective excitations in liquid metals may be - different from that in rare gas liquids - rather solid like. The first convincing experiment was performed by Copley and Rowe /20/ on liquid rubidium close to its melting point. They found peaks at finite frequencies in $S(Q,\omega)$ at Q-values corresponding to atomic distances. An example of their results is illustrated in Fig.1o.

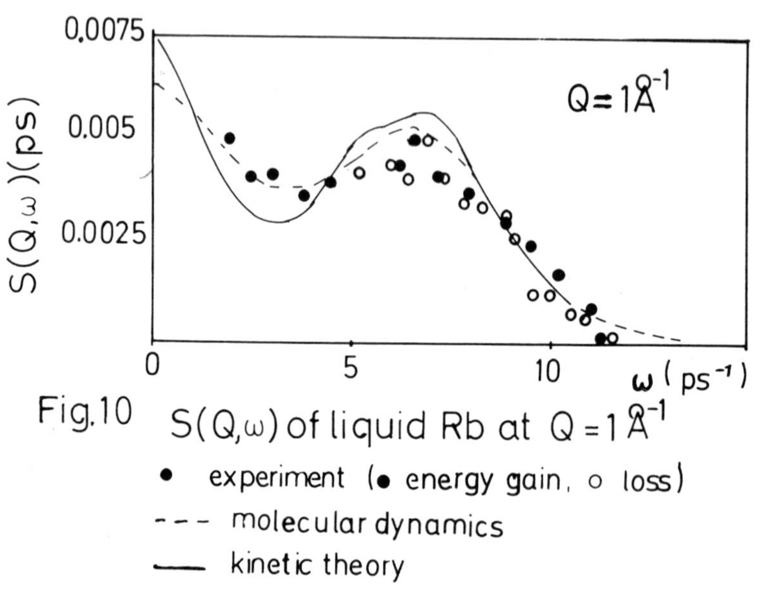

Fig.10 $S(Q,\omega)$ of liquid Rb at $Q = 1\ \text{Å}^{-1}$
• experiment (• energy gain, o loss)
--- molecular dynamics
—— kinetic theory

The peaks, changing their position with Q were found to persist up to Q of about 1 Å^{-1}, which is about 2/3 towards the first maximum of $S(Q)$. This particular structure of $S(Q,\omega)$ has been verified by molecular dynamics calculations /21/. A three peak structure of $S(Q,\omega)$ can be understood qualitatively with viscoelastic models if a Q-dependent relaxation time is introduced. Satisfying descriptions are now also achieved by microscopic kinetic theories /22/.The ω-values of the peaks plotted as a function of Q - the familiar way to illustrate phonon dispersions in solids - is shown in Fig.11

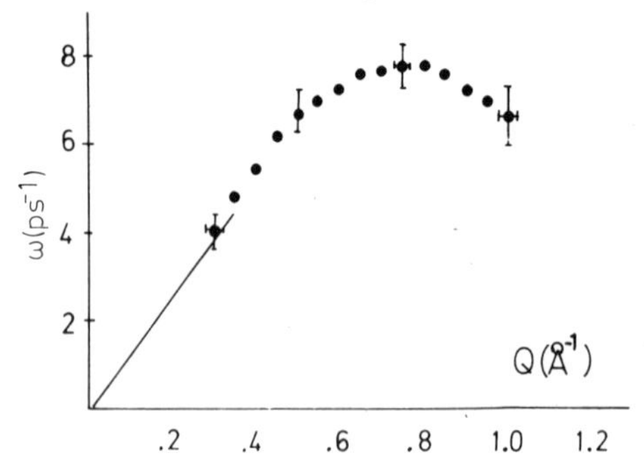

This curve is rather close to the average dispersion of longitudinal phonons in solid rubidium close to the melting point.

Recent pressure dependent measurements on liquid rubidium at the high flux reactor /23/ confirm the earlier results ar normal pressure. With the higher resolution available these measurements indicate that the upper Q-limit for propagating longitudinal modes is even higher (~1,2 $Å^{-1}$) in agreement with model predictions.

With increasing pressure up to 1,3 kbar corresponding to a density increase of 6 % a general increase in mode frequencies was found, which is qualitatively understandable with the simple theory of electron screening in metals.

Similar structures have been seen recently in $S(Q,\omega)$ data of liquid lead /24/ but are far less pronounced in liquid aluminium /25/, liquid copper /26/ and liquid gallium /27/.

Although the microscopic reason for these persisting modes is not completely clear it is believed that they are due to the special form of the effective pair potential in liquid alkali metals having a relative soft core. But Lewis and Lovesey /28/ concluded from a series of computer simulations for different potential shapes that the potential also has to be fairly harmonic around its minimum. A similar argument was put forward by Sjögren /29/.
Systematic measurements on liquid metals having effective potentials of different shape would certainly help to advance a more detailed understanding of the requirements for persisting high frequency modes.
From the data existing so far it can be concluded that in liquid metals propagating longitudinal collective excitations may exist to fairly high Q-Values. However with increasing Q the relaxation time ($\sim 10^{-12}$sec) decreases below the inverse excitation frequency thus preventing the propagation.
Continuing interest persists also in understanding details of the behaviour of transverse collective modes in liquid metals. Microscopic theories predict high frequency transverse modes at high Q-values. Unfortunately neutrons are only good probes for longitudinal excitations. Details on transverse modes probably can be investigated only via their coupling to longitudinal modes by very accurate measurements of $S(Q,\omega)$ and comparison of such data with proposed models.

c) Density dependence of $S(Q,\omega)$

From the discussion in section V on the density dependence of the static structure factor it is suggestive to try to get deeper insight in the dynamic behaviour of liquid metals by studying the pressure dependence of $S(Q,\omega)$. In this way in principle information on the time-dependent triplet correlation function may be obtained. A time-dependent analog to equations 21.) and 22.) was given by Egelstaff et al./30/ and can be written

$$\left.\frac{\partial S(Q,\omega)}{\partial p}\right|_T = \int e^{i[\underline{Q}(\underline{r}-\underline{r}')-\omega t]} \frac{\partial G(\underline{r}',0;\underline{r},t)}{\partial p} d(\underline{r}-\underline{r}') \qquad 32.)$$

with

$$\rho k_B T \frac{\partial G(\underline{r}',0;\underline{r},t)}{\partial p} = \int [G_3(\underline{r}', 0;\underline{r}'',0;\underline{r},t) - G(\underline{r}',0;\underline{r}'',0) \cdot G(\underline{r}',0;\underline{r},t)] \, d\underline{r}''$$

where the triplet correlation function $G_3(\underline{r}',0;\underline{r}'',0;\underline{r},t)$ is the probability of finding atoms at $(\underline{r}',0)$, $(\underline{r}'',0)$ and $(\underline{r},t)$.

It has been shown /31/ that at short t or large ω equation reduces to

$$\left.\frac{\partial \ln S(Q,\omega)}{\partial p}\right|_T = \left.\frac{\partial \ln S(Q)}{\partial p}\right|_T \qquad 33.)$$

If experimentally deviations from equation 33.) could be detected they would give information on longtime collision effects. The experiment on the pressure dependence of $S(Q,\omega)$ of liquid rubidium /23/ was designed for such studies. In this experiment $S(Q,\omega)$ was measured at twelve points in (p,T)-space covering a temperature range of 15oK and a pressure range of 1.3 kbar.

Besides the general information on the structure of $S(Q,\omega)$ mentioned already it is hoped that not only accurate ρ- and T-dependent $S(Q,\omega)$ data but also more accurate $S(Q)$-data will be made available. The short discussion on the Placzek corrections in section II may have indicated that finally the only unambigous way to accurate $S(Q)$ is to measure $S(Q,\omega)$ accurately.

Such data will allow to extract a pair potential from the T-dependence of $S(Q)$ at constant density which may be a more unique way /32/ than the one described in section IV. Similarly from the density dependence of $S(Q)$ at constant temperature density effects on the potential may be studied and information on the static triplet correlation function g_3 may be extracted. Finally the density dependence of $S(Q,\omega)$ may lead to some information on the time dependent triplet correlation function.

References

/1/ P.A.Egelstaff,"An Introduction to the Liquid State",Academic Press, London(1967)

/2/ J.E.Enderby,"Physics of Simple Liquids",North-Holland, Amsterdam (1968)

/3/ see e.q. S.A. Rice and P. Gray,"The Statistical Mechanics of Simple Liquids", Wiley, New York (1965)

/4/ L.S. Ornstein and F.Zernike, Physik Z. 19, 134 (1918)

/5/ N.S. Ashcroft and J.Leckner, Phys. Rev. 145, 83 (1966)

/6/ J.D. Weeks, D.Chandler, and H.C. Andersen, J.Chem.Phys. 54, 5237 (1971)

/7/ N.K. Ailawadi, D.E. Miller, and J. Naghizadeh, Phys.Rev.Lett. 36, 1494 (1976)

/8/ W.M. Shyu and G.D. Gaspari, Phys.Rev. 170, 687 (1968)

/9/ W. Freyland, W.Gläser and F. Hensel, Ber.Bunsenges. Phys. Chem., in press

/10/ R.Evans and W. Schirmacher, J.Phys.C.II, 2437 (1978)

/11/ P. Schofield, Proc. Phys. Soc. 88, 149 (1966)

/12/ J.Yvon, "Correlations and Entropy in Classical Statistical Mechanics", Pergamon Press Oxford (1966)

/13/ P.A. Egelstaff, D.I. Page and C.R.T. Heard, J.Phys. C4, 1453 (1971)

/14/ e.g. J.R.D. Copley, Computer Physics Comm. 7, 289 (1974)

/15/ M.W. Johnson, B. Mc Coy, N.H. March, Phys. Chem. Liq. 6, 243 (1977)

/16/ N.H. March and M.P. Tosi, "Atomic Dynamics in Liquids", MacMillan London (1976)

/17/ K. Carneiro, Phys. Rev. A14, 517 (1976)

/18/ A. Rahman, Phys. Rev. A11, 2191 (1975)

/19/ J.B. Suck, Thesis Unversity of Karlsruhe (1975)

/20/ J.R.D. Copley, J.M. Rowe, Phys. Rev. A9, 1656 (1974)

/21/ A. Rahman, Phys. Rev. A9, 1667 (1974)

/22/ A.Zippelius and W. Götze, Kinetic theory for the coherent scattering function $S(q,\omega)$ of classical liquids, Phys.Rev. A17, 414 (1978)

/23/ P.A. Egelstaff, W. Gläser, R.McPherson, J.B. Suck, A.Teitsma, unpublished

/24/ J.R.D. Copley, B. Dorner, O. Söderström, J.B. Suck, unpublished

/25/ O.J. Eder, B. Kunsch, J.B. Suck, unpublished

/26/ S.Hagen, Thesis University of Karlsruhe (1973)

/27/ U. Löffler, Thesis University of Karlsruhe (1973)

/28/ J.W.E. Lewis and S.W. Lovesey, J.Phys. C10, 3221 (1977)

/29/ L. Sjögren, J.Phys. C12, 425 (1979)

/30/ P.A.Egelstaff, Ann. Rev. Phys. Chem. 24, 159 (1973)

/31/ L.Groome, K.Gubbins, J.Dufty, Phys.Rev. A13, 437 (1976)

/32/ S.K. Mitra, P. Hutchinson, P.Schofield, Phil. Mag. 34, 1087 (1976)

OBSERVATION OF SHORT-RANGE ORDER+

H. Ruppersberg

Fachbereich Angewandte Physik, Universität des
Saarlandes, 66oo Saarbrücken, W. Germany

ABSTRACT. Chemical short range order has been studied in liquid and solid ^7Li alloys and in amorphous $Ti_{0.6}Ni_{0.4}$. The theoretical background and some results are briefly discussed.

Close to the tripel-point in liquid metals the atoms are arranged similar to hard spheres (HS) in a dense random packing for which the structure factor S(q) can approximately be calculated following the Percus-Yevic (PY) assumption (1). If we now have a liquid alloy of A and B atoms which are mixed in the proportions x_A and $x_B=1-x_A$ the question arises whether, here again, the atoms are arranged as in a HS-system. If the atoms are of equal size and if as in a mixture of isotopes of the same element there is no special chemical interaction, then, obviously, there will be a random occupation of the sites of the liquid HS like structure by A and B atoms. Accordin to Ashcroft and Langreth (2), the three partial structure factors of such a mixture can approximately (PY) be calculated, also if the A- and B-spheres are of different size. The entropy of mixing solutions with randomly distributed A and B atoms corresponds to the entropy term of an ideal solution. But it is well known that these so-called regular solutions are very exceptional. In real liquids the entropy of mixing shows more or less strong positive or negative deviations from the ideal term. From this we can conclude that the simple picture of a random HS mixture does not hold in general. The excess entropy might be due to a non-random distribution of the A- and B-particles over the sites of the liquid structure yielding preferred self- or hetero-coordination.These two cases are still quite simple because they represent what one may call

+This work has been supported by the Deutsche Forschungsgemeinschaft and by the Institut Max von Laue, Paul Langevin, Grenoble, France.

"substitutional" alloys for which the global structure is identical to the structure of a pure liquid metal except for possible size effects. The situation is almost the same as in the case of homogenious disordered solid-solutions like CuAu and the structure should be described by indicating the global structure plus something analogous to the short-range order (sro) parameters. The next possible case is what one might call an "interstitial" alloy which will be formed if a relatively small amount of small particles is dissolved in a solvent of large particles of which the atoms remain arranged as in a pure liquid. That is exactly the Polk-Bernal model. Besides these relatively simple structures one can imagine more complicated cases in which, because of directed bonding, the global structure or the arrangement of the solvent in the case of an interstitial alloy, is no longer HS like. For example, there may be formation of polyhedra, like in crystalline, liquid and amorphous quartz, or of chains which exist in amorphous Se and its alloys.

It may be assumed, and it has been proved experimentally in two or three cases that the special bonding which yield a special chemical order in the crystalline phase is not altered on melting so that a similar chemical sro exists in the liquid. Because in the melt and in the amorphous phases the atoms are not forced on lattice sites, the structure of liquid alloys will possibly be simpler and not as varied as for the solid alloys. On the other hand, because of experimental problems and because for amorphous phases there is no three dimensional information available as for a single crystal, it will be very difficult to obtain the same quantitative information about chemical sro for liquids as for a solid. It is, however, very important to get as much information as possible about this sro because it is intimately related to many of the physicochemical properties and there is strong evidence that a special sro in the liquid is a necessary condition for the formation of a glass on rapid quenching.

Theory

In special cases the presence of sro becomes directly visible in the coherently scattered intensity per atom $I^c(q)$, which is obtained from a diffraction experiment. In this paper the following definition of the total structure factor has to be adopted:

$$(1)\quad S(q) = I^c(q) / \langle b^2 \rangle$$

where b is the coherent scattering amplitude and $\langle b^2 \rangle = x_A b_A^2 + x_B b_B^2$. To discuss sro effects it is advantageous to subdivide $S(q)$ into three partial "number concentration"-structure factors $S_{mn}(q)$ (with m,n = N,C) which have been introduced by Bhatia and Thornton (3):

$$(2)\quad S(q) = \{\langle b \rangle^2 S_{NN}(q) + 2\langle b \rangle \Delta b\, S_{NC}(q) + \Delta b^2 S_{CC}(q)\} / \langle b^2 \rangle .$$

where $\Delta b = b_A - b_B$ or:

(3) $S(q) = a\,S_{NN}(q) + b\,S_{NC}(q) + (1 - a)S_{CC}(q)/x_A x_B$

The physical meaning of the $S_{mn}(q)$ may best be elucidated starting from the definition of the number-density fluctuation $\delta n_i(\underline{R})$ of i atoms (i = A,B) which occurs at a given time at the place $\underline{R}$ (3):

(4) $\delta n_i(\underline{R}) = \sum_j \delta(\underline{R} - \underline{R}_j^i(\underline{R}) - x_i \rho_o$

$\underline{R}_j^i$ is the place of the j-th i-atom and $\int_{\Delta V} \sum \delta(\underline{R} - \underline{R}_j^i))d^3R$ is the number of i-atoms which are in the volume element ΔV.
$x_i \rho_o \Delta V = x_i(N/V)\Delta V$ is the mean number of i-atoms in this volume.
Thus, $\delta n_i(\underline{R})$ denotes the i-particle density fluctuation.
$\delta n(\underline{R}) = \delta n_A(\underline{R}) + \delta n_B(\underline{R})$ is the fluctuation of the total number-density and its Fourier Transform (FT) is given by:

(5) $N(\underline{q}) = \int e^{-i\underline{q}R}\,\delta n(\underline{R})d^3R$

$S_{NN}(\underline{q})$ is related to the FT of $\delta n(\underline{R})$ by

(6) $S_{NN}(\underline{q}) = N^{-1} \langle N^x(\underline{q}) \cdot N(\underline{q})\rangle = N^{-1}\int\int e^{i\underline{q}(\underline{R}-\underline{R}')}\,\delta n(\underline{R})\delta n(\underline{R}')d^3R d^3R'$

and with the distance vector $\underline{r} = \underline{R} - \underline{R}'$ we obtain:

(7) $S_{NN}(\underline{q}) = \int e^{i\underline{qr}}(\int \delta n(\underline{R})\ \delta n(\underline{R} - \underline{r})\ d^3R)d^3r$

which is the FT of the distance correlation of the total number-density fluctuations. Thus, $S_{NN}(\underline{q})$ describes the global structure, a fact which can also be derived from Eq. (2): if an experiment is performed for which $b_A = b_B$ which is, for example, the case if one studies a MgAl alloy by X-ray diffraction, than $\Delta b = 0$ and at the right site of Eq. (2) remains only the term with $S_{NN}(\underline{q})$. But, evidently in that case the radiation cannot distinguish between A and B atoms and it recognizes only the overall arrangement of the atoms. In the case of an isotropic substance, like a crystalline powder or an amorphous phase, $S_{NN}(\underline{q})$ depends only on the absolute value of q. For $x_i \to 1$, $S_{NN}(q)$ approaches the structure factor of the pure i-component. In terms of the Faber-Ziman partial pair correlation functions, the FT of $S_{NN}(q)$ is given by:

(8) $FT\{S_{NN}(q) - 1\} = (2\pi^2)^{-1}\int q(S_{NN} - 1)\sin(qr)dq = r(\overline{\rho(r)} - \rho_o)$

with

(9) $\overline{\rho(r)} = x_A\rho_{AA} + x_B\rho_{BB} + 2x_A\rho_{AB}$

Eq. (8) defines the operator FT { }.

To discuss the remaining S_{mn} functions, we introduce the function $\delta x(\underline{R})$, which describes the local fluctuation of concentration (3):

$$(10)\quad \delta x(\underline{R}) = \rho_o^{-1}\,(x_B \delta n_A(\underline{R}) - x_A \delta n_B(\underline{R}))$$

so that if the $\delta n_i(\underline{R})$ change in proportion to the respective mean concentration x_i, then $\delta x(\underline{R}) = 0$, as it should. The FT $C(\underline{q})$ of $\delta x(\underline{R})$ is defined analgous to Eq. (5) and $S_{NC}(\underline{q})$ is given by: $S_{NC}(\underline{q}) = \mathrm{Re}\langle N^x(\underline{q}) \cdot C(\underline{q})\rangle$ which is the FT of the distance correlation between concentration- and density-fluctuations. $S_{NC}(q)$ oscillates around zero and vanishes for $x_i \to 0$. With FT { } defined in Eq. (8) we obtain:

$$(11)\quad \mathrm{FT}\,\{S_{NC}(q)\} = r\,((\rho_{AA} + \rho_{AB}) - (\rho_{BA} + \rho_{BB}))$$

which vanishes if the global surrounding of A atoms is the same as for B atoms. Finally, $S_{CC}(\underline{q})$ is given by: $S_{CC}(\underline{q}) = N\langle C^x(\underline{q}) \cdot C(\underline{q})\rangle$ which corresponds to the FT of the distance correlation of concentration fluctuations. $S_{CC}(q)$ oscillates around $x_A x_B$ and is identical to this value if there are no such fluctuations, i.e. if the atoms are distributed at random. The term $\Delta b^2 S_{CC}(q)$ in Eq. (2) is well known for the case of solid crystalline disordered substitutional solutions where it has been called "Laue diffuse scattering" (LDS). Through its FT the LDS is related to the Warren sro-parameters as will be shown by the next equations (4):

$$(12)\quad \mathrm{FT}\,\{S_{CC}(q)/x_A x_B - 1\} = r\rho_{CC}(r) \text{ with}$$

$$(13)\quad r\rho_{CC}(r) = r(x_B(\rho_{AA} + \rho_{AB}) + x_A(\rho_{BB} + \rho_{BA}) - \rho_{BA}/x_A)$$

which becomes in the case $S_{NC} = 0$:

$$(14)\quad r\rho_{CC}(r)_{S_{NC}=0} = r\,(\overline{\rho(r)_{S_{NC}=0}} - \rho_{BA}(r)/x_A)$$

with $\rho(r)$ given in Eq. (9). For random distribution, $\rho_{BA}(r) = x_A\overline{\rho(r)}$ and $\rho_{CC}(r)_{S_{NC}=0} = 0$, it becomes negative for preferred hetero-coordination and vice versa. $\rho_{CC}(r)$ corresponds exactly to the $\rho(r)$ which has been published in 1944 by Wilchinsky (5) for polycrystalline Cu_3Au and from which the Warren sro-parameters α_p are obtained according to:

$$(15)\quad \alpha_p = Z_p^{-1} \int_{r_p-\varepsilon}^{r_p+\varepsilon} 4\pi\, r^2 \rho_{CC}(r)_{S_{NC}=0}\, dr$$

r_p is the mean distance from a given atom to the p-th coordination shell which is calculated from the lattice parameter and the crystal structure. The actual distance is between $r_p - \varepsilon$ and $r_p + \varepsilon$. Z_p is the number of atoms in the p-th shell. For liquid alloys we have no well defined coordination shells and to obtain the α one needs an appropriate model of the liquid structure. It is, however, very instructive to discuss directly the $\rho_{CC}(r)$-curves and to compare them for example with the corresponding curves of the solid phase. Before doing so, we have to discuss another very important property of the $S_{CC}(q)$ curve, namely its value for $q = 0$.

It has been shown by Bhatia and Thornton (3) that

(16) $RT/S_{CC}(0) = \partial^2G/\partial x^2 = E^{xs} + RT/x_Ax_B$

G is the Gibbs free energy, and $E^{xs} = - RT\ \partial \ln\gamma_A/\partial x_B^2$ is the "excess stability function" which has been introduced by Darken (6). γ is the activity coefficient. E^{xs} is negative in segregating systems and $RT/S_{CC}(0)$ becomes zero at the critical point of a miscibility gap. E^{xs} is positive for systems with a tendency for compound formation. From the normalization equation $\int q^2(S_{CC}(q)-x_Ax_B)dq = 0$, follows that $S_{CC}(q)$ necessarily oscillates around x_Ax_B if $E^{xs} \neq 0$. From this it has been concluded by Thompson (7) that there has to be some chemical sro whenever E^{xs} is different from zero. Komarek (8) studied thermodynamic properties and the phase diagrams of binary alloys consisting of main group metals and semi-metals. Amongst 405 binary systems only 50 showed no tendency toward compound formation or no miscibility gap. This demonstrates that chemical sro is a very general property of liquid and certainly also of amorphous alloys. However, if we wish to know exactly the nature of this sro we have to measure the total S_{CC} curve.
In the general case, the total structure factor S(q), calculated according to Eq. (1) from the results of a diffraction experiment, is a superposition of the three weighted $S_{mn}(q)$ functions, Eq. (2). Already for crystalline powders, where $S_{NN}(q)$ consists of sharp Bragg-peaks and of some thermaldiffuse scattering, it is very difficult to separate the diffuse $S_{CC}(q)$ scattering from the rest of S(q). For non-crystalline phases also $S_{NN}(q)$ has very broad peaks and an ad hoc separation is impossible. For special alloys, however, there is a chance to measure $S_{CC}(q)$ directly by a diffraction experiment. If in Eq. (2) $\langle b \rangle$ is equal to zero, then $S_{CC}(q)=x_Ax_BS(q)$. The condition $\langle b\rangle = x_Ab_B + (1-x_A)b_B = 0$ yields the composition of this special alloy which is called "zero alloy":

(17) $x_A^o = b_B/(b_B - b_A)$

$0 < x_A^o < 1$ only if b_A and b_B have different signs which means that the coherent scattering from A and B atoms occurs with a relative phaseshift of π. This happens in the case of neutron scattering. Most of the isotopes have a positive b. Negative b-values are reported for ^1H, ^7Li, ^{nat}Ti, ^{nat}Mn, ^{62}Ni, ^{152}Sm and ^{162}Dy. If these elements or isotopes mixed with a positive scatterer at the proportion given by Eq. (17) yield a homogeneous alloy, then a neutron scattering experiment leads directly to $S_{CC}(q)$. Such a zero alloy experiment has been proposed already in 1967 by Krivoglaz (9) to study solid Li alloys. But, as far as I know, at that time this proposition has not been realized. Since 1973 we have been studying the following systems at the Institut Laue-Langevin in Grenoble: ^7Li with Na, Ba, Sr, Ca, Mg, Ag and Pb which is a series of alloys ranging from a system with a miscibility gap (LiNa) to a system with strong tendency towards compound formation (LiPb). Recently, also a metallic glass, namely TiNi with a composition close to x_A^o has been investigated (14).

Results

The NaLi phase diagram is dominated by a region of two immiscible liquids. The consolute point occurs at $T_c = 577 \pm 2$ K and at $x_{Li} = 0.64 \pm 0.02$ which is very close to the zero alloy composition $x^o_{Li} = 0.61$. In Fig. 1 $S_{CC}(q)/x_Ax_B = S(q)$ is shown at 13 K above T_c (11). This curve is very different from a HS curve. It is dominated by strong small angle scattering from which with the help of an Ornstein-Zernike plot of curves measured at different temperatures the value $S_{CC}(0)/x_Ax_B = 2.1\ 10^{-3}T\ T_c^{1.1}(T - T_c)^{-1.1}$ was obtained yielding S(q) in Fig. (1) of about 80. Beyond the small angle region S(q) is very flat. It practically contains only two further ripples with an amplitude of 0.1 which are located at the q-values of the first peak of S(q) of pure Li and Na respectively. $4\pi r^2\rho_{CC}(r)$, Fig. 2, is positive over large distances indicating extended clusters with preferred self-coordination. The small peaks at 3 and 4 Å are due to preferred Li-Li and Na-Na distances respectively. If it is assumed that each atom has ten nearest neighbours, at 590 and 725 K α_1-values of about 0.5 and 0.3 result from Eq. (15). From this follows that at these temperatures a Li atom has a spatial average of 8 and 7 Li atoms as nearest neighbours instead of 6 in the case of a random distribution. On approaching T_c from 590 K, the $\rho_{CC}(r)$ curve will remain almost invariant at small distances. The range of the sro, however, will diverge at T_c. Because the volume per atom varies almost linearily between pure Li and pure Na, it should be concluded that the overall structure of the alloys remain HS like with some disturbances due to the size effect in the enriched clusters.

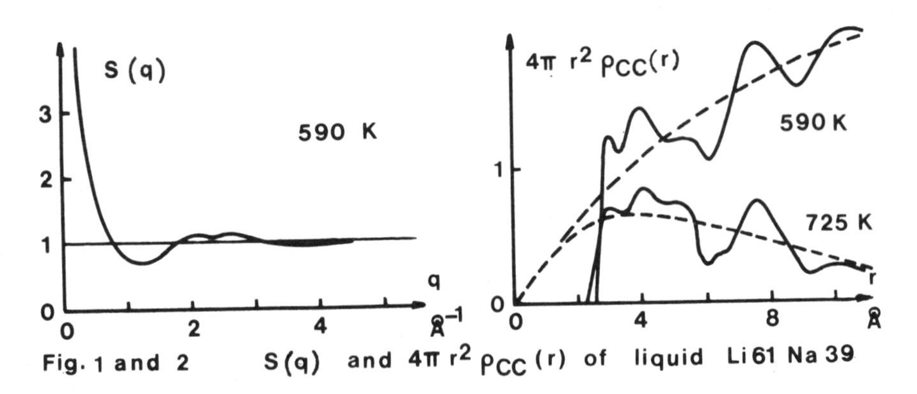

Fig. 1 and 2 S(q) and $4\pi r^2 \rho_{CC}(r)$ of liquid Li61 Na 39

Fig. 3 shows $S(q)=S_{CC}(q)/x_Ax_B$ of liquid $Li_{0.68}Ca_{0.32}$ (15). Because S(0) is close to one, E^{xs} is small, demonstrating that the solution is nearly regular. The $4\pi r^2\rho_{CC}(r)$-curve, Fig. 4, indicate preferred self-coordination at $r \approx 3$ Å and $r \approx 4$ Å and hetero-coordination at r ≈ 3.5 Å, which, however, is a pure size effect: because the diameters of Li and Ca are 3.1 and 3.9 Å, the aforementioned distances are distances between atoms in neighbouring LiLi, CaCa and LiCa pairs, respectively. The dotted curve in Fig. 3 has been calculated (PY) for a hard sphere system. The excellent agreement with the measured curve demonstrates that there is no special chemical interaction between Li and Ca atoms. For LiMg the contrary has been observed by Herbstein and Averbach (16) who measured the X-ray patterns of the solid alloys. They found preference for hetero-coordination, a reduction of the lattice parameter, and non-additive nearest neighbour distances. Preliminary results from a neutron diffraction study of the solid and liquid zero alloy $Li_{0.7}Mg_{0.3}$ (bcc β-phase) (15) are shown in Fig. 5. S(O) < 1 and the peak at about 1.5 $Å^{-1}$ is characteristic for preferred hetero-coordination: it corresponds to a super-lattice peak of an ordered

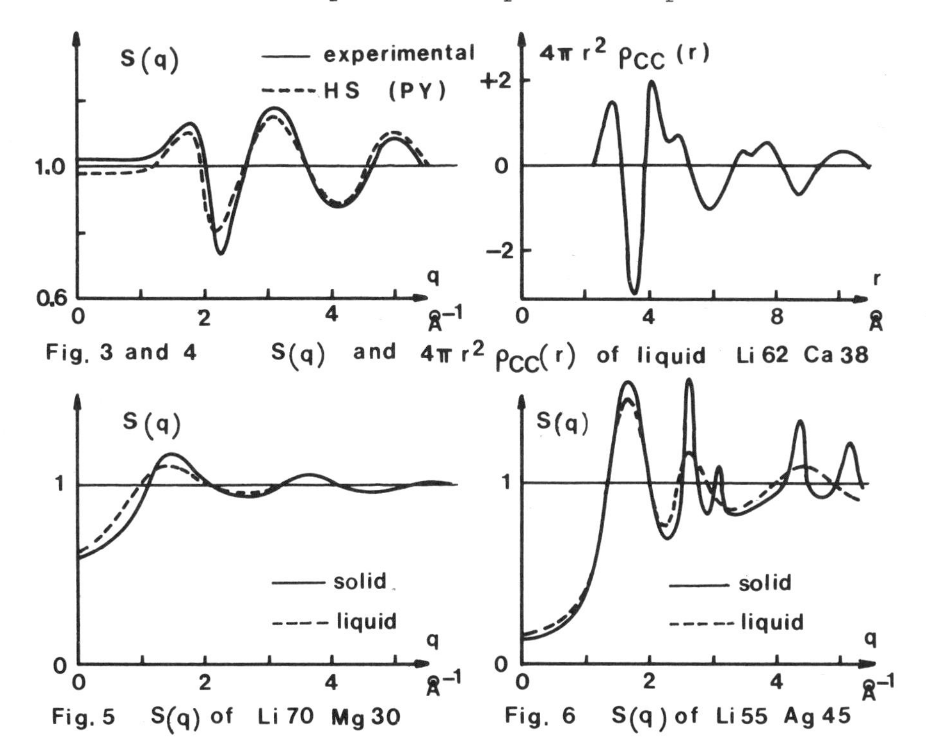

Fig. 3 and 4 S(q) and $4\pi r^2 \rho_{CC}(r)$ of liquid Li 62 Ca 38

Fig. 5 S(q) of Li 70 Mg 30

Fig. 6 S(q) of Li 55 Ag 45

crystalline alloy. Upon melting, the general behaviour of $S_{CC}(q)$ does not change. Thus, in the liquid we have the same type of sro as in the crystalline phase and I think that there is no doubt that the same would be true for an amorphous phase prepared by rapid quenching of this liquid.

LiAg is another alloy system for which a comparison of the sro in the solid disordered α-phase and the liquid phase was possible (10), however, not exactly at the zero-alloy composition but for $Li_{0.55}$

$Ag_{0.45}$ for which the values of a and b, Eq. (3) are 0.13 and -1.34 respectively. On melting the Bragg peaks disappear and transform into the broad S_{NN}-peaks of the liquid phase. This effect is clearly visible in Fig. 6 at about 2.5 $Å^{-1}$. As in the case of LiMg, the broad sro peak in S(q) of the solid phase at about 1.7 $Å^{-1}$ remains almost unaltered on melting. At lower temperatures it transforms into a superlattice line. Thus, there is here again a strong relation between the sro in the disordered solid α-phase and in the liquid. From a study of liquid LiAg alloys in the composition range between 1 and 55 at % Ag (13) we have the impression that there is no significant variation of the type of sro with composition in spite of the six different intermetallic phases which are formed on solidification. It has already been mentioned in the first part of this paper that there are good reasons to assume that the chemical arrangement of atoms might be simpler in the liquid than in the solid. The global structure of liquid AgLi alloys, once again, is HS like. This has been demonstrated by calculating $S_{NN}(q)$ approximately from a combination of X-ray and neutron scattering data and by comparing it with a PY-curve (13). A very impressive agreement is shown in Fig. 7. The solid line is the approximate partial structure factor a_{LiAg} which has been determined by the aforementioned experiments.

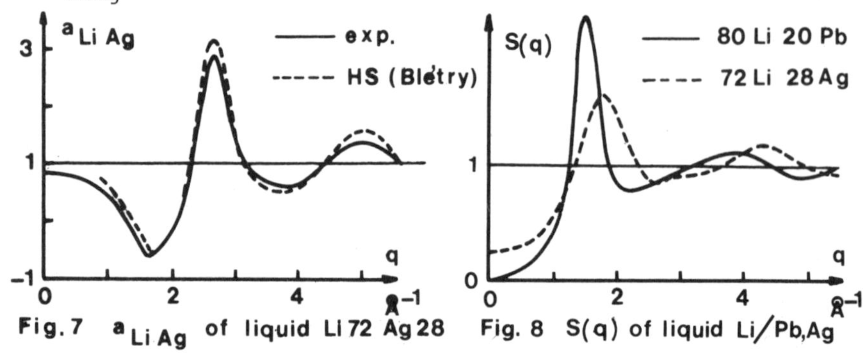

Fig. 7 $a_{Li\,Ag}$ of liquid Li 72 Ag 28   Fig. 8 S(q) of liquid Li/Pb,Ag

The dotted line has been obtained by Blêtry (17) from a computer simulated HS model with superimposed sro. A good agreement was also observed for a_{LiLi} and a_{AgAg}. In LiPb the zero alloy composition $x^{o}_{Pb} = 0.2$ happens to coincide exactly with the stoeciometric composition Li_4Pb. The corresponding $S_{CC}(q)/x_A x_B$ curve (4) is shown together with the curve of the liquid LiAg-zero alloy in Fig. 8. S(O) of LiPb is still smaller than S(O) of LiAg and S(q) has a much sharper first peak. The results in r-space are presented in Fig. 9. The negative peaks at about 3 Å indicate preference for unlike nearest neighbours. In $Li_{0.8}Pb_{0.2}$ the Pb atoms have probably only Li-atoms as nearest neighbours. It is interesting to note that the sharp peak in $S_{CC}(q)$ of $Li_{0.8}Pb_{0.2}$ yields a very long range sro in this case. From the combination of X-ray and neutron diffraction data, $S_{CC}(q)$ could approximately be calculated for $Li_{0.5}Pb_{0.5}$ and the sro came out to be much more rapidly damped. The chemical sro with preference for hetero-coordination in alkali alloys is

due to a charge transfer which renders the bonding at least partially ionic. Schirmacher (18) could derive physicochemical properties of these alloys in assuming an interaction between the particles which is the sum of a metallic potential and a screened Coulomb potential. From this, one might conclude that the long range of the sro of $Li_{0.8}Pb_{0.2}$ is due to an especially weak screening at the stoeciometric composition.
A distinct sro has also been observed (14) in amorphous NiTi alloys. The X-ray and neutron diffraction curves of $Ni_{0.4}Ti_{0.6}$ are shown in Fig. 10. The weighting factors of Eq. (3) are for X-rays (q = 0) and neutrons a = 0.986/0.088 and b = -0.485/-1.154, respectively. Due to the difference in (1-a), Eq. (3), the strong S_{CC}-peak in the neutron curve is only visible as a small prepeak in the X-ray scattering pattern.

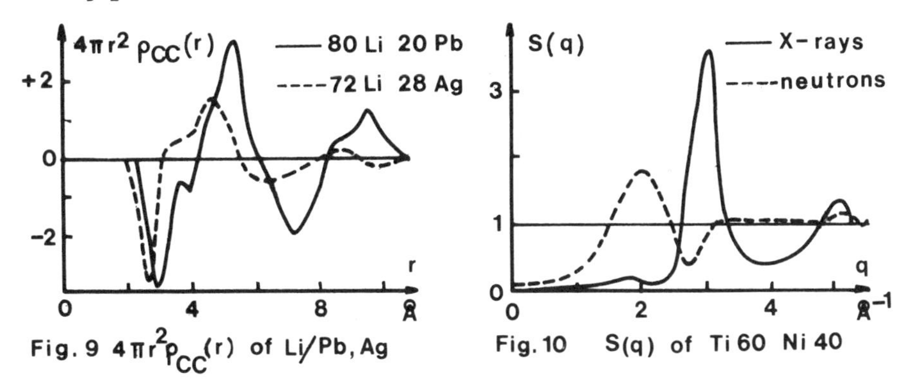

Fig. 9 $4\pi r^2 \rho_{CC}(r)$ of Li/Pb, Ag

Fig. 10 S(q) of Ti 60 Ni 40

Finally it should be noted that in binary mixtures of elements with positive b, the value of (a-1), Eq. (3), is in general quite small which makes it difficult to detect sro effects by neutron diffraction or to calculate $S_{CC}(q)$ from isotopic substitution experiments. If b of one of the constituants, say A, is comparatively small, one can study the sro in dilute solutions of B in A. For example, 5% Ni in V is a zero alloy with (a-1) = 1, and for LiPb-alloys investigated with X-rays it amounts to about 0.5 and 0.9 for 0.1 and 3 at % Pb, respectively.

References

1. N.W. Ashcroft and J. Lekner, Phys. Rev. 145, 83 (1966)
2. N.W. Ashcroft and D.C. Langreth, Phys. Rev. 156, 685 (1967)
3. A.B. Bhatia and D.E. Thornton, Phys. Rev. 1 B, 3004 (1970)
4. H. Ruppersberg and H. Egger, J. Chem. Phys. 63, 4095 (1975)
5. Z. Wilchinsky, J. Appl. Phys. 15, 806 (1944)
6. L.S. Darken, Trans. AIME 239, 80 (1967)
7. J.C. Thompson, J. Phys. (Paris), Colloque C4 (1974) 367
8. K.L. Komarek, Ber. Bunsen Ges. Phys. Chem. 80, 936 (1977)
9. M.A. Krivoglaz, "Theory of X-Ray and Therman-Neutron Scattering by Real Crystals", Plenum Press, N.Y. (1969)
10. H. Reiter, H. Ruppersberg and W. Speicher, Inst. Phys. Conf. Ser. 30, 133 (1977)
11. H. Ruppersberg and W. Knoll, Z. Naturforsch. 32a, 1374 (1977)
12. H. Ruppersberg, Metall, in press (1979)
13. H. Reiter, Doctor Thesis, Saarbrücken (1979)
14. H. Ruppersberg, D. Lee and C.N.J. Wagner, to be published
15. H. Ruppersberg, not published
16. F.H. Herbstein and B.L. Averbach, Acta Met. 4, 407 (1956) and 4, 414 (1956)
17. J. Blétry, Z. Naturforsch. 33a, 327 (1978)
18. W. Schirmacher, private communication

EXPANDED METALS

F. HENSEL

Institute of Physical Chemistry, University of Marburg,
D-3550 Marburg, Germany

ABSTRACT. A brief review is presented of the continuous metal-nonmetal transition observed in expanded fluid metals like Hg,Cs, Rb and K. Recent experimental information, especially thermodynamic data, is summarized and discussed.

1. INTRODUCTION

The electrical and thermodynamic properties of metallic liquids over a wide range of temperatures and pressures have received considerable attention in recent years. This interest originates, in part, from the fact that a favourable combination of physical properties such as high latent heats of vaporization, high thermal conductivity and high temperature liquid range makes liquid metals attractive as working fluids for many purposes. In spite of this technological interest many fundamental properties of fluid metals, including the equation of state data, are not known with sufficient certainty. Reliable theoretical predictions of such data are still lacking. This lack of progress in the face of the growing interest in such data is mainly due to two causes. Because of the large cohesive energies of metals, experimental data relevant to the quation of state have to be determined at very high temperatures and relatively high pressures. The liquid range of metallic fluids extends to very high temperatures which is demonstrated by Fig.1. Only the critical data for Hg [1], [2], Cs [3], Rb [3], K [4] and Na [5] have been measured with static experimental methods. The values for the heavy metal molybdenum have recently been determined by Fucke and Seydel [6] in a transient experiment with exploding wires.

metal	T_c (°C)	p_c (bar)
Hg	1490	1530
Cs	1740	115
Rb	1820	130
K	1950	155
Na	2230	250
Mo	14000	5700

Fig.1. Critical data of metals

The main difficulty for a theoretical prediction of thermodynamic data, including the critical data, is the absence of an adequate interatomic-potential function for the entire liquid range, i.e. over a large range of densities. The difficulty is essentially due to the fact that regardless of the way in which the interionic forces in a metal are described they must reflect the screening of the ionic charges by the electron gas and must therefore be always implicit functions of the electron density, i.e. of the ion number density. In addition it is well known that in an expanded fluid metal in the limit of a very low density the metallic properties disappear. This transition from metal to nonmetal implies that the type of chemical binding or at least of interatomic interaction changes, i.e. however we describe the interatomic forces in a fluid metal, the description must change with density. This is in contrast to nonconducting liquids like Ar, for which to a first approximation the thermodynamic properties are usually described by reference to a single pair-potential at all densities.

The aim of the present paper is to give a short review of recent experimental results with particular emphasis on their relevance to the following questions: Where in the phase diagram does the metal-nonmetal transition occur? Does it necessarily coincide with the thermodynamic transition from liquid to vapour across the saturation curve? It is not long since the subject "expanded metals" has been reviewed [7], [8]. Therefore I will restrict the following to some basic points and very recent mostly unpublished results.

2. EXPANDED MERCURY

Because liquid Hg has a relatively low critical temperature, extensive experimental results including dc-conductivity [1], [2], [9], thermoelectric power [10], [11], Hall effect [12], Knight shift [13], optical absorption [14], optical reflectivity [15], sound velocity [16], specific heat [17] and equation of state data [1], [9] are available over a wide range of density. All these properties establish the occurence of a metal-nonmetal transition induced by a density change in the range between 11 and about $9 g/cm^3$. There Hg exhibits a rapid variation of its thermodynamic and transport properties. The following noteworthy observations are specifically important for the present understanding of the electronic structure of expanded fluid Hg:

The electrical conductivity, σ, decreases by 8 orders of magnitude as the density d is reduced from $13.6 g/cm^3$ to $2 g/cm^3$. At high densities between $13.6 g/cm^3$ and $11 g/cm^3$ σ varies from 10^{+4} to $3 \cdot 10^{+3}$ $ohm^{-1}cm^{-1}$. Here the electron mean free path, L, is larger than the interatomic distance, a, and the Hall constant R_H has the free electron value. σ, R_H and the thermoelectric power S can be described by Ziman's "Nearly Free Electron" (NFE) model. For d smaller than $11 g/cm^3$ L becomes comparable with a. It is certainly not likely that the NFE model will hold here where the electron ion interaction is so strong. Further expansion of Hg from $11 g/cm^3$ to $9 g/cm^3$ reduces σ from 3000 $ohm^{-1}cm^{-1}$ to 300 $ohm^{-1}cm^{-1}$, makes R_H increase markedly above its free electron value and leaves the Knight shift K about constant. There is an apparent contradiction between K, σ and R_H between 11 and $9 g/cm^3$. σ and R_H can be understood within the framework of the random phase model [18] with the assumption that the density of states at the fermi energy $N(E_F)$ decreases rapidly in this range, whereas the constancy of K is consistent with the assumption that the s-component of N(E) is constant. Detailed discussions of this problem have recently been published [19], [7].

There may be some controversial opinions about the interpretation of the Hg data in the density range between $11 g/cm^3$ and $9 g/cm^3$, but there remains certainly no doubt that fluid Hg shows the behaviour of a disordered semiconductor for densities smaller than $9 g/cm^3$. This is consistent with all the existing data mentioned above, but the most convincing evidence stems from very recent and very accurate simultaneous measurements of σ and d as a function of temperature and pressure [9]. With these data it has become possible to derive $(E_C - E_F)$ as a function of density from equations (1) and (2) by plotting $\ln\sigma$ versus $1/T$.

$$\sigma = \sigma_o \exp\left(- \frac{(E_C - E_F)}{kT}\right) \qquad (1)$$

$$S = k/e\left[\frac{E_c - E_F}{kT} + 1\right] \quad (2)$$

(E_C denotes the mobility edge in the conduction band). If one assumes that $(E_C-E_F) = E_o - ßT$ is a linear function of T at constant volume then the slope of a plot of $\ln\sigma$ against $1/T$ is E_o/k and the intercept on the σ-axis is $\sigma_o \cdot \exp(ß/k)$. E_o-values obtained from a plot of $\ln\sigma$ versus $1/T$ at constant volume for d smaller than 8.4 g/cm^3 are shown in Fig.2. Values of ß can be obtained independently from S using eq.(2) which should apply for d smaller than $8g/cm^3$ [10]. The calculated ß's together with the corresponding σ_o-values are: For d = 6.8, 7.2, 7.6, $8.0g/cm^3$ ß = $7.53\cdot10^{-4}$, $6.75\cdot10^{-4}$, $5.29\cdot10^{-4}$, $3.56\cdot10^{-4}$ eV/K and σ_o = 140, 110, 140, 180 $ohm^{-1}cm^{-1}$, respectively. The corresponding (E_C-E_F) values are included in Fig.2. The noteworthy facts are: σ_o is within the accuracy of the present experiment independent of d and close to the value predicted by Mott for σ at a mobility edge. The extrapolation of E_o and (E_C-E_F) suggests that both quantities vanish at a density of about $8.8g/cm^3$, which considerably exceeds the thermodynamic critical density of $5.3g/cm^3$. There is no correlation between the metal-nonmetal transition and the critical point of the liquid-vapour phase transition.

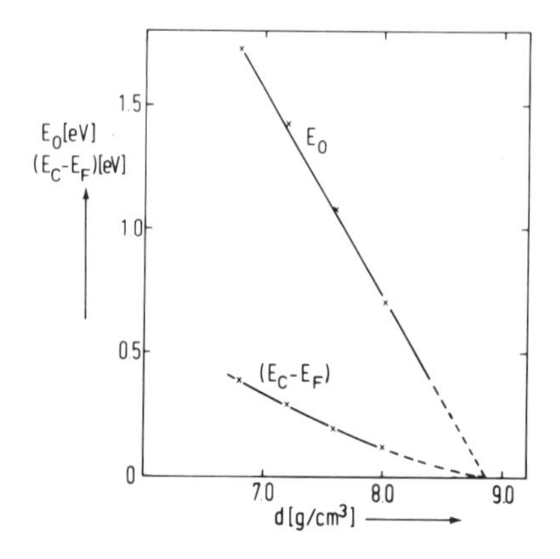

Fig.2. Activationenergies as a function of density.

We turn now to the equation of state data, from which different derivatives of the free energy such as isothermal compressibility X_T and thermal expansion coefficient α_p can be calculated as a function of density. The results for α_p and X_T are shown in Fig. 3. The most interesting result seems to be, that both quantities rise quite sharply as the density falls below $9g/cm^3$ and it is surely tempting to speculate that the turnover in the slopes of

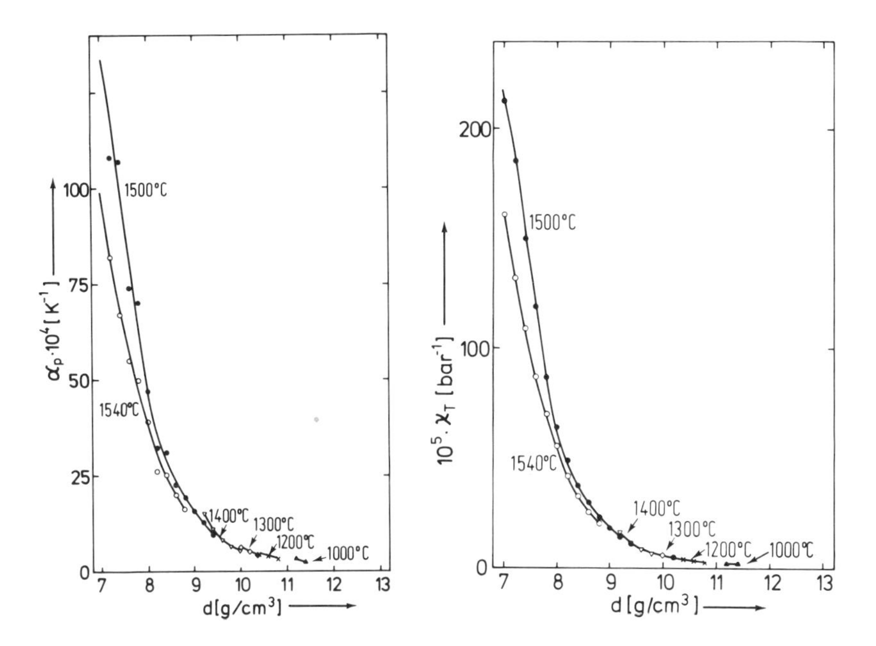

Fig.3. α_p and χ_T of Hg as a function of density

α_p and χ_T coincides with the metal-semiconductor transition. It must be pointed out, however, that these two quantities diverge as the critical point is approached in all substances and it is not easy to separate these two effects. On the other side a metal-semiconductor transition implies a change from metallic to other interatomic forces and therefore changes in thermodynamic properties. But there is no theory available which deals simultaneously with thermodynamic and electronic properties of low density metals. The difficulty is essentially due to the fact that exchange and correlation amoung valence electrons at low densities are not understood. In the absence of a definite theory one must at present resort to a meanfield type of analysis for the equation of state of expanded metals. The underlying idea of such treatment is that the liquid consists of hard-sphere particles immersed in a uniform potential U which is a smooth function of volume V at constant T and which provides the cohesion that the hard sphere system otherwise lacks. The thermodynamic relation

$$p = T \cdot \left(\frac{\partial S}{\partial V}\right)_T - \left(\frac{\partial U}{\partial V}\right)_T \qquad (3)$$

enables $(\partial U/\partial V)_T$ to be calculated from the experimental results [9]. Fig.4 shows $(\partial U/\partial V)_T$ for Hg along the saturated vapour curve.

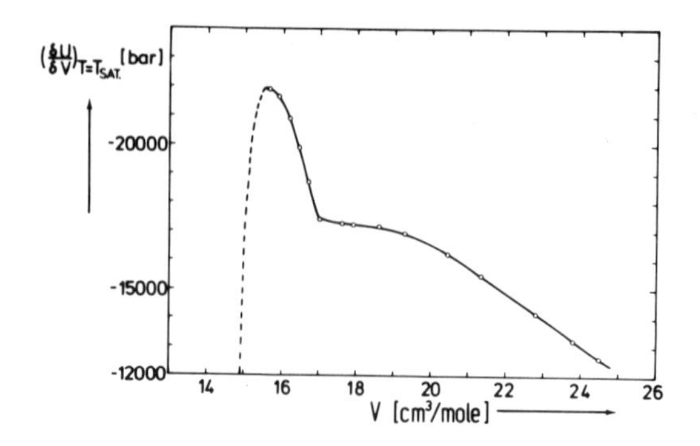

Fig.4. Inner pressure of liquid Hg along the saturation curve

The obvious difference between the behaviour of $(\partial U/\partial V)_T$ of Hg and that of nonmetallic fluids [20] is the quite sharp break in the slope $(\partial^2U/\partial V^2)_T$ of the Hg-curve if one approaches a density of about 11g/cm³ ($L \sim a$). Again this is not surprising as U is mainly the cohesion energy of the system and is therefore expected to transform from a metallic type to the form for nonconducting liquids.

The effect of a few atomic percent of In on the conductivity of Hg in the metal-nonmetal transition range was first studied by Zillgitt et al.[21]. Examples of the effect at different pressures (left side) or at different number densities (right side) are given in Fig.5. These data are taken from the work of Schönherr [22]. They demonstrate that a comparison between pure Hg and

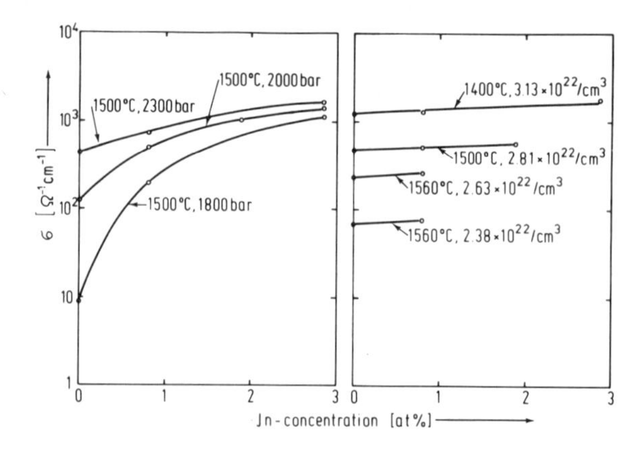

Fig.5. σ of dilute Hg-In mixtures as a function of In-concentration at constant pressure and constant number density.

dilute In-Hg solutions is only reasonable if both the temperature and the number density of atoms are constant. It is obvious from the left side of Fig.5 that a very large increase of σ with increasing In-concentration occurs near the metal-nonmetal transition density of pure Hg. Indeed, In increases the density at constant pressure and the excess volume of mixing becomes considerably large near the transition density of 9g/cm^3 of pure mercury. This is not surprising because the compressibility χ_T of the solvent Hg increases rapidly near 9g/cm^3. The close relation between χ_T of the pure solvent Hg and the excess volume of mixing V^E is demonstrated by Fig.6.

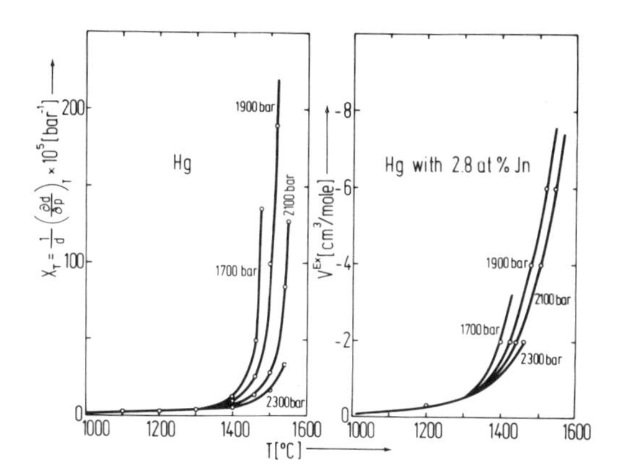

Fig.6. Excess volume of mixing and compressibility of pure Hg as a function of temperature.

3. EXPANDED ALKALI METALS

For an array of one-electron atoms, whether crystalline or not, a metal-nonmetal transition can only occur as a consequence of the interaction between the electrons [23]. The formal treatment is due to Hubbard [24]. He finds that if the distances between the atoms exceed a critical value, a splitting of the conduction band into two bands occurs. The lower band is occupied and the upper band is empty and the material is a nonmetal with an activation energy for conduction. In numerous papers it has been pointed out by Mott, that in a disordered expanded system of monovalent atoms with a mean interatomic distance near that for the metal-nonmetal transition, the density of states can be represented by two overlapping Hubbard bands, so that the position of the Fermi energy is in the minimum of the density of states. If the depth of the minimum is large enough, the states around the Fermi energy should become localized; i.e. a mobility gap should be formed. This gives a gradual transition to a nonmetallic state which experimentally should be very similar to the band crossing transition observed for divalent mercury.

From recent measurements of the electrical conductivity σ and the absolute thermoelectric power S for fluid Cs [25], [3], Rb [26] and K [4] it can be suggested that in the fluid alkali metals NFE-behaviour exists over a wide density range down to about 50% of the normal melting point density. At this density the electron mean free path L is approximately equal to the reciprocal wavevector k_F at the Fermi edge, the temperature coefficient of σ changes sign from negative to positive and the Hall coefficient R [27] exhibits nearly the free electron value down to this density. Therefore the Ziman (NFE)-model should apply for the electrical transport properties of the fluid alkali metals in this density range. This has very recently been proved experimentally for fluid rubidium by measurements of the density [26], the electrical conductivity [26] and the structure factor S(Q) [29], as a function of temperature and pressure. It is found [28] that the density dependence of σ can be well represented by the (NFE)-model in the density range between 1.4g/cm^3 and 0.9g/cm^3, i.e. that σ is mainly determined by the number of free valence electrons, the configuration of the scatterers S(Q) and the scattering properties of the screened positively charged ions.

Very recent data by Franz [3] of the simultaneously measured conductivity and density of fluid Cs extent to the critical range. It is obvious from Fig.7 that at the critical point σ is about 200-300 ohm^{-1}cm^{-1} which has been proposed by Mott to be the "minimum metallic conductivity". This is consistent with the ob-

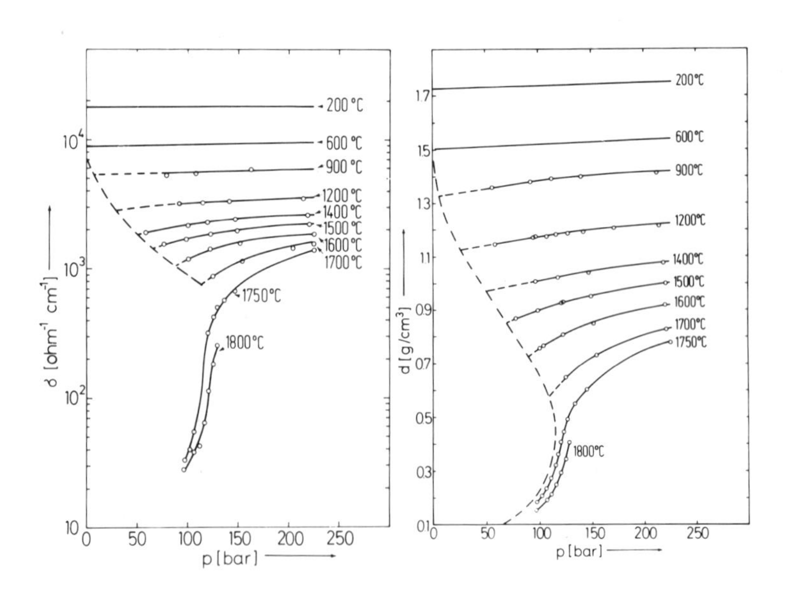

Fig.7. Conductivity and equation of state of fluid Cs.

served linear dependence of the logarithm of the conductivity on the thermoelectric power with a slope of about e/k [30] which is obtained at σ smaller than 300ohm^{-1}cm^{-1}. Hence it cannot be excluded that for the alkali metals the thermodynamic critical point will coincide with the metal-nonmetal transition. It must be pointed out, however, that the situation is more complicated and that one has to prove whether a monovalent vapour, such as Cs, is indeed monovalent. The diatomic molecules Cs_2 are quite stable and their existence in dense Cs vapour close to the critical point has recently been concluded by Freyland from measurements of the magnetic susceptibility of expanded Cs [31].

REFERENCES

1. F.Hensel and E.U.Franck, Ber.Bunsenges.phys.Chem. 70,1154 (1966).
2. I.K.Kikoin and A.R.Sechenkov, Physics Metals Metallogr. 24, 5 (1967).
3. G.Franz, Thesis, University of Marburg (1980).
4. W.Freyland and F.Hensel, Ber.Bunsenges.phys.Chem. 76, 347 (1972).
5. V.S.Bhise, Thesis, Columbia University, New York (1976).
6. U.Seydel and W.Fucke, J.Phys.F. 8, L157 (1978).
7. N.E.Cusack in Metal-Non-Metal Transitions in Disordered Systems, ed. L.R.Friedman and D.P.Tunstall, Edinburgh, 1978.
8. F.Hensel, Liquid Metals 1976, ed. by Evans and Greenwood, Bristol 1967.
9 G.Schönherr, R.W.Schmutzler and F.Hensel, Phil.Mag. (1979) in press.
10. R.W.Schmutzler and F.Hensel, Ber.Bunsenges.phys.Chem. 76, 531 (1972).
11. L.J.Duckers and R.G.Ross, Properties of Liquid Metals, Proc. 2nd Int.Conf.on Liquid Metals, Tokyo, Taylor and Francis, London, 1972, p.365.
12. U.Even and J.Jortner, Physic.Rev.Letters 28, 31 (1972).
13. V.El-Hanany and W.W.Warren, Physic.Rev.Letters 34, 1276 (1975).
14. H.Uchtmann, F.Hensel, Phys.Letters 53A, 239 (1975).
15. H.Ikezi, K.Schwarzenegger, A.L.Simons, A.L.Passner and S.L. McCall, Phys.Rev. B18, 2494 (1978).
16. M.Inutake, K.Suzuki and S.Fujiwaka, Proc.XIVth Int.Conf.on Phenomena in Ionized Gases, 1979.
17. R.W.Schmutzler, Habilitationsarbeit, University of Marburg 1979.
18. N.K.Hindley, J.non-cryst.Solids 5, 17, 31 (1970).
19. L.F.Mattheis and W.W.Warren Jr., Phys.Rev. B16, 624 (1977).
20. A.F.M.Barton, The Dynamic Liquid State, (Longman Group Ltd., 1977).

21. M.Zillgitt, R.W.Schmutzler and F.Hensel, Phys.Lett. A39, 419 (1972).
22. G.Schönherr, Thesis, University of Marburg 1978.
23. N.F.Mott, Phil.Mag. 6, 287 (1961).
24. J.Hubbard, Proc.Roy.Soc.(London) A281, 401 (1964).
25. H.P.Pfeifer, W.Freyland, F.Hensel, Phys.Letters 43, 111 (1973).
26. H.P.Pfeifer, W.Freyland and F.Hensel, Ber.Bunsenges.phys.Chem. 83, 204 (1979).
27. U.Even and W.Freyland, J.Phys.F. 5, L104 (1975).
28. R.Block, J.B.Suck, W.Gläser, W.Freyland and F.Hensel, Liquid Metals 1976, ed. by Evans and Greenwood, The Institute of Physics, Bristol and London 1976.
29. W.Freyland, F.Hensel and W.Gläser, Ber.Bunsenges.phys.Chem. (1979) in press.
30. W.Freyland, H.P.Pfeifer and F.Hensel, Proc. 5th Int.Conf.Amorphous and Liquid Semiconductors, ed. J.Stuke and W.Bredig, Taylor and Francis, London 1974.
31 W.Freyland, Phys.Rev. 1979, in press.

LIQUID SEMICONDUCTORS

F. Hensel

Institute of Physical Chemistry, University of Marburg,
D-3550 Marburg, Germany

ABSTRACT. A brief summary is presented of very recent experimental results observed in different fluid systems exhibiting semiconducting properties, with special reference to fluid selenium and sulphur at very high temperatures and the liquid ionic binary alloy CsAu.

1. INTRODUCTION

The purpose of this article is to describe recent experimental work on the electrical, optical and thermodynamic properties of fluid conductors whose electronic properties are different from those of metals and are similar to those of semiconductors. Such fluids are often called "liquid semiconductors". The study of these materials has received remarkable interest during the last 15 years. It is the subject of various books [1], [2], review articles [3], [4] and conference proceedings [6], devoted to selected topics. Hence any attempt at a comprehensive review of the subject here would be unnecessary because of the recent surveys that exist. Instead I have selected for attention those new experimental results which have not adequately reviewed elsewhere or which are still unpublished; e.g. high temperature and high pressure studies of the elemental liquid semiconductors Te, Se and S and their mixtures; and selected properties of ionic liquid alloys which are composed of two metallic elements and which are nonmetallic at definite stoichiometric compositions (like CsAu, Li_3Bi or Mg_3Bi_2). For those topics omitted on purpose from the present article, especially the liquid alloys of metals with the chalcogen elements S, Se and Te, the reader is referred to the more extensive articles [1], [3], [4].

2. ELEMENTAL SEMICONDUCTORS

The group VI A chalcogen elements exhibit a rich variety of physical properties in their liquid state ranging from the insulating properties of oxygen and the metallic properties of polonium. In between are the liquid semiconductors sulphur, selenium and tellurium. Especially liquid selenium and tellurium and their liquid mixtures have been studied over a wide range of temperatures and pressures by many investigators both in regard to their electrical properties [7], [8], [9] and their molecular structure [10], [11]. Their behaviour is consistent with the assumption that the chain-like structure of solid tellurium, selenium and of the solid solutions of both consisting of mixed chains persists to a large extent in the melt at temperatures close to the melting point. This assumption leads to a well-developed energy gap and semiconducting properties. However if the temperature is increased, the chains should dissociate and decrease in length; the decrease of the chain length goes in the direction of a completely random atomic configuration which is typical for metals i.e. a semiconductor to metal transformation is expected to occur with increasing temperature. This proposal is consistent with recent measurements of the electrical conductivity of fluid tellurium [9], selenium [8] and Se-Te mixtures [7], up to very high temperatures and pressures. Data are shown in Fig.1 for Se. The limitations of the "semicon-

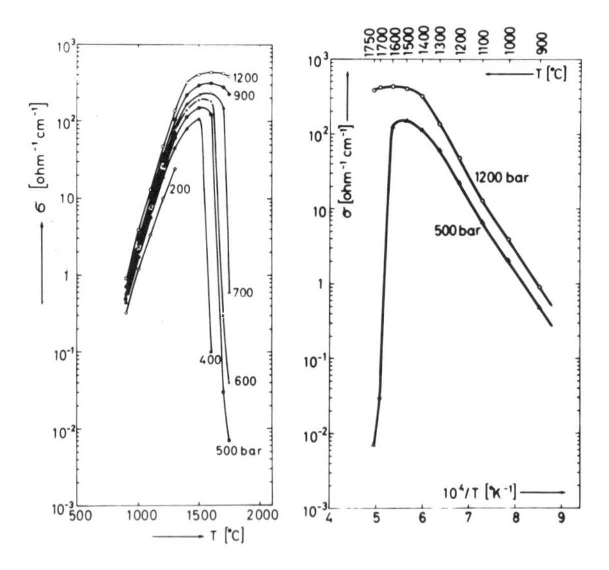

Fig.1. Conductivity isobars versus temperature and reciprocal temperature at supercritical pressures (p_c = 380 bar).

ductor model"for selenium become apparent at high temperatures.The strong increase in the conductivity in orders of magnitudes reaches a maximum. Values of about 100 $ohm^{-1}cm^{-1}$, in the region of the so-called "minimum metallic conductivity", are observed for temperatures above 1300°C and pressures above 600 bars. The hypothesis of an increase in the coordination number in the region of the minimum metallic conductivity in fluid Se is in accordance with an anomaly in the equation of state in this range [12] which can be directly seen from Fig.2. In contrast to the behaviour of normal

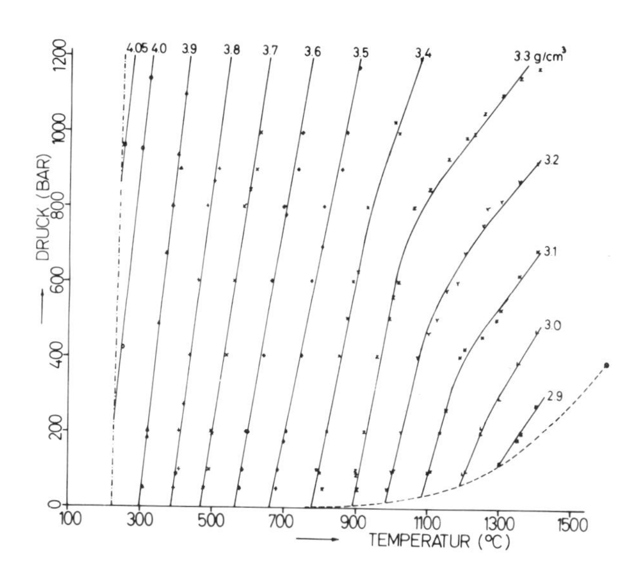

Fig.2. Isochores of fluid selenium.

liquids the isochores of selenium tend to bend by raising the temperature and thereby lowering the density. This probably indicates that close to the kinks in the isochores a transformation to a more dense fluid occurs. In order to compare the behaviour of selenium with that of sulphur Fig.3 [12] gives a plot of the molar volume of both at a fixed ratio of p/p_c as a function of the ratio of T/T_c (p_c = critical pressure, T_c = critical temperature). At about 0.8 T_c the curve for Se flattens out and starts raising again close to the critical point. The unusual flat region demonstrates the change of the short range order in liquid Se. Obviously the average coordination number increases with increasing temperature leading to a more dense packing despite of the competing thermal expansion. A minimum in the thermal expansion coefficient is also observed for Se-Te alloys. The minimum shifts towards higher temperatures with increasing Se-content. This shift is certainly related to the larger stability of the bonds within the selenium chain-molecules. The transformation to a more metallic state occurs at higher temperatures for selenium

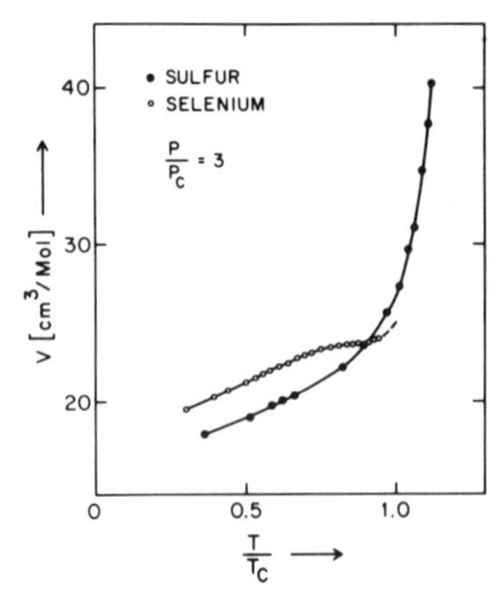

Fig.3. Molar volume of fluid selenium and sulphur as a function of the reduced temperature.

than for tellurium. For selenium the maximum conductivity is observed near the critical temperature of the liquid-vapour phase equilibrium, whereas for tellurium the maxima occur far beyond the critical temperature. The binding of the covalent S-S bond in sulphur ring- or chain-molecules is much stronger than the Se-Se bond. Thus indications for metallic conduction in sulphur have been reported only at very high pressures of more than 100 kbars at high temperatures. The most striking feature of liquid sulphur is the spontaneous polymerization that occurs at about 160°C. Below this temperature liquid sulphur is mainly composed of eight-atom rings; above 160°C a fraction of the rings polymerize into long chains with a mean chain length of about 10^6 atoms. At high temperatures the mole fraction of polymer increases, but the average chain length decreases [13]. The two ends of each chain give rise to unpaired localized electrons as static susceptibility [14] and ESR measurements [15] have shown. Very recent optical absorption measurements [16] demonstrate that the first optical transition for the free-radical chains occurs at about 1.3 eV. Fig.4 shows the absorption spectra of liquid sulphur in the energy range 0.5-3.5 eV for temperatures between 140°C and 1000°C at pressures slightly higher than the corresponding vapour pressures. The predominant feature of the absorption curves is the expected red shift of the edge with increasing temperature and the weak absorption bands in the energy range between 1-1.5 eV which increase very rapidly with temperature. The intensities and the positions of these bands have been determined from the area under the absorption band after deconvolution of the overlapping exponential edges and bands. Since the intensities of the absorption bands have the same temperature dependence as the ESR signal, it has been concluded that both features have the same physical origin, i.e. they are connected with the chains which possess unpaired electron spin states at the chain ends. The role of these

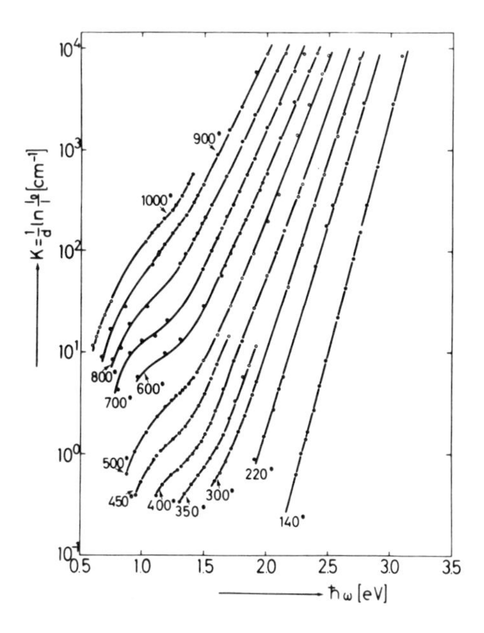

Fig.4. Optical absorption edges of liquid sulphur at different temperature T [°C].

free-radical chain-ends for the electrical conductivity of liquid sulphur - a problem of special interest in the light of the current theoretical discussion of the electrical properties of glassy and liquid chalcogenides - has recently been studied by Edeling et al.[17]. Conductivity data σ are shown in Fig.5 as a function of the reciprocal temperature. The dominant feature of the

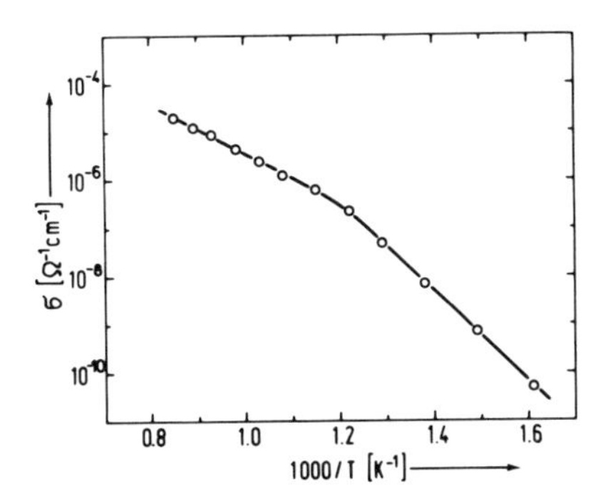

Fig.5. Conductivity of liquid sulphur as a function of reciprocal temperature.

(log σ, 1/T) curve is the relatively sharp break in the slope at a temperature T_C of about 550°C, corresponding to activation energies E_A = 1.05 eV for $T > T_C$ and E_A = 1.9 eV for $T < T_C$.

Assuming that the dangling bonds form acceptor states within the bandgaps at an energy ε one can easily show [17] that the fermi energy E_F is related to the concentration of dangling bonds N^* by eq.(1)

$$E_F - E_V = (\varepsilon - E_V)/2 + (kT/2) \ln (2N_V/N^*) \qquad (1)$$

where N_V is about $N(E_V) \cdot kT$ for carriers at the mobility edge E_V. The temperature dependence of N^* can be given in the analytic form

$$N^* = A \exp (- E^*/kT) \qquad (2)$$

with E^* = 0.8 eV. Eqs.(1) and (2) can be compared with $E_A = E_F - E_V$ = 1.05 eV of σ for $T > T_C$. Neglecting terms of the order of kT, it is seen that

$$E_A \approx (\varepsilon - E_V + E^*)/2 \qquad (3)$$

and with E^* = 0.8 eV, the value $\varepsilon - E_V$ = 1.3 eV is obtained in excellent agreement with the value of the position of the optical absorption band.

For $T < T_C$, E_A of σ is increased by a factor of about 2. The most probable cause of this variation in E_A is the inevitable presence of a very small number N_X of monovalent impurities x in liquid sulphur [13]. Under such circumstances, liquid sulphur behaves like a doped and compensated semiconductor [17] with (N^*+N_X) available defect levels and N^* acceptors. $N_X \approx 6 \cdot 10^{15} cm^{-3}$ - i.e. less than 1 p.p.m. of monovalent impurities - are sufficient to explain the shape of the experimental (σ, 1/T) curve.

3. IONIC ALLOYS

There has recently been a considerable amount of work on properties of liquid binary alloys which one would expect to be metallic and which behave more like semiconductors or molten salts. Although the pure metals like, e.g. Li, Cs, Au, Bi and Sb are liquid metals for which Ziman's theory of nearly free electrons is valid the liquid mixtures Mg-Bi, Li-Bi, Cs-Sb and Cs-Au have quite a low conductivity σ in the neighbourhood of compositions satisfying simple chemical valence requirements. At the same time the temperature dependence of σ is positive. This is strong indication that the formation of chemical compounds is responsible for this phenomenons as first pointed out by Wagner [18] for the liquid Mg-Bi system. The essential, distinguishing feature between metals on the one hand and semiconductors and ionics on the other concerns the way in which the electronic energy levels of the atoms merge into bands in condensed phases. If any band is partly full at the absolute zero of temperature, the substance is a metal. If all bands are full or empty, the material is a nonmetal. This classification does not distinguish between semiconductors and ionics. Ionics are semiconductors with band gaps larger than about 3 eV at 25°C and the electrical conductivity is directly related to the diffusive motion of charge carriers of atomic or molecular dimensions under conditions of thermal excitation.

The very interesting group of liquid electrical conductors is formed by those compounds which are at the borderline between electronic semiconductors and ionics. This includes compounds like e.g. CuJ and the intermetallic compounds Li_3Bi and CsAu which have room temperature gaps of 2.8 eV, 1.4 eV and 2.6 eV, respectively. The electronegativity differences of the constituent atoms are intermediate. Those compounds exhibit often features of both electronic and ionic conduction depending on the magnitude of the gap. The process of melting can be accompanied by a transition from a solid electronic semiconductor to a liquid ionic conductor. A typical example is CsAu for which the electrical conductivity σ is shown in Fig.6. For all samples studied the conductivity decreases on melting from values between 100 and 20 $ohm^{-1}cm^{-1}$ in the solid depending on the deviations from the stoichiometry to a value of about 3 $ohm^{-1}cm^{-1}$ in the liquid. This marked decrease of σ on melting indicates that the electrical transport mechanism in liquid CsAu is very different from that in the corresponding solid. One can contrast the behaviour of CsAu with that of other solid semiconductors which mostly show a more or less large increase of σ on melting. Two main groups may be distinguished. In the first group of semiconductors, such as Ge, Si, InSb etc., σ becomes metallic on melting. In the second group to which semiconductors like Se, In_2Se_3, Ga_2Te_3 etc. belong a transition from a solid to a liquid semiconductor is observed on melting accompanied usually by an increase in σ. It is evident from the results in Fig.6 that

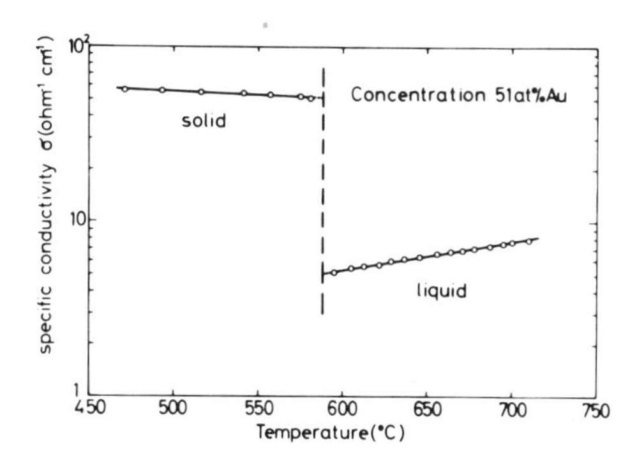

Fig.6. Conductivity of solid and liquid CsAu near the stoichiometric composition.

CsAu does not belong to one of these groups. For semiconductors with a high degree of ionic binding and room temperature gaps between 2.5 eV and 3.0 eV the solid liquid phase transition coincides with a transition from a solid semiconductor to a liquid mainly ionic conductor. Electromigration experiments show that liquid CsAu conforms to Faraday's law [19]. The electronic contribution to σ is very small due to the large energy gap in the liquid. This is evident from the spectral dependence of the absorption coefficient K of solid and liquid CsAu shown in Fig.7 [20]. The absorption edges exhibit the expected red shift with increasing temperature. The lower part of Fig.7 shows the temperature dependence of the position of the optical gap. It is a plot of the photon energy corresponding to $K \approx 10^4 cm^{-1}$ as a function of temperature. There occurs a discontinuous, abrupt change in the energy of 0.8 eV on melting; such a discontinuous change between 0.6 and 0.8 eV is often observed at the melting point of salts [21]. The data available at present establish the occurrence of a gradual metal-nonmetal transition induced by concentration changes in liquid Cs-Au mixtures. The minimum in the transport properties

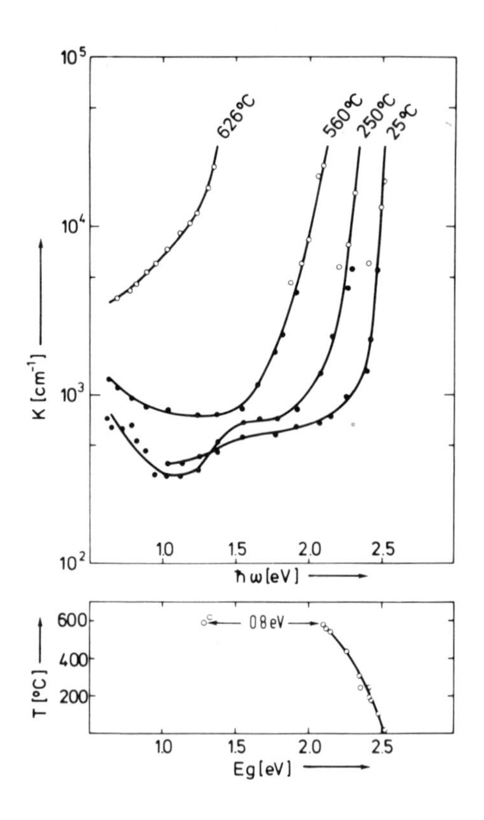

Fig.7. Optical absorption in solid and liquid CsAu.

occurs in the vicinity of a composition which corresponds to a liquid consisting of Cs^+ and Au^- ions according to the experimental evidence mentioned above. It seems, therefore, reasonable to interpret the experimental results by assuming that bound states associated with the formation of CsAu ionic assemblies remove electrons from the conduction bands as the composition is changed between the pure metals and CsAu. Then the liquid alloys are assumed to be a binary mixture of Cs and CsAu for concentrations smaller than 50 at% Au or Au and CsAu for concentrations larger than 50 at% Au, respectively. This implies that the electrical transport properties of the alloys in the whole composition range are dependent on the nature of the electronic constitution in CsAu, i.e. that the main features of the chemical bonds between Cs and Au atoms remain unaltered over the whole concentration range.

The most direct evidence for the validity of this assumption comes from the study of the concentration dependence of the molar volume V_m (Fig.8, left side) [22]. Fig.8 shows on the right side the partial molar volumes $\overline{V}$ of Cs and Au as a function of the mole fraction x of Au [23]. It is obvious that for dilute solutions in the metallic range (i.e. $x < 0.3$ and $\sigma > 10^3 ohm^{-1}cm^{-1}$ $\overline{V}_{Cs}$ is nearly independent of the concentration and close to the value of pure liquid Cs, whereas $\overline{V}_{Au}$ tends to large negative values approching $-107\ cm^3/mole$ for x values close to zero. The latter value is consistent with the picture that even in very dilute solutions an electron transferred from Cs pairs up with the valence electron of Au and is localized. The attractive forces between the unlike charges and the repulsive forces between the like

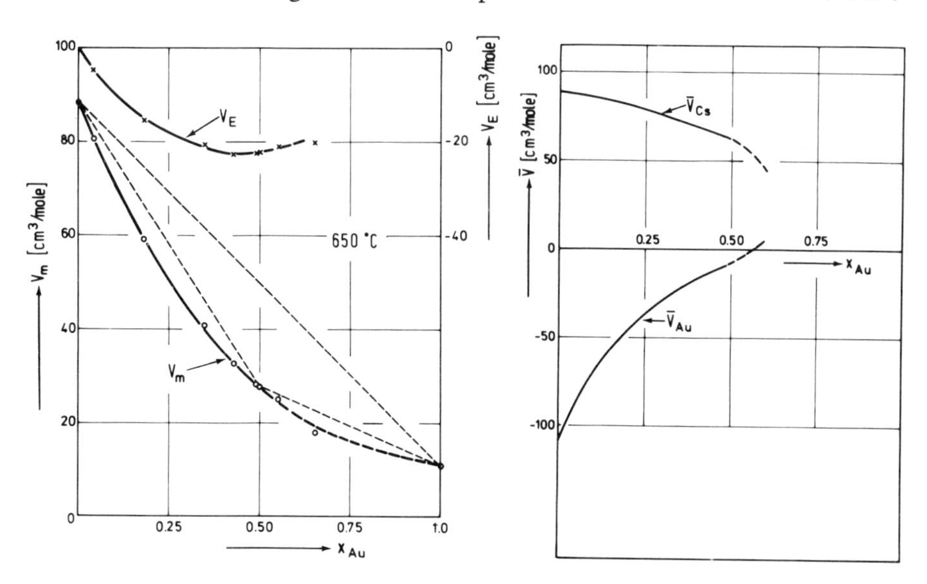

Fig.8. Molar volume V_m, excess-volume V_E and partial molar volumes in liquid Cs-Au alloys.

charges lead to short range ordering such as that which occurs in molten salts.

REFERENCES

1. M.Cutler, Liquid Semiconductors, Academic Press, New York, 1977.
2. V.M.Glazov, Liquid Semiconductors, Plenum Press, New York, 1969.
3. J.E.Enderby, in Amorphous and Liquid Semiconductors, ed. J.Tauc, Plenum Press, New York, 1974.
4. W.W.Warren, in Advances in Molten Salt Chemistry 4, 1979.
5. J.E.Enderby, in Metal-Non-Metal Transitions in Disordered Systems, ed. L.R.Friedman and D.P.Tunstall, Edinburgh, 1978.
6. F.Hensel, in Amorphous and Liquid Semiconductors, Proc.7th Int.Conf., Edinburgh, ed. W.Spear, 1977.
7. J.C.Perron, Advanc.in Phys. 16, 657 (1967).
8. H.Hoshino, R.W.Schmutzler, W.W.Warren, F.Hensel, Phil.Mag. 33, 255 (1976).
9. H.Endo, H.Hoshino, R.W.Schmutzler, F.Hensel in Liquid Metals 1976, ed. R.Evans and D.Greenwood, Inst.Phys.,Bristol,1977.
10. B.Cabane and J.Friedel, J.Phys.(Paris) 32, 73 (1977).
11. H.Thurn and J.Ruska,J.Noncryst.Solids 22, 331 (1976).
12. R.Fischer, R.W.Schmutzler, F.Hensel, 8th Int.Conf.on Amorphous and Liquid Semiconductors, Boston, 1979.
13. B.Meyer, Chem.Rev. 76, 367 (1976).
14. J.A.Paulis, C.H.Massen, P.van der Leeden, Trans-Faraday Soc. 58, 474 (1962).
15. D.C.Koningsberger, Thesis Eindhoven (1971).
16. G.Weser, F.Hensel, W.W.Warren, Ber.Bunsenges.phys.Chem. 82, 588 (1978).
17. M.Edeling, R.W.Schmutzler, F.Hensel, Phil.Mag.39,547 (1979).
18. C.Wagner, Thermodynamics of Alloys, Addison-Wesley, Reading, MA (1952).
19. K.D.Krüger, R.W.Schmutzler, Ber.Bunsenges.phys.Chem. 80, 816 (1976).
20. P.Münster, W.Freyland, Phil.Mag. (1979), in press.
21. E.Mollwo, Z.Phys. 124, 118 (1948).
22. A.Kempf and R.W.Schmutzler, to be published.
23. R.W.Schmutzler, Habilitationsschrift, University of Marburg, 1979.

3.2 Amorphous and glassy systems

METALLIC GLASSES - PREPARATION, UNIQUE PROPERTIES, APPLICATIONS AND EXPERIMENTAL RESULTS

H.-J. Güntherodt

Institut für Physik, Universität Basel
CH-4056 Basel, Switzerland

1. INTRODUCTION

The field of metallic glasses has grown rapidly during the last few years. The main interest was initiated by possible technical applications based on outstanding mechanical, magnetic and chemical properties. However, there is also a great deal of new and exciting physics involved. The research in this field has pointed towards several similarities in common with dielectric glasses and liquid metals. This brief review deals mainly with the preparation, unique properties, applications and experimental results on the physical properties such as electron transport, magnetic and optical properties. Other topics such as ionic structure, glass formation, electron spectroscopy, mechanical and low temperature properties have been reviewed by other talks at this Institute. More recent general reviews are covered by the authors own publications[1,2,3] and the references therein.

2. PREPARATION

Ordinary glasses are formed at relatively low cooling rates; in fact, glass formation is in many cases difficult to avoid in cooling from the liquid state. If the cooling rate is increased, it is possible to induce glass formation in many materials which are normally crystalline. Metallic glasses are formed by a cooling rate of 10^6 degrees per second, which can be achieved by two main techniques: splat-cooling and melt-spinning. There are several other techniques which are described in the literature.[1]

In general amorphous metals can be prepared by a variety of methods: 1. Evaporation, 2. Sputtering, 3. Chemical deposition, 4. Electrodeposition, and finally, 5. Rapid quenching from the liquid state. Amorphous metals prepared by the last method are the so-called metallic glasses or glassy metals. At present, the following families of metallic glasses are known: T-N (e.g. $Fe_{80}B_{20}$,$Pd_{80}Si_{20}$), N-N (e.g. $Mg_{70}Zn_{30}$,$Ca_{70}Mg_{30}$), T_L-T_E (e.g. $Ni_{60}Nb_{40}$,$Ta_{55}Ir_{45}$), RE-N (e.g. $La_{70}A\ell_{30}$), RE-MN (e.g. $La_{80}Au_{20}$), RE-T_L (e.g. $Gd_{70}Co_{30}$,$Gd_{70}Ni_{30}$) and U-T (e.g. $U_{70}Fe_{30}$); where T : transition metal, N: polyvalent metal, T_L : late transition metal, T_E : early transition metal, RE: rare earth metal, MN: monovalent noble metal, and U: Uranium.

3. UNIQUE PROPERTIES

Metallic glasses show an unique combination of specific metallic and glassy properties. They have high mechanical strength combined with ductility, remarkable corrosion resistance, and they show soft-magnetic behavior and high mechanical hardness. Table 1 gives an overview by comparing several properties of crystalline metals, dielectric glasses and metallic glasses. This crude comparison has to be taken with some reservations.

Table 1 Metallic glasses show an unique combination of properties from the metallic and glassy state.

	METALS	GLASSES	METALLIC GLASSES
STRUCTURE	CRYSTALLINE	AMORPHOUS	AMORPHOUS
BONDING	METALLIC	COVALENT	METALLIC
STRENGTH	NOT IDEAL	IDEAL	IDEAL
DEFORMATION	GOOD,DUCTILE	BAD,BRITTLE	GOOD,DUCTILE
HARDNESS	SMALL	LARGE	LARGE
FRACTURE LIMIT	HIGH	LOW	HIGH
CORROSION	ATTACKED	RESISTANT	RESISTANT
OPTICS	NON-TRANSPARENT	TRANSPARENT	NON-TRANSPARENT
TH,EL.CONDUCTIVITY	HIGH	LOW	HIGH
MAGNETISM	SEVERAL PROPERTIES, PARA-, FERRO-MAGNETIC	NON-MAGNETIC	SEVERAL PROPERTIES, MAINLY SOFT-MAGNETIC

4. APPLICATIONS

The mechanical properties (high strength, hardness, wear resistance and ductility) make metallic glasses useful for reinforcing filaments in plastics, rubber and cement. High-pressure and wear resistant components, and all kinds of springs can be made.

The magnetic properties (soft-magnetic with very low coercive forces and high permeabilities) can lead to applications in all kinds of transformers, shielding materials, and tape-recording heads.

The electrical properties (large resistivity comparable with well-known resistance alloys and nearly zero or negative temperature coefficient) of metallic glasses make them suitable for electrical resistors, for low temperature heating wires, and for resistance thermometers.

Several metallic glasses are extremely corrosion resistant and this may offer many applications in chemistry, biophysics and medicine.

There are other promising applications such as hydrogen storage etc. Of great interest are the microcrystalline alloys produced by the melt-spinning process. The brazing foils are already used in several fields.

5. EXPERIMENTAL RESULTS

5.1 Electrical Transport

The experimentally observed magnitude, temperature dependence, and composition dependence of the electrical resistivity of metallic glasses at high temperatures ($T \gtrsim T_D$; T_D: Debye temperature) are comparable to liquid metals data. Beck[4] has shown at this Institute that the extended Ziman theory can describe these liquid state results including rare earth metals. The negative temperature coefficient of resistivity (NTCR), observed in both the liquid and amorphous state, plays a key role in the theoretical description. Along these lines, the Evans, Greenwood, and Lloyd[5] formulation of the Ziman liquid metal theory was used as a first starting point to understand the electrical resistivity of amorphous metals at high temperatures. A more general formalism of this diffraction model was developed by several authors to describe the electrical resistivity in the range $T < T_D$. The predictions of the diffraction model are in excellent agreement with experimental findings in a variety of amorphous metals.[6]

However, alternative theories, which have somewhat changed by the years, are often involved to explain the low temperature behavior, in particular the NTCR's. These alternative theories do not give concrete numerical values. Nowadays, the main objections to the diffraction model have been reduced to the ideas of Cochrane et al.[7] and of Tsuei.[8] Nevertheless, the relation of those ideas to the metallic glass problem has been questioned by Black and Gyorffy.[9]

Indeed, a possible breakdown of the diffraction model in view of new experimental results has to be discussed. These results are the rather unexpected NTCR's of glassy alloys belonging to the groups of T_E-T_L, T_L-RE, and RE-MN. The first two groups also show NTCR's in the liquid state. In contrast, the liquid alloys of the last group show a positive temperature coefficient in the liquid state. The apparent inconsistency of these NTCR's with the diffraction model can be replaced by the fact, that $2k_F$-values of the alloys cannot simply be extrapolated from the values of the pure components. The alternative, that a simple extrapolation procedure yields the correct $2k_F$-values in these alloys and the diffraction model fails, seems at present quite unlikely.

5.2 Magnetic properties

This subject is highly divided into two fields. The first one deals with the ferromagnetic transition metal (Fe,Co,Ni) based alloys of group T-N. These represent the potential technological significant alloys. The current work in this field is devoted to our understanding of both their fundamental and applications oriented properties. The second field mainly deals with the fundamental properties of ferro-, para- and diamagnetic metallic glasses such as Gd-Co(RE-T), Pd-Zr($T_E - T_L$), Mg-Zn(N-N), Pd-Si(T-N) and of dilute alloys. They are not important for any applications, but provide a good opportunity to explore further details of magnetism in metallic glasses. A recent review[10] deals mainly with amorphous alloys other than glassy metals. However, the basic principles might be extended to the glassy state. This field is also covered by other talks at this Institute.

5.3 Optical properties

In general, the optical reflectivity of metallic glasses has to be measured in a large energy range. From these data the optical constants ε_1, ε_2 can be calculated by using a Kramers-Kronig analysis. The optical conductivity is most suitable to distinguish the two significant energy ranges: a) the intraband (Drude) range, where values for the resistivity and the number of conduction electrons can be deduced, and b) the range of interband transitions, where the observed structures can be related to the electronic band structure. Structures in the optical data correspond in the crystalline state to transitions from a filled to an empty state at selected points in k-space, obeying k-conservation. However, in the amorphous state, where the k-conservation fails, the interband absorption reflects more or less the density of states. Therefore, the optical spectroscopy data are in close relation to the results obtained by electron spectroscopy.[11] This relation was established for glassy Pd-Si alloys[12] and is under investigation for alloys of the T_E-T_L group. The varying d-band splitting of those alloys is also seen by comparison of the optical reflectivities of glassy Fe-Zr and Pd-Zr alloys.

ACKNOWLEDGEMENTS

The content of this contribution is based on the work of my collaborators. T. Gabriel takes care of the preparation. The section on electrical transport is part of the PhD thesis of M. Liard. Dr. M. Müller informed me about his magnetic measurements. The optical reflectivity measurements were performed by Dr. K.P. Ackermann and R. Lapka. The part on applications reflects stimulating discussions with several groups in the industry. The author is very grateful for all the collaborations and discussions. Financial support of the Swiss National Science Foundation, the "Kommission zur Förderung der wissenschaftlichen Forschung", the "Eidgenössische Stiftung zur Förderung Schweizerischer Volkswirtschaft durch wissenschaftliche Forschung" and the Research Center of Alusuisse is gratefully acknowledged.

REFERENCES

1. H.-J. Güntherodt in "Festkörperprobleme" (Advances in Solid State Physics), Vol.XVII, ed. by J. Treusch (Vieweg, Braunschweig), p.25, 1977.
2. H.-J. Güntherodt, H. Beck, P. Oelhafen, K.P. Ackermann, M. Liard, M. Müller, H.U. Künzi, H. Rudin and K. Agyeman, Nato Advanced Study Institute "Electrons in Disordered Metals and at Metallic Surfaces", Ed. by P. Phariseau,

B.L. Gyorffy and L. Scheire (Plenum, New York), 1979.
3. H.-J. Güntherodt et al., Critical Assessment Article "Metallic Glasses", J. of Phys. F: Metal Phys. 1980.
4. H. Beck, this book.
5. R. Evans, D.A. Greenwood, and P. Lloyd, Phys. Lett. A35, 57 (1971).
6. L.V. Meisel and P.J. Cote, Phys. Rev. B17, 4652 (1978).
7. R. Cochrane, R. Harris, J. Ström-Olsen, and M. Zuckermann, Phys. Rev. Letters 35, 676 (1975).
8. C.C. Tsuei, Sol. State Commun. 27, 691 (1978).
9. J.L. Black and B.L. Gyorffy, Phys. Rev. Letters 41, 1595 (1978).
10. R.W. Cochrane, R. Harris, and M.J. Zuckermann, Phys. Reports 48, 1 (1978).
11. P. Oelhafen, this book.
12. A. Schlegel, P. Wachter, K.P. Ackermann, M. Liard, and H.-J. Güntherodt, Solid State Commun. 31, 373 (1979).

FORMING ABILITY AND THERMAL STABILITY OF METALLIC GLASSES

E. Hornbogen, I. Schmidt

Institut für Werkstoffe, Ruhr-Universität Bochum, Germany

1. INTRODUCTION

Thermal stability as well as glass forming ability will be discussed in this paper. We are dealing with relatively complex properties of metallic materials which are not identical, but their origin is interrelated. Glass forming ability refers to the property of the molten metal to reach low temperatures without crystallization. The existence of a substrate is unavoidable. It provides not only heat transfer from the melt, but also an interface at which heterogeneous nucleation can occur. Thermal stability of a metallic glass refers to its behaviour during heating from low temperatures or during isothermal aging. In both cases a good thermal stability or glass forming ability is caused by a low crystallization rate. In addition, changes in structure and properties of a glass before crystallization may be included into the term "thermal stability".

The crystallization temperature T_c and the glass transition temperature T_g are not exclusively determined by thermodynamical equilibria, but also by kinetic conditions. They are therefore not independent of cooling rate or aging treatment. A glass can be obtained if the molten alloy is cooled faster than a critical cooling rate to a temperature at which isothermal crystallization requires a very long period of time. Good glass forming ability shows an alloy for which this critical cooling rate is not trespassed for relatively large cross sections and which can be heated far above ambient temperature without crystallization (Fig. 1).

2. THERMODYNAMIC STABILITY OF THE GLASS STRUCTURE

The major cause for the stability of the amorphous structure is the entropy of mixing. As a consequence of the rather disordered glass structure this term becomes more significant for the reduction of Free Energy than in crystalline solids. In addition, several conditions exist, which provide further reduction of Free Energy at defined compositions of the alloy. These conditions have been discussed a great deal in the literature and therefore it is sufficient only to summarize them briefly:

a) The atomic size ratio is important in a random densest packing of hard spheres. The number of holes can be determined, which provide space for smaller atoms. In a RDPHS-structure these atoms, for example boron or carbon, are completely encaged by larger atoms like iron. It is found that for a little less than 20 % all of the larger holes are filled. This in turn coincides with the fact that many glass-forming alloys are found in this compositional range (1,2,3).

b) A different structure is characterized by a composition which allows a maximum concentration of ordered clusters in the liquid, a little above the melting temperature. For certain stochiometric compositions the interaction energy between different atomic species provides additional thermodynamic stability, which can be determined from measurements of anomalies of physical properties of the liquid above the melting temperature. The existence of many earth-alkaline-metal base glasses (for example Ca) has been predicted and found, following this structural feature (4,5).

c) Observations of a negative temperature coefficient of electrical resistivity (6) have stimulated considerations of the concentration of valancy electrons e/i. In certain alloys maximum crystallization temperatures have been found for alloys with a little less than two conduction electrons pro atom (6,7).

$$2 \left| k_F \right| \simeq K_P \qquad (1)$$

The maximum of the structure factor due to the closest neighbour is K_P. With the model of nearly free electrons equation 1 is fulfilled for $^e/i \simeq 1.7$. k_F is the wave vector of the electrons with Fermi energy. This rule is similar to those by which the stability of Hume-Rotherey-phases is predicted. The interaction of the electrons with the structure factor leads to a reduction of the density of states at this critical value (8,9,10).

All three principles may have a justification to explain partially the behaviour of limited groups of alloy. In no case they can be generalized; for each of these models examples can be found, which contradict the predictions.

From ceramic and polymeric materials it is known that covalent bonding and elongated shape of molecules are additional factors which explain the good glass forming abilities of these materials (Tab. 1). The structure of these glasses differs much from that of metallic glasses, mainly due to the different types of bonding.

At present, it is not quite clear, whether all metallic glasses form a unique structural species. For continuous series of metal and semiconductor atoms like Al-Si continuous sequences from RDPHS to random network structure (RN) have been found (11,12). It is not unlikely that pure metals and alloys of similar atoms like Fe-Ni or Fe-Cu will form microcrystalline structures (13). This is not easy to confirm directly, because crystallization has progressed much in these materials at ambient temperature . The microcrystalline microstructure found under these conditions can be explained as having originated from an original ultra-microcrystalline structure. This structure is characterized by zones of closely packed atoms with an average diametre of < 10 nm, surrounded by a structure similar to high angle grain boundaries. In this case the transition from microcrystalline to glassy is continuous. For micro-crystallites which contain in average 20 atoms the portion of grain boundary structure surpasses 50 %. In Fig. 2 a schematic survey is given for the possible structural species of metallic glasses. These varieties may be of importance if the mechanism of nucleation of crystalline phases is to be understood.

The glass structures which are obtained by any method of solidification usually are imperfect, i.e. they do not represent the (metastable) lowest Free Energy condition of a glass. Many changes in mechanical, magnetic and electrical properties occur under conditions at which crystallization has not yet started (14,15,16,17). The following structural changes must be responsible for these phenomena, which usually are associated with a very low effective activation energy and a decrease in free volume .

A. Healing out of structural defects possibly continuum dislocations, vacancies or tangling bonds.

B. Segregation of small solute atoms into favorable positions and short distance rearrangement of solute atoms.

C. Short range ordering or short range decomposition.

It may be mentioned that the segregation behaviour of high angle grain boundaries can serve in some cases as a two dimensional analogon which may help to understand what happens during aging the three dimensional glass (18).

3. THE RELATIVE STABILITY OF GLASS AND CRYSTALLINE PHASES

It is not sufficient to limit the discussion to the absolute stability of the glass phase. At least equally important is the relative stability, i.e. the difference of Free Energy between the various stable or metastable crystalline phases (Fig. 3).

The well known relation between the Free Energy diagram and the phase diagram is shown for stable and metastable equilibria of a simple eutectic system. The Free Energy difference $\Delta F = F_a - F_c$, between the amorphous state and one or two crystalline phases is identical with the "driving force" for crystallization. When ΔF is small the amorphous structure can be relatively resistent to crystallization. It is evident that the composition of ΔF_{min} is not necessarily identical with $F_{a\ min}$.

There exist three basic modes of crystallization which lead towards a new metastable or stable equilibria (Fig. 4) (19).

I homogeneous (or polymorphous) crystallization
II simultaneous (or eutectic) crystallization
III heterogeneous (or primary) crystallization

The reactions I and II can produce stable or metastable equilibria, while reaction III always leads to a metastable mixture of glass and crystal.

I s $a \rightarrow \alpha$
I m $a \rightarrow \alpha_m$

II s $a \rightarrow \alpha + \beta$
II m $a \rightarrow \alpha_m + \beta_m$
III m $a \rightarrow a' + \alpha_m$

Partial information on the isothermal Free Energy diagrams can be obtained from the phase diagram, which is usually well known. A full knowledge of F-x-diagrams for all temperatures or complex alloys is rarely available. Nevertheless even semi-quantitative diagrams provide important information on the necessary conditions under which certain crystallization reactions can possibly take place. To answer the question, which of the reactions really occurs, rate of nucleation and growth have to be considered. The composition for the transition from homogeneous to eutectic crystallization (reactions I to II) for example can be determined by comparing the velocities of the two reaction fronts (Fig. 5, equ. 8 to 12), i.e. the particular reaction is found which proceeds most quickly. The heterogeneous crystallization reaction (III m) can take place in the total range of I s and II. Its occurence is favored by easy nucleation. The velocity of its reaction front is determined by long range diffusion and consequently, it is always slow. All the reactions which lead to metastable phases, will be followed by secondary reactions.Examples are:

$$a' + \alpha_m \rightarrow \beta + \alpha$$

crystallization of the residual glass which surrounds crystals which had formed by heterogeneous crystallization.

$$\alpha_m \rightarrow \alpha + \beta$$

decomposition of a metastable homogeneous crystal.

$$\alpha_m + \beta_m \rightarrow \alpha + \beta$$

attainment of complete segregation in an incompletely segregated eutectic.

4. NUCLEATION OF CRYSTALLINE PHASES

Nucleation in crystalline solids is reasonably well understood - even in the rather complex cases of defect solid solutions in which metastable phases form (20,21). The situation in glasses seems to be simpler. An additional problem is an uncertainty about the diffusion coefficient, because the activation energy of diffusion is not independant of temperature in the range of the glass transition T_g (Fig. 6). In addition, the following aspects have to be considered to understand the nucleation phenomena which play a role in connection with glass formation.

Usually the metal is brought into touch with a substrate to provide heat flow from the liquid. This places an interface at the site of maximum undercooling (Fig. 7). As a consequence conditions for heterogeneous nucleation are favorable. The energy balance is modified by the contribution of interfacial energy σ_{as} to the energy which is required for nucleation (Fig. 8) (22,23):

$$\Delta F(i) = -A\ \Delta f_{ac}\, i + B\ \sigma_{ac}\, i^{2/3} - C\ \sigma_{as}\, i^{2/3} \qquad (2)$$

If the interfacial energy between glass and crystal σ_{ac} is large, the contribution of the surface of the substrate can reduce considerably the critical size of a nucleus i_N (Fig. 9a) and increase the rate of nucleation at this surface.

$$i_N(T) \sim \frac{T_m (B \cdot \sigma_{ac} - C \cdot \sigma_{as})}{\Delta H_m\ \Delta T} \qquad (3)$$

A less well known modification of the classical nucleation theory, which evidently has to be considered, is the existence of a minimum nucleus size $i_{N\,min}$. For large undercooling $\Delta T/T_m$ i_N cannot approach zero, but only the size of the elementary cell of the crystalline phase which has to be formed:

$$i_N = i_N(T) + i_{N\,min} \qquad (4)$$

For small undercooling ΔT $i_{N\,min}$ can be neglected. For large undercooling crystal structures with a small number of atoms in the elementary cell, like bcc and fcc, will form faster than those with large cells like intermetallic σ-phases. A theory of nucleation which includes this lower limit, has not yet been formulated in detail. There is however, much evidence for the fact that crystalline phases with simple, small elementary cells form much more rapidly than those with complicated large cells (Fig. 9b).

For the equation that describes crystallization in a tTc-diagram, three special features have always to be considered, if crystallization is observed at extremely high undercooling:

$$t_N = t_o \exp\frac{\Delta F_N\ (T, \sigma_{as}, i_a)}{RT} \exp\frac{Q_a\ (T)}{RT} \qquad (5)$$

ΔF_N can be reduced depending on surface energy and the

material of the substrate (Equ. 3).
ΔF_N will not be reduced to zero for complicated large elementary cells of the crystalline phase (Equ. 4).
Q_a (and also D_{oa}) the diffusion coefficient of the undercooled liquid does not follow a simple function for its temperature dependence (Fig. 6).

In addition, the diffusion coefficient of the metallic glass depends very much on composition and details of the glass structure (Fig. 6). The activation energy of diffusion in the glass Q_a is not necessarily smaller than in a comparable crystal Q_α (Tab. 2). This can be understood by the fact that metalloid atoms can be encaged by metal atoms for example in the $T_{80}M_{20}$- glasses. They must leave the cage for long range diffusion.

To approach the reality of the possibilities for the modes of crystallization during the cooling procedure used for the production of glasses, several principle cases must be considered (Fig. 10). For a given alloy and thickness of the ribbon or splat a maximum cooling rate exists at the surface of the substrate, while away from it, the cooling rate reaches its minimum. From a tTc-diagram which contains these limiting cooling rates and the periods of time for the start of crystallization, the principle microstructures can be derived (Fig. 1, 10,11,12,22):

a) homogeneous glass
b) crystallization starts exclusively at the interface
c) crystallization starts exclusively in the interior
d) crystallization both at the surface and inside the material
e) crystallization starts exclusively at the outer free surface
f) no nucleation, but growth of crystals after rapid heating of a surface layer.

The time dependence of heterogeneous or homogeneous nucleation (Equ. 5) and the growth characteristics of the reactions (see chapter 5) determine whether crystallization occurs and which process dominates. For formation of a glass the minimum cooling rate must be greater than that required for the onset of all possible reactions $t_{N\,min}$ (Fig. 5 and 10).

$$t_{N\,min} > \frac{T_m - T_g}{dT/dt} \tag{6}$$

In RN- or RDPHS-structures individual nucleation of crystals in the interior must occur, which is usually not ideally homogeneous (Equ. 2) depending on whether the undercooled liquid contains inclusions or other frozenin inhomogeneities. The situation is different for a microcrystalline structure. In this case a very large number of preformed nuclei exists. They simply have to grow. If other factors (for example a high diffusion coefficient D_a) are favorable, the rate of crystallization can be extremely high. This may explain why pure metals and many alloys crystallize into an ultra fine grain structure already far below ambient temperature, and therefore are regarded as non-glass forming materials (Fig. 13,14).

The other extreme is an alloy in which no nucleation takes place inside the undercooled liquid. Crystallization starts exclusively at the interface. In this case crystallization is complete, if the reaction front will travel through the thickness d of the ribbon or splat without being frozen in or encountering individually nucleated particles. For isothermal growth the minimum velocity v_{min} is given by the relation

$$t_{Nc} > t_{Nd} + \frac{d}{v_{min}} > t_c \qquad (7)$$

t_{Nc} is the time required for the individual nucleation[x)] ahead of the reaction front, t_{Nd} for the individual nucleation. For $t < t_C$ no crystallization can occur and a glass is formed.

5. VELOCITY OF DISCONTINUOUS REACTIONS

Equation 7 contains the velocity of propagation of a reaction front which has been initiated at a surface. The models of such reactions have been shown in Fig. 5. The velocity is determined by the driving force i.e. by the Free Energy change in the reaction front (Fig. 3,4, 5) and a mobility factor which is determined by the diffusion conditions in this front (24).

[x)] Nomenclature:
Heterogeneous nucleation - at the surface or the interphase with the cooling substrate.
Individual nucleation - Formation of nuclei in the interior of the glass at inclusions or as yet unknown frozen in discontinuities in the glass structure.
Homogeneous nucleation - Formation of nuclei at statistical fluctuations caused by thermal agitation in a perfect glass structure.

$$v = \Delta f \cdot m = \frac{1}{A} \cdot \frac{\partial F}{\partial x} \cdot m \qquad (8)$$

For the three reactions which are shown in Fig. 5 the following equations can be derived by analysing the diffusion conditions (24,25,26):

$$v_I = f_c \frac{v_m}{RT} \cdot \frac{D_a}{b} \qquad (9)$$

for the homogeneous crystallization,

$$v_{II} = (f_c - f_i) \frac{v_m}{RT} \cdot \frac{D_a}{S} \qquad (1o)$$

for the eutectic crystallization.

For the primary crystallization a comparable discontinuous mechanism which leads to a constant velocity is impossible. Long range diffusion must occur perpendicular to the reaction front. As a consequence the velocity decreases with time t or path of travel x.

$$v_{III} = \frac{\partial x}{\partial t} \propto \frac{D_a}{x}, \qquad x \propto (D_a t)^{1/2} \qquad (11a,b)$$

A primary crystallization reaction will slow down after it has been initiated by individual nucleation or at a surface. Therefore this mechanism can dominate crystallization only by copius individual nucleation inside the undercooled liquid.

To explain the glass forming ability and the microstructure in the as-crystallized state, not only the velocity but also the type of reaction which takes place has to be determined. This is shown for the reactions I and II (Fig. 4) for the assumption that the time required to nucleate both reactions is equal or negligible (Equ. 7). Then the principle can be applied that the reaction occurs for which the velocity is maximum, for example:

$$v_{II} < v_I = \max. \qquad (12)$$

The condition $v_I = v_{II}$ determines the composition at which the homogeneous crystallization reaction is replaced by decomposition (Fig. 15). On this basis the behaviour of many important glass forming alloys can be understood. As an example may serve the different behaviour of Fe-B- and Fe-C-based alloys (Fig. 16)(27-30). It is well known that it is easier to obtain relatively thick amorphous ribbons with Fe-B-alloys than with Fe-C, in spite of the fact that both are alloys of eutectic composition and the behaviour of the solute in the melt is similar (31). The major difference between the two alloy systems becomes evident from Figs. 16, 17, 18. In the Fe-C-system a sequence of metastable simple-structured solid solutions exist in a wide range of concentrations, while boron destabilizes these phases (Tab. 3). As a consequence crystallization must be associated with decomposition for most Fe-B melts while Fe-C can crystallize rapidly by homogeneous crystallization (Fig. 19). The favorable effect of boron on the glass forming ability seems to be not only due to a stabilization of the glass phase, but also to enforcing a reaction which requires decomposition, therefore long range diffusion and a lower velocity of the reaction front.

6. CRITICAL COOLING RATES - GLASS FORMING ABILITY

It has been shown in the previous chapter that the ability to freeze-in the liquid structure cannot be explained by an unique property. In addition it has to be distinguished between the behaviour of the material during cooling from a liquid and while a glass is reheated to temperatures at which diffusional rearrangements become rapid.

During cooling from the liquid state the material is exposed to quite a special set of conditions:
1. The melt is overheated to a temperature $T_1 = T_M+\Delta T_1$.
2. The melt is brought into touch with the substrate which is at a low temperature $T_2 = T_M-\Delta T_2$.
3. The material of the substrate is characterized by a surface energy σ_{sa} and the surface morphology which influences the heterogeneous nucleation of the crystalline phases.
4. Heat conductance through the substrate determines the shape and time dependence of the temperature gradient inside the liquid (Fig. 7).

As a consequence of those conditions a number of basic types of microstructures can be found after cooling (Fig. 10). An explanation can be given by a discussion of the tTc-diagram for homogeneous and heterogeneous nucleation together with the cooling rates, which decrease with increasing distance from the substrate (Fig, 7,20).

The following necessary conditions can be derived for the relations between applied cooling rates and critical cooling rates to interpret the different situations:

$$\dot{T}_{max} > \dot{T}_{het} \quad \dot{T}_{min} > \dot{T}_{ind} \qquad (13a)$$

$$\dot{T}_{max} < \dot{T}_{het} \quad \dot{T}_{min} > \dot{T}_{ind} \qquad (13b)$$

$$\dot{T}_{max} > \dot{T}_{het} \quad \dot{T}_{min} < \dot{T}_{ind} \qquad (13c)$$

$$\dot{T}_{max} < \dot{T}_{het} \quad \dot{T}_{min} < \dot{T}_{ind} \qquad (13d)$$

$$\dot{T}_{min} < \dot{T}_{het} \quad \dot{T}_{min} > \dot{T}_{ind} \qquad (13e)$$

$$\dot{T}_{max} << \dot{T}_{het} \qquad (13f)$$

It is evident that glass forming ability is a complex property which involves the effects from the production method (i.e. especially properties of the substrate) in addition to the intrinsic properties of the material which is to be characterized (see Fig. 1o).

7. CRYSTALLIZATION TEMPERATURE - THERMAL STABILITY

An identical situation does not exist for the crystallization temperature (Fig. 21). For isothermal heat treatments it can be supposed that temperature gradients, and in addition the interface with the substrate, are absent. As a consequence the favorable conditions for heterogeneous nucleation of discontinuous reaction at this interface are no longer fulfilled. It leaves, however, still the possibility of heterogeneous nucleation at the free surface , which indeed is sometimes observed. In general reactions will start by individual nucleation in the interior of the glass.

Another difference is due to the fact that the relative stability of glass and crystalline phases (chapter 3) may have changed due to the higher undercooling ΔT_2. An example for this behaviour is the formation of metastable ε- or γ-solid solutions during cooling of Fe-C alloys, as comparedto the most stable α-phase if an Fe-C glass is heated to temperatures below 700 ^{o}C (Fig. 22) (33).

The macroscopic property changes which are caused by reheating are often characterized by the crystallization temperature T_c. This is not an equilibrium property, as for example transformation temperatures, which are based on thermodynamic equilibria. T_c can only be defined using a tTc-diagram (Fig. 23). For an isothermal heat treatment it is the temperature at which crystallization starts after defined period of time. For continuous heating experiments a constant heating rate has to be applied, if a reproduceable value for the crystallization temperature is to be obtained. If not the start is to be characterized, but for example, the maximum rate crystallization, it is unavoidable to investigate the micromechanisms of crystallization in addition to measurements of macroscopic physical properties. This is especially true if the reaction proceeds by more than one stage.

8. COMBINATION OF FACTORS WHICH FAVOR FORMATION AND THERMAL STABILITY OF METALLIC GLASSES

Favorable conditions in this sense are characterized by:

$$\dot{T}_{het} \to \min, \quad \dot{T}_{ind} \to \min, \quad \dot{T}_c \to \max \qquad (14a,b,c)$$

It has been discussed in the preceeding chapters that these conditions are effected by the crystallization behaviour. In some glass forming systems a factor is dominating which in others is less significant. This situation has led to frequent confusion especially for attempts to find unique physical property responsible for good glass forming ability. It follows a list of factors that can be expected to contribute to the limit in which a non-crystalline structure is obtained and preserved:

1. Thermodynamic stability of the non-crystalline structures at certain compositions, which may be due to electron density, atomic size ratio or local order.
2. Structural type of the glass. A microcrystalline structure requires no nucleation, while RDPHS- and RN-structure species do have to form nuclei.
3. Properties which retard nucleation are:

a) high surface energy $\sigma_{a\alpha}$
b) low undercooling $T_M - \Delta T_2$ (eutectic composition)
c) low diffusion coefficient D_a (Q_a high) corresponds to high viscosity of the amorphous phase
d) large size of the elementary cell of the crystalline phase α
e) unfavorable conditions for heterogeneous nucleation

4. Low growth velocities of discontinuous reactions:

a) no stable or metastable phase of the same composition as the undercooled liquid
b) large differences in composition of phases which form by an eutectic reaction
c) avoidance of any discontinuous reactions (see 3e)
d) low diffusion coefficient (see 3c)

The different glass forming abilities of Fe-C and Fe-B

alloys can serve as an example to illustrate the combination of some of these factors.

1. Everything which is known about properties of liquid Fe-C and Fe-B alloys does not provide an indication for differences of the thermodynamic stability of the non-crystalline phases. Both alloys should possess favorable conditions for stability of the non-crystalline phases.

2. Concentrated solutions of both alloys are of the RDPHS-type. For dilute solutions and pure iron a transition ot the microcrystalline structure can be expected, which explains the absolute loss of glass forming ability in this range of composition.

3. The best glass forming conditions are observed in both alloy systems in the environment of the eutectic composition. The crystalline phases which can form are different in both alloys. With its α-, γ-, ε-phases the Fe-C system provides in a wide range of compositions stabl or metastable small elementary cell solid solutions. In the Fe-B alloys of the same compositional range, these phases do not exist. A metastable phase like $Fe_{23}B_6$ which could nucleate in a narrow composition range is associated with a much larger elementary cell as compared to the simple solid solution. Therefore nucleation of crystalline phases is more difficult in Fe-B- than in Fe-C- alloys of comparable composition.

4. This difference becomes even more pronounced if discontinuous reactions induced by heterogeneous nucleation in the interface are considered. All Fe-C alloys can crystallize very rapidly by easy nucleation and fast growth of homogeneous crystalline solutions. In the Fe-B alloys an eutectic reaction is required under comparable conditions. The nucleation of at least one of the crystalline phases is much less probable than that of solid solutions. Even if a reaction front would have developed, long range diffusion is required which allows only low velocities (Equ. 10). This seems to be a major cause for the good glass forming ability of Fe-B alloys.

5. More complex alloys can be only understood by observing the effects of additional alloying elements on all the factors that influence crystallization. If the stability of an Fe-C-base glass is to be increased this can be easily done by the addition of boron and some other metalloid elements like P, S and the consequent induction of decomposition (Fig. 19). The stabilizing effect of transition metal atoms with less filled d-shells (Cr, Mo, W) should have a different origin (Fig. 18). Measurements of the effect of third elements on the activity coefficients of carbon in iron indicate a change in different direction, if the electron density

of iron is decreased or increased. In spite of the lower tendency for decomposition of carbon, the glass forming ability is improved if these elements replace iron. This effect must be due to the stabilisation of the amorphous structure itself, due to the formation of associates of carbon with this type of third element. This indicates the direction in which alloys with optimum properties should be found. If new glass forming alloys are sought or known ones to be improved it is not sufficient to consider only the relation between the composition and the thermodynamic stability of the non-crystalline phase alone. Equally important are the properties which control nucleation and growth of the crystalline phases. They may vary rather independently of the absolute stability of the non-crystalline state.

9. SUMMARY

Thermal stability implies the resistance of a liquid to crystallization during cooling and an existing glass to crystallization during heating.

Thermal stability is effected by:
1. Thermodynamical stability and the structure of the non-crystalline phase;
2. the relative stability of non-crystalline and metastable or stable crystalline phases that exist below the melting temperature;
3. the mode of nucleation within the limits of heterogeneous in the interface of and homogeneous inside the undercooled liquid;
4. the mode of crystal growth with fast propagation without decomposition and slow propagation associated with decomposition.

The glass forming ability of alloys can only be understood if all these factors are taken into consideration simultaneously.

Acknowledgement:

The support of this work by the Deutsche Forschungs-Gemeinschaft (DFG, HO 325/11) is greatfully acknowledged.

LIST OF SYMBOLS

symbol	dimension	explanations
A	m^3	geometric factor equ. 2
A'	m^2	area
a	-	amorphous phase = glass
α	-	stable crystalline phase
α_m	-	metastable crystalline phase
b	m	atomic distance
B	m^2	geometric factor equ. 2
C	m^2	geometric factor equ. 2
D_a	m^2s^{-1}	diffusion coefficient in the am. phase
e	-	number of electrons
F	J	Free Energy
ΔF	J	Free Energy in a fluctuation with a structure of a more stable phase
ΔF_N	J	activation energy for nucleation
$f=\Delta f$	Jm^{-3}	any driving force
f_{ac}	Jm^{-3}	driving force for crystallization
f_i	Jm^{-3}	energy required for the formation of interfaces
ΔH_m	J	heat of melting
i	-	number of atoms
i_N	-	number of atoms in a nucleus
m	$m^4J^{-1}s^{-1}$	mobility of a crystallization front
Q_a	$Jmole^{-1}$	activation energy for diffusion in the amorphous phase
RT	$Jmole^{-1}$	thermal energy
S	m	lamellar spacing
σ_{ac}	Jm^{-2}	interfacial energy - glass/crystal
σ_{as}	Jm^{-2}	interfacial energy - glass/substrate
T	K	any temperature
ΔT	K	undercooling
T_1	K	temperature of overheating above T_m
T_2	K	temperature of undercooling below T_m
T_m	K	melting temperature
T_c	K	crystallization temperature
T_g	K	glass transition temperature
$\dot{T}=\frac{dT}{dt}$	Ks^{-1}	any cooling or heating rate (c.r., h.r
$\dot{T}_{ind}$	Ks^{-1}	critical c.r. to avoid homogeneous or individual nucleation
$\dot{T}_{het}$	Ks^{-1}	critical cooling rate to avoid heterogeneous nucleation at the surface
$\dot{T}_{min}$	Ks^{-1}	minimum c.r. inside the liquid ribbon
$\dot{T}_{max}$	Ks^{-1}	maximum c.r. at the interface with the substrate
t	s	any time

t_o	s	time constant in equ. 5
t_N	s	time until onset of nucleation
t_{Nc}	s	time for homogeneous or individual nucleation inside the material
t_{Nd}	s	time for initiation of a discontinuous reaction at the surface
t'	s	critical period of time used to define T_c
v_m	m^3	molar volume
v	ms^{-1}	velocity of a reaction front
v_I	ms^{-1}	v. for homogeneous (or polymorphous) crystallization
v_{II}	ms^{-1}	v. for simultaneous (or eutectic) crystallization
v_{III}	ms^{-1}	v. for heterogeneous (or primary) crystallization
x		coordinate parallel to direction of crystallization front

REFERENCES

1. J.D. Bernal, Nature 185 (1960) 68
2. J.D. Bernal, Proc. Roy.Soc. 280A (1964) 299
3. D.E. Polk, Acta Met. 20 (1972) 485
4. B. Predel, Ber. d. Bunsenges. Bd. 80, Nr. 8 (1976)695
5. F. Sommer, G. Duddek, B. Predel, Z. Metallkde. 69 (1978) 587
6. H.J. Güntherodt, Adv. in Solid State Physics XVII, Vieweg (1977)
7. H.J. Güntherodt, Verhandl. DPG 13 (1978) 225
8. S.R. Nagel, J. Tauc, Phys. Rev. Lett. 35 (1975) 380
9. S.R. Nagel, J. Tauc, 2. Conf. Rapidly Quenched Metals, Massachusetts (1975) p. 337
10. H. Beck, R. Oberle, 3. Conf. Rapidly Quenched Metals, Brighton (1978) p. 416
11. P.H. Mangin, G. Marchal, B. Rodmacq, Chr. Janot, Phil. Mag. 36 (1977) 643
12. P.H. Mangin, M. Piecuch, G. Marchal, C. Janot, J.Phys.F.: Metal Phys. 8 (1978) 2085
13. E. Hornbogen, J.Mat.Sci. 13 (1978) 666
14. A. Pampillo, D.E. Polk, Mat.Sci.Eng. 33 (1978) 275
15. T. Egami, T. Ichikawa, Mat.Sci.Eng. 32 (1978) 293
16. T.Egami, Mat.Sci.Eng. 13 (1978) 2587
17. T. Egami, Mat.Res.Bull. 13 (1978) 587
18. E. Hornbogen, Proc. ICSMA-5 Aachen 1979, 10P06 p. 1337
19. U. Köster, Crystallization and Decomposition of Amorphous Germanium Alloys, Westdeutscher Verlag, Opladen (1976) 2604

20. A.C. Zettelmoyer ed., Nucleation, Marcel Dekker, New York 1969, p. 309
21. E. Hornbogen, Fortschr. Miner. 50 (1973) 270
22. D. Turnbull, J.C. Fisher, J.Chem.Phys. 17 (1949) 71
23. J.W. Cahn, Acta Met. 4 (1956) 449
24. K. Lücke, Z.Metallkde. 52 (1961) 1
25. C. Zener, Trans AIME 167 (1946) 550
26. D. Turnbull, K.N. Tu, Scripta Met. 1 (1967) 173
27. R.C. Ruhl, M. Cohen, Trans AIME 245 (1969) 241
28. P.G. Boswell, G.A. Chadwick, J. Mat.Sci. 11 (1976) 2287
29. I. Schmidt, E. Hornbogen, Z.Metallkde. 69 (1978) 221
30. U. Herold, U. Köster, Z.Metallkde. 69 (1978) 326
31. H. Schenk, E. Steinmetz, Stahleisen-Sonderberichte 7 (1968) 9
32. B.F. Oliver, US Steel Corporation, Research Center Rep.No. 966 (1961)
33. P.H. Shingu, K. Shimomura. K. Kobayashi, Mat.Sci. Eng. 23 (1976) 183
34. T. Masumoto et.al., 3. Conf.Rapidly Quenched Metals, Brighton (1978)p. 265
35. W. Felsch, Z.Physik 195 (1966) 201
36. I.W. Donald, H.A. Davies, 3. Conf.Rapidly Quenched Metals, Brighton (1978)p. 273
37. U. Köster, private communication
38. H. Warlimont, Z.Metallkde. 69 (1978) 212

Material	Type of bonding	Q_η ($kJmol^{-1}$)	Glass forming tendency
SiO_2	mainly covalent	5oo	++
organic	covalent H-bond	12o	++
Na	metallic	~ 4	--
Fe-B		>4	+

TAB. 1. TENDENCY FOR GLASS FORMATION AS A FUNCTION OF THE NATURE OF BONDING. Q_η : ACTIVATION ENERGY FOR VISCOUS FLOW IN THE STABLE LIQUID STATE.

Alloy	Q_1 ($kJmol^{-1}$)	Q_2	
α -Fe-C	8o		1: long range rearrangement
α -Fe_{sd}	22o		2: short range rearrangement
a -Fe-C	~15o		sd: self diffusion
a -$Fe_{75}P_{15}C_{1o}$	22o	1oo	
a -$Fe_{4o}Ni_{4o}P_{14}B_6$	26o	135	
a -Fe	<1oo		

TAB. 2. ACTIVATION ENERGY FOR DIFFUSION IN A GLASS (A) AND A CRYSTAL (α) (38) .

k	in Fe
<o.1	B
o.2	C,P
o.3	Ni,Ti
o.4	Ge
o.5	Au,Zr
o.6	Si,Cu,Sn
o.9	V,Cr,Co,Al

TAB. 3. EQUILIBRIUM SEGREGATION COEFFICIENT IN FE. $k = X_S/X_L$ (32).

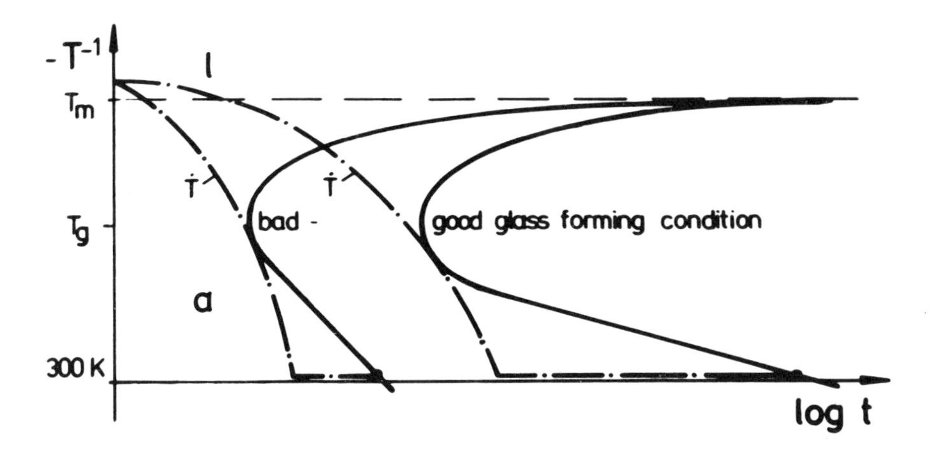

FIG. 1. TIME-TEMPERATURE-CRYSTALLIZATION -DIAGRAM INDICATING THE START OF CRYSTALLIZATION AND THE CORRESPONDING COOLING RATES (DOTTED LINES).

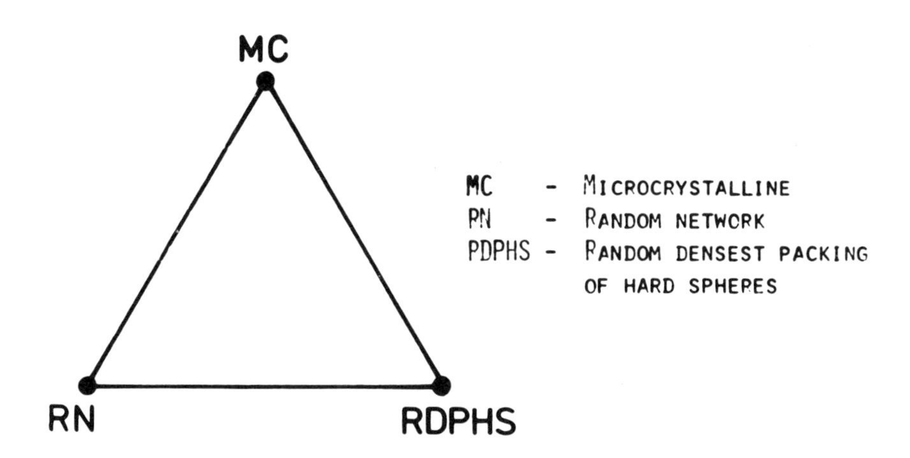

FIG. 2. POSSIBLE STRUCTURAL SPECIES OF METALLIC GLASSES

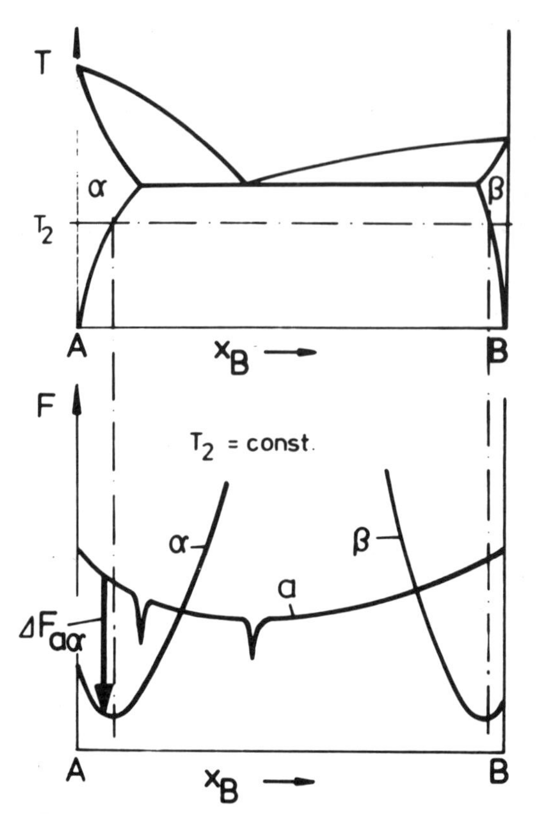

FIG. 3. CONNECTION BETWEEN PHASE-DIAGRAM AND FREE-ENERGY-DIAGRAM

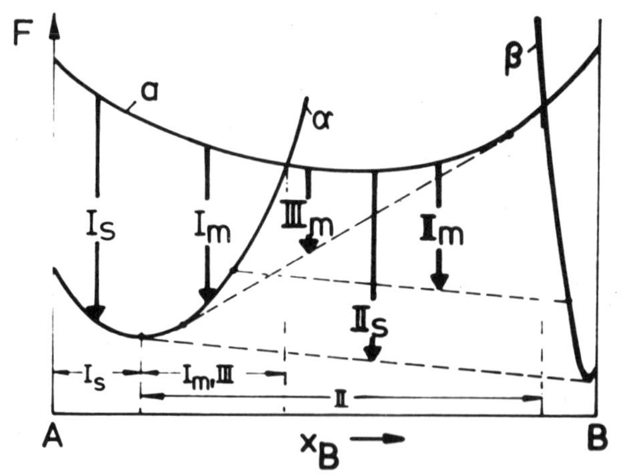

FIG. 4. CONCENTRATION RANGES FOR THE POSSIBLE OCCURRENCE OF CRYSTALLIZATION MECHANISMS.

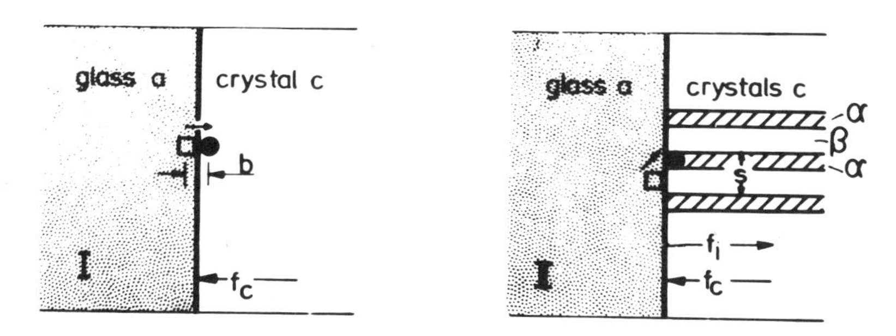

FIG. 5. DIFFUSION CONDITIONS IN THE REACTION FRONT OF MECHANISM I AND II

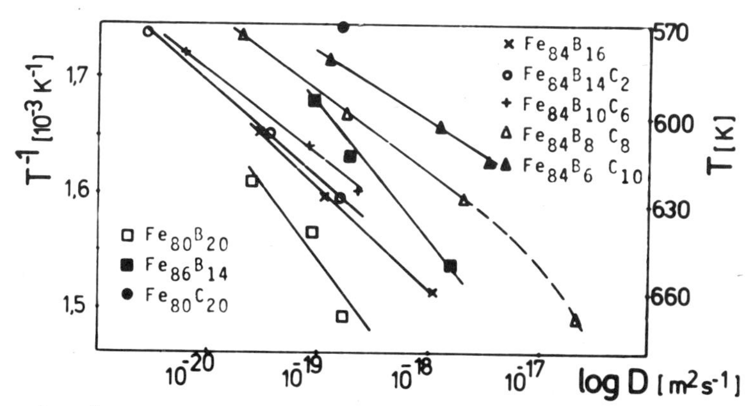

FIG. 6. TEMPERATURE DEPENDENCE OF THE DIFFUSION COEFFICIENT OF VARIOUS METALLIC GLASSES (37).

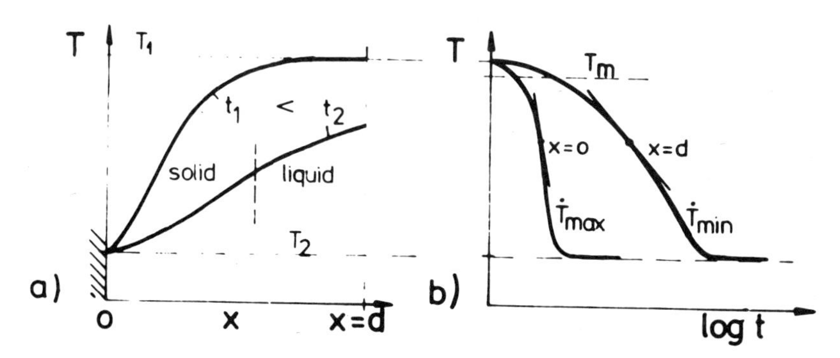

FIG. 7. SCHEMATIC REPRESENTATION OF THE COOLING CONDITIONS OF A RIBBON THICKNESS D: A) TEMPERATURE GRADIENCE FOR TWO PERIODS OF TIME T_1, T_2 B) COOLING RATES FOR TWO EXTREME POSITIONS .

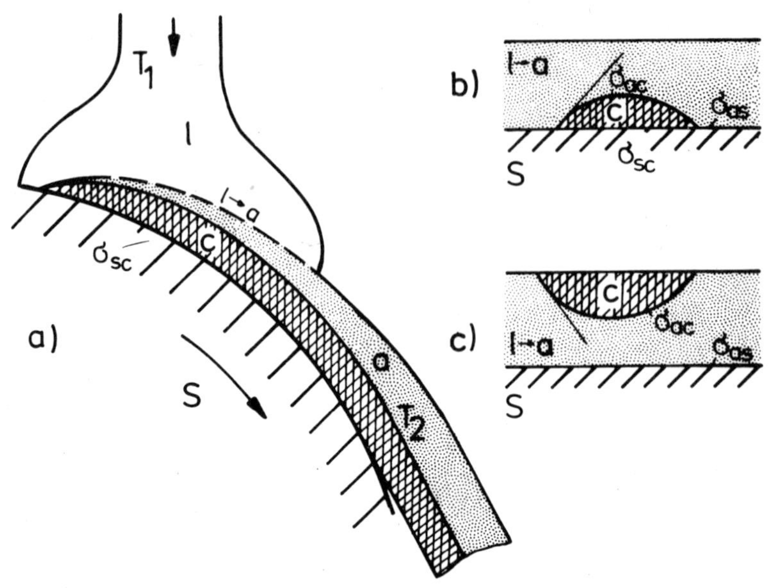

FIG. 8. A)SITUATION FOR A LIQUID COOLED BY A ROTATING SUBSTRATE FOR THE CASE THAT HETEROGENEOUS NUCLEATION AT THE INTERFACE IS NOT INHIBITED, B)HETEROGENEOUS NUCLEATION AT THE INTERFACE AS, C)HETEROGENEOUS NUCLEATION AT THE FREE SURFACE.

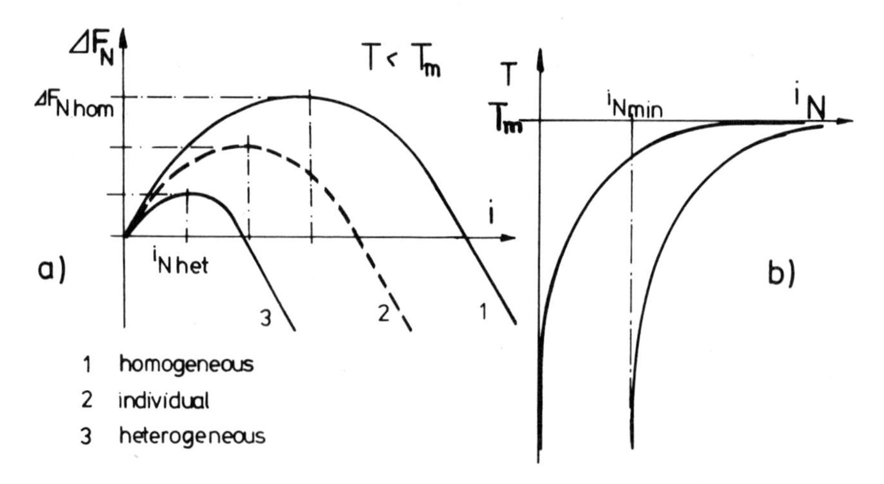

FIG. 9. A)ACTIVATION ENERGY F_N AND CRIT. NUCLEUS SIZE FOR DIFFERENT TYPES OF NUCLEATION B)CONCEPT OF A MINIMUM CRIT. NUCLEUS SIZE IDENTICAL WITH THE SIZE OF THE ELEMENTARY CELL .

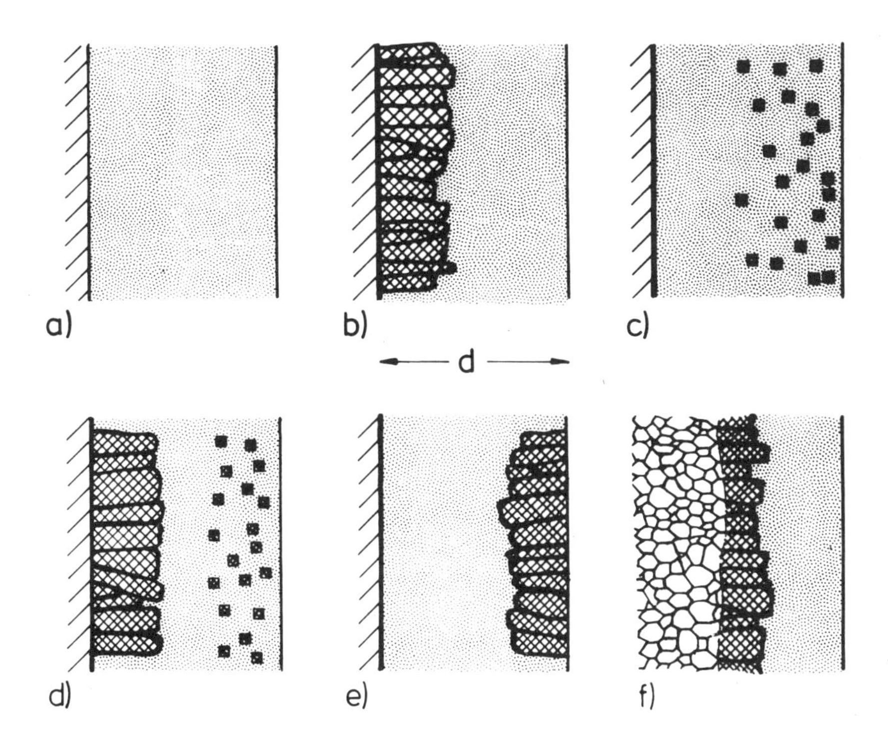

Fig. 1o Survey of types of structure, which are obtained for alloys with different nucleation behaviour .

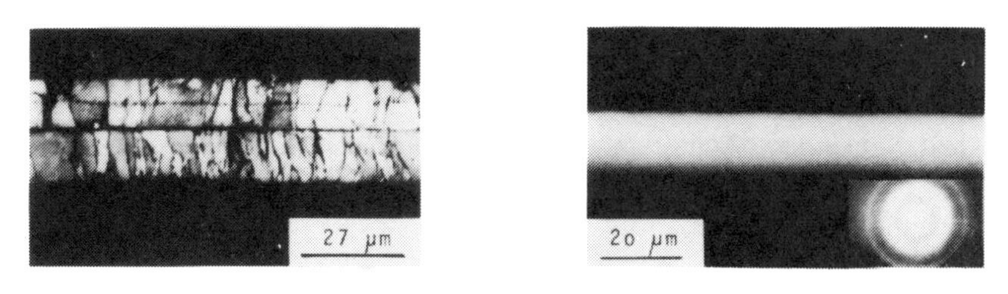

Fig. 11. Microscopic analysis of two splat-cooled alloys a) formation of ε-phase by homogeneous crystallization of $Fe_{81}C_{19}$, b) glass formation of $Fe_{75}Mo_{9}C_{16}$.

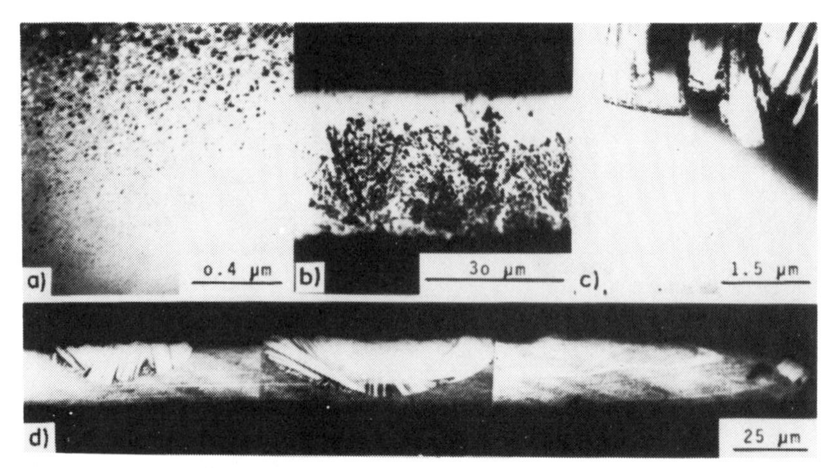

FIG. 12 PARTIALLY CRYSTALLINE STRUCTURES OBTAINED IN ALLOYS WITH DIFFERENT CRYSTALLIZATION BEHAVIOUR: A) INDIVIDUAL NUCLEATION IN $Fe_{59}Cr_{29}C_{12}$ B) NUCLEATION AT THE INTERFACE IN $Fe_{71}Si_{12}C_{17}$ C,D) HOMOGENEOUS CRYSTALLIZATION OF $Fe_{75}Mo_{9}C_{16}$ NUCLEATING AT THE FREE SURFACE .

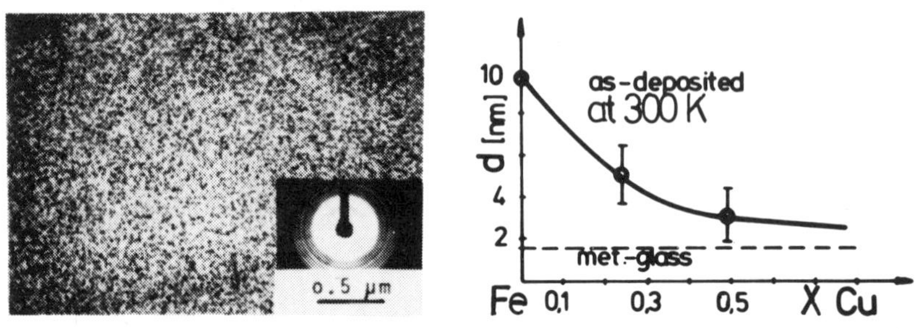

FIG. 13. ULTRAFINE GRAIN STRUCTURE OF VAPOUR-DEPOSITED Fe-Cu ALLOYS. A) TEM OF $Fe_{75}Cu_{25}$, BCC STRUCTURE B) MEASUREMENT OF AVERAGE GRAIN DIAMETRE AS A FUNCTION OF Cu CONTENT .

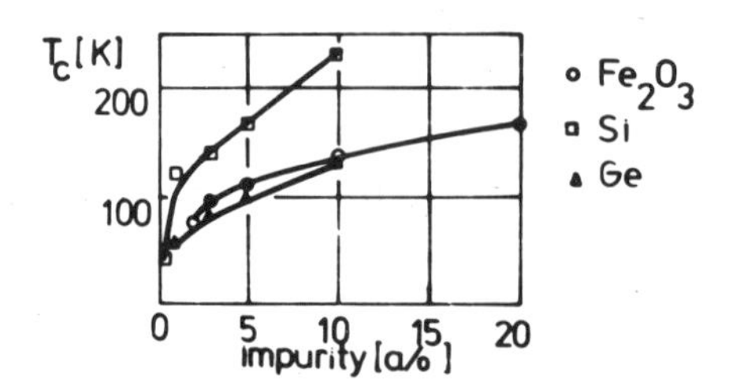

FIG. 14. CRYSTALLIZATION TEMPERATURE OF Fe-FILMS AS A FUNCTION OF IMPURITY CONTENT (35) .

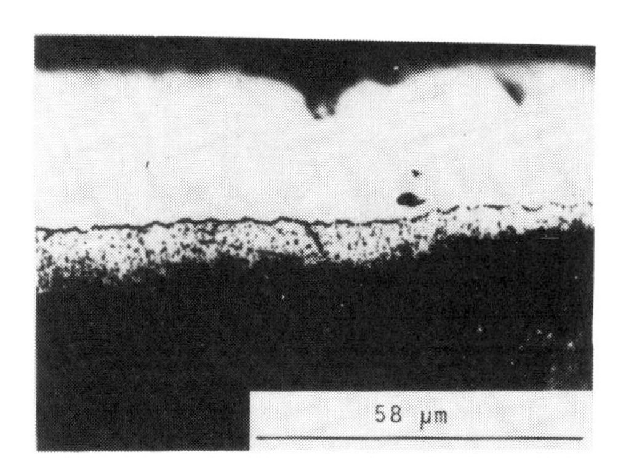

FIG. 15 THE FRONT OF HOMOGENEOUS CRYSTALLIZATION (I) IS PRECEDING THE FRONT OF THE COMPETING DECOMPOSITION REACTION (II) IN $Fe_{61}Cr_{26}C_{13}$ (SEE FIG. 4).

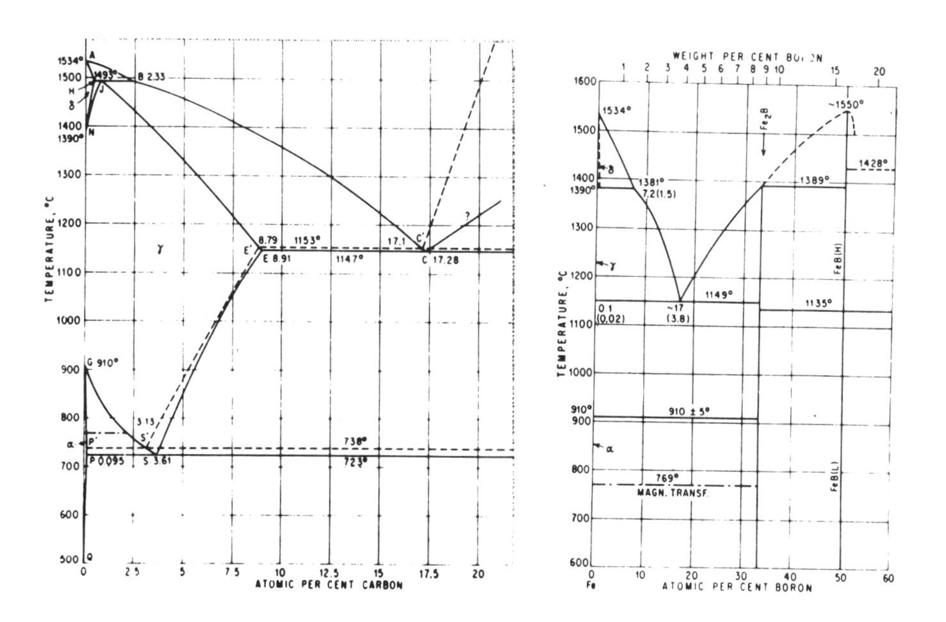

FIG. 16. PARTIAL FE-C - AND FE-B -PHASE DIAGRAM .

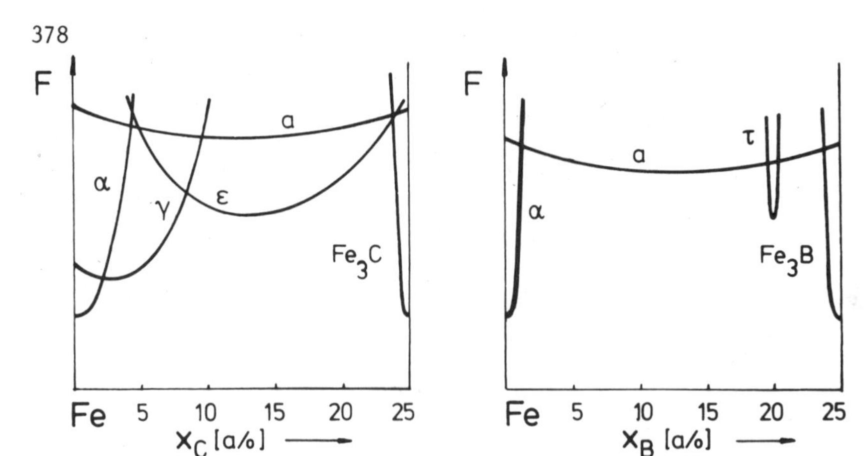

FIG. 17. SEMI-SCHEMATIC FREE-ENERGY - DIAGRAMS FOR SOLID Fe-B - AND Fe-C -ALLOYS .

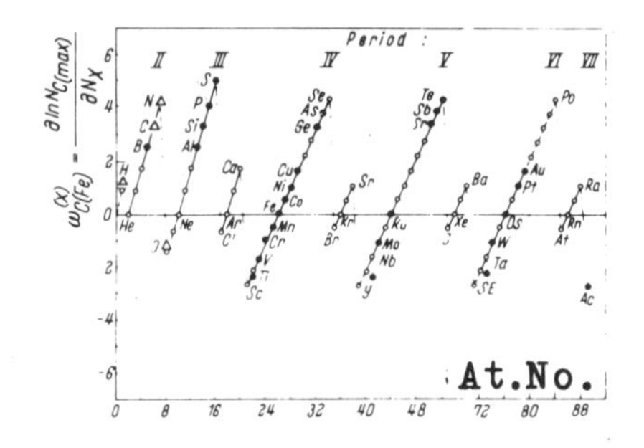

FIG. 18. ACTIVITY PARAMETERS OF C IN Fe-C-X MELTS EFFECTED BY THIRD ALLOYING ELEMENTS (31) .

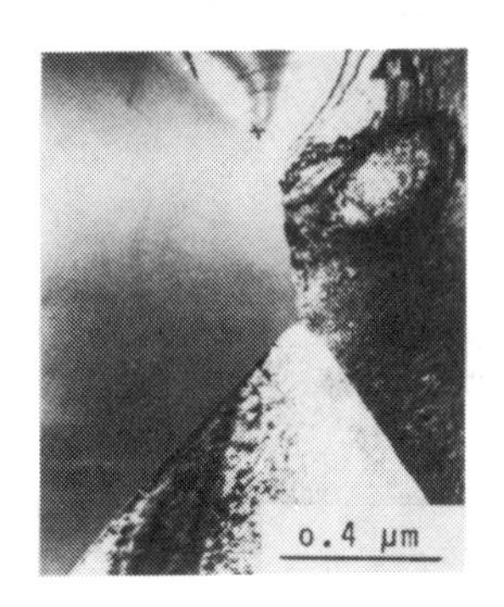

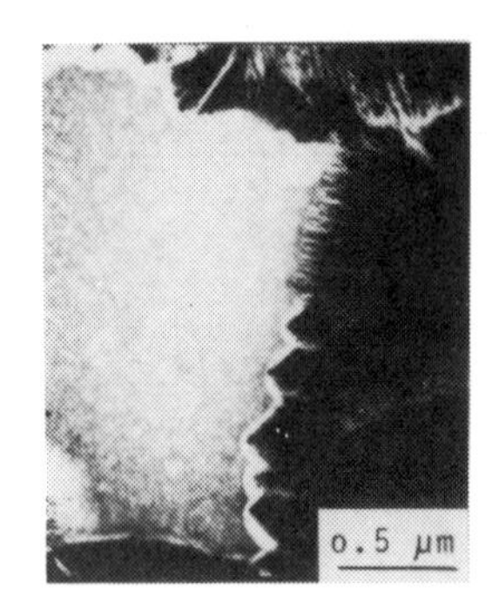

FIG. 19. MICROSCOPIC ANALYSIS OF CRYSTALLIZATION MECHANISM IN Fe-ALLOYS: A) $Fe_{81}C_{19}$ (HOMOGENEOUS) B) $Fe_{80}B_6C_{14}$ (EUTECTIC) .

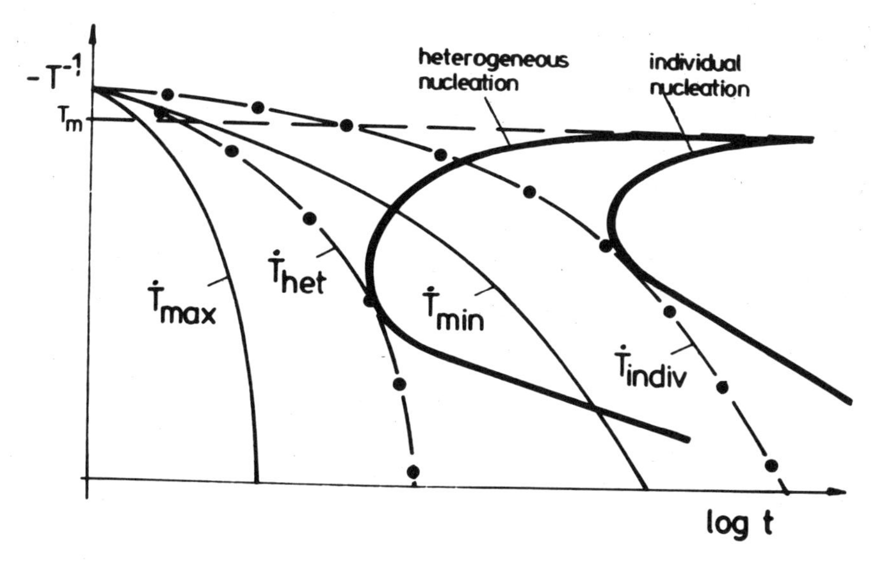

FIG. 20. SCHEMATIC TTC -DIAGRAM, WHICH CAN BE USED TO EXPLAIN THE DIFFERENT POSSIBILITIES OF CRYSTALLIZATION BEHAVIOUR (SEE FIG. 10, 12, EQU. 13).

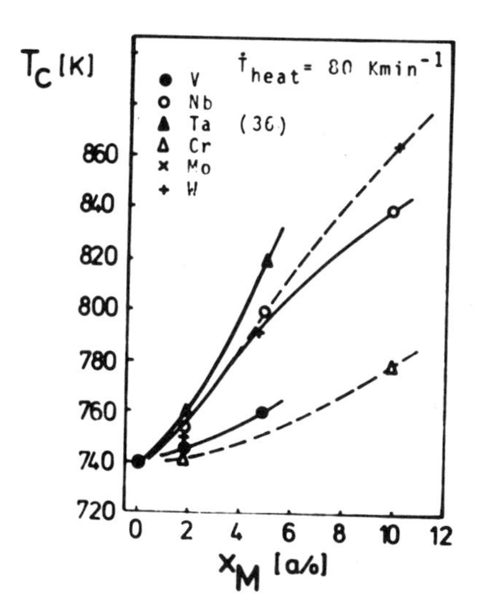

FIG. 21. EFFECT OF REPLACEMENT OF FE BY A THIRD ELEMENT M ON THE CRYSTALLIZATION TEMPERATUPE DETERMINED BY DTA IN $Fe_{83-x}M_xB_{17}$.

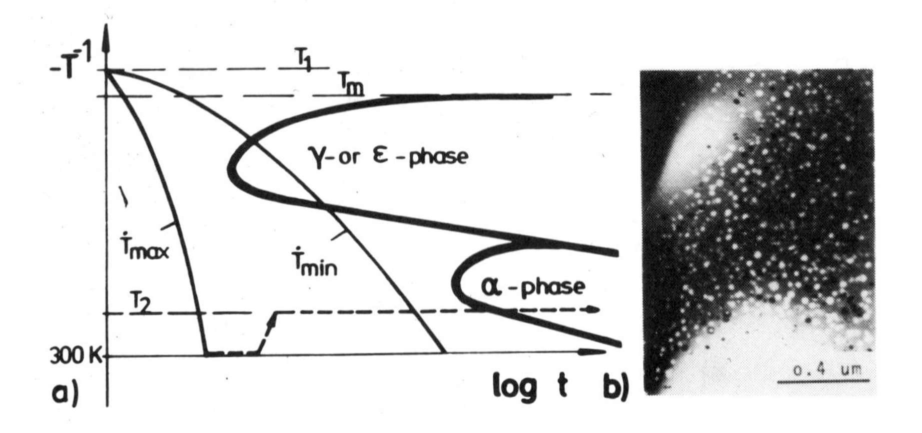

FIG. 22. FORMATION OF BCC α-FE INSTEAD OF HCP ε (SEE FIG. 19.) BY RE-HEATING FROM THE GLASS CONDITION: A) TTC-DIAGRAM B) TEM OF INDIVIDUAL NUCLEATION OF α-FE (600 K / 5 MIN).

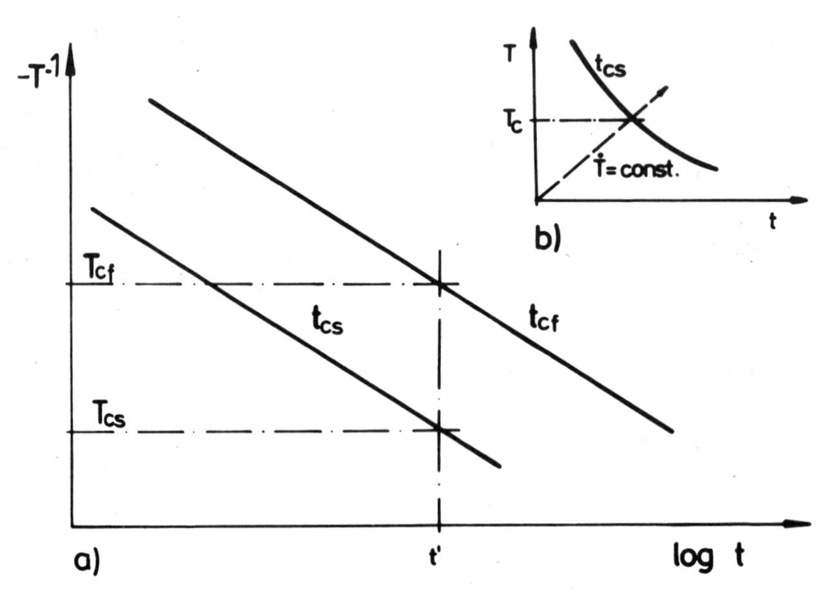

FIG. 23. DEFINITION OF THE CRYSTALLIZATION TEMPERATURE T_C: A) USE OF A TTC-DIAGRAM AND ISOTHERMAL HEAT TREATMENT TO DETERMINE T_{CS} AND T_{CF} (START OR FINISH OF CRYST.) B) LINEAR TTC-DIAGRAM WHICH CAN BE USED TO DEFINE T_C FOR A CONSTANT HEATING RATE.

NONMAGNETIC PROPERTIES OF THE GLASSY STATE

A. C. Anderson

Department of Physics and Materials Research Laboratory
University of Illinois at Urbana-Champaign
Urbana, Illinois 61801

1. INTRODUCTION

The nonmagnetic, low-temperature properties of crystalline metals can be explained in terms of thermal phonons and "conduction" electrons, which are, respectively, excitations of the lattice system and the electronic system. Thermal phonons and conduction electrons are also important in glassy metals at low temperatures. But in glassy metals there exists, in addition, a spectrum of highly localized excitations which apparently is common to all amorphous materials. This paper will examine the limited information which is available concerning these localized excitations.

Most of the low-temperature data on amorphous materials has been obtained from nonmetallic glasses. The reason is threefold. First, the quite recent interest in glassy metals has developed as a small portion of a more general interest in amorphous and disordered materials. Second, many measurements on glassy metals are encumbered by a contribution from the conduction electrons. And third, many measurements are extremely difficult to perform on the sample geometries generally available for glassy metals, i.e., ribbons or splats. Since most experimental effort has been directed to the nonmetallic glasses, Section 2 will discuss information, attesting to the existence of localized excitations, which has been gleaned from dielectric glasses. Section 3 briefly

---

*This work has been supported by the National Science Foundation under Grant DMR-77-08599 and the Department of Energy under Contract EY-76-C-02-1198.

describes the only useful theoretical model which is currently available, and the various attempts which have been made to experimentally test this model. Finally, in Section 4, the relevant measurements on metallic glasses will be presented and the results compared with the behavior of dielectric glasses and the theory.

The reader should be aware of two facts concerning this paper. First, the bibliography is not intended to be exhaustive. In general the most recent references are cited. The author makes no apologies for the disproportionate number of references to work from his own laboratory.

Second, the presentation is intended to be logical rather than historical. Only the following brief historical perspective will be offered. By the 1960's, various "anomalies" were recognized in the low-temperature behavior of amorphous materials. As one example, the thermal conductivity of many glasses, plastics, glues and greases used for cryostat construction were found to have nearly the same thermal conductivity below ≈ 1 K, which is $\approx 10^{-4}T^2$ (W/cm K). Nevertheless, these low temperature properties attracted little attention until a paper by Zeller and Pohl [1] appeared in 1971 which suggested that the various anomalies might have a common origin. This paper catalyzed the current broad interest in the low-temperature behavior of amorphous materials.

2. EVIDENCE FOR LOCALIZED EXCITATIONS

The presence of various excitations is most apparent in the specific heat of a material. Figure 1 shows the specific heat of two amorphous materials, namely, fused silica [2] and a polymer [3]. The specific heat, C, has been divided by the cube of the temperature, T, so as to emphasize the departure of C from the phonon contribution C_{ph}. The phonon contributions for the two glassy materials, indicated by the dashed lines in Fig. 1, were calculated in the Debye approximation using longitudinal and transverse acoustic velocities measured below 1 K.

It is obvious in Fig. 1 that, with decreasing T, the specific heat of the two amorphous materials increases greatly over the phonon contribution. This behavior indicates the presence of an additional set of excitations which contribute a specific heat $C_{ex} = C - C_{ph}$. At 0.1 K, $C_{ex} \simeq 100\ C_{ph}$. Furthermore, C_{ex} is approximately proportional to T, indicating that the energy spectrum n(E) of the excitations is nearly independent of energy E for $E/k \lesssim 1$ K (k is the Boltzmann constant). Finally, it is to be noted that the behavior depicted in Fig. 1 is true for all amorphous materials, i.e., the density of excitations n(E) is roughly the same for all glassy dielectrics independent of chemical composition or thermal history.

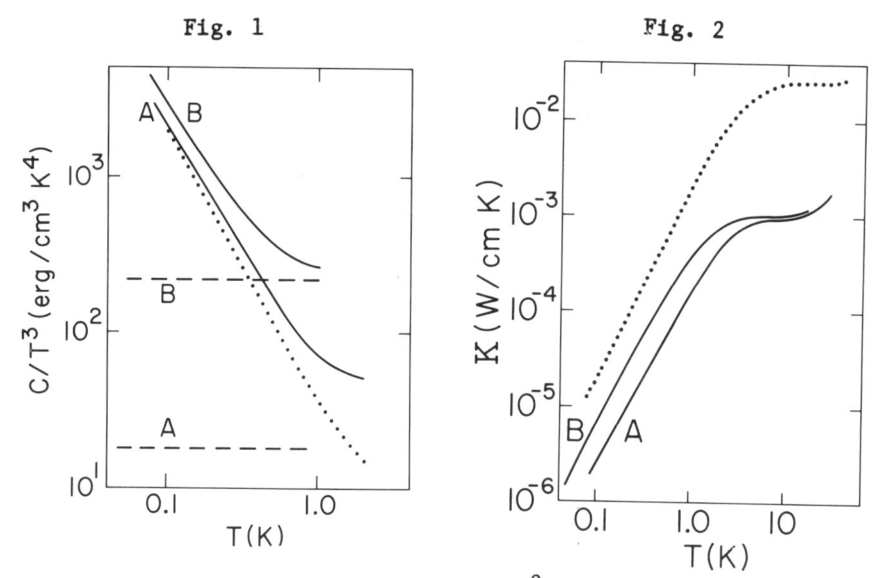

Fig. 1 Specific heats C, divided by T^3, of (A) fused silica and (B) a polymer are represented by solid lines. The dashed lines represent the calculated phonon contributions for these two amorphous materials. The dotted curve is C/T^3 for crystalline Na β-alumina which contains planes of high disorder.

Fig. 2 Thermal conductivity κ of (A) fused silica and (B) an amorphous polymer. The dotted curve is for crystalline Na β-alumina.

Additional information can be obtained from thermal transport measurements. Figure 2 shows the thermal conductivity κ of the same fused silica [2] and amorphous polymer [3] used to obtain Fig. 1. Three features are of interest, as they are common to all glassy materials. First, κ is proportional to T^2 below ≈ 1 K. Second, there is in the range ≈ 5-20 K a "plateau" which has a weaker temperature dependence. Third, the magnitudes of κ, at a given temperature, are similar.

The role played in the thermal conductivity, by the excitations which dominate the specific heat below 1 K, has been determined by measuring κ in both silica glass and polymer samples containing large numbers of microscopic holes [4]. The holes served to scatter the heat carriers, and the mean free path of the heat carriers

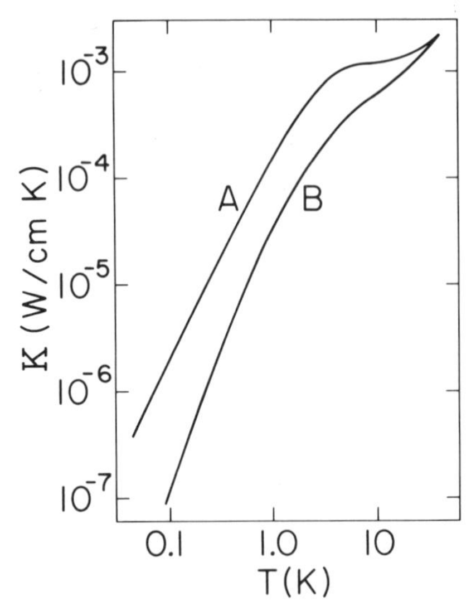

Fig. 3 Thermal conductivity κ of (A) silica glass and (B) silica glass containing 3 x 10^6 cm^{-2} of 5 x 10^{-4} cm diameter holes. Holes in a glassy polymer produced the same effect. Curve B has been corrected upward for the material "removed" by the holes, see Ref. 4.

could be deduced by measuring (in a scanning electron microscope) the average distance between holes. The results are shown in Fig. 3. The presence of holes produces both a reduction in κ and a change in temperature dependence. The results are in excellent agreement with a theoretical calculation, using no adjustable parameters, which assumes that only phonons transport heat. Thus, even though the specific heat of the excitations is a factor of $\approx$ 100 larger at 0.1 K than that of the phonons, the excitations make a negligible contribution to κ at 0.1 K. In brief, the excitations are highly localized.

The presence of the holes reduces κ even in the region of the plateau, see Fig. 3. As explained in Ref. 4, this reduction indicates that bulk amorphous materials act as a low-pass filter for phonons, that whereas phonons having a frequency below $\approx 10^{11}$ Hz have mean free paths of order 10^{-2} cm, those above $\approx 10^{11}$ Hz have very short mean free paths of order the phonon wavelength [2] ($\approx 10^{-7}$ cm). As a consequence, the phonons responsible for heat conduction in a glass at 20 K have a wavelength a factor of $\approx$ 10 larger than the phonons which transport heat in a crystal at 20 K. Measurements of phonon diffraction [4] and the thermal Kapitza resistance [5] provide independent evidence that phonons of frequencies greater than $\approx 10^{11}$ Hz do not transport heat, at least for temperatures below $\approx$ 30 K.

Having established that only thermal phonons transport heat in nonmetallic glassy materials, we may return to Fig. 2 and inquire whether it is the localized excitations which scatter the phonons and thereby produce the characteristic $\kappa \approx 10^{-4}T^2$ (W/cm K) below 1 K and the plateau above 1 K. Unfortunately, it is not possible from C_{ex} (Fig. 1) to deduce n(E) for E/k $\gtrsim$ 1 K since the phonon contribution C_{ph} may begin to dominate the specific heat for T $\gtrsim$ 1 K. Hence, it is not known if there are a sufficient number of localized excitations to produce the plateau in κ.

For T $\lesssim$ 1 K, however, it is probable that the localized excitations do cause the phonon scattering. Support for this speculation is derived from additional measurements on the fused silica and the polymer, used in Figs. 1 and 2, in which the atomic arrangement was altered. Neutron irradiation of fused silica [2] increased κ by $\approx$ 30%, but reduced C_{ex} by $\approx$ 30%. This may be interpreted as a reflection, in both properties, of a 30% reduction in n(E). Increasing the cross-link density of the polymer [3] increased C_{ex} by 50% and decreased κ by 50%. Again, an increase of 50% in n(E) would account for the changes observed in both C_{ex} and κ (as well as other properties [3]). It has been suggested that the thermal conductivity is not related to scattering by the localized excitations, [6,7] but no experimental evidence supports this proposal. Hence, in light of the above results, it will be assumed in the remainder of this paper that the localized excitations do in fact scatter phonons.

Influence of the localized excitations is also observed in other properties, such as thermal expansion [8], optical absorption [9] and emission [10], nuclear [11] and electron [12] magnetic resonance, ultrasonic [13] and dielectric [14,15] dispersion and absorption, and in ultrasonic [16] and dielectric [17,18] transient behavior. In some cases, because of problems in sample preparation, certain properties may not be readily accessible to measurement. Nevertheless, when all measurements for a given material are considered, it may be stated that evidence for the presence of localized excitations has been found in every amorphous material which has been studied. (As one example, see Ref. 19 and papers cited therein for problems encountered with amorphous arsenic).

Before proceeding to a theoretical model for the excitations, it should be noted that the localized excitations can survive in a highly crystalline environment. Figures 1 and 2 include data [20] from the crystalline superionic conductor Na β-alumina. The same broad energy spectrum of localized excitations is present [12,21, 22]. The excitations are here associated with the highly mobile Na ions which lie on monatomically thick planes in this microscopically layered material. A $\approx$ 20% nonstoichiometric excess of Na plus the attendent charge-compensating defects make the planes highly disordered. This disorder presumably creates the localized excitations. Unfortunately, how much "disorder" is required to produce glass-like properties has not been determined.

3. TESTS OF A PHENOMENOLOGICAL MODEL

In 1972, P. W. Anderson, et al. [23] and W. A. Phillips [24] independently proposed a tunneling-states model to explain the low temperature behavior of amorphous materials; that is, the localized excitations arise through the process of quantum mechanical tunneling. It is often objected that entities as large as atoms or molecules should have a vanishingly small probability of tunneling in a solid. However, it is now well documented [25,26] that such tunneling can occur.

In the tunneling-states model, it is assumed that some atomic-size entity resides in a potential energy double-well as depicted in Fig. 4. Here a symmetrical well is used for convenience. The entity of mass m may execute the same zero-point motion about either position x_1 or x_2. This degeneracy in energy is lifted by the possibility of tunneling through the barrier of width $\Delta x = x_2 - x_1$ and height ΔV. The energy spectrum of the tunneling entity then consists of a ground state and a single excited state at energy $\varepsilon \propto \exp-[\Delta x(2m\,\Delta V)^{1/2}/\hbar]$. (It is assumed that other energy levels have energies $>> \varepsilon$). Since, in a random structure like a glass, one might expect a range of ΔV and Δx, there should also occur a range in ε. Excitations of energy ε contribute a Schottky peak to the specific heat at $T \approx \varepsilon/k$. A uniform range in ε would contribute a superposition of peaks giving $C_{ex} \propto T$ as observed in glasses. Also, an incident phonon of frequency ω could be absorbed by a tunneling-state of energy $\varepsilon = \hbar\omega$ and then be reradiated in a different direction. This would constitute a resonant scattering mechanism responsible for the small thermal conductivity of glasses. The strength of the interaction, γ, between phonon and localized excitation is an additional parameter in the model.

There would also be asymmetrical wells as shown in Fig. 5. Now the difference in energy, E, between the excited state and ground state becomes $E = \sqrt{\varepsilon^2 + \xi^2}$. Also, the ground state is localized near position x_2, the excited state near x_1. If the tunneling entity finds itself in the excited state at x_1, the probability of tunneling to the ground state (with creation of a phonon of energy E) depends on the size of the barrier. Since a range in barrier sizes is expected, there should also occur a range in relaxation times τ_1, the time spent in the excited state.

As a consequence of the range in τ_1, a specific heat measurement must occur over a time interval $\Delta t > \tau_{max}$, where τ_{max} is the longest relaxation time in the sample. For $\Delta t < \tau_{max}$, only some fraction of the total C_{ex} would be measured. Also, in the thermal conductivity, phonons of frequency ω would be scattered only by

Fig. 4

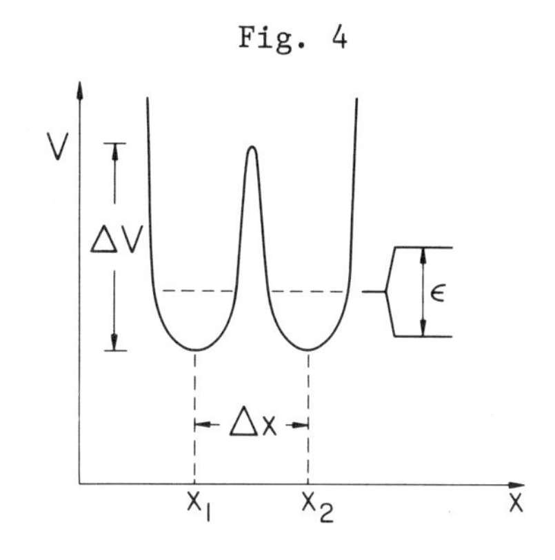

Fig. 5

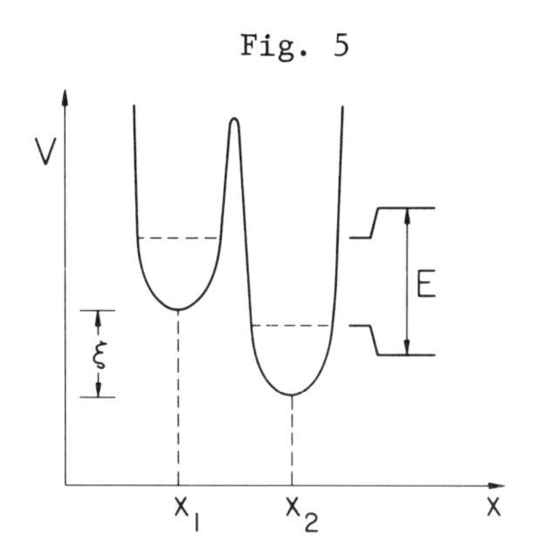

Fig. 4 Potential energy V of some entity in a glass as a function of a generalized coordinate x. ε represents the net energy difference between the ground state and the excited state.

Fig. 5 Potential energy V of some entity in a glass as a function of a generalized coordinate x. E represents the net energy difference between the ground state and the excited state for this asymmetrical potential.

those localized excitations having, roughly, $\tau_1 \lesssim \tau$. For $\tau_1 > \tau$ the excitation does not have an opportunity to respond to the passing acoustic wave. Hence only a fraction of the total density of excitations $n(E,\tau_1)$ should be effective in reducing the thermal conductivity of a glassy material.

The variations in ΔV and Δx have not been calculated from first principles for any material. Hence, a phenomenological approach is used. A spectrum $n(E,\tau_1)$ of localized excitations is selected which can provide agreement with experimental data. An appropriate trial spectrum is indicated in Fig. 6. To fit the specific heat data, the E dependence of $n(E) = \int n(E,\tau_1)\, d\tau_1$ must be roughly a constant, $n(E) = 10^{33}$ erg^{-1} cm^{-3}, for E/k ranging from ≈ 0.01 K to 10 K. The τ_1 dependence of $n(E,\tau_1)$ can be investigated by performing the specific-heat measurement within a time interval $\Delta t < \tau_{max}$. An early attempt to observe the τ_1 dependence in the specific heat was rather inconclusive. [27,2] A more recent measurement [28], using $\Delta t \gtrsim 10^{-6}$ sec, indicated <u>no</u> τ_1 dependence of $n(E,\tau_1)$. This is not consistent with other measurements discussed below, and represents a serious test of the tunneling-states model.

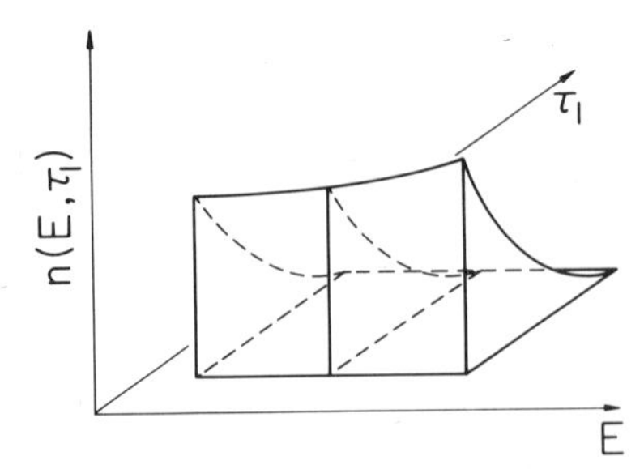

Fig. 6 Schematic distribution of localized excitations $n(E,\tau_1)$ as a function of their energy E and relaxation time τ_1. Very roughly, the ranges in E/k and τ_1 which have been measured lie between 0.01 and 10 K and 10^{-6} and 1 sec, respectively.

On the other hand, this "fast" heat capacity measurement involves phonons and localized excitations far from thermal equilibrium. A definitive analysis of data from such a nonequilibrium system is not, as yet, possible.

If n(E) has been obtained from specific-heat measurements, the thermal conductivity can be calculated from the tunneling-states model provided γ, the phonon coupling constant, is known. The parameter γ has been measured independently as described later in this paper. Substitution of the measured n(E) and γ in the model gives a calculated $\kappa \propto T^2$ as observed experimentally, but with a magnitude a factor of $\approx$ 10 too small. In brief, not all localized excitations interact with the thermal phonons. There must be a τ_1 dependence. The τ_1 dependence depicted in Fig. 6 is adequate to provide an excellent fit to both the specific heat and the thermal conductivity data. The magnitude of γ, $\approx$ 1 eV, appears to be roughly the same for all amorphous dielectrics if it is assumed that the τ_1 dependence of $n(E,\tau_1)$ is roughly the same from material to material [29].

The adequacy of the model may be further tested by a measurement of the ultrasonic dispersion. Figure 7 shows the acoustic velocity dispersion for the same fused quartz and polymer samples used in Figs. 1 and 2. The solid lines are calculated [2,3] using the tunneling-states model and parameters consistent with the κ and C_{ex} data. The agreement is satisfactory, the model is useful.

Referring again to Fig. 6, it was hoped that the number of parameters needed to specify $n(E,\tau_1)$ would be small. As indicated previously, n(E) was initially believed to be independent of E, at least for 0.01 K < E/k < 1 K. However, it now appears that n(E) increases with increasing E. It has also been argued that n(E) = 0 at small E because the magnitude of ΔV in Fig. 4 is never infinite,

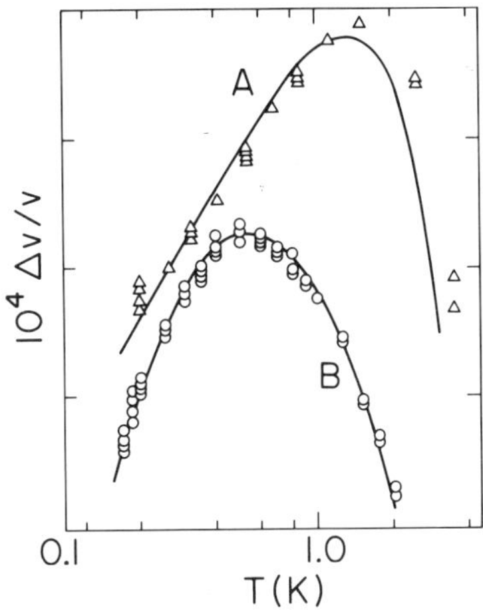

Fig. 7 Variation of transverse ultrasonic velocity, v, with temperature for (A) fused silica (5×10^7 Hz) and (b) an amorphous polymer (10^7 Hz). The zero of the vertical scale is arbitrary. Longitudinal waves have a similar temperature dependence.

and hence the tunnel splitting ε is never zero [30, 3]. However, the localized excitations have been probed directly in ultrasonic absorption measurements [13] to energies as low as $E/k = 7 \times 10^{-3}$ K. Hence any gap in n(E) must lie at still smaller energies.

A complicated E and τ_1 dependence of $n(E,\tau_1)$ does not invalidate the tunneling-states model as it presently exists. A complicated $n(E,\tau_1)$ only limits the utility of the model because of the large number of variable parameters that must be determined experimentally. In addition, there may be more than one distribution $n_i(E,\tau_1)$ for a given material. For example, there may exist in silica glasses a density $n_2(E,\tau_1)$ for localized excitations associated with OH^- impurities and an independent $n_1(E,\tau_1)$ for all other locatized excitations [31]. Finally, the phonon coupling parameter γ might contain some E dependence.

With any number of variable parameters available in $n(E,\tau_1)$, a fit of the model to the specific heat, thermal conductivity and ultrasonic data of a given glassy material, as discussed above, is generally possible, and hence such a fit does not provide a definitive test of the tunneling-states model. It might seem desirable to test the model directly, that is, by changing the mass of the tunneling entity, or the size or shape of the barrier through which it tunnels, and then attempting to observe a corresponding change in E. Unfortunately, this approach has not been successful [29,32-35] since any shift in energy is obscured by the broad spectrum of energies n(E) already existing.

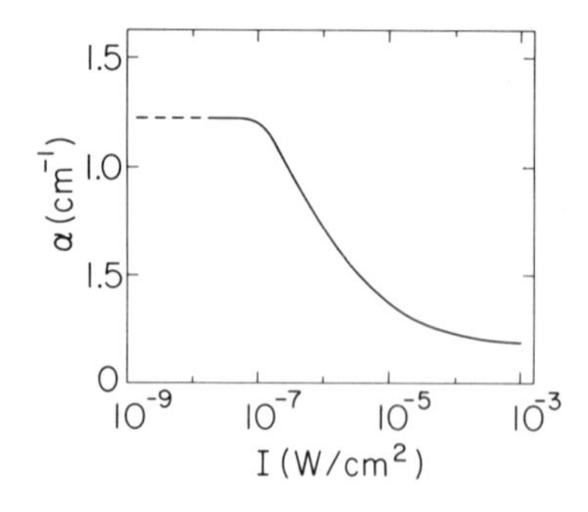

Fig. 8 Variation of ultrasonic attenuation α of a borosilicate glass versus the intensity I of the applied acoustic wave at 0.5 K and 10^9 Hz. (Ref. 13).

The tunneling-states model does have a distinguishing characteristic. The localized excitations are represented by a collection of two-level states as discussed previously. The presence of a two-level state can be probed in several ways. Here it may be helpful to keep in mind that a magnetic dipole with spin 1/2 in a magnetic field is also a two-level system. Phenomena associated with the spin-1/2 system, such as behavior observed in nuclear magnetic resonance measurements, should occur in an analogous fashion in a glass. However, in the case of the glass, the "resonance" is extremely inhomogeneously broadened in energy as indicated by n(E).

Application of an ultrasonic stress field of frequency ω causes transitions to the excited state for those localized excitations having energy $E = \hbar\omega$. The transitions remove energy from the ultrasonic wave producing a measurable attenuation. However, if the wave is too intense, the two energy levels will become equally populated and the attenuation will decrease. (In nuclear magnetic resonance this effect is referred to as saturation, in optics as self-induced transparency). The effect has been measured [13], see Fig. 8, and was one reason Phillips [24] selected a two-level tunneling state to represent the localized excitations.

The application of an appropriate sequence of ultrasonic pulses produces an ultrasonic analog of the electromagnetic spin echo of a magnetic system or the photon echo of an optical system. (These phonon echoes should not be confused with reflections from surfaces of a sample, which are also referred to as echoes). Knowledge of the amplitude and time duration of the applied acoustic pulses provides a direct measure of the phonon coupling parameter γ for those localized excitations having energy $E = \hbar\omega$. See Ref. 16 for more details.

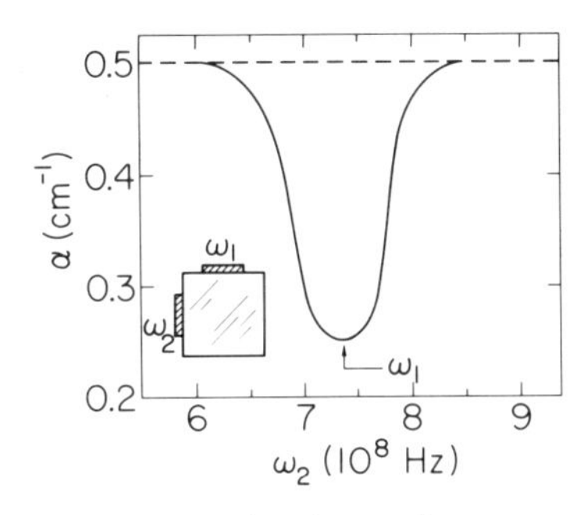

Fig. 9 Variation of attenuation α_2 at 0.5 K of a weak ultrasonic pulse of frequency ω_2 after prior application of a strong pulse at frequency $\omega_1 = 7.4 \times 10^8$ Hz. The inset shows, schematically, an arrangement of the two ultrasonic transducers. (Ref. 13).

Another pulsed ultrasonic experiment is depicted in Fig. 9. The transducer at the top supplies an intense pulse at frequency ω_1 which saturates (equally populates) those localized excitations having energy $E_1 = \hbar\omega_1$. The second transducer immediately thereafter supplies a weak pulse of frequency ω_2. As shown by the data [13] of Fig. 9, if $\omega_1 = \omega_2$ the attenuation of the second pulse is small since the first pulse had already saturated the excitations having $E_1 = \hbar\,\omega_1 = \hbar\,\omega_2$. If ω_2 is very different than ω_1, the attenuation is normal and large. The important feature is that as ω_2 approaches ω_1, the attenuation begins to decrease. This decrease indicates that the energy injected into the system at ω_1 diffuses rapidly to other localized excitations with energies somewhat different than E_1. The time constant for this process to occur is referred to as τ_2 in analogy with magnetic spin systems. τ_2 concerns the exchange of energy between localized excitations whereas τ_1 concerns the exchange of energy between the localized excitations and thermal phonons. The "hole burning" experiment of Fig. 9 demonstrates that $\tau_2 \ll \tau_1$, hence the localized excitations interact quite strongly with each other [13,36], they are not isolated in the glassy matrix. In addition, a satisfactory explanation of these data [36,37] requires a broad spectrum of τ_1 as indicated in Fig. 6.

If all or some fraction of the localized excitations are associated with an electric-dipole moment, the previously mentioned experiments can be repeated using electric rather than elastic fields. In particular, dielectric dispersion [14], dielectric attenuation [15], dielectric saturation [22], and electric "echoes" [17,18] have all been measured in glassy materials. The tunneling-states model provides a reasonable explanation of these phenomena with the introduction of one additional parameter, namely, the effective electric-dipole moment associated with the localized excitations.

In summary, essentially all measurements on the low temperature properties of nonmetallic glassy materials can be explained qualitatively by using the theory of tunneling states to model the localized excitations. To make a quantitative comparison with the theory requires that all experimental measurements used to obtain the phenomenologically-fit parameters be made on the same sample, since quantitative differences can exist between different samples of the same material [2,3]. Unfortunately, very little data has been obtained in this manner. Experience to date suggests that the tunneling-states model is adequate to quantitatively explain nearly all available data provided $n(E,\tau_1)$ and, perhaps, $\gamma(E)$ are not arbitrarily restricted to some highly simplified, but convenient, mathematical dependence on E and τ_1.

Showing that a model is adequate does not prove that the theory is correct. Nevertheless, should the present model be superseded, it is obvious that the new theory must provide most of the ingredients found in the tunneling-states model. Eventually, of course, it would be desirable to have a microscopic description of a localized excitation in each glassy material.

In closing this Section, it should be added that alternative models have been proposed to explain the low-temperature properties of amorphous materials [38-40]. However, unlike the tunneling-states theory, these models have provided no guidance to the experimentalist.

4. THE LOW TEMPERATURE PROPERTIES OF GLASSY METALS

Glassy metals originally became of interest to the author because the mass density of the glassy state differs from that of the crystalline state by only $\approx$ 2%, whereas in fused silica, for example, the difference is $\approx$ 20%. There was reason to expect that the concentration of localized excitations might be smaller in a compacted material [41]. As Fig. 10 shows, however, the phonon mean free path in glassy PdSi alloy is very similar to that of fused silica [42], that is, the density of localized excitations has not been decreased significantly by the more "compact" structure of the metal. The same statement applies to other glassy metals, whether quenched from the melt or electrodeposited [42].

Of course, a measurement of thermal conductivity in a metal gives the total conductance contributed by both phonons and conduction electrons. The electronic part, κ_{el}, may be calculated by substituting the electrical resistivity ρ, measured at low temperatures, into the Wiedemann-Franz relation $\kappa_{el} = 2.45 \times 10^{-8}$ (T/ρ) (W/cmK). Subtraction of κ_{el} from the measured thermal conductivity gives the phonon contribution for PdSi plotted in Fig. 10. Authors attempting to utilize other techniques have suggested that this approach is not valid because of possible magnetic impurity effects.

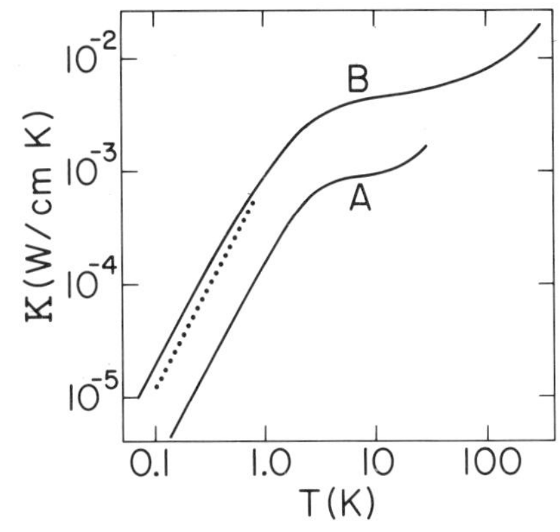

Fig. 10 Phonon thermal conductivity κ of (A) fused silica and (B) glassy PdSi alloy. Although the κ of these two materials differ by a factor $\approx$ 10, the phonon mean free paths are nearly identical for $T \lesssim 2$ K. The dotted line represents the κ of superconducting glassy ZrPd alloy (T_c = 2.5 K, Ref. 43).

However, in glassy PdSi (and most other samples) magnetic effects are too insignificant to appreciably alter the thermal transport.

The contribution from the conduction electrons is also removed if the glassy metal can be taken into the superconducting state. This is possible for glassy ZrPd, the thermal conductivity data [43] are included in Fig. 10. The specific heat of this sample was also measured, and was found to be linear in T and to correspond to a density of localized excitations n(E) similar to that found in other glasses [43].

In addition to C_{ex} and κ, ultrasonic attenuation [44], ultrasonic velocity dispersion [44], and saturation of ultrasonic attenuation [44,45] of glassy metals also have the same behavior as observed in amorphous dielectrics. Clearly, localized excitations exist in glassy metals in the same manner as in glasses and polymers. But there is one major difference. The relaxation times of the localized excitations, as deduced from ultrasonic measurements, are orders-of-magnitude shorter than in fused silica [44,45]. It is assumed that this fast relaxation is related to the presence of conduction electrons [44].

If electrons do interact with the localized excitations, then some effect of the localized excitations should be observed [46] in the low temperature electrical resistivity ρ. Unfortunately, any such effect is masked by the strong electron scattering from the atomic disorder (the electrical residual-resistivity-ratio is essentially unity) and by electron scattering from magnetic states or impurities. A very small, and questionable, influence of localized excitations on ρ has been reported [47] for glassy PdSi, where a 50% reduction in the density of localized excitations, produced by a mild heat treatment, caused a slight change in the temperature dependence of ρ below 1 K.

There is a second quantitative difference between glassy metals and amorphous dielectrics. In a dielectric, the strength of the interaction between phonons and localized excitations as deduced from thermal conductivity measurements was found to be the same as that from ultrasonic measurements. In glassy metals this appears not to be true, the interaction found from thermal conductivity being a factor of $\approx$ 10 stronger than that for the ultrasonic technique [5,42]. A satisfactory explanation of this difference has yet to appear [42].

The number of low temperature measurements on glassy metals is small. Additional experimental (as well as theoretical) work is needed. However, the difficulty of such measurements should be borne in mind. Low-temperature determinations of specific heat, thermal conductivity, ultrasonic behavior, etc., are very difficult for the most readily fabricated samples, namely, quench-cooled ribbons and splats.

5. CONCLUDING REMARKS

Several questions remain unanswered. For example, is the "plateau" in the phonon thermal conductivity curves of Figs. 2, 3, and 10 caused by the localized excitations or by another mechanism [48,49]? In the glassy metals, how strong is the interaction between the conduction electrons and the localized excitations? Indeed, what is the microscopic nature of the localized excitations? Why is the density n(E) so broad in energy, and why is n(E) so nearly the same for essentially all amorphous (and at least some highly disordered crystalline) materials?

A partial answer to the last question has been suggested [50]. Consider a homogeneous material containing several isolated tunneling defects. The energy splitting of all defects is then the same, and is represented as E_0 in Fig. 11. Next, increase the density of defects until one interacts with the strain field produced by another. The narrow spectrum at E_0 of Fig. 11 then broadens. If the defects are sufficiently concentrated, two interesting consequences appear. First, n(E) becomes independent of E at small E and, second, n(E) at small E becomes independent of the concentration of defects. Hence the broad spectrum in energy of n(E), and the uniformity of n(E) between different materials, could result if the basic "defects" in the glass are strongly interacting. The maximum specific heat of fused silica has been calculated from this theory using interaction parameters consistent

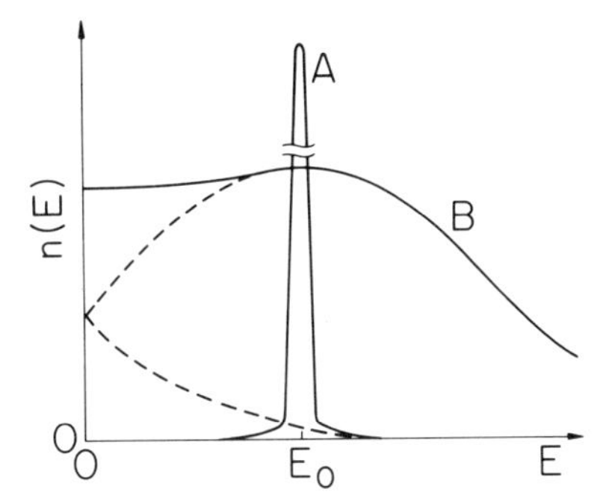

Fig. 11 Schematic distribution n(E) of energy levels of "defects" for (A) a homogeneous material in which the defects do not interact and (B) a material in which the defects interact quite strongly via their elastic strain fields. Only a single energy level at energy E_o has been assumed for the isolated defect. The dashed lines have been added to produce curve B for $E < E_o$.

with the measurements of τ_2 (Sec. 3). The calculated upper bound on C_{ex} is a factor of 7 larger than the measured C_{ex}. The smaller measured C_{ex}, the slight positive temperature dependence of C_{ex}, and the small but finite variation of C_{ex} between different glassy materials suggest that the concentration of "defects" is generally not sufficient to produce the limiting case where n(E) becomes independent of E.

The very speculative nature of the previous paragraph, and the fact that even this speculation has avoided any suggestion of a microscopic or atomic description of a localized excitation, serves to indicate how little has been learned thus far from the extensive theoretical and experimental effort directed to the low-temperature behavior of amorphous materials.

REFERENCES

1. R. C. Zeller and R. O. Pohl, Phys. Rev. B4, 2029 (1971).
2. T. L. Smith, P. J. Anthony, and A. C. Anderson, Phys. Rev. B17, 4997 (1978).
3. D. S. Matsumoto, C. L. Reynolds, and A. C. Anderson, Phys. Rev. B19, 4277 (1979).
4. M. P. Zaitlin and A. C. Anderson, Phys. Rev. B12, 4475 (1975).
5. D. S. Matsumoto, C. L. Reynolds, and A. C. Anderson, Phys. Rev. B16, 3303 (1977).
6. G. J. Morgan and D. Smith, J. Phys. C7, 649 (1974).
7. D. Walton, Phys. Rev. B16, 3723 (1977).
8. G. K. White, Phys. Rev. Lett. 34, 204 (1975).
9. M. A. Bosch, Phys. Rev. Lett. 40, 879 (1978).

10. V. Khiznyakov (private communication).
11. M. Rubenstein and H. A. Resing, Phys. Rev. B13, 959 (1976).
12. S. R. Kurtz and H. J. Stapleton, Phys. Rev. Lett. 42. 1773 (1979).
13. S. Hunklinger and W. Arnold, in Physical Acoustics, ed. by W. P. Mason and R. N. Thurston (Academic, New York, 1976) Vol. 12, p. 155, and papers cited.
14. M. von Schickfus, S. Hunklinger, and L. Piche, Phys. Rev. Lett. 35, 876 (1975).
15. M. von Schickfus and S. Hunklinger, Phys. Lett. 64A, 144 (1977).
16. J. E. Graebner and B. Golding, Phys. Rev. B19, 964 (1979).
17. L. Bernard, L. Piche, G. Schumacher, J. Joffrin, and J. Graebner, J. Phys. (Paris) 39, L126 (1978).
18. M. V. Schickfus, B. Golding, W. Arnold, and S. Hunklinger, J. Phys. (Paris) 39, C6-959 (1978).
19. G. E. Jellison, G. L. Peterson, and P. C. Taylor, Phys. Rev. Lett. 42, 1413 (1979).
20. P. J. Anthony and A. C. Anderson, Phys. Rev. B14, 5198 (1976); 16, 3827 (1977).
21. P. J. Anthony and A. C. Anderson, Phys. Rev. B19, 5310 (1979).
22. U. Stron, M. von Schickfus, and S. Hunklinger, Phys. Rev. Lett. 41, 910 (1978).
23. P. W. Anderson, B. I. Halperin, and C. M. Varma, Phil. Mag. 25, 1 (1972).
24. W. A. Phillips, J. Low Temp. Phys. 7, 351 (1972).
25. V. Narayanamurti and R. O. Pohl, Rev. Mod. Phys. 42, 201 (1970).
26. V. I. Goldanskii, Nature 279, 109 (1979).
27. W. M. Goubau and R. A. Tait, Phys. Rev. Lett. 34, 1220 (1975); J. Heinrichs and N. Kumar, Phys. Rev. Lett. 36, 1406 (1976).
28. R. B. Kummer, R. C. Dynes, and V. Narayanamurti, Phys. Rev. Lett. 40, 1187 (1978).
29. A. C. Anderson, in Fast Ion Transport in Solids, ed. by P. Vashishta, J. Mundy, C. Milendres, and A. Rahman (North Holland, New York) to be published.
30. J. C. Lasjaunias, R. Maynard, and M. Vandorpe, J. Phys. (Paris) 39, C6-973 (1978).
31. T. L. Smith and A. C. Anderson, Phys. Rev. B19, 4315 (1979).
32. C. N. Hooker and L. J. Challis, in Phonon Scattering in Solids (La Documentation Francaise, Paris, 1972), p. 364.
33. R. B. Stephens, Phys. Rev. B14, 754 (1976).
34. T. L. Smith, J. R. Matey, and A. C. Anderson, Phys. Chem. Glasses 17, 214 (1976).
35. L. E. Wenger, K. Amaya, and C. A. Kukkonen, Phys. Rev. B14, 1327 (1976).
36. J. L. Black and B. I. Halperin, Phys. Rev. B16, 2879 (1977).

37. W. Arnold, C. Martinon, and S. Hunklinger, J. Phys. (Paris) 39, C6-961 (1978).
38. H. P. Baltes, Solid State Commun. 13, 225 (1973).
39. L. S. Kothari and Usha, J. Non-Cryst. Solids 15, 347 (1974).
40. W. H. Tanttila, Phys. Rev. Lett. 39, 554 (1977); and references cited therin.
41. J. R. Matey and A. C. Anderson, J. Non-Cryst. Solids 23, 129 (1977).
42. J. R. Matey and A. C. Anderson, Phys. Rev. B16, 3406 (1977); 17, 5029 (1978).
43. J. E. Graebner, B. Golding, R. J. Schutz, F.S.L. Hsu, and H. S. Chen, Phys. Rev. Lett. 39, 1480 (1977).
44. B. Golding, J. E. Graebner, A. B. Kane, and J. L. Black, Phys. Rev. Lett. 41, 1487 (1978).
45. P. Doussineau, P. Legros, A. Levelut, and A. Robin, J. Phys. (Paris) 39, L-265 (1978).
46. J. L. Black and B. L. Gyorffy, Phys. Rev. Lett. 41, 1595 (1978).
47. S. B. Dierker, H. Gudmundsson, and A. C. Anderson, Solid State Commun. 29, 767 (1979).
48. M. P. Zaitlin and A. C. Anderson, Phys. Status Solidi B71, 323 (1975).
49. D. P. Jones, N. Thomas, and W. A. Phillips, Philos. Mag. B38, 271 (1978).
50. M. W. Klein, B. Fischer, A. C. Anderson, and P. J. Anthony, Phys. Rev. B18, 5887 (1978).

LOW TEMPERATURE PROPERTIES OF METALLIC GLASSES

Gunther von Minnigerode

I. Physikalisches Institut der Universität Göttingen
Bunsenstraße 9, D-3400 Göttingen

INTRODUCTION

Metallic glasses are prepared by ultra-rapid quenching from the melt or the vapor of metal alloys. The non-crystalline state in the produced thin filaments or films may be described by the model of a dense random packing of spheres of different sizes|1|.
By reason of the lack of long range order such a state is often called briefly "amorphous" or "glassy". The electronic structure of crystalline metals is mainly determined by the short range order and only details of the Fermi surfaces etc. are sensitive to the long range order. Therefore, metallic glasses are metals with several nearly free electrons per atom. The low temperature properties of metallic glasses are governed by the characteristic features of the electronic system in an extremely disordered metal|2|.

ELECTRICAL CONDUCTIVITY

The lack of long range order in amorphous metals means a very short and nearly temperature independent mean free path of the electrons. Therefore, the d.c.resistivity ρ of amorphous metals, even at low temperatures, is comparable with the high resistivity of the molten alloy, typically 100-300 $\mu\Omega$ cm. Moreover the wave-vector of the conduction electrons $\vec{k}$ is no longer a good quantum number. The very weak temperature coefficient of the electrical resistivity (TCR) $\equiv (1/\rho)\ (d\rho/dT)$ in the amorphous state may be of a negative or positive sign depending on the type and content of the metallic glass and the measuring temperature|3,4|. In many cases the TCR of the glass in a wide range of higher temperatures corresponds with that of the liquid alloys|5|. Such a behavior

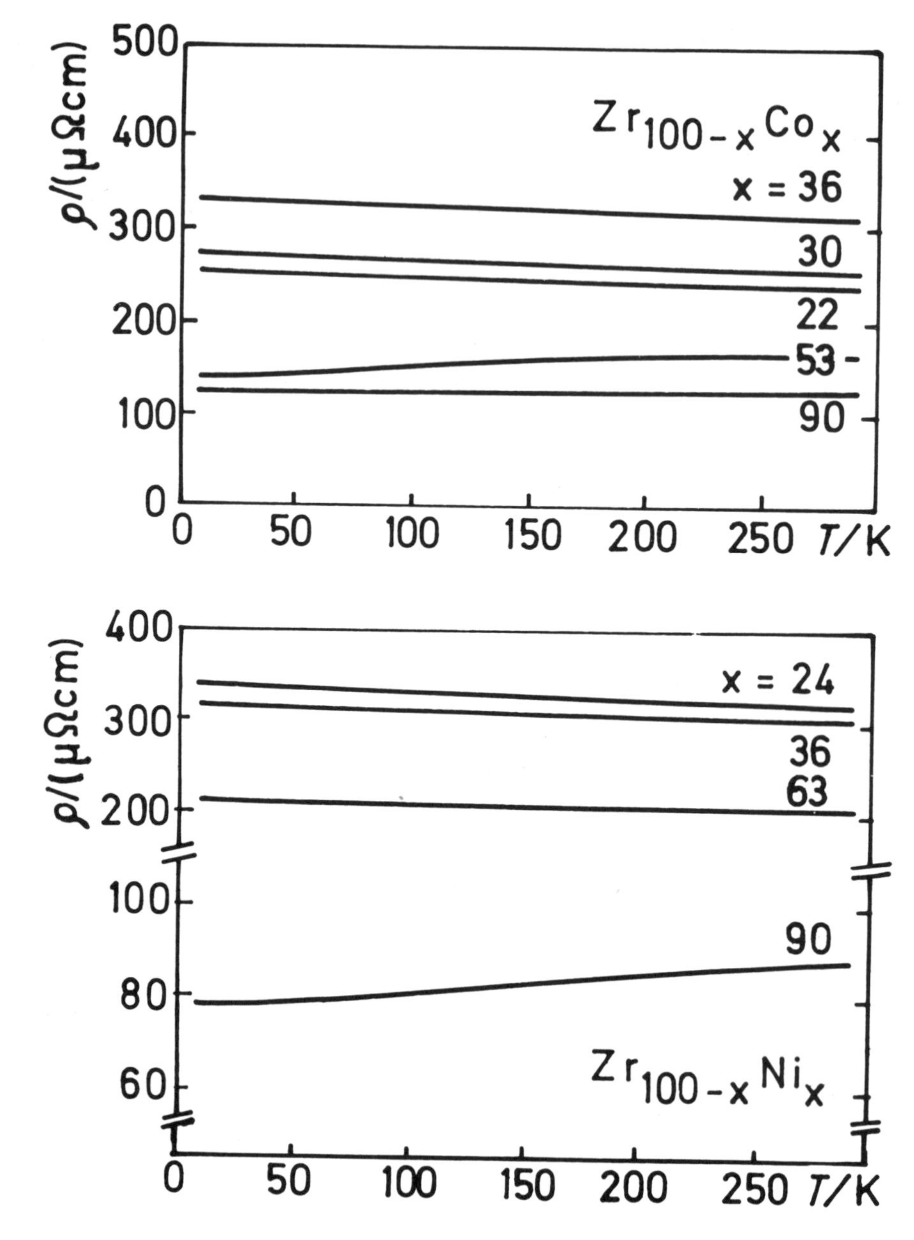

Fig.1. Electrical resistivity of metallic glasses based on Zr. (Buschow et al, Phys. Rev. B19,3843,1979).

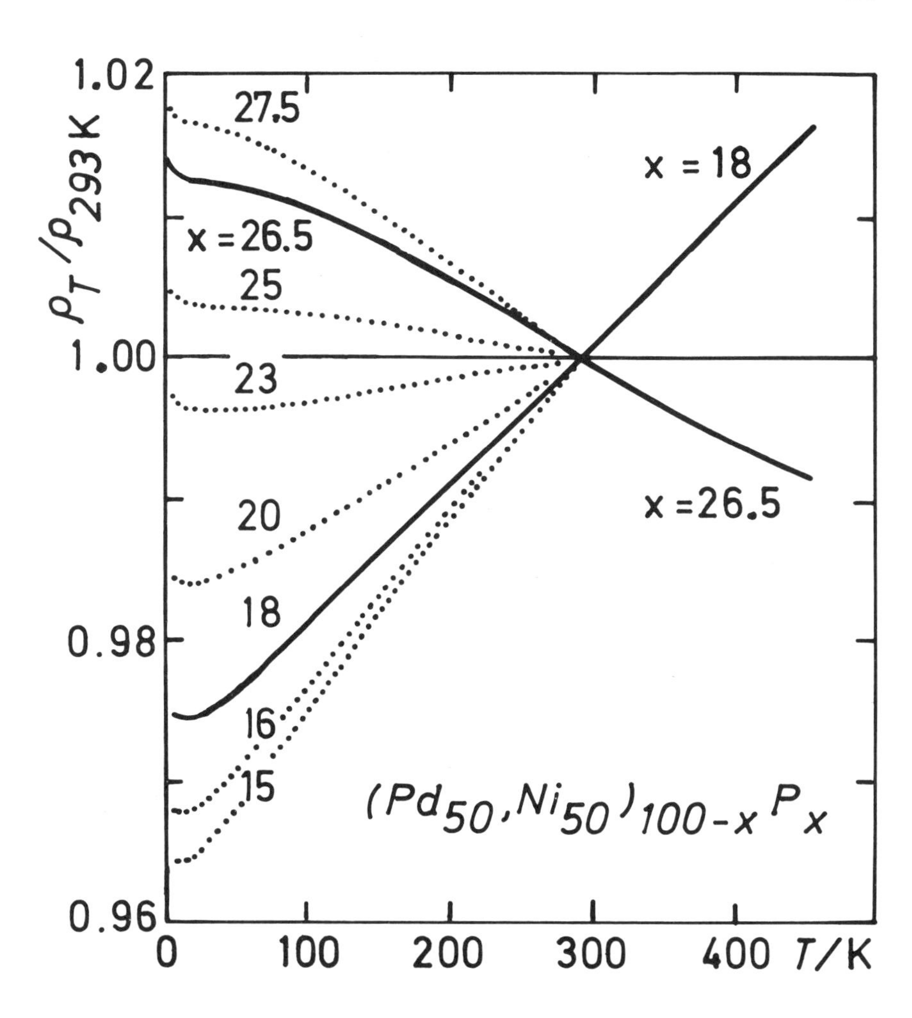

Fig.2. Normalized electrical resistivity versus temperature for $(Pd_{50},Ni_{50})_{100-x}P_x$ glasses. (reference |4|).

is predicted by the extended Ziman theory |6|. In this theory of pseudopotential scattering of nearly free electrons, the temperature dependence of ρ is primarly determined by the temperature dependence of the structure factor S $(q \approx 2k_F)$. In the case, where $2k_F$ nearly coincides with scattering vector Q_p of the main peak of the structure factor, TCR is negative. This is a direct consequence of the broadening of the interference function with increasing temperature, which lowers the values near the center of the main peak (TCR<O) but raises all values further away (TCR>O). In this approach ρ is nearly linear in T at temperatures above 1/2 the Debye temperature and saturates at low temperatures.

However, in many amorphous metals at low temperatures, the resistivity increases rapidly as the temperature is lowered and saturates only below the liquid helium regime. Together with a positive TCR at higher temperatures, resistivity minima take rise|4|. This behavior is very similar to the Kondo effect in crystalline dilute magnetic alloys. Nevertheless there has been considerable dispute in the recent literature regarding the functional form of the resistivity data and what mechanism is responsible for the low-temperature anomaly. The dispute is nourished by the fact that the spin-flip Kondo scattering requires spins in zero effective magnetic field. It is not surprising that this condition is fulfilled for those amorphous metals that are nonmagnetic, but have a dilute concentration of magnetic impurities, which may originate also accidentally by impure materials. But a puzzling aspect stems from the fact that anomalous low-temperature resistivity minima have been observed in very concentrated ferromagnetic glasses. Moreover, in several of the amorphous metals the low-temperature resistivity anomaly is not influenced by strong magnetic fields (~50kOe), except for a shift due to a magnetoresistance. For these reasons, it has been postulated that the resistance anomaly does not have a magnetic origin but is an intrinsic property of the amorphous structure. Cochrane et al |7| have made the suggestion that the anomaly arises from the electron scattering from two-level tunneling systems. A Kondo-like treatment of this model yields $\rho(T) = \rho(o) - C \cdot \ln(T^2+\Delta^2)$, where C is a constant and Δ is the temperature characteristic of the splitting between the two levels. But recent calculations |8| have pointed out that there are deviations from the Kondo Hamiltonian, that the calculated values are three orders of magnitude too small to fit the experimental data, and that the sign of the anomaly should be reversed.

If the structural or tunneling model does not explain the anomaly, the question for the mechanism still remains. Lately Grest and Nagel |9| made the intriguing suggestion of a modified Kondo mechanism based on a wide distribution of effective fields P(H) in some ferromagnetic glasses, which was first suggested by Sharon and Tsuei |10| in order to fit their Mössbauer data. They assumed

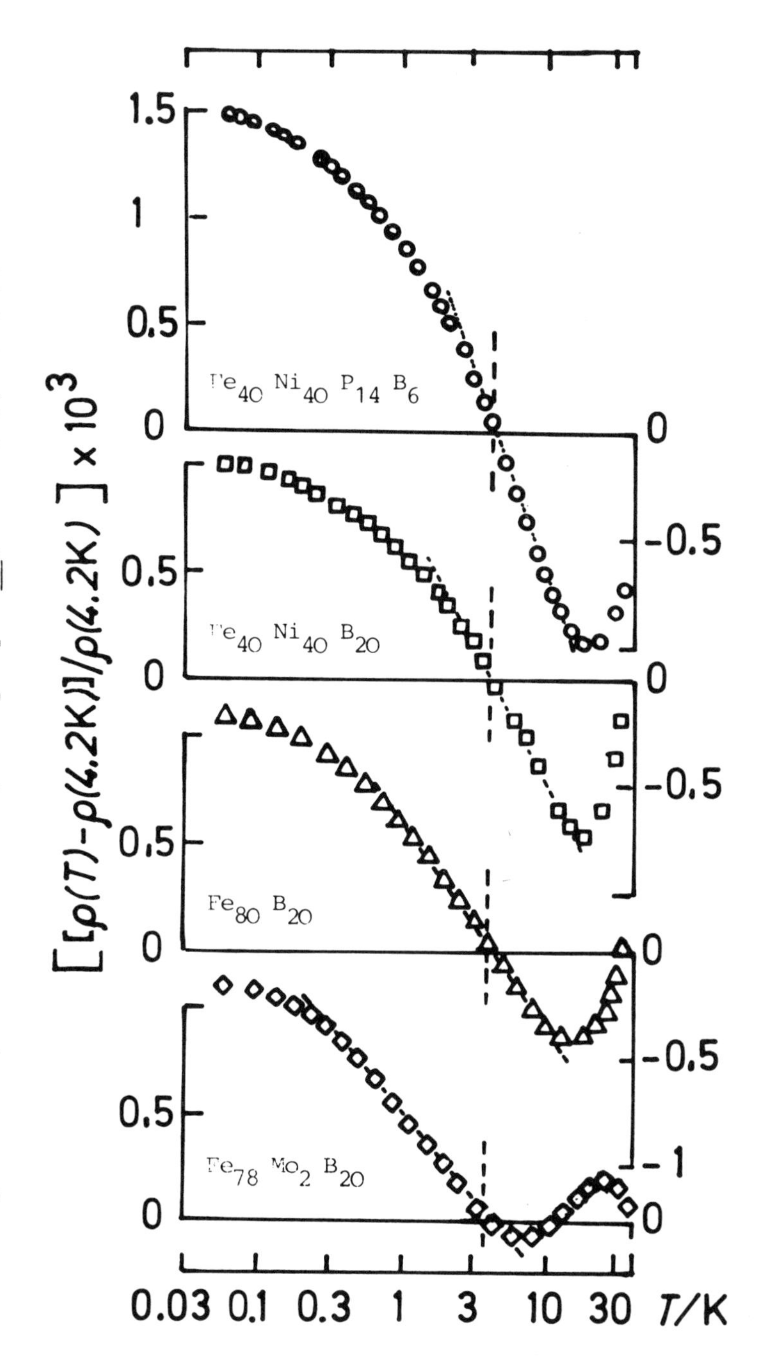

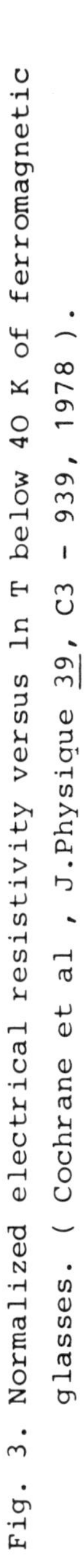

Fig. 3. Normalized electrical resistivity versus ln T below 40 K of ferromagnetic glasses. (Cochrane et al , J.Physique 39, C3 - 939, 1978).

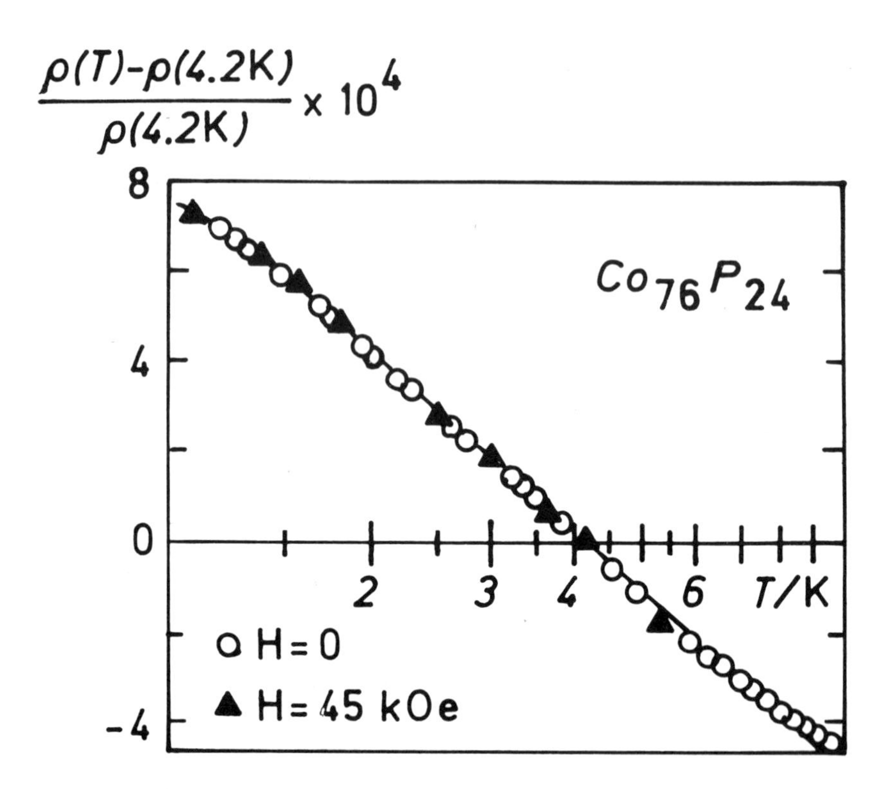

Fig. 4. Normalized electrical resistivity change versus ln T below 10 K. Magnetic fields up to 50 kOe cause a positive magneto-resistivity at high fields but do not change the temperature dependence of the resistivity of amorphous ferromagnetic $Co_{76}P_{24}$ at low temperatures. (Cochrane et al , J. Phys. F, 7 , 1799, 1977).

a P(H) which was peaked at some high positive field (200-300kOe) but had a sizable tail extending to H=0. The spins in the sites of very small fields participate in spin-flip scattering. The calculations of Grest and Nagel indicate that direct-exchange and RKKY interactions do not produce such a tail in P(H). However, if one takes into account the superexchange interactions between those next-nearest-neighbor magnetic atoms which are separated by a metalloid atom, there will be a long flat tail in P(H) that extends to large negative fields (-100kOe). This distribution can explain why the Kondo scattering persists in ferromagnetic glasses and in applied magnetic fields. The change of the distribution by additional alloying offers a wide field of influence on the resistivity anomaly.

At low temperatures most alloys including many strongly magnetic ones show a positive magnetoresistivity at high fields |4|. As already mentioned, this contribution is usually only weakly temperature dependent. But there are significant magnetic field effects on the resistivity minimum in some rare earth alloys related to the onset of magnetic order |11-12|.

Finally it should be mentioned that even the onset of crystallization causes a significant change in the TCR |13|.

ELECTRON-PHONON INTERACTION AND SUPERCONDUCTIVITY

In amorphous metals the wavevector conservation law in the electron-phonon interaction is weakened. Therefore, this interaction is enhanced |14|. The softening of the low-frequency phonon modes may be even more important for the enhancement of superconductivity |15|. However, in transition metal superconductors both of these effects may be overcompensated by a decrease of the electronic density of states at the Fermi level by the smearing of sharp peaks in the electronic density of states function of d-electrons |16|. In La-based glasses a reduction of the hybridization of 5d with 4f states causes a weakening of the electron-phonon interaction |13,17|.

Because of the very short mean free path of the electrons, amorphous superconductors are properly described by the extreme "dirty limit" of the microscopic theory. Amorphous superconductors are thus extreme type II materials with large values of H_{c2} and a very small critical current density in the presence of an applied magnetic field as a consequence of missing fluxpinning centers |18|.

THERMAL CONDUCTIVITY

The measurements provide the total thermal conductivity of phonons and electrons: $k = k^{ph} + k^{e}$. The electronic contribution to the thermal conductivity in the normal state k_n^e obeys the Wiedemann-Franz

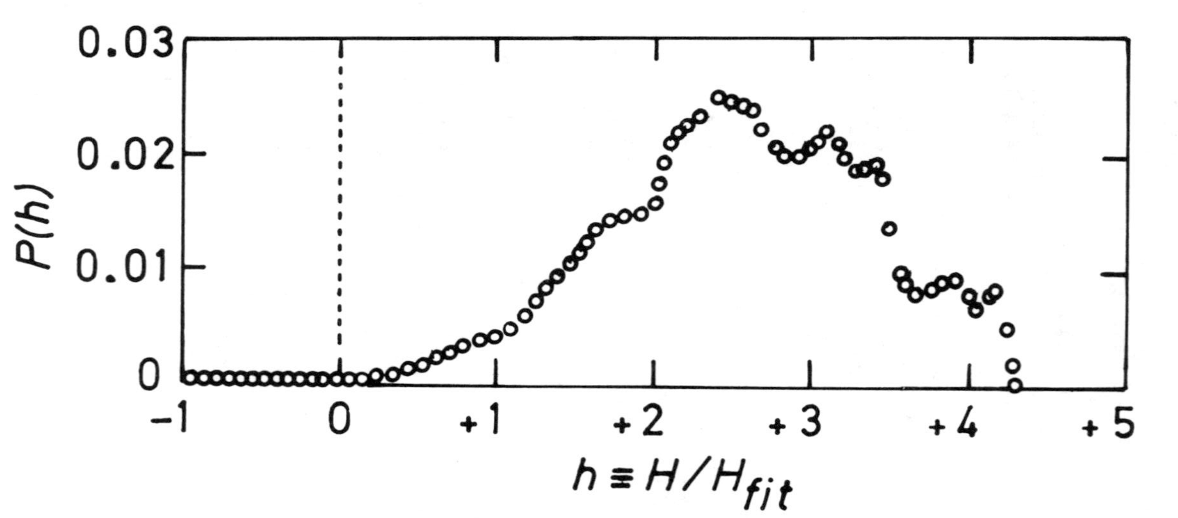

Fig. 5. Calculated effective-field distribution at a magnetic atom which interacts with its neighbors via RKKY, direct exchange, and superexchange. The radial distribution function for the $Fe_{80}P_{20}$ glass, $2k_F$ = 3 Å, and J_{nn} = 1.2 are used. No metalloid-metalloid nearest neighbors are allowed. A data point is plotted at a field h only if there are at least 20 ppm in a bin of width Δh = 0.05. A random value between J_{SE}= -0.45 and -0.65 is used for 60% of time and J_{SE}= 0 otherwise. Mössbauer data are consistent with $H_{fit} \simeq$ 100kOe. Note that the distribution has a long flat tail that extends to large negative values of H. (reference |9|).

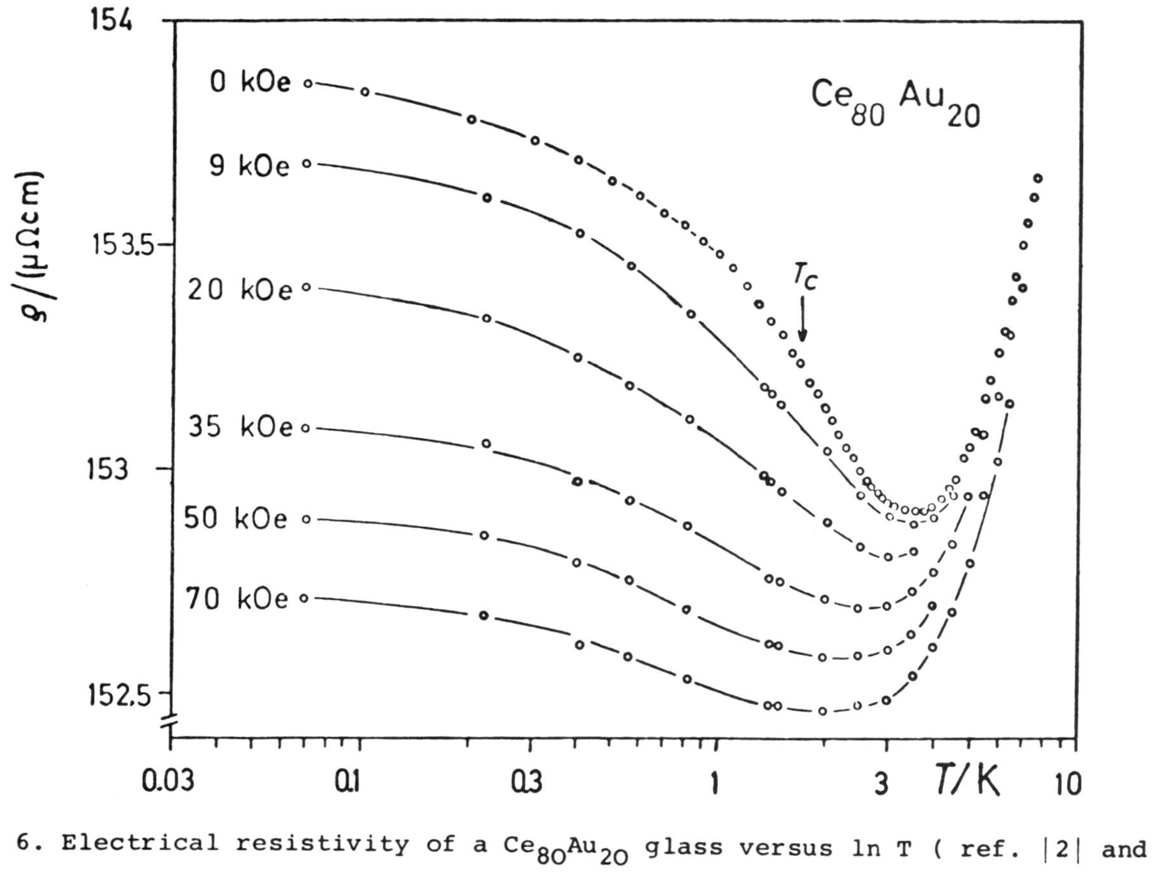

Fig. 6. Electrical resistivity of a $Ce_{80}Au_{20}$ glass versus ln T (ref. |2| and |33|).

law $k_n^e = L_o T/\rho$, where L_o = Lorenz number = $2.45 \cdot 10^{-8} W \cdot Ohm/K^2$, since the electron-defect scattering is predominant |19|. In the superconducting state the electronic contribution to the thermal conductivity k_s^e decreases rapidly as the temperature is reduced below the transition temperature T_c, and may be calculated following reference |19|. We are interested in the thermal conductivity contributed by the phonons. Therefore, the electronic contribution is substracted off: $k^{ph} = k - k^e$. The phonon thermal conductivities of glassy metals have been measured in this way by different authors |20,21,22|. The magnitudes and the temperature dependences are close to those found for insulating glasses: A nearly T^2 dependence below ≈ 1K, a weakly temperature dependent part at some degrees (often marked as the plateau), and a precipitous increase above the plateau. The phonon scattering which causes the relatively small thermal conductivity is supposed to be dominated by localized centers which are intrinsic to amorphous materials. Such centers are provided by a model which assumes that atoms or groups of atoms reorient via quantum-mechanical tunneling. These two-level tunneling systems failed to be observed in the electrical conductivity, but acoustical phenomena predicted by the theory of tunneling states have been observed too |20,23|.

Even though the tunneling theory of the localized low-energy excitations has achieved notable success, the microscopic nature of the entity which undergoes tunneling remains uncertain. In this context it is surprising that the density of the two-level systems and their coupling to acoustic phonons in a densely packed disordered metal are so similar to those properties in the much more loosely packed network structures of dielectric glasses. Moreover the low-energy excitations have also been observed in an amorphous Pb,Cu-film, a simple metal without metalloids and covalent bonds |21|.

Finally, an extra contribution linear in T in the heat capacity of the superconducting $Zr_{70}Pd_{30}$ -glass has been observed providing the most direct evidence for such low-energy excitations |22|. The plateau in the phonon thermal conductivity curve vs. temperature comprises two contributions, that from low-frequency phonons plus that from high-frequency phonons. Moreover the fast rise of $k^{ph} = \frac{1}{3} C_D \cdot \ell^{ph} \cdot \bar{v}$ above the plateau (faster than T^3 and most clearly shown in reference |21|) could be due to the specific heat C_D of the propagating phonons rising faster than T^3. These aspects will be discussed in the next paragraph.

SPECIFIC HEAT

As already mentioned, measurements of the specific heat C_p between 0.1 and 10K on the superconducting metallic glass $Zr_{70}Pd_{30}$ (T_c =2.53K) exhibit an approximately linear term in C_p below 0.2K |22|

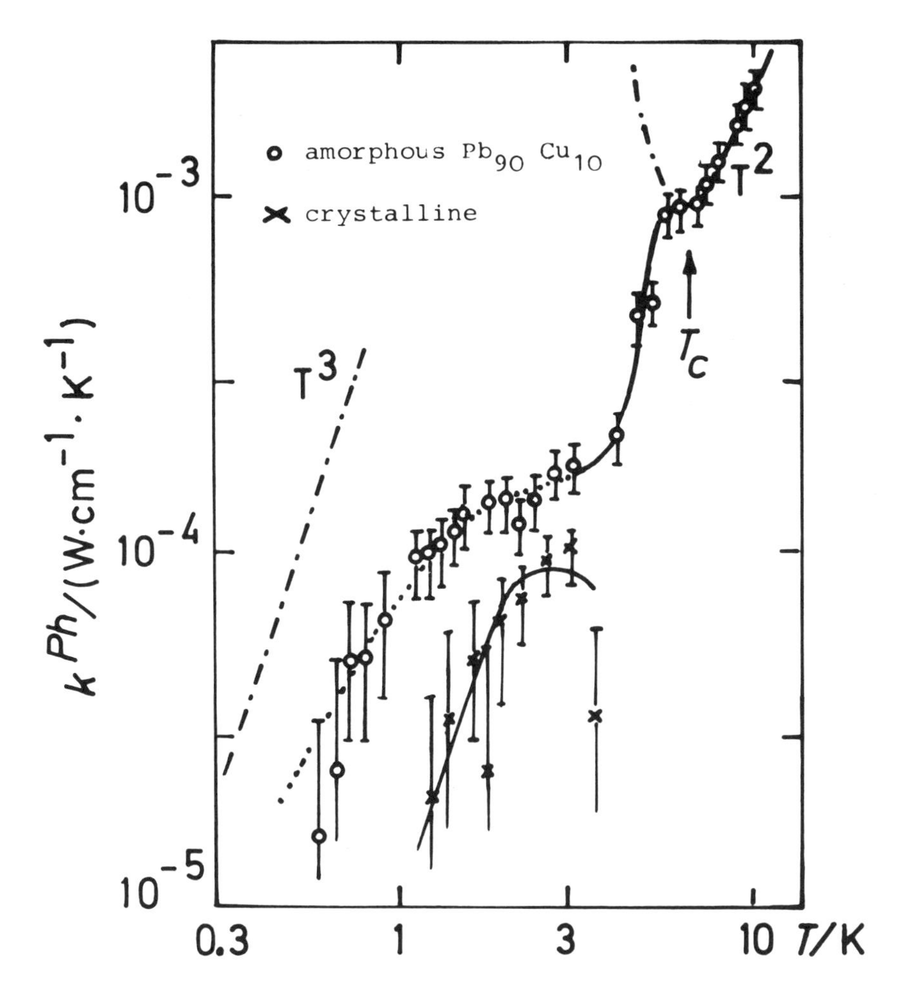

Fig. 7. Phonon thermal conductivity in superconducting amorphous and crystalline $Pb_{90}Cu_{10}$ films versus temperature (bilogarithmically). The solid lines are intended as a visual guide. The dashed-dotted lines give a lower bound of the thermal conductivity in a strong-coupling superconductor limited by phonon-electron scattering and size effect respectively (reference |21|).

The specific heat varies over more than four orders of magnitude between 0.1 and 10K. The data were fitted above T_C=2.53K by the expression

$$C_p = A\cdot T + B\cdot T^3 + C\cdot T^5$$

The coefficient A incorporates the combined contributions from the electronic specific heat, electron-phonon interactions, and contributions from low-energy configurational excitations (as already discussed in the last paragraph). The second and third term describe the phonon specific heat. Below T_C the data were fitted by

$$C_p = DT + BT^3 + \text{contributions of electronic quasiparticles.}$$

The first term may be the simplest representation of the extra specific heat at the lowest temperatures resulting from the configurational excitations (or normalconducting inclusions?). The same phonon T^3 term was taken and the T^5 term is neglected at the low temperatures. For the electronic quasiparticle contributions see reference |22|.
The following coefficients were determined:
$A=4.664\ mJ/(mol\cdot K^2)$; $B=0.334\ mJ/(mol\cdot K^4)$; $C=8.78\cdot 10^{-5}\ mJ/(mol\cdot K^6)$; $D=0.102\ mJ/(mol\cdot K^2) = 2.187\%$ of A
The linear specific heat below T_C is similar to that of typical insulating glasses, for which the coefficient $D \approx (0.024-0.360)\ mJ/(mol\cdot K^2)$. Moreover the experimental upper limit of any crystalline normalconducting phase is 0.1%. Therefore, the anomalous specific heat of this superconducting metallic glass at the lowest temperatures is due to the intrinsic defects characteristic of the amorphous state. If the measured value of D is interpreted as arising from two-level tunneling states, a density of states of $n_o = 6D/(\pi^2\cdot k_B^2) = 0.51\cdot 10^{23}\ mol^{-1}(eV)^{-1}$ is obtained. Finally the relatively large and positive C coefficient of the T^5 term should be mentioned which makes the main difference to the specific heat of the crystalline material in the normal state. This term reaches 1% of the value of the T^3 at about 6K.

Different authors measured the specific heat of Pd-Si and Pd-Si-Cu alloys, which are easy glass formers. Golding et al |24| fitted their data of a $Pd_{77.5}Si_{16.5}Cu_6$ glass in the temperature range 1.8-3.6K with the following coefficients (in the same nomenclature as before: $A=1.21\ mJ/(mol\cdot K^2)$; $B=0.122\ mJ/(mol\cdot K^4)$; $C=3.7\cdot 10^{-3}\ mJ/(mol\cdot K^6)$). With crystallization A increases by 5%, B decreases by 34%, and C decreases by 70%. The mass density of the glass of $10.52\ g\cdot cm^{-3}$ (at room temperature) increases only by 1.6% with crystallization. Longitudinal and transverse sound velocities were measured by MHz-pulse techniques. A large softening of long-wavelength transverse acoustic phonons in the glass, relative to the crystall, is observed. The large change in transverse sound velocity v_T ($1.797\cdot 10^3\ m\ s^{-1}$ in the glass, $2.068\cdot 10^3\ m\ s^{-1}$ in the crystall, $\Delta v_T = 15\%$) corresponds to a weakening of the shear elastic modulus (35%) and a decrease of the Debye temperature

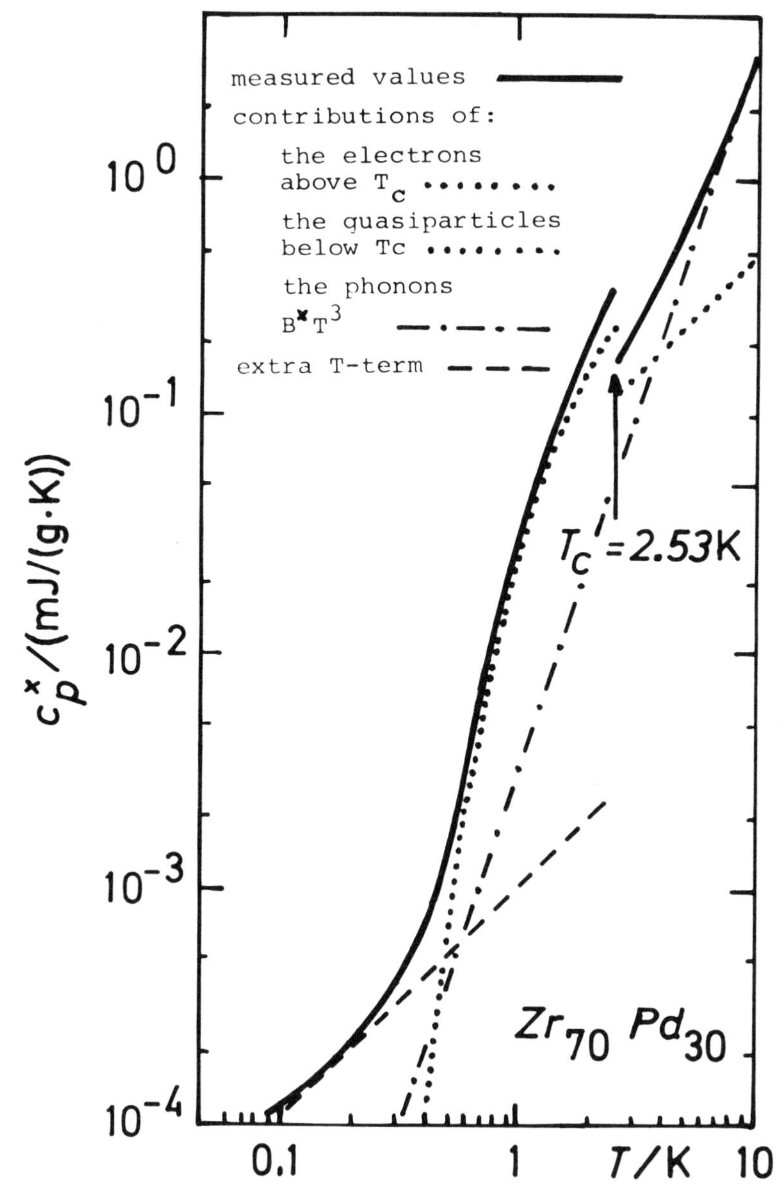

Fig. 8. Specific heat of the superconducting $Zr_{70}Pd_{30}$ glass (bilogarithmically),
$c_p = c_p^{x} \times 95.86$ g/mol , (reference |22|).

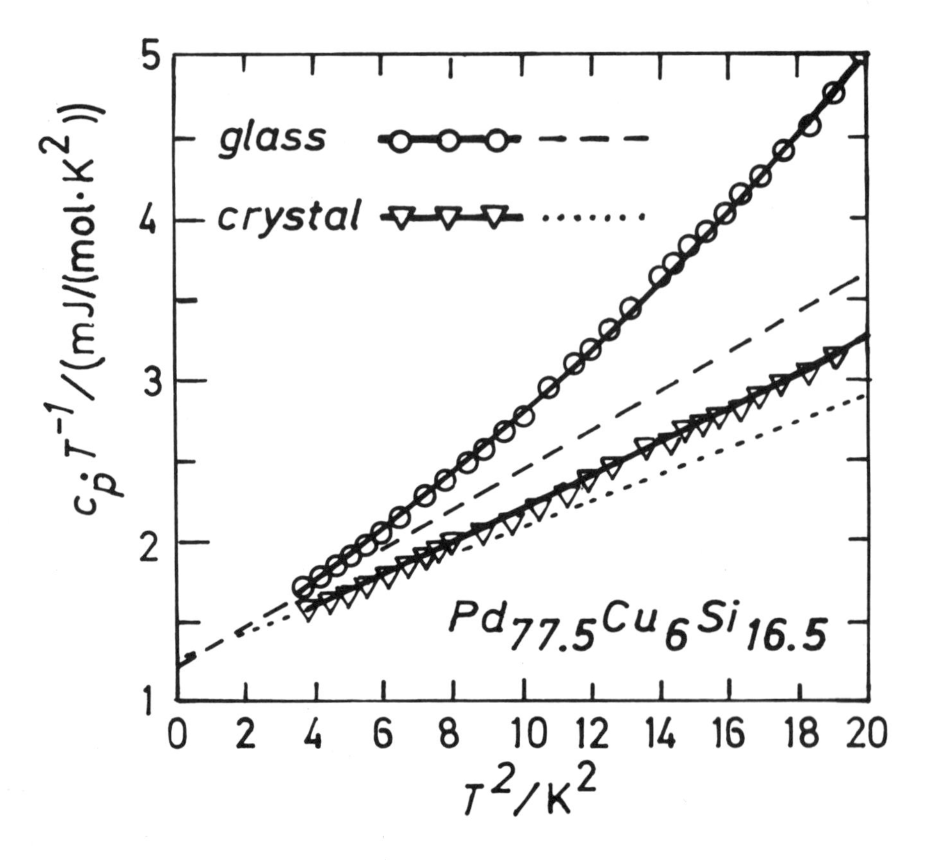

Fig. 9. Specific heats c_p of $Pd_{77.5}Cu_6Si_{16.5}$. Plot of c_p/T versus T^2. The straight lines give the contribution of $A + B \cdot T^2$ for the glass (dashed line) and the crystall (dotted line) in reference |24|.

(250K in the glass, 288K in the crystall). The reduced resistance to shear is supposed to arise from the possibility of nonuniform changes in the local atomic density in the slightly looser structure of the glass |25|. From the measured coefficient A the dressed density of electronic states at the Fermi surface may be calculated $N^{*}(E_F) = (1+\lambda)\cdot N(E_F) = 3A/(2\pi^2\cdot k_B{}^2) = 1.547\cdot 10^{23}$ states/(spin·mol eV). By assuming 1 free electron per atom of Pd or Cu and 4 free electrons per atom of Si one obtains z = 1.495 electrons per atom and a bare density of states at the Fermi surface $N_o(E_F) = V_{mol}\cdot 2\pi m_o\cdot(3z\cdot N_L/\pi\cdot V_{mol})^{1/3}/\hbar^2 = z^{1/3}\cdot 0.729\cdot 10^{23}$ states/(spin·mol·eV) = $0.834\cdot 10^{23}$ states/(spin·mol·eV). In the Debye T^3 approximation for $T \ll \Theta$ the coefficient B is given by $B = 234\ N_L\ k_B/\Theta^3 = 1{,}944\cdot 10^6\ mJK^{-1}/\Theta^3$. The Debye temperature Θ calculated from the measured value of B is Θ = 252K. This value is in good agreement with Θ = 249K calculated by $\Theta = (6\pi^2 N_L/V_{mol})^{1/3}\cdot$ $\cdot(\hbar\langle v\rangle/k_B)$ from the measured sound velocities v_T and v_L with $\langle v\rangle = 1.44\ v_T\left[2+(v_T/v_L)^3\right]^{-1/3}$ (see reference 24).

More recently the same glass $Pd_{77.5}Si_{16.5}Cu_6$ has been studied by Lajaunias et al |25|. They measured the specific heat between 30 mK and 2K. Below 80mK, the specific heat is dominated by a hyperfine T^{-2} term, almost independent of an amorphous or crystalline structure (the coefficient of this term is $3.3\cdot 10^{-4}$ mJ·K·mol^{-1}). The coefficient of the T term between 0.2 and 1K (A=1.29 mJ mol^{-1}K^{-2}) is about 6,6% larger than the value in reference |24| (whereas the mass density is enhanced only by 1.1%), and decreases a little (within the error) with crystallization.

Measurements of specific heats where taken by Massalski et al |26| in a number of binary Pd-Si and ternary Pd-Si-Cu amorphous alloys. By the fitting of the data as before, at temperatures between 1.5K - 4.2K, the coefficient A of the T term was determined as a function of z with increasing Si content. The values of A (1.15 - 1.3 mJ mol^{-1}K^{-2}) are not proportional to $z^{1/3}$, as expected for free electrons, but show a maximum at z = 1.57. These authors calculated even a smaller bare density of electronic states using a smaller V_{mol} = 6.983cm^{3}mol^{-1} from x-ray data (instead of V_{mol}= 8.664cm^{3}mol^{-1} from mass density used here). The main contribution for the ratio $N^{*}(E_F)/N_o(E_F)$, which is 1.85 in our calculation, may be the phonon enhancement factor $(1+\lambda)$. But contributions of the d-band or effective thermal masses of the electrons have also to be considered. In any case there is no indication of a minimum in the density of states in the range of the amorphous state in this alloy system, as proposed by Nagel and Tauc |27|.

A series of amorphous alloys of $La_{100-x}\ Ga_x$ with x = 16 to 28 have been studied by Shull et al |28|. The heat capacities, superconducting transition temperatures, electrical resistivities and upper critical fields have been measured. Debye temperatures, densities of states, and electron-phonon coupling constants have

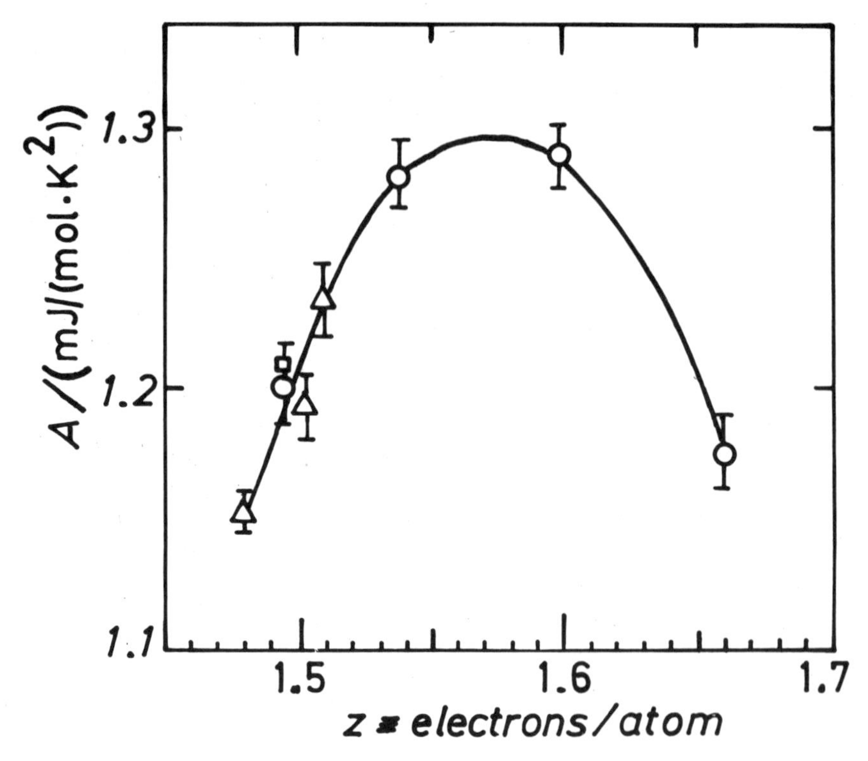

○ $Pd_{100-x}\ Si_x$ with ($16.5 \leq x \leq 22$).

△ $Pd_{100-y-x}\ Cu_y\ Si_x$, ($3 \leq y \leq 9$),($16 \leq x \leq 17$).

□ $Pd_{77.5}\ Cu_6\ Si_{16.5}$ Golding et al. reference |24|.

Fig. 10. Coefficient of the T-term in the specific heats of Pd-Si and Pd-Cu-Si glasses as a function of z ≡ electrons/atom. (reference |26|).

been evaluated from these measurements. The data C_p/T are plotted as a function of T^2 and fitted above T_c by a straight line with the condition that at T_c the entropy of the superconducting state and the normal state is the same. The coefficients A and B are determined by the intercept and the slope of this line. At T_c there is a jump ΔC_p in the specific heat. The ratio $\Delta C_p/AT_c$ varies between 1.37 and 1.97 (compared with 1.43 for a weak-coupling BCS-superconductor). With increasing x the transition temperatures decrease |2|, the Debye temperatures (calculated from B) increase from 110K to 118K, and the densities of electronic states at the Fermi surface (calculated from A) cover a maximum of $N^*(E_F) = 8.6\cdot10^{23}$ states/(spin·mol·eV) at x = 24. These values of the density of states are consistent with those determined by the slope of the $H_{C2}(T)$-curve at T_c. The electron-phonon coupling constant λ is calculated using the McMillan expression of T_c using $\mu^*=0.1$ and the measured values of T_c and Θ. The estimated values of λ decrease from 0.85 to 0.75 with increasing x. The bare densities $N_o(E_F)$ were calculated with the mass density and under the assumption that La and Ga each contribute 3 free electrons per atom. The ratio $N^*(E_F)/(1+\lambda)N_o(E_F) \equiv m^*/m_o \approx 2.2$ is substantially larger than 1.

We avoid the discussion of magnetic contributions to the low temperature specific heat, whereas it is an interesting field. But some measurements of the specific heat of amorphous films of simple metals quenched from the vapour shall be mentioned |29,30|. These measurements are tricky because of the small mass of the film compared to that of the substrate. A large softening of the lattice has been observed by an enhancement of the lattice specific heat by a factor of 3.5 for amorphous $In_{80}Sb_{20}$ |29|, and a factor of 6 for amorphous $Pb_{70}Bi_{30}$ |30|. Down to 0.5K no indication for a linear T-term in the superconducting region due to low energy excitations of the lattice has been found. But the anomaly may be hidden under a rather large T^3-term even at 0.5K due to the remarkably low Debye temperature of Θ =50K in amorphous $Pb_{70}Bi_{30}$ and Θ = 61K in amorphous $In_{80}Sb_{20}$.

MAGNETIC PROPERTIES

Transition metal-metalloid glasses based on Fe or Co are at low temperatures very soft ferromagnets. Their properties and possible applications are reviewed in reference |31|. The fundamental magnetic properties of rare-earth based amorphous alloys are of special interest for the low temperature physicist |2|. Very large fields are required at low temperatures to saturate the magnetization. Anomalous hysteresis loops and associated magnetic after-effects are observed. These experiments are explained with a random anisotropy model of local crystal field effects acting on the non-zero orbital contributions to the magnetic moment of

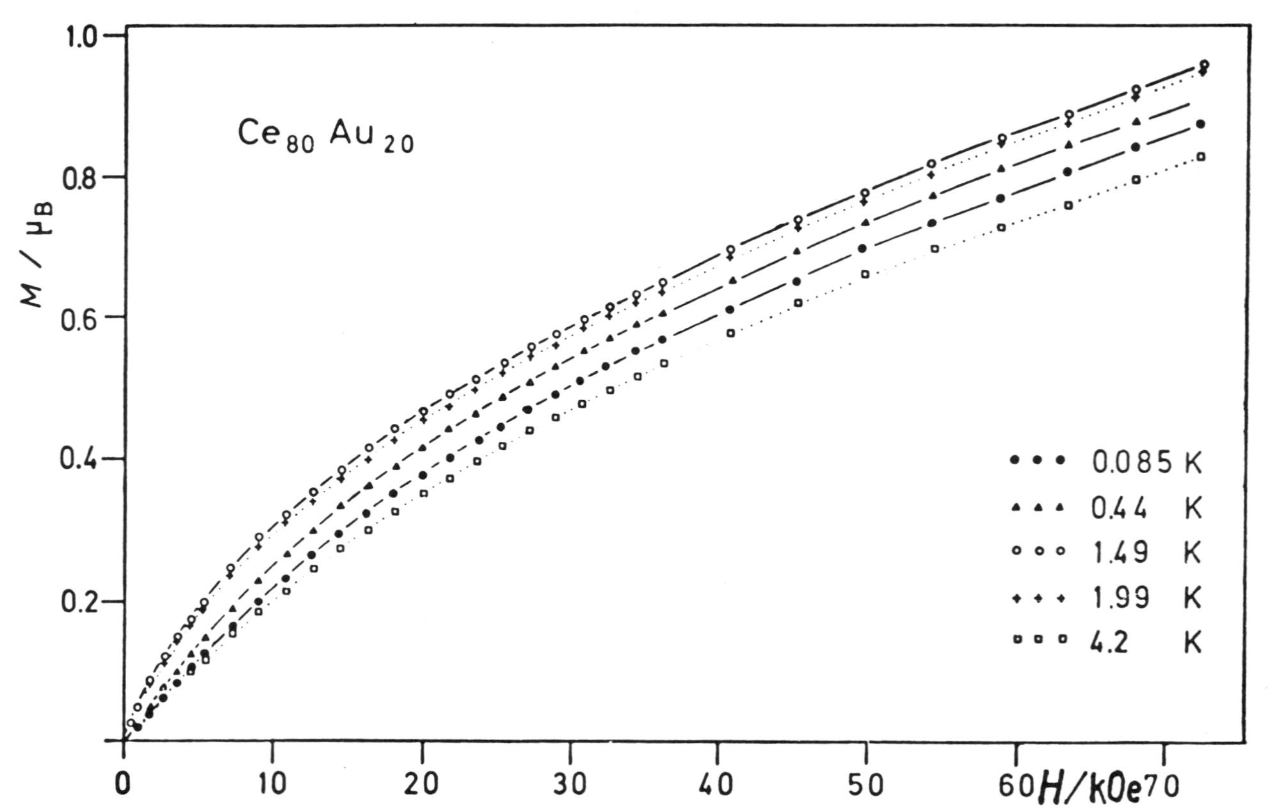

Fig. 11. Magnetization per Ce-atom in units of μ_B (saturation value = 2.14) as a function of the magnetic field above and below the magnetic ordering temperature T_c = 1.7 K |2 and 33|.

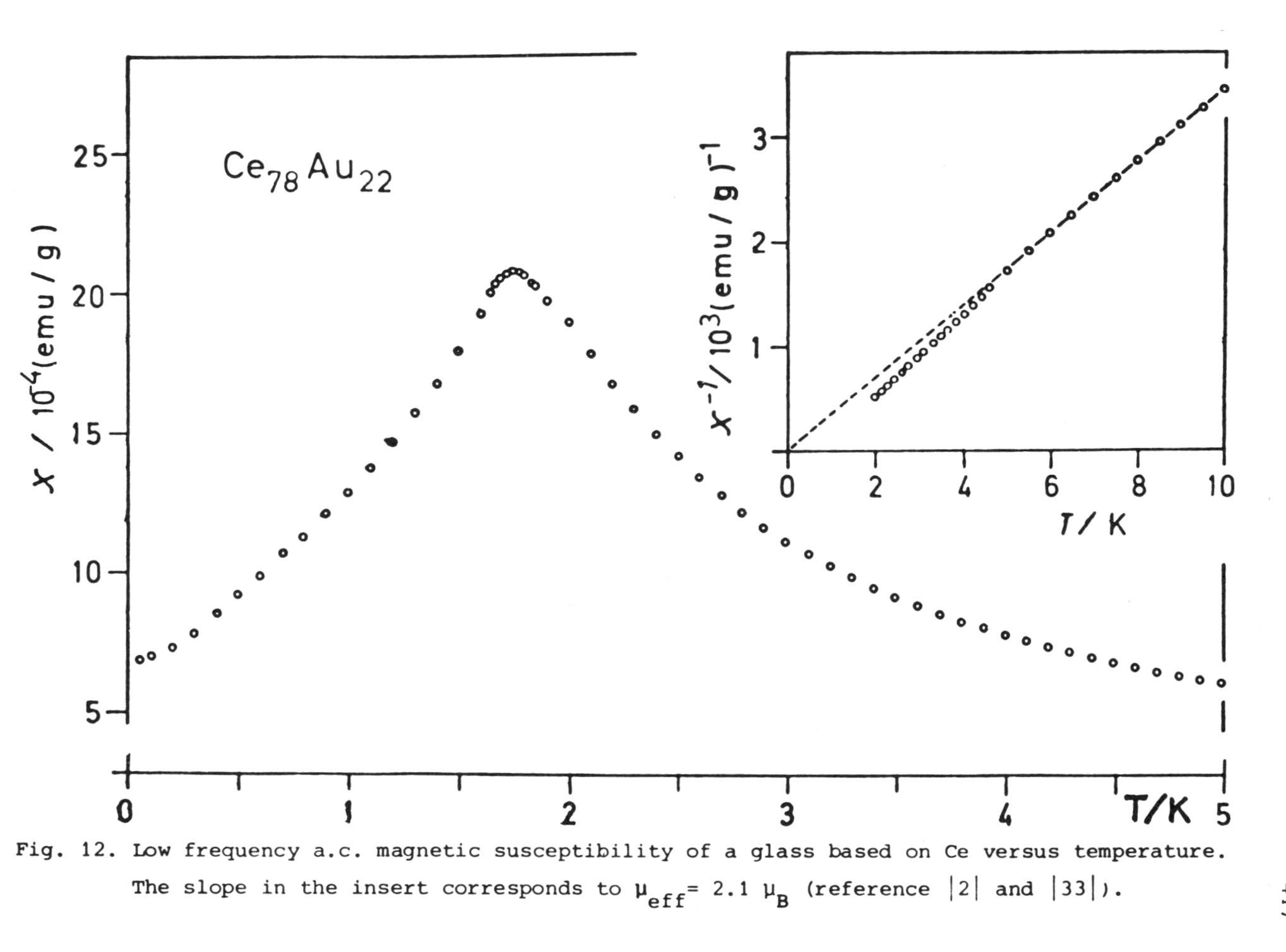

Fig. 12. Low frequency a.c. magnetic susceptibility of a glass based on Ce versus temperature. The slope in the insert corresponds to $\mu_{eff} = 2.1\ \mu_B$ (reference |2| and |33|).

the rare-earth ions|32|. The magnetic structure of e.g. amorphous $Ce_{80}Au_{20}$ at intermediate temperatures is that of a "speromagnet", with the Ce-moments pointing in the randomly distributed directions of easy magnetization. Very large external magnetic fields are required to rotate the Ce-moments into the direction of the applied field, as illustrated by measurements of the magnetization. Magnetic susceptibility measurements at rather low temperatures demonstrate a magnetic ordering below T_C=1.7K. The nature of the "ordered" state below T_C is still open, it may be similar to a random antiferromagnet or to a spin glass|33|.

CONCLUSIONS

Some low temperature properties of metallic glasses have been reviewed. The point of main interest is the existence of localized low-energy lattice excitations in amorphous metals and their influence on the transport properties.

We gratefully acknowlege the support of our own work with amorphous metals by the SFB 126 and the help by the OSV-planning-group in typewriting the manuscript.

REFERENCES

1. Rapidly Quenched Metals III, ed. by B. Cantor, The Metals Society, London, 1978.
   See for example: F. Spaepen, Vol. 2, p. 253, and references therein.

2. G.v.Minnigerode and K. Samwer, "Electrons in Metallic Glasses", to be published in Revue de Chimie Minerale.

3. H.J. Güntherodt and H.U. Künzi, in Metallic Glasses, ed. by J.J. Gilman and H.J. Leamy, American Society for Metals, Ohio, 1978, p. 247.

   D. Korn, H. Pfeifle, and G. Zibold, Z. Physik 270, 195 (1974).

4. R.W. Cochrane, J. Physique 39, C6-1540 (1978), and references therein.

5. H.J. Güntherodt, in Advances in Solid State Physics XVII, ed. by J. Treusch, Vieweg, Braunschweig, 1977, p. 25.

6. S.R. Nagel, Phys. Rev. B16, 1694 (1977)

   K. Froböse and J. Jäckle, J. Phys. F, 7, 2331 (1977)

   P.J. Cote and L.V. Meisel, Phys. Rev. Lett. 40, 1586 (1978), and Vol. 2, p. 62 in reference|1|, and references therein.

Y. Waseda and H.S. Chen, Phys. Stat. Solidi (b) 87, 777 (1978).

R. Harris, M. Shalmon, and M. Zuckermann, Phys. Rev. B18, 5906 (1978).

7. R.W. Cochrane, R. Harris, J.O. Ström-Olsen,and M.J. Zuckermann, Phys. Rev. Lett. 35, 676 (1975).

8. J.L. Black, B.L. Gyorffy, J. Physique 39, C6-941 (1978), and Phys. Rev. Lett. 41, 1595 (1978).

J.L. Black, Verhandl. DPG (VI), 14,375 (1979), and private communications together with J. Jäckle.

9. G.S. Grest and S.R. Nagel, Phys. Rev. B19, 3571 (1979).

10. T.E. Sharon and C.C. Tsuei, Phys. Rev. B5, 1047 (1972).

11. R. Azomoza, A. Fert, I.A. Campell, and R. Meyer, J. Phys. F, 7, L 327 (1977).

S.J. Poon, J. Durand, and M. Yung, Solid State Commun. 22, 475 (1977).

12. U. Ernst, W. Felsch, and K. Samwer, to be published.

13. P.M. Nast and K. Samwer, to be published.

14. G.Bergmann, Phys. Rep. 27C, 159 (1976).

15. K.H. Bennemann and J.W. Garland, in Superconductivity in d- and f-band metals, ed. by D.H. Douglass, AIP, New York, 1972, p. 103.

G. Kerker and K.H. Bennemann, Z. Physik 264, 15 (1973).

16. W.L. Johnson, Vol. 2, p. 1 in reference |1|.

17. K.Agyeman, R. Müller, and C.C. Tsuei, Phys. Rev. B19, 193 (1979).

18. W.L. Johnson, J. Appl. Phys. 50, 1557 (1979).

19. H. von Löhneysen and F. Steglich, Z. Physik B29, 89 (1978).

20. J.R.Matey and A.C. Anderson, Phys. Rev. B16, 3406 (1977).

21. H. von Löhneysen and F. Steglich, Phys. Rev. Lett. 39, 1205, (1977).

22. J.E. Graebner, B. Golding, R.J. Schutz, F.S.L. Hsu, and H.S. Chen, Phys. Rev. Lett. 39, 1480 (1977).

23. G. Bellessa, P. Doussineau, and A. Levelut, J. Physique Lett. 38, L-65 and L483 (1977).

24. B. Golding, B.G. Bagley, and F.S.L. Hsu, Phys. Rev. Lett. 29, 68 (1972).

25. D. Weaire, M.F.Ashby, J. Logan, and M.J. Weins, Acta Met. 19, 779 (1971).

26. T.B. Massalski, U. Mizutani, K.T. Hartwig, and R.W. Hopper, Vol. 2, p. 81 in reference |1|.

27. S.R. Nagel and J. Tauc, Phys. Rev. Lett. 35, 380 (1975), and in Rapidly Quenched Metals II, ed. by N.J. Grant and B.C. Giessen, MIT Press, Cambridge, Mass. 1976, Vol. 1, p. 337.

28. W.H. Shull and D.G. Naugle, Phys. Rev. Lett. 39, 1580 (1977).

W.H. Shull, D.G. Naugle, S.J. Poon, and W.L. Johnson, Phys. Rev. B18, 3263 (1978).

29. A.Combert and S. Ewert, Z. Physik B25, 173 (1976).

30. G. Kämpf and W. Buckel, Z. Physik B27, 315 (1977).

31. C.D. Graham and T. Egami, Vol. 2, p. 96 in reference |1|.

32. R.W. Cochrane, R. Harris, and M.J. Zuckermann, Phys. Rep. 48, 1 (1978).

33. U. Ernst, W. Felsch, and K. Samwer, to be published in Proc. ICM 79, München.

AMORPHOUS METALS

INVESTIGATED BY MÖSSBAUER SPECTROSCOPY

U. Gonser and H.-G. Wagner

Fachbereich Angewandte Physik, Universität
des Saarlandes, 66oo Saarbrücken, W. Germany

Ever since it was realized that metals can be produced in the amorphous state (1,2) there have been numerous investigations into a variety of these materials. A breakthrough was achieved with the discovery of the $T_{80}M_{20}$ alloys (T = Fe, Co, Ni, Mo; M = B, C, N, P, etc.).It was found that the amorphous state can be stabilized significantly by high valence elements of small size which are known to occupy interstitial lattice sites in metals. The high stability is obtained in the composition range of 2o at % metalloid which coincides with extraordinarily deep eutectics. The phase diagrams of Fe-B, Co-B, Ni-B, Fe-P, Co-P and Ni-P are shown in Fig. 1. This unique feature of the $T_{80}M_{20}$ amorphous alloys combined with similarities in their properties has been interpreted as an indication that in the quenching process the liquid structure of the eutectic is preserved rather than the various phases appearing in the phase diagrams. Ever since it was found that amorphous metals could be produced in a continuous process, with the advent of roller quenching techniques, big national research programs have been under way and industrial involvement has become quite impressive. The first commercial devices using amorphous metals are now available. The events of recent years may be considered as one of the rare cases where technology has by-passed science, that is, a material has been put to important applications while the central question of its structure is still uncertain. Even worse, the concentrated research effort has led to some controversies. Three of these controversies will be

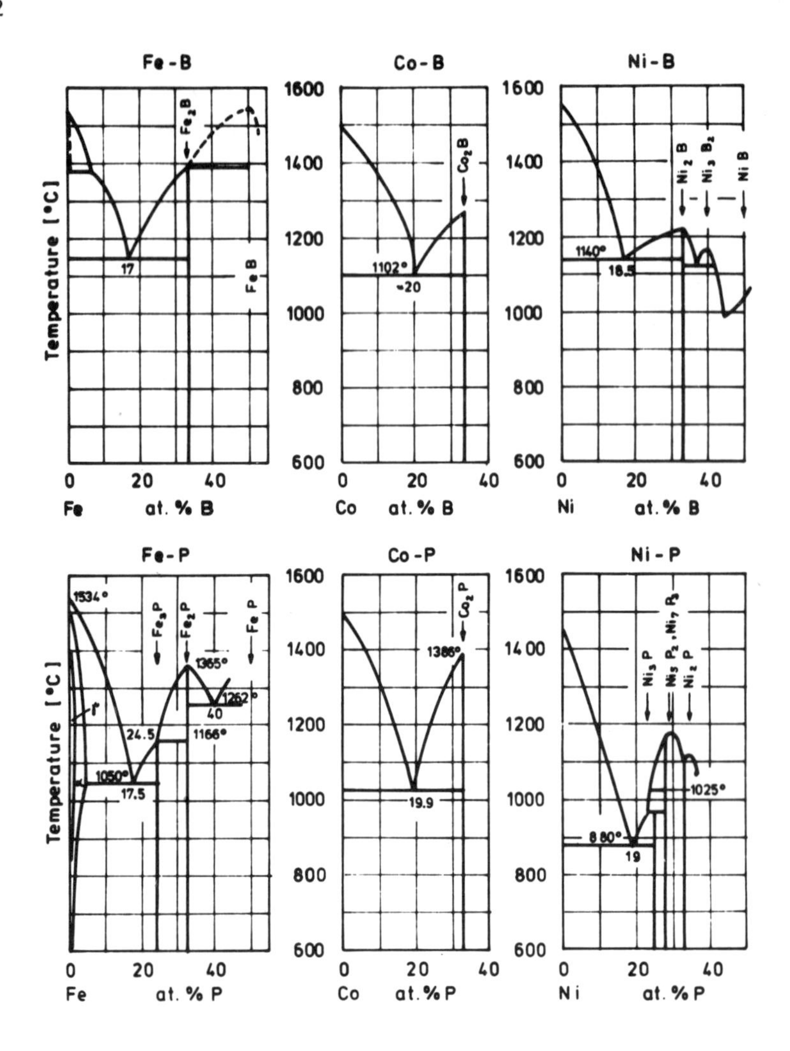

Fig. 1: Phase diagrams of Fe-B, Fe-P, Co-B, Co-P, Ni-B and Ni-P (taken from Hansen and Anderko, Constitution of Binary Alloys)

listed in this article.

The first controversy seems a rather formal one: "amorphous" versus "glass". It is interesting to note that in recent years there has been a shift in the terminology from amorphous toward glass. A contributing factor to this transition might be found in the human nature of scientists: it seems it is not so desirable for them to be associated with the term amorphous, they prefer the word glass which sounds more interesting, transparent, well-understood. Particularly, here in Zwiesel - the "Glass-City" - we might say: glass is beautiful!

The problem of nomenclature is connected with the question: which term describes the material best? Even the trade names seem to reflect this controversy: Metglas® (USA), Amomet® (Japan) and Vitrovac ® (Germany). The Greek origin of "amorphous" means lacking structure, and we use the word to express our ignorance or inability to define anything. In contrast glasses are usually defined: transparent glasses as undercooled liquids, or spin glasses, Wigner glasses, Stoner glasses etc. They might be considered released from the ill-defined amorphous pool. At present the term "amorphous metals" seems appropriate because it reflects the state of the art. One may assume that well-defined expressions will be coined as soon as we have deeper understanding of the nature of amorphous metals.

The methods used for studying the structure of amorphous alloys were mainly diffraction techniques employing x-rays and later neutrons and electrons. While it has become clear from the resulting interference and radial distribution functions that there is no long range crystalline order, the precise nature of atomic arrangements in units with diameters less than about 20 Å is still rather speculative.

The second controversy concerns these speculations, that is, the best and most appropriate model. For many years there was strong support for dense random packing of hard spheres (DRPHS) as a first order approximation to the structure, while recent years have seen the emergence of new models such as molecules, molecular units, tetrahedron packing, short-range order, microcrystalline structure, quasi-crystalline structure on the basis of locally distorted off-stoichiometric lattice etc.

The three descriptions: microcrystallites, network, (directional bonds) and dense random packing of hard spheres (DRPHS) can be regarded as the extremes or corner points of the amorphous state (see also article by Hornbogen and Schmidt). In the triangular representation (Fig. 2) one might find a position for any amor-

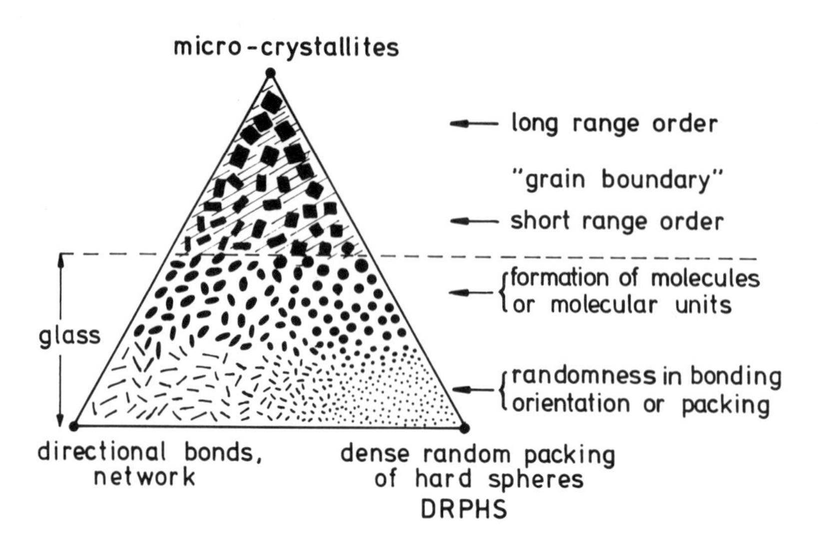

Fig. 2: Schematic representation of the amorphous state

phous metal. A line is drawn which separates the lower part of the triangle where the term glass seems appropriate. Here the material is characterized by continuous models. In contrast, in the upper part of the triangle the material is represented by discontinuous models, that is, some kind of "grain boundaries" are essential separating the micro-crystallites - or fragments of it - from each other (shaded area).

The reason for the continuing disagreement over the structure is the limited spatial resolution of the diffraction techniques. It was therefore an obvious step to start using microscopic methods sensitive to the chemical and topological short range arrangement. Results of one such method, the Mössbauer effect, will be discussed in this paper.

The Mössbauer effect probes the local environment of each resonant nucleus through the hyperfine interactions between nuclei and their electron shell. Starting from earlier work by Tsuei et al. (3) and Cochrane et al. (4), there have been numerous Mössbauer investigations of amorphous metals (see MEDI (5)). Unfortunately those concerned with the question of structure, instead of deciding the issue, merely reflect the different views already in existence. To understand why Mössbauer spectra of the same material produce different interpretations it is necessary to have a look at typical spectra of amorphous alloys containing Fe and to consider how these data are processed. Fig. 3 shows a Mössbauer spectrum of $Fe_{80}B_{20}$ (Metglas® 2605) taken at room temperature with a 25 mCi ^{57}Co in Rh source. It is important to point out that the overall shape of this spectrum is typical for all ferromagnetic amorphous metals containing Fe, i.e. almost independent of the constituents. We observe a broad magnetic spectrum with a very small asymmetry. There is a general agreement that this constitutes a distribution of magnetic hyperfine interactions with a possible fluctuation of isomer shifts and/or the existence of small quadrupole splittings.

The relative line intensities indicate a preferred orientation of spins in the ribbon plane. At first glance fitting theoretical spectra based on structural models to these experimental data seems a hopeless task. There are, however, two approaches which have both been successfully employed. One is to fit a continuous hyperfine field distribution to the data, assuming all other parameters such as line width, isomer shift etc. known (6,7). Apart from this rather unrealistic assumption the distribution analysis involves an expansion of the distribution function as a Fourier series producing all the well known truncation problems. It is therefore very dubious whether this method can give any reliable information on local environments.

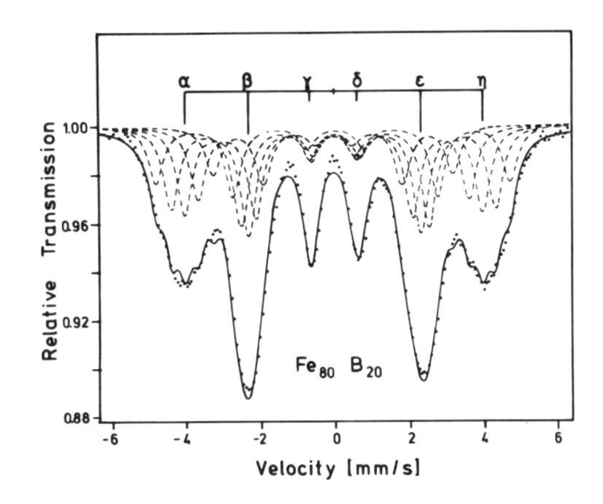

Fig. 3: Mössbauer spectrum of $Fe_{80}B_{20}$

The second approach - which we introduced (8) - tries to decompose the broad spectrum into a set of subspectra. The problem with this method is that, of course, the number of subspectra used is arbitrary. We found, however, that for all amorphous metals of the type $T_{80}M_{20}$ which we investigated, five subspectra would always produce a good fit. Important in our analysis is the fact that the relative intensities of the five subspectra correspond closely to the frequency of occurrence of five nearest neighbour configurations predicted by Bernal in his model of dense random packing of hard spheres (9,1o). Straight lines were obtained by plotting the hyperfine fields H_i of the subspectra versus C_o (Fig. 4). This held true for spectra of amorphous metals ($T_{80}M_{20}$) with various compositions for the metal T as well as the metalloid M, taken at different temperatures between 4.2 K and room temperature, under external fields up to 50 kOe and with the application of external stress up to 125 kp/mm^2. We therefore feel justified in assuming that the five subspectra used in our fitting correspond to the five most probable nearest neighbour configurations of Bernal's model. Bernal developed his geometrical model in order to explain the liquid state. The similarities of the Mössbauer spectra in conjunction with the field distribution analysis as well as the unique eutectics in the phase diagrams - almost independent of chemical constituents - indicates that the Bernal model is generally applicable and describes well the frozen-in structure of amorphous metals in the eutectic region.

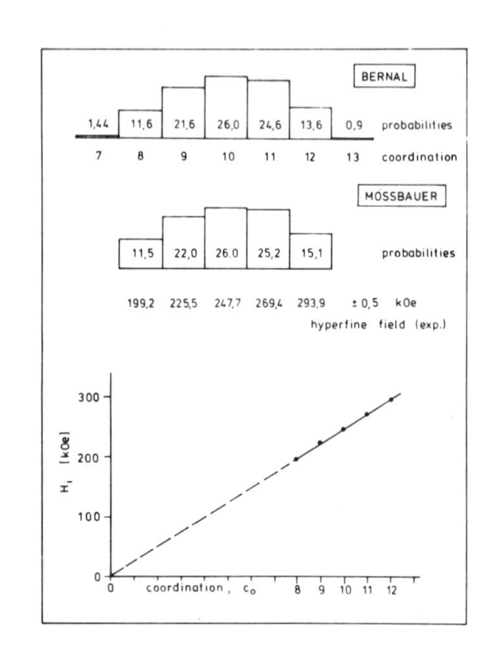

Fig. 4: Close contact coordination C_o and probabilities of occurrence of five hyperfine fields, H_i, determined from the subspectra. H_i vs C_o.

Other fitting attempts have been made to obtain information on the structure of amorphous metals, for instance, by assuming a short range order of a cementite-type structure. Obviously, we have now come to the third controversy: the problem is whether the distribution of hyperfine fields is mainly due to the perturbation by the metalloid atoms (M) while the metal atoms occupy "unique" lattice sites of a quasi- or microcrystalline arrangements or is governed by the various nearest neighbours transition metal atoms (T) in a kind of Bernal structure.

Besides the atomistic structure another interesting issue is the investigation of magnetic properties of amorphous alloys in view of their great potential as soft magnetic materials. The magnetic structure is usually discussed in terms of the domain structure or - closely connected with this - the spin texture. In crystalline ferromagnetic materials normally the magnetic crystalline anisotropy governs the orientation of the spins within the domains, that is, along the easy directions of magnetization. In amorphous metals the domain patterns are determined by three types of anisotropy energies: shape anisotropy, magnetoelastic coupling energy and structure anisotropy (11,12). Essentially two characteristic domain patterns have been observed in amorphous metals: broad stripes with a width of about 25 μm and patches of maze or fingerprint type patterns with a smaller domain width of about 3 - 5 μm. Considering positive magnetostriction only, the latter type of domain pattern can be related to regions with compressive stresses producing closure domains. This is schematically shown in Fig. 5.

The spin response to external magnetic fields can be studied by observing the angular dependence of the hyperfine interaction as evidenced in the relative line intensities. The ratio of the intensities of the lines for $\Delta m = 0$, I_2 (I_5) and $\Delta m = \pm 1$, I_3 (I_4) is

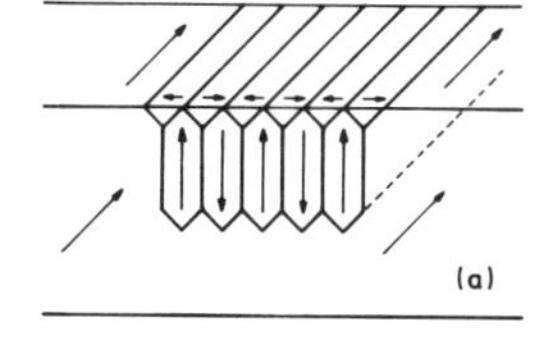

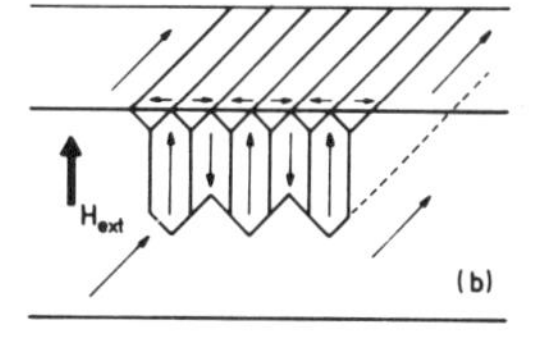

Fig. 5: Domain structures (see text)

$$I_2/I_3 = 4 \sin^2 \Theta / 1 + \cos^2 \Theta$$

where Θ denotes the angle between the internal magnetic field and the propagation direction of the γ-rays. In the case of a preferred spin orientation (domain structure or spin texture) the angle in the equation represents an average angle $\bar{\bar{\Theta}}$. Fig. 6 shows a sequence of spectra of the $Fe_{80}B_{20}$ Metglas® taken at different external fields perpendicular to the ribbon plane and parallel to the propagation direction of the γ-rays (13). In the fitting we assumed that all subspectra exhibit and reflect the same spin orientation. The angle $\bar{\bar{\Theta}}$ is schematically indicated on the left side of Fig. 6. Because of the negative hyperfine interaction on Fe the internal magnetic field shrinks with increasing external field. Asymmetries in the spectra were rather small and the values obtained for the quadrupole splittings were zero within the error limits. The relative intensities I_2/I_3 versus external fields are plotted in Fig. 7 for an as-quenched sample and one annealed at 63o K for 25 min.

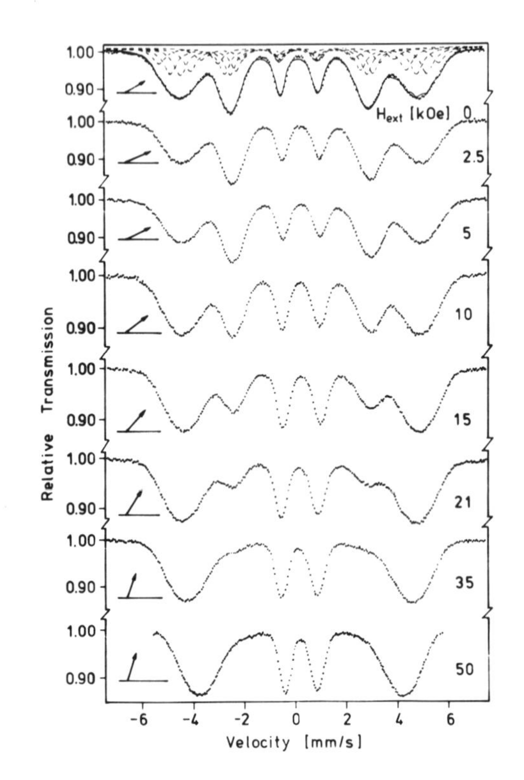

Fig. 6: Mössbauer spectra of $Fe_{80}B_{20}$ in longitudinal magnetic fields H_{ext}

For comparison, results on -Fe are also included. Surprisingly, small external fields turn the spins into the ribbon direction, away from H_{ext}, rather than aligning them. This might be explained by assuming that some of the fingerprint domains are converted to stripe domains. We have observed a similar effect in transverse magneto-resistivity measurements.

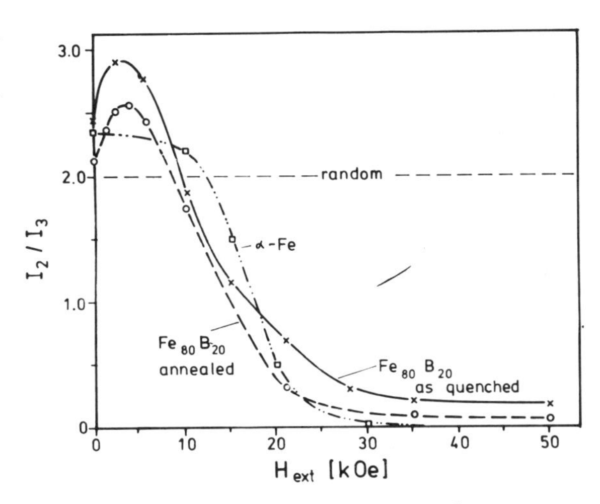

Fig. 7: Ratio of relative line intensities I_2/I_3 vs. H_{ext}

It is also quite remarkable that even external fields of 5o kOe do not cause the lines corresponding to transitions with $\Delta m = 0$ to disappear ($I_2/I_3 > 0$). This indicates that saturation magnetization and alignment of all spins has not been accomplished. A possible explanation for this was given by Kronmüller et al. (11). The existence of quasi dislocations with an estimated density of $1o^{13}/cm^2$ leads to an inhomogeneous spin arrangement near these defects (Fig. 8). Such quasi dislocations might be the result of mass density fluctuations due to the rapid cooling process as schematically shown in Fig. 9. Keeping the outer contour constant a two dimensional dense random packing of circles is disturbed by the addition and the removal of a linear chain of 7 discs.

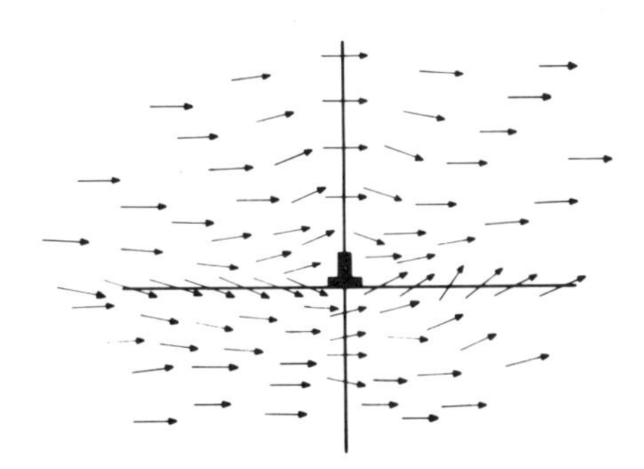

Fig. 8: Schematic representation of spins in the vicinity of a quasi dislocation

Fig. 9 a and c show possible resulting configurations with the overlap describing compressive stresses and the lack of contact representing tensile stresses. The application of an external tensile stress leads to an increase in the observed hyperfine field (14) as depicted in Fig. 1o.

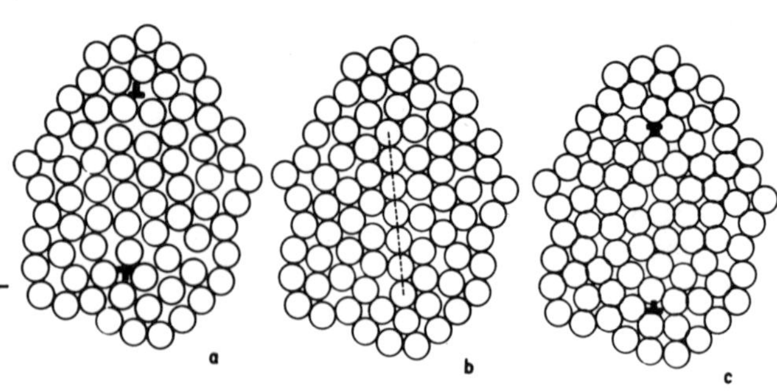

Fig. 9: Model of amorphous structure with quasi dislocation (see text)

This might be explained by an increase in the exchange interaction with increasing atomic separation, in accordance with the Bethe-Slater curve.

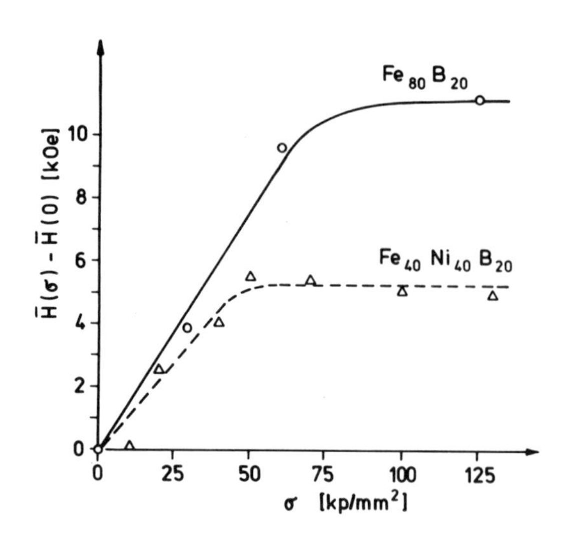

Fig. 1o: Change in hyperfine fields with external tensile stress

Linearly polarized γ-rays were used to study the spin texture and its change with applied tensile stress (14, 15,16); The source was a transversely magnetized α-Fe foil containing about 5o mCi ^{57}Co. The four lines with $\Delta m = \pm 1$ are polarized perpendicular to the two lines with $\Delta m = 0$. Moving the six source lines (designated by A, B, C, D, E, F) over the broad absorber lines (α, β, γ, δ, ε, η) we expect a 36 lines spectrum. The line positions are easily calculated by adding or subtracting the corresponding line positions of source and absorber while the transition probabilities and polarizations of the corresponding transitions determine the relative intensities.

For the parallel case ($H_S \parallel H_A$) and the perpendicular case ($H_S \perp H_A$) of magnetization in source and absorber we expect 2o and 16 lines respectively, which are plotted in the centre of Fig. 11. In the upper spectrum of Fig. 11 the ribbon direction R and the source magnetic field H_S were parallel ($H_S \parallel R$) whereas in the lower spectrum they were perpendicular to each other ($H_S \perp R$). The two spectra are rather similar, however on closer inspection differences in the relative line intensities become evident, indicating a preferred orientation (texture) of the spins in the ribbon di-

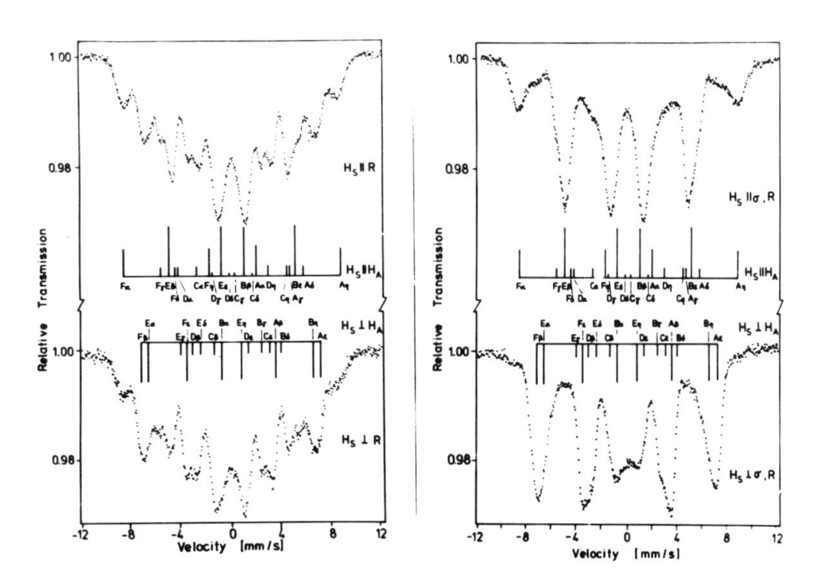

Fig. 11: Fig. 12

Mössbauer spectra of amorphous $Fe_{40}Ni_{40}P_{14}B_6$. A transversely magnetized Co^{57}-α-Fe foil was used as the source which had its direction of magnetization parallel (upper spectra) and perpendicular (lower spectra) to the ribbon direction. The spectra in Fig. 12 were taken under applied tensile stress.

rection. If a tensile stress σ is applied to the absorber ($\sigma \parallel R$) the spectrum is changed considerably (see Fig. 12). Now the spectra for the arrangements $H_S \parallel \sigma$, R and $H_S \perp \sigma$, R tally well with the corresponding stick diagrams indicating that the applied stress has aligned the spins.

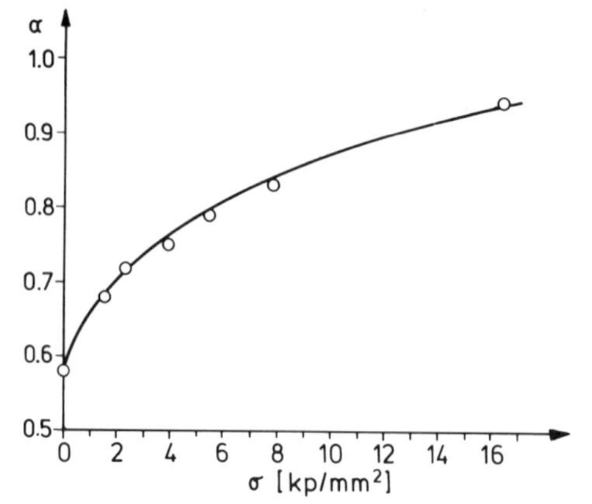

Fig. 13: Polarization parameter α vs. external tensile stress σ

From the relative line intensities of Mössbauer spectra obtained with linearly polarized γ-rays and tensile stress σ, the degree of alignment can be deduced (14). A polarization parameter α is introduced which is zero or one in the limits of full polarization of the spins, that is, either parallel (one) or perpendicular (zero) to σ, which is in the ribbon direction. The value of o.5 corresponds with random orientation. In Fig. 13 α is plotted as a function of σ. It should be possible to obtain information on the anisotropy constant K from this curve, as the ease of alignment is proportional to λ/K (17), with λ the linear magnetostriction constant of the material.

With this discussion we hope to have demonstrated the important contribution the Mössbauer effect can make to the study of amorphous metals. While it alone cannot decide the issue of the structure of these materials, further studies of amorphous systems with varying alloying concentration and different metalloids should eventually, in conjunction with the results of other experimental methods, answer the question of the structure of amorphous metals.

Acknowledgement

The financial support of the Deutsche Forschungsgemeinschaft is gratefully acknowledged.

REFERENCES

1. Brenner and Riddel
   J. Research Nat. Bur. Standards, 39, 385 (1947)
2. W. Buckel, R. Hilsch
   Z. f. Physik 138, 1o9 (1954)
3. C.C. Tsuei, G. Longworth, S.C.H. Lin
   Phys. Rev. 17o, 6o3 (1968)
4. R.W. Cochrane, R. Harris, M. Plischke, D. Zobin, M.J. Zuckermann
   Phys. Rev. B12, 1969 (1975)
5. J. Stevens, V. Stevens
   Mössbauer Effect Data Index
6. T. Kemeny, I. Vincze, B. Fogarassy, S. Arajs
   Phys. Rev. B (in press)
7. J.M. Dubois, G. Le Caer, A. Amamon, U. Herold
   To be published in J. de Physique
8. U. Gonser, M. Ghafari, H.-G. Wagner
   Proc. Int. Conf. Mössbauer Spectroscopy, Bucharest Vol. 1, 159 (1977) and J. Magn. and Magn. Materials 8, 175 (1978)
9. J.D. Bernal
   Nature 185, 68 (196o)
1o. J.D. Bernal
   Proc. Roy. Soc. A28o, 299 (1964)
11. H. Kronmüller, M. Fähnle, M. Domann, H. Grimm, R. Grimm, B. Gröger
   J. Magn. and Magn. Materials (in press)
12. M. Takahashi, T. Suzuki, T. Miyazaki
   Japan J. Appl. Phys. 16, 521 (1977)
13. H.-G. Wagner, U. Gonser, A. Schertz
   3. Intern. Conf. Rapidly Quenched Metals, Brighton 2, 333 (1978)
14. U. Gonser, H. Fischer, M. Ghafari, H.-G. Wagner, R.S. Preston
   ICM München (1979)
15. H. Fischer, U. Gonser, R.S. Preston, H.-G. Wagner
   J. Magn. and Magn. Materials 9, 336 (1978)
16. U. Gonser
   J. de Physique (in press)
17. R.C. O'Handley, C.-P. Chou
   J. Appl. Phys. 49 (3), 1659 (1978)

MECHANICAL PROPERTIES OF METALLIC GLASSES

H.U. Künzi

Institut für Physik, Universität Basel
CH-4056 Basel, Switzerland

INTRODUCTION

The mechanical behavior of materials certainly a very interesting subject in the physics of the solid state has been considerably enlarged by the discovery of metallic glasses. Even though the first metallic glasses were prepared some 20 years ago it was not realized until the beginning of this decade that with respect to future applications some metallic glasses exhibited outstanding mechanical properties.[1,2] Among these properties we find excellent mechanical strength and hardness. Apart from these technological aspects these new materials also offer a great scientific interest. The absence of any periodicity in the arrangement of their atoms gives rise to interesting questions in comparison with crystalline metals.

It is well-known from numerous investigations in crystalline metals that the mechanical properties are highly dependent on the shortrange order and the presence of lattice defects or impurities. This seems to be true also for the amorphous materials with an almost equal sensitivity. This fact drastically contrasts with the behavior of other physical properties as e.g. the electrical resistivity. In crystalline metals the electrical resistivity is known to be highly sensitive to deviations from a perfect order, in the amorphous state, however, the electrical resistivity merely reflects the disordered structure and is almost independent to small variations in such a structure. In the case of glassy metals the investigation of structure sensitive properties such as the mechanical properties which at least can give indirect information on the shortrange order are of great value as classical diffraction experiments do not have sufficient sensitivity to determine the local arrangements of atoms. It is therefore hoped that the mechanical properties may help to give further insight into these problems.

The aim of this contribution is to review the basic mechanical behavior of metallic glasses and to discuss their relevance to structure, atomic dynamics and other properties. According to the different types of stress-strain behavior exhibited by metallic glasses the paper will be devided in different sections.

In common with all other solids, metallic glasses behave in an essentially elastic manner at low temperatures and low stress levels. The elastic constants relevant for this mode of deformation will be discussed in the first section. The second section concerns the anelastic behavior. This time dependent elastic behavior is a consequence of stress induced relaxations operating within the material and in many cases has its origin in thermally activated atomic jumps. At these temperatures the stress-strain behavior is still fully reversible. At higher temperatures close to the glass transition the mechanical deformation becomes more and more irreversible. Viscous flow which is a characteristic mode of deformation for all glassy materials at high temperatures starts to dominate. As viscous flow also the plastic deformation is irrecoverable. Its occurence is however confined to the relatively high stress levels prior to fracture. This mode of deformation is characteristic for crystalline metals and seems to distinguish metallic from non-metallic glasses which usually fail by brittle fracture.

2. ELASTICITY OF METALLIC GLASSES

For crystalline materials the number of independent elastic constants that fully describes the system depends on the symmetry properties of the crystal. In the case of an amorphous structure which at least in a fully relaxed state is randomly disordered and therefore isotropic the number of independent elastic constants is two. However, as the metallic glasses are in configurational frozen states a non-randomly distributed directional topological or compositional short range order may be frozen in. Such a directional character in the structure may result from preparation or any other directional treatment. Non-isotropic elastic constants have indeed been observed in a number of ferromagnetic glassy alloys.[3-5] For reasons not yet sufficiently understood these materials may show an uniaxial magnetic anisotropy. As metallic glasses are magnetically soft the magnetoelastic coupling may be large and thus is the cause of a substantial mechanical anisotropy. For non-magnetic metallic glasses, however, no anisotropic effects have yet been reported and consequently their elastic behavior is fully described by two elastic constants.

The experimental determination of Young's modulus and the shear modulus is quite simple. We have the choice between a variety of methods. With statical methods the elastic deformation is directly measured as a function of the applied load. With resonance methods the elastic constants are determined from resonant frequencies of a sample. In this case the geometrical form should be simple and the dimensions have to be known very accuratly as they enter in the expressions for the resonant frequencies. In the absence of this knowledge these methods still allow for a very high relative accuracy which may be important in investigations where we study the influence of external parameters as the temperature, the magnetic field, disolved gases (variation of external pressure), heat treatments and others on the elastic constants. Another quite often used method is the determination of the sound velocity with ultrasonic pulse techniques. At not too high frequencies (wave length $\gg$ lateral dimensions) and with wire or ribbon shaped samples this method is independent of geometrical dimensions except for the length along which the pulse travels. Infact all these methods give reliable results and have been used during the past years for the determination of the elastic constants in metallic glasses.[6-11]

As the elastic constants of materials do not depend very strongly on the structure, metallic glasses show values which are of about the same magnitude as comparable crystalline metals. Table 1 shows that for metallic glasses the Young's modulii vary as in crystalline metals between the values for elastically hard metals as Co and the elastically soft metals as Mg.

A rather interesting result is, however, obtained when we directly compare the elastic constants of the alloy before and after crystallization. Young's modulus increases by typically 30% and the shear modulus by about 40% . It should be remembered that crystallization in most cases leads to phase separation and complex microstructures. But in spite of this fact the spread about the given mean values are relatively small. In view of the small density differences between the crystalline and glassy alloys (typically 1-2 %) the observed variations of 30 and 40% are rather large. In Quarz for comparison the shear stiffness changes by 38% on crystallization, but at the same time the density increases by about 20% .

Table 1 Comparison of Young's modulii between crystalline and glassy metals. (a = ref.12).

Crystal	$E\left[{}^{N}/m^{2}\right]$	Glass	$E\left[{}^{N}/m^{2}\right]$
Co	$20.6 \cdot 10^{10}$	$Co_{70}Fe_{5}Si_{15}B_{10}$	$19.0 \cdot 10^{10}$
		$Fe_{80}B_{20}$	16.9 [a]
Cu	12.3	$Nb_{50}Ni_{50}$	14.1 [a]
Ti	10.8	$Cu_{50}Ti_{50}$	9.86[a]
Mg	4.4	$Mg_{70}Zn_{30}$	3.5

It has been shown as early as 1971 by Weaire et al.[13] that this relative elastic softness of metallic glasses has its origin in the amorphous nature of their atomic structure. Compared to an ideal crystal, atomic sites in an amorphous structure are not unique. Therefore, under the action of an external stress a local arrangement of atoms may change its configuration (internal displacements) while still remaining perfectly amorphous. Such rearrangements may give a contribution to the strain and thus reduce the elastic stiffness.

3. ANELASTICITY OF METALLIC GLASSES

While discussing the elastic properties we did not bother about how the deformation proceeded in time. For a given stress we rather were interested in the equilibrium value of the deformation as given by Hooke's law. On a close examination, however, all solids show in addition to an immediate elastic response a time dependent creep and the equilibrium value of deformation is achieved only after a certain time. The same is true after releasing the load. A deformation is called anelastic, if after a loading and deloading cycle the deformation is fully reversible, i.e. the shape of the sample is left unchanged.

In many cases this creep behavior can be described by a single relaxation time τ. Instead of Hooke's law the stress-strain behavior of the anelastic solid is then described by

$$J_R \sigma + J_U \tau \dot{\sigma} = \varepsilon + \tau \dot{\varepsilon}$$

where J_R and J_U are the relaxed and unrelaxed compliances (= reciprocal modulii). The anelastic behavior[14] is caused by stress-induced relaxation processes operating within the material. The mechanical stress may therefore be used as a sensitive experimental probe to investigate the kinetics of such relaxations.

The choice of the measuring method depends on the length of the relaxation time involved in the relaxation process. For long relaxation times $\tau > 1s$ the anelasticity can be studied directly, by measuring the time dependence of the creep. (elastic after-effect).[6,15] For shorter relaxation times $1 < \tau < 10\mu s$ the anelasticity is measured indirectly by the internal friction (energy dissipation) of an oscillating sample[16] and at even shorter relaxation times by the ultrasonic attenuation.[17]

There are several physical effects that can contribute to the anelasticity of solids. In metallic glasses up to now relaxations due to the thermoelastic effect, the magnetomechanical coupling (ΔE effect) and structural rearrangements involving atomic jumps have been observed.

3.1 The thermoelastic effect

The thermal expansion of matter when heated is a well-known manifestation of the thermoelastic interaction. The inverse phenomena, namely heating of matter by applying a stress or a strain, is an alternative expression of the same effect. Its contribution to internal friction in flexurally vibrating reeds arises from transverse thermal currents which flow as a consequence of the non-homogeneous temperature distribution within the sample, caused by thermoelastic heating and cooling of the sample. As Zener[8] has demonstrated the internal friction Q^{-1} due to this effect can be described by a simple Debye peak.

$$Q^{-1} = \frac{\alpha^2 ET}{c_p} \frac{\omega\tau}{1+\omega^2\tau^2}$$

α is the linear thermal expansion coefficient, E Young's modulus, c_p specific heat and T the absolute temperature. The peak damping occurs at the frequency $\omega_o = 1/\tau$ where τ is the relaxation time for the transverse heat exchange which in turn depends on the thickness d of the sample and the thermal diffusivity of the material under consideration $\tau = d^2/\pi^2 D$.

As melt quenched metallic glasses have a thickness of about 50µm the thermoelastic damping becomes dominant in the lower kilocycle range. The thermal diffusivity and conductivity can be directly obtained from the frequency ω_o, whereas the peak height is a sensitive measure for the thermal expansion coefficient α.

Figure 1 shows the thermoelastic peak in a glassy $Cu_{50}Zr_{50}$ alloy. Points represent experimental results, obtained by exciting the fundamental as well as overtones in samples of different length and hence different fundamental frequencies. The continuous curve shows the fitted Debye peak. The room temperature values obtained for the linear thermal expansion coefficient was $\alpha = 8.2 \cdot 10^{-6}\ {}^oK^{-1}$ and for the thermal diffusivity was $D = 2.6 \cdot 10^{-6}\ m^2/s$. The thermal conductivity λ calculated from D $\lambda = 6.4\ {}^W/m\ {}^oK$ is almost twice as large as the electronic contribution calculated from the Wiedemann-Franz law using the free-electron value for the Lorenz ratio.

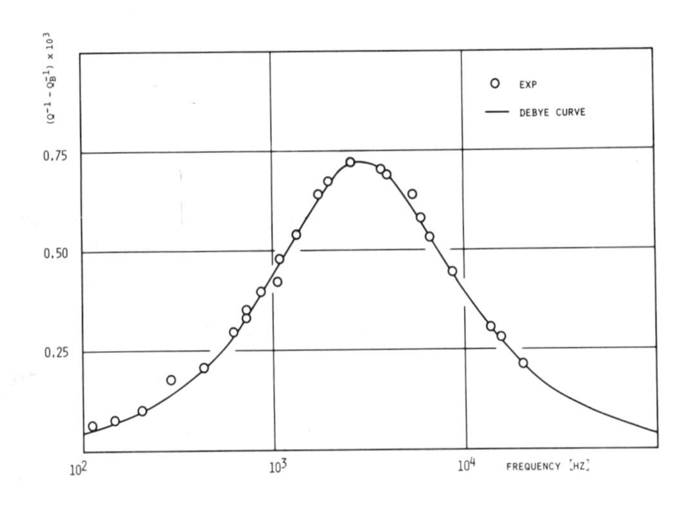

Fig.1: Thermoelastic peak in glassy $Cu_{50}Zr_{50}$.

3.2 Magnetoelastic behavior

The basic parameter that describes the magnetomechanical coupling is the magnetostriction given by the change of the geometrical dimensions of a sample as a function of the magnetic field or alternatively as the variation of the magnetization induced by a mechanical stress. At low frequencies in the fully relaxed state this phenomena may give rise to large changes of Young's modulus as a function of the applied field (ΔE-effect). In particular it has been found in recent years that due to the magnetical softness this effect may be very large in ferromagnetic metallic glasses. M.A. Mitchell et al.[19] reported the substantial value of $\Delta E/E = 2.26$ for the maximum ΔE-effect in a $Fe_{71}Co_9B_{20}$ glass. This means that the magnetostrictive strain is more than twice the purely elastic strain.

At higher frequencies ($f \gtrsim 100$kHz depending on thickness of sample) stress induced eddy currents prevent the magnetostrictive strain from being in phase with the elastic strain, the ΔE-effect decreases with the complementary appearance of an internal friction peak.[20]

Since in metallic glasses the ΔE-effect is large and in addition can be monitored by comparatively low fields, these materials could be used as variable ultrasonic delay lines. A further interesting feature of the magnetoelastic coupling is that in certain materials the magnetostriction can exactly compensate for the thermal expansion[21] (invar effect).

3.3 Structural relaxations

Under the influence of a mechanical stress various microstructural changes may occur in crystalline as well as amorphous solids. When such rearrangement give additional contributions to the strain at constant stress or, alternatively give rise to a stress relaxation at constant strain they are subject of the time dependent elasticity. The kinetics of such relaxations are in many cases well described by thermally activated atomic movements. It should, however, be noted that at low temperatures also tunneling processes may be dominant, as it is for instance the case for the low lying excitations (two level system) characteristic for most glassy materials.[17,22]

For a thermally activated relaxation the relaxation time τ is given by the Arrhenius law

$$\tau = \tau_o e^{E/kT}$$

where E and τ_o are the activation energy and pre-exponential factor respectively, k is Boltzmann's constant and T is temperature. The internal friction in the presence of structural relaxation centers with a well-defined activation energy and τ_o is again as for the thermoelastic relaxation described by a Debye peak.

In the case where instead of a single relaxation time τ , we have a distribution of relaxation times (slightly dissimilar relaxation centers) which may be due to a distribution in the activation energies and/or τ_o values, the internal friction still shows a peak which, however, will be broader than a simple Debye peak. In amorphous materials with a random structure we might in fact intuitively expect this later situation to occur. The value of the relaxation time itself when given by the Arrhenius law depends very strongly on the temperature and instead of measuring the internal friction at constant temperature as a function of frequency, the Debye peak may be observed in the temperature dependence of Q^{-1} at a constant frequency. Such peaks have indeed recently been observed in several metallic glasses.[23-27] Figure 2 shows an example for a $Co_{35}Y_{65}$ glass. The measurement had been carried out at the fundamental as well at three overtones and the fact that the peak shifts with frequency indicates the thermally activated nature of this relaxation. The peak damping accurs when $\omega\tau = 1$. This allows to determine τ at the temperatures where the peaks occur.

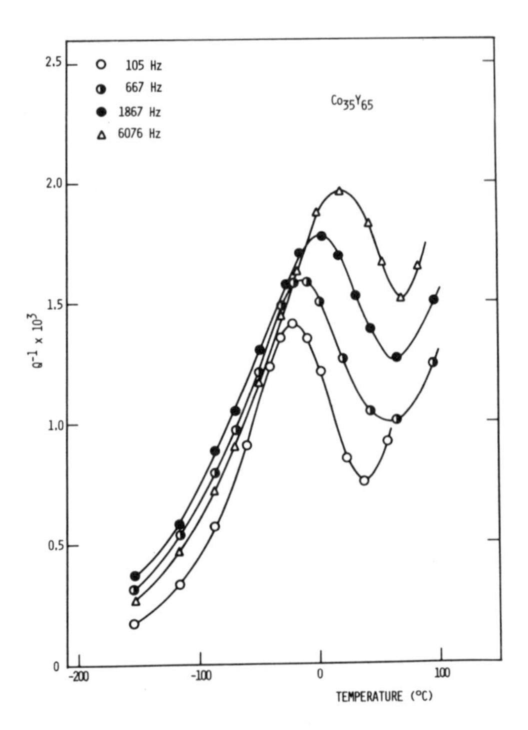

Fig.2: Thermally activated internal friction peak in a glassy $Co_{35}Y_{65}$ alloy.

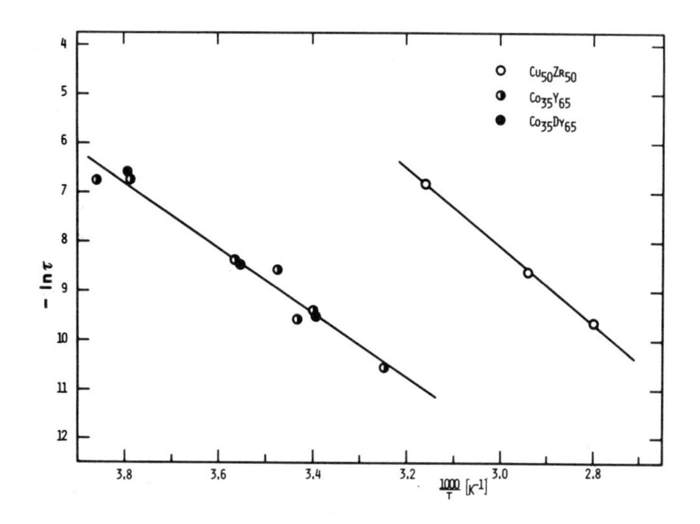

Fig.3 Arrhenius representation of the relaxation times.

Figure 3 shows the Arrhenius representation of these relaxation times for $Co_{35}Y_{65}$ as well as for $Cu_{50}Zr_{50}$ and $Co_{35}Dy_{65}$ which show similar peaks.

Table 2 Activation energies and the pre-exponential factors τ_o .

	T_p [°K]	τ_o [sec]	E [eV]	Ref.
$Cu_{50}Zr_{50}$	314	$1.4 \cdot 10^{-14}$	0.69	26
$Co_{35}Y_{65}$	260	$2.2 \cdot 10^{-14}$	0.56	26
$Co_{35}Dy_{65}$	262	$7.4 \cdot 10^{-15}$	0.59	26
$Nb_{75}Ge_{25}$	260	$3.0 \cdot 10^{-14}$	0.55	24
$Pd_{80}Si_{20}$	200	-	(0.4)	27

τ_o values of order 10^{-14} sec found for the metallic glasses are consistent with relaxations involving single jumps of an atomic specie. These results drastically contrast with the peaks occuring near 250 °K in $Fe_{40}Ni_{40}P_{14}B_6$ and $Fe_{32}Ni_{32}Cr_{14}P_{12}B_6$ [25] with activation energies of 1.47 and 1.08 eV respectively and hence if interpreted as Debye peaks with pre-exponential factors of $1.6 \cdot 10^{-32}$ and $6 \cdot 10^{-26}$sec. Such values are of course far byond any vibrating period of an atom in a solid and in fact do not seem to find a simple explanation.

In the case of $Nb_{75}Ge_{25}$[28] and $Pd_{80}Si_{20}$ it is well-established that the peaks are due to disolved hydrogen.

In $Cu_{50}Zr_{50}$, $Co_{35}Dy_{65}$ and $Co_{35}Y_{65}$ it is not yet known whether the relaxation centers are impurity related or intrinsic. Measurements made on samples that had been stored in a hydrogen atmosphere and with oxygen doped samples did not give different results. Another interesting feature of the peaks in the above three alloys is that in contrast to the findings in $Nb_{75}Ge_{25}$ and $Pd_{80}Si_{20}$ they do not decrease after annealing treatments below the glass transition.

Independent of the specific nature of the jumping atoms it should also be pointed out that for an amorphous structure the peaks are rather sharp. In fact they are not simple Debye peaks but rather have a width of about two to three times the width of such a peak. If we were to explain this width with a continuous spectrum in the activation energies alone, a distribution width of ±10% around the meanvalues given in Table 2 would be enough. It seems therefore evident that the short range order of metallic glasses cannot be described by a fully random arrangement of atoms.

Financial support of the "Kommission zur Förderung der wissenschaftlichen Forschung" and of the Swiss National Science Foundation is gratefully acknowledged.

REFERENCES

1. J.J. Gilman, J. of Appl. Phys. 46 (1625) 1975.
2. L.A. Davies in "Metallic Glasses", American Society for Metals, Metals Park, Ohio, 1978.
3. B.S. Berry and W.C. Pritchet, Phys. Rev. Letters 34 (1022) 1975.
4. B.S. Berry and W.C. Pritchet, J. of Appl. Phys. 47 (3295) 1976.
5. B.S. Berry and W.C. Pritchet, AIP Conf. Proc. 34 (292) 1976.
6. B.S. Berry, in ref.2.
7. C.-P. Chou, L.A. Davies and N.C. Narasimhan, Scripta Met. 11, (417) 1977.
8. H.S. Chen and J.T. Krause, Scripta Met. 11 (761) 1977.
9. H.S. Chen, J. Appl. Phys. 49 (3289) 1978.
10. S. Tyagi and A.E. Lord, Jr., J. Non-Cryst. Solids 30 (273) 1978.
11. C.-P. Chou, L.A. Davies and R. Hasegawa, J. of Appl. Phys. 50, (3334) 1979.

12. L.A. Davies, C.-P. Chou, L.E. Tanner and R. Ray, Scripta Met. 10 (937) 1976.
13. D. Weaire, M.F. Ashby, J. Logan, M.J. Weins, Acta Met. 19 (799) 1971.
14. A.S. Nowick and B.S. Berry, Anelastic Relaxation in Crystalline Solids, Academic Press, New York, 1972.
15. H. Kimura, T. Murata and T. Masumoto, Sci. Rep. RITU, A26 270 1977.
16. B.S. Berry and W.C. Pritchet, IBM Res. Develop., July 1975 (334).
17. G. Weiss, W. Arnold, K. Dransfeld and H.-J. Güntherodt, in press.
18. C. Zener, Phys. Rev. 53 (90) 1938.
19. M.A. Mitchell, J.R. Cullen, R. Abbundi, A. Clark and H. Savage, J. Appl. Phys. 50 (1627) 1979.
20. K.I. Arai, N. Tsuya, M. Yamada, IEEE Trans. of Magnetics MAG-12 No 6 (936) 1976.
21. F. Fukamichi, M. Kikuchi, S. Arakawa and T. Masumoto, Sol. State Comm. 23 (955) 1977.
22. S. Hunklinger and W. Arnold, Physical Acoustics Vol.12, p.155, ed. by W.P. Mason and R.N. Thurston, Academic Press, New York, 1976.
23. B.S. Berry, W.C. Pritchet and C.C. Tsuei, Phys. Rev. Letters 41 (410) 1978.
24. B.S. Berry and W.C. Pritchet, "Rapidly Quenched Metals III" Proc. of 3rd Int. Conf. on "Rapidly Quenched Metals, Brighton, Vol.2 (21) 1978.
25. H.N. Yoon and A. Eisenberg, J. of Non-Cryst. Solids 29 (357) 1978.
26. H.U. Künzi, K. Agyeman and H.-J. Güntherodt, Sol. State Comm. 32 1979.
27. H.U. Künzi and K. Agyeman, Proc. of 3rd European Conf. on Internal Friction and Ultrasonic Attenuation in Solids, Manchester, 1979.
28. G.S. Cargill III: This volume.

ELECTRONIC STRUCTURE OF METALLIC GLASSES STUDIED BY ELECTRON SPECTROSCOPY [1]

P. Oelhafen

Institut für Physik, Universität Basel
CH-4056 Basel, Switzerland

ABSTRACT. A wide variety of metallic glasses containing an early (T_E) and late (T_L) transition metal or a transition metal and normal metal or metaloid (N) have been studied by electron spectroscopy (UPS,XPS and AES). The metallic glasses were prepared by rapid quenching from the liquid state using a piston and anvil apparatus operating in high vacuum in the 10^{-8} Torr range. The measurements were performed with a combined UPS/XPS/AES spectrometer (LHS 10 , Leybold-Heraeus).

Valence bands of T_L-T_E glasses

The valence bands of a large number of glassy alloys with different alloy concentrations (Cu-Zr,Ni-Zr,Co-Zr,Fe-Zr,Pd-Zr,Pt-Zr,Rh-Zr, Nb-Ni,Ta-Ni and Cu-Ti) have been studied by UPS with excitation energies of 21.2 and 40.8 eV . The most important general feature of the photoemission valence band spectra is the d-band splitting and the shift of the d-states of T_L to higher binding energies. As a consequence the valence bands are dominated by two peaks related to the d-states of the two constituents. The largest d-band binding energy shift of about 1.7 eV was found in the glassy Pd-Zr.[2] In T_L-T_E alloys with T_L from a given series the d-band shift is decreasing with decreasing number of groups between T_L and T_E in the periodic table. A correlation between the glass forming ability of the alloy and the d-band binding energy shift was observed: the bigger the binding energy shift the better the glass forming ability. The d-bands of the alloy components show a concentration dependence: the d-band peak binding energy of T_L is decreased with increasing content of T_L (towards the value of pure metal d-band centroid position) and its half-width is increased.

A comparison of valence band spectra of glassy alloys ($Cu_{60}Zr_{40}$, $Pd_{35}Zr_{65}$) with corresponding spectra of crystalline systems (Cu_3Zr_2, $PdZr_2$) shows that the d-band splitting and d-band binding energy shift is not a specific property of the amorphous alloys but is also found in the crystalline phase. The d-band binding energies are essentially the same in each phase but the peak shape is changed from a Gaussian-like peak in the amorphous phase to a peak with more structure in the crystal phase.

Core level spectroscopy (CLS)

Important information may be obtained from core level spectroscopy: the core level binding energy E_B of an alloy constituent compared to the pure component is related to the charge transfer occuring on alloying and the core level line shape is related to the local density of states near the Fermi level E_F. The measured core level binding energy shifts are generally small ($|\Delta E_B| < 0.6$ eV) in the systems studied except in the case of Pd-Zr where a Pd 3d core level shift of 1.4 eV was measured. This indicates that the net charge transfer is generally small in the metallic glasses. The core level line shapes of the T_L are highly asymmetrical in the pure metal (except in Cu alloys) and become symmetrical in the alloys due to a strong decrease of the local density of d-states at E_F. For the T_E a decrease or increase of the local density of states at E_F can be observed by a more symmetrical (Ta 4f in Ni-Ta and Nb 3d in Ni-Nb) or more asymmetrical (Zr 3d in Ni-Zr, Co-Zr, Fe-Zr and Ti 2p in Cu-Ti) core level line shape. The change of local density of states of T_E on alloying may be explained by a shift of the corresponding d-band to lower binding energies.

Valence bands and core levels of T_L-N glasses

Two systems of T_L-N glasses (Pd-Si and Fe-B-Si) have been studied. The valence bands are dominated by the d-bands of T_L which are slightly shifted towards higher binding energies. The Pd d-band shift increases with increasing Si content.[3] The Pd-Si valence band exhibit a shoulder at E_F which can be attributed to Pd d-states. The core level binding energy shifts are small ($|\Delta E_B| < 0.5$ eV) in the two systems. Again the Pd 3d core levels become highly symmetrical in the alloy due to the strong decrease of the local density of states in contrast to the Fe-B-Si glass in which the Fe $3p_{3/2}$ line asymmetry is almost unchanged on alloying Fe with B and Si.

REFERENCES

1. This subject will be thoroughly discussed in "Glassy Metals I" Ed. H.-J. Güntherodt and H. Beck, Topics in Current Physics, Springer Verlag Berlin, Heidelberg, New York, 1980.
2. P. Oelhafen, E. Hauser, H.-J. Güntherodt and K.H. Bennemann, Phys. Rev. Lett. 43. 1134 (1979).
3. P. Oelhafen, M. Liard, H.-J. Güntherodt, K. Berresheim and H.D. Polaschegg, Solid State Commun. 30, 641 (1979).

PRESSURE DEPENDENCE OF THE ELECTRICAL RESISTANCE OF SOME METALLIC GLASSES

G. Fritsch*, H. Schink. J. Willer, E. Lüscher

Physik-Department E13, Technische Universität München
8046 Garching, Germany

ABSTRACT. We report on measurements of the electrical resistance as a function of pressure and temperature. The metallic glasses examined are $Fe_{40}Ni_{40}P_{14}B_6$ (Metglas** 2826), $Fe_{32}Ni_{36}Cr_{14}P_{12}B_6$ (Metglas 2826A) as well as $Fe_{80}B_{20}$ (Metglas 2605). Data were taken between 1.5 K and 400 K and between zero pressure and about 100 kbar. We propose a qualitative explanation of the results.

INTRODUCTION

The electrical resistivity of amorphous alloys has been reported in literature for quite a lot of systems as a function of temperature /1/. However, pressure data are scarce /2/. Theoretically, the results of nonmagnetic metallic glasses are understood quite well based on the Ziman theory. In the magnetic case, especially in the glasses containing Fe, Ni, Cr, Pand B, there is a resistivity minimum which shifts dramatically with composition. No detailed and consistent explanation of this behaviour is given by now. Since temperature variation gives the combined effect of temperature and a small volume change due to thermal expansion, it seems important to measure the volume effect alone by pressure variation in order to get additional information on the scattering mechanism.

---

*Present address: ZWE Physik, Hochschule der Bundeswehr, 8014 Neubiberg, Germany

**Trademark by Allied Chemical

EXPERIMENTAL DETAILS

The samples were cut from thin ribbons (length: 1.5 mm, thickness about 50 μm and width 0.1 to 0.2 mm). The high pressure technique and the experimental set-up is described elsewhere /3/. The pressure calibration was done in situ with a thin lead-strip, monitoring the superconducting transition temperature. The width of the transition gives information on the pressure homogeneity within the cell. This figure is included in the error-bars for the pressure. Data were taken when relaxing the pressure, in order to avoid irreversible changes in the sample and in the experimental geometry. The zero pressure runs were made in a different sample holder and with different sample geometry. No magnetic field was applied. A lowering of the crystallization temperature is not to be expected when applying pressure /4/.

RESULTS AND DISCUSSION

The electrical resistance of the three Metglasses measured is given in Figures 1 to 4. The occurance of a resistivity minimum is clearly seen in Fig. 1. Its shift with pressure is reproduced together with the variation of the resistance ratio R/R_{min} at various temperatures (Figs. 2, 3 and 4). Whereas the dependence on temperature looks very similar in all cases - apart from the location of the minimum - the pressure coefficient dR/dP shows positive and negative values. The data reveal also that dR/dP is almost independent from temperature.

Since we applied no corrections for the change of the sample geometry with pressure, we measure directly the behaviour of the electrical resistivity /5/ but without the pre-integral factors, speaking in terms of the Ziman formulation.

We propose, in accordance with the suggestion made by Gudmundsson et al. /6/, that the total resistivity can be represented as the sum of a potential scattering term and a magnetic contribution. In addition we assume the following: Whereas the potential part should be described by a Ziman-type formula, i.e. should show almost zero temperature dependence, the magnetic part should be splitted into two contributions. One in the range below T_{min} and one above, since T_{min} exhibits a complicated pressure dependence, depicted in Figs. 2 to 4.

The low temperature part should be related to the Kondo-effect. In order to support this conjecture, we offer the following arguments:

i) A certain number of spins feels zero field environment, since there is a broad distribution of internal magnetic fields /7/.

ii) T_{min} seems to scale inversely with T_c, the Curie temperature. Taking T_c as a measure of the coupling strength between the spins, the probability of zero field conditions should be reduced for high T_c. Then the number of "free" spins reduces, therefore the magnitude of the Kondo-effect too, lowering in turn T_{min}.

iii) The temperature and pressure variation is fitted well by the equation $\rho_0(P) - \rho_1(P) \cdot [\ln T/T_K(P)]^2$ in case of 2826A where enough data points are available to make a fit meaningful. $\rho_1(P)$ is reduced when the pressure is increased, showing directly the decrease of the number of spins involved as anticipated. $T_K(P)$ is also decreasing with rising pressure. ($T_K(0) \approx 0.5$ K for 2826A). $\rho_0(P)$, the potential scattering part decreases strongly with pressure, giving the main contribution to dR/dP in this range.

The high temperature part above T_{min} should have a positive temperature coefficient. This statement may be derived from the fact, that the potential scattering contribution should be nearly a constant. Hence, we fitted the data to an equation of the form $\bar{\rho}_0(P) + \bar{\rho}_1(P) \cdot T + \bar{\rho}_2(P) \cdot T^2$.

From our analysis the following picture emerges: There is a background resistivity (ρ_0, $\bar{\rho}_0$, $\bar{\rho}_1$) nonmagnetic by assumption, with a small positive temperature coefficient ($\approx 1 \cdot 10^{-4}$ μΩm/K). It accounts for about 70 - 80 % of the pressure variation, the pressure coefficients being $(1/\rho_0) \cdot d\rho_0/dP, \cdot 10^{-4}$ kbar^{-1}): $-(2.5 \pm 0.5)$ (2826A), $+(1.5 \pm 0.5)$ (2826) as well as $-(8.6 \pm 0.5)$ (2605). A quadratic term $\bar{\rho}_2 T^2$ is superimposed above T_{min}, a Kondo like term below T_{min}. Their pressure coefficients are (kbar^{-1}): $-(1.4 \pm 0.3) \cdot 10^{-3}$ ("Kondo", 2826A); $-(6 \pm 0.6) \cdot 10^{-2}$ ("T^2", 2826) and $+(2 \pm 0.2) \cdot 10^{-2}$ ("T^2", 2605). It can be shown that the pressure dependence of $\rho_1(P)$ and $\bar{\rho}_2(P)$ is responsible for the shift of $T_{min}(P)$.

The positive T^2-term may be explained by the following models for the scattering mechanisms:

i) s-d-scattering, as discussed by Mott /8/ or

ii) scattering by magnons /8/.

No studies concerning the pressure dependence of those mechanisms are published in literature. Hence, the state of affairs has to be left undecided. A more detailed version of this work will be published elsewhere /9/.

ACKNOWLEDGEMENT. This work was done in collaboration with Prof. M. Kalvius' group at the Technische Universität Munich. We like to thank Profs. H.-J. Güntherodt, J. Jäckle, K. Bennemann and G. von Minnigerode for valuable discussion.

REFERENCES

1. H.-J. Güntherodt, Adv.Solid State Phys. 17, 25 (1977)
2. D. Lazarus, Solid State Comm. 32, 175 (1979)
3. J. Willer, J. Moser, J.Phys. E 12, 886 (1979)
4. M. Cedergren, G. Bäckström, J.Non-Cryst. Solids 30, 69 (1978)
5. G. Fritsch, E. Lüscher, NATO-ASI Liquid Metals and Alloys, edts. E. Lüscher, H. Coufal, (Proc. to be published)
6. H. Gudmundsson, K.V. Rao, A.C. Anderson, NATO-ASI Liquid Metals and Alloys, edts. E.Lüscher, H.Coufal (Proc. to be publ.)
7. C.L. Chien, D. Musser, E.M. Gyorgy, R.C. Scherwood, H.S. Chen, F.E. Luborsky, J.W. Walter, Phys.Rev. B20, 283 (1979)
   G.S. Grest, S.R. Nagel, Phys.Rev. B19, 3571 (1979)
8. N.F. Mott, Phil.Mag. 26, 1249 (1972), Adv. in Phys. 13, 325 (1964)
9. G. Fritsch, H. Schink, J. Willer, E. Lüscher (to be published)

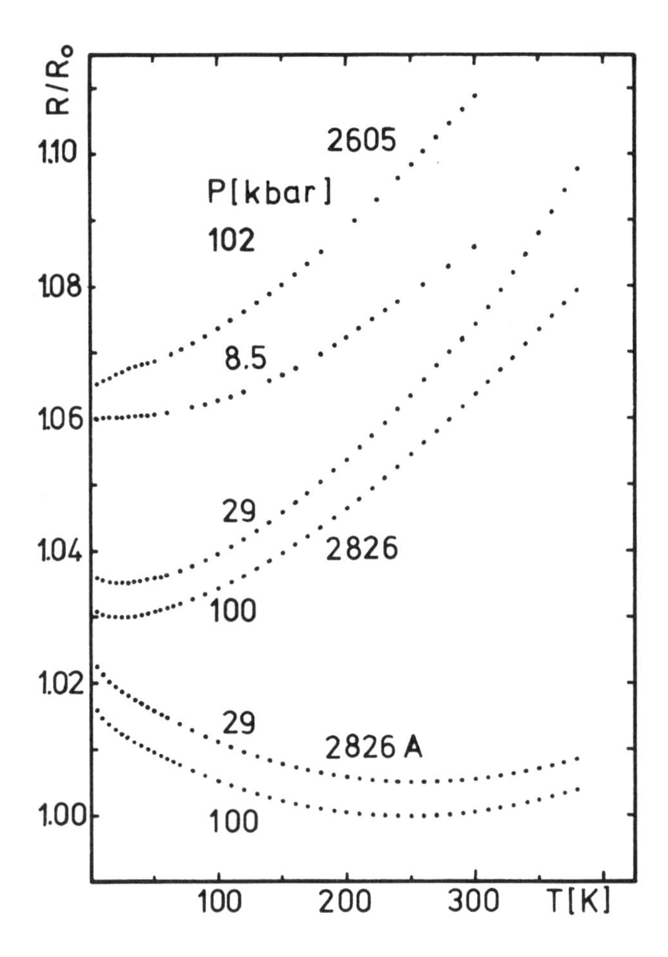

Fig. 1. Electrical resistance as a function of temperature T at various pressures P. The data are normalized with respect to the value at T_{min} and shifted upward in order to avoid overlap. R_{min} is about 1.35; 1.6 and 1.57 μΩm for the Metglasses 2826, 2826A and 2605, resp.

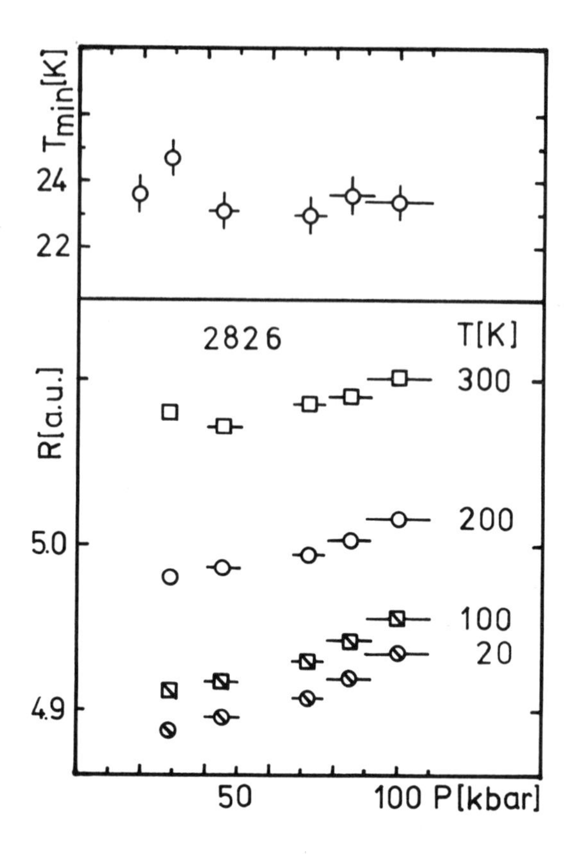

Fig. 2. The pressure dependence of T_{min} and of the resistance R at various temperatures for Metglas 2826.

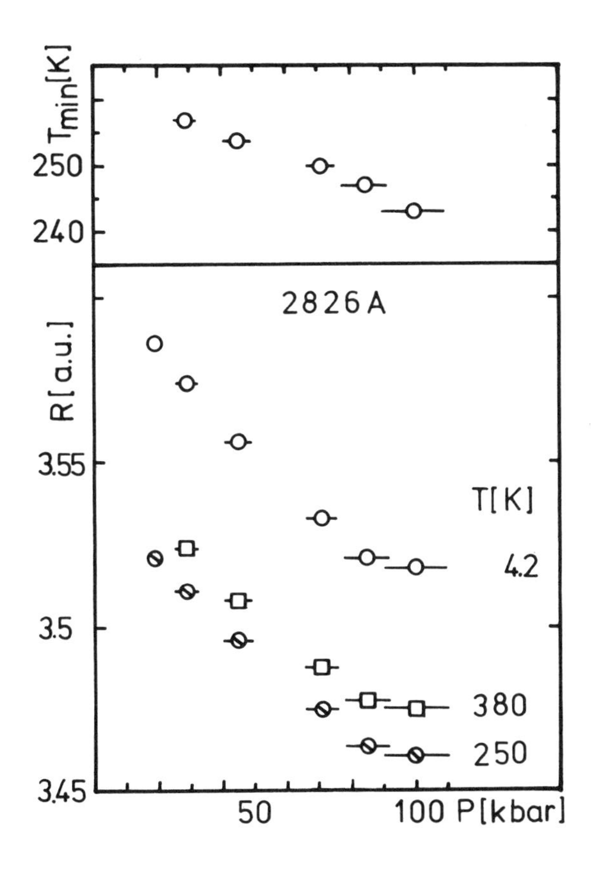

Fig. 3. Same as Fig. 2, except for Metglas 2826A.

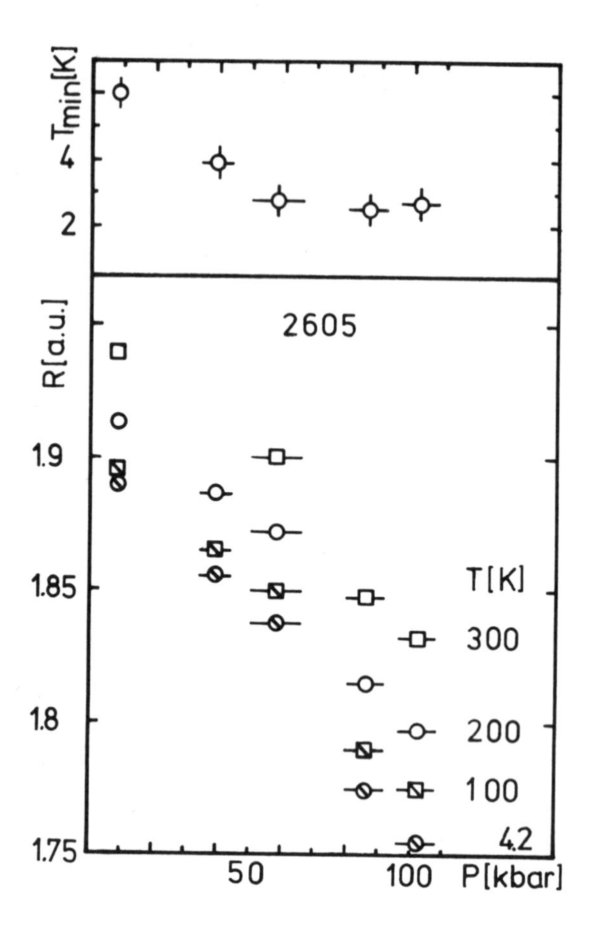

Fig. 4. Same as Fig. 2, except for Metglas 2605.

EXAFS: POSSIBILITIES, ADVANTAGES AND LIMITATIONS FOR THE INVESTIGATION OF LOCAL ORDER IN METALLIC GLASSES

R. Haensel, P. Rabe, G. Tolkiehn and A. Werner

Institut für Experimentalphysik der Universität Kiel, D 2300 Kiel 1, F.R. Germany

ABSTRACT. Throughout this book many examples for the close relationship between the macroscopic properties and the local structure of liquid and amorphous metals and alloys are given. Complementing the conventional X-ray, electron and neutron scattering techniques the analysis of the "Extended X-ray Absorption Fine Structure" (EXAFS) has been developed to a reliable method for structural analysis. Although the phenomenon is known since the thirties of the century, its application for the determination of local geometrical parameters has only been introduced about ten years ago. The development of the method has been greatly stimulated by extensive theoretical work, e.g. more refined calculations of complex scattering amplitudes and by the availability of synchrotron radiation as an intense X-ray continuum source. Two aspects are discussed in this paper: i) The basic principles of EXAFS and the usual data evaluation techniques demonstrated for crystalline Fe; ii) Applications on amorphous systems, the modifications of the data evaluation due to nonsymmetric pair distributions and the comparison with other methods for structure determination.

1. INTRODUCTION

Since many decades X-rays are used in the investigation of the geometrical and the electronic structure of matter. Whereas diffraction studies are strongly related to the geometrical structure, absorption and its secondary effects are normally used to probe the electronic structure. EXAFS combines these two fields, investigating a scattering phenomenon to deduce geometrical structure informations by means of absorption measurements.

The roots of the method go back to the thirties when Kronig (1) tried to understand the origin of the fine structure observed in the energy range up to several hundred eV above the absorption edges. As an example Fig.1 (upper curve) shows the K-absorption spectrum of metallic Fe.

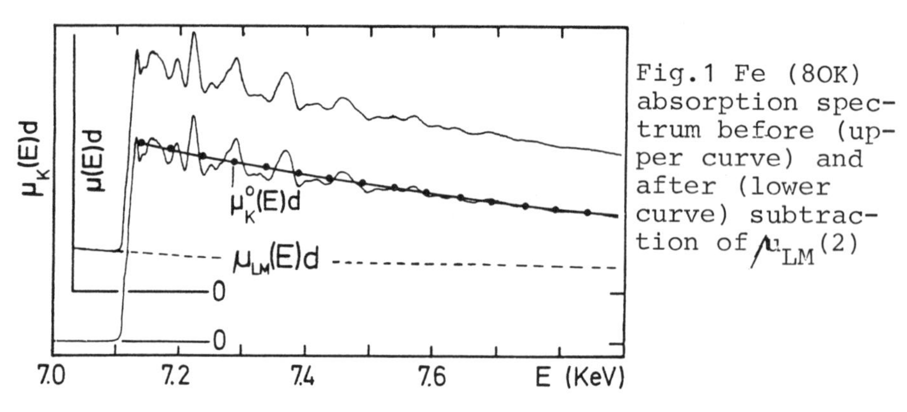

Fig.1 Fe (80K) absorption spectrum before (upper curve) and after (lower curve) subtraction of μ_{LM} (2)

EXAFS can be described as an electron diffraction process, where the electron source is the absorbing atom. The final state wave function of the excited state is formed from the outgoing photoelectron wave and the parts of it, which are backscattered from the surrounding atoms. Outgoing and backscattered waves interfere near the excited atom. According to Fermi's Golden Rule the photoabsorption coefficient $\mu(E)$ depends on the density of final states $\rho(E)$ and transition matrix element $M(E)$

$$\mu(E) \sim \rho(E) \cdot |M(E)|^2 \qquad (1)$$

E is the photon energy. The term $\rho(E)$ has a monotonic behaviour except in the region near the edge. The matrix element $M(E)$ represents the overlap of the initial and final state wave functions at the central atom. This overlap depends on the relative phase of outgoing and scattered wave and can be changed by tuning the wavelength (kinetic energy) of the photoelectron wave.

We express the absorption coefficient μ by

$$\mu = \mu_K^o(1 + \chi) + \mu_{LM} \qquad (2)$$

where μ_K^o is the monotonous absorption coefficient due to the excitation of the K-electrons of an isolated atom and μ_{LM} that of the weaker bound electrons (see Fig.1). χ represents the modulating part of the absorption coefficient, specific for the EXAFS modulation. About ten years ago Lytle, Sayers and Stern (3) demonstrated the usefulnes of EXAFS to probe the local geometrical structure in matter. By their efforts they initiated great improvements of the theoretical background for a quantitative interpretation of the EXAFS data (4-10). A special impetus was given to the method by the availability of synchrotron radiation as an intense X-ray continuum source. The brightness of a synchrotron radiation source is superior to the bremsstrahlung intensity of a conventional X-ray generator by several orders of magnitude (11). A typical experimental setup is described in ref.(12).

2. BASIC CONSIDERATIONS

As has been mentioned above EXAFS originates from a modulation of the transition matrix element $M(E)$ caused by the scattering of the photoelectron wave. The kinetic energy of the photoelectrons E_{kin} is

$$E_{kin} = E - E_o \qquad (3)$$

where E is the photon energy and E_o the binding energy. For the following it is convenient to introduce the wave number k which is calculated from

$$k = \sqrt{\frac{2m}{\hbar^2} E_{kin}} \qquad (4)$$

The EXAFS above K-edges for unoriented samples can be written as (4-10)

$$\chi(k) = \sum_j A_j(k) \cdot \sin(2kR_j + \alpha_j(k)) \quad (5a)$$

with

$$A_j(k) = -\frac{1}{k}\frac{N_j}{R_j^2} \cdot |f_j(\pi,k)| \cdot e^{-R_j/\lambda} e^{-2\sigma_j^2 k^2} \quad (5b)$$

and

$$\alpha_j(k) = 2\delta_1(k) + \arg f_j(\pi,k) \quad (5c)$$

The summation is carried out over all coordination shells j. N_j is the number of identical atoms in the scattering shell j at the distance R_j; $/f_j(\pi,k)/$ is the backscattering amplitude characteristic for the scattering atom; $\alpha_j(k)$ is the scattering phase, consisting of the phase shift arg $f_j(\pi,k)$ due to the potential of the scatterer and the phase shift $\delta_1(k)$ of the central atom. The first exponential in eq.(5b) describes the finite range of the photoelectron wave, the second exponential is a Debye-Waller factor and describes the disorder (dynamical (thermal) or static (structural)) with mean square relative displacements σ_j^2. The $\chi(k)$ determined from the absorption spectrum of Fe is shown in Fig.2. The original data have been reduced as discussed in detail elsewhere (13).

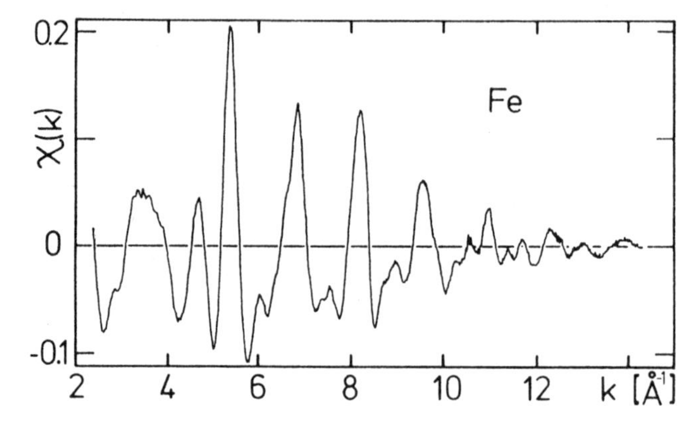

Fig.2: $\chi(k)$ extracted from the Fe absorption spectrum in Fig.1 (2)

In the following we will discuss in more detail the different ingredients of the EXAFS formula in eq.(5).

a) The argument of the sine term: It contains the information about distances R_j between the central atom and the surrounding scatterers. The frequency of $\chi(k)$ in k-space is modified by the k-dependent parts of the phase $\alpha_j(k)$. Its exact knowledge therefore is obviously important for the absolute determination of distances from EXAFS data. These phase shifts mainly depend on the atomic number of the central and the scattering atoms. The chemical bonding character only plays a minor role (chemical transferability)(13,14).

Two ways are possible to obtain the phase shifts $\alpha_j(k)$. They can simply be determined by adding calculated values of $\delta_1(k)$ and arg $f(\pi,k)$. Examples obtained from Lee and Beni (7) are shown in Fig.3.

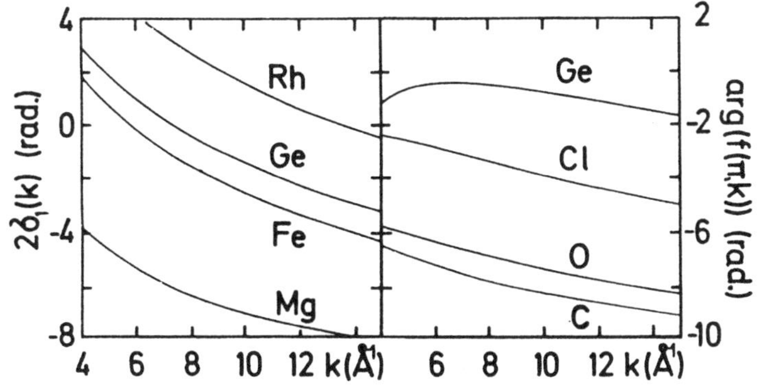

Fig.3: Scattering phases taken from Lee and Beni (7)

Alternatively the chemical transferability can be used to obtain the phase from EXAFS measurements of a crystallographically well known compound. Utilizing these phase shifts in systems of unknown geometry yields typical accuracies in the determination of bond lengths of ± 0.02 Å. Changes of interatomic distances for atom pairs in comparable chemical environments due to phase transitions (e.g. amorphous-crystalline (15)) or to applied external forces (e.g. pressure (16)) may yield even higher accuracies. In special cases accuracies of ± 0.003 Å have been obtained (17).

b) The amplitude: The k-dependence of the backscattering amplitude $|f_j(\pi,k)|$ is specific for the scattering atom. Calculated values for $|f_j(\pi,k)|$ are shown in Fig.4 (10). For light elements like F $|f_j(\pi,k)|$ is a monotonically decreasing function for increasing k.

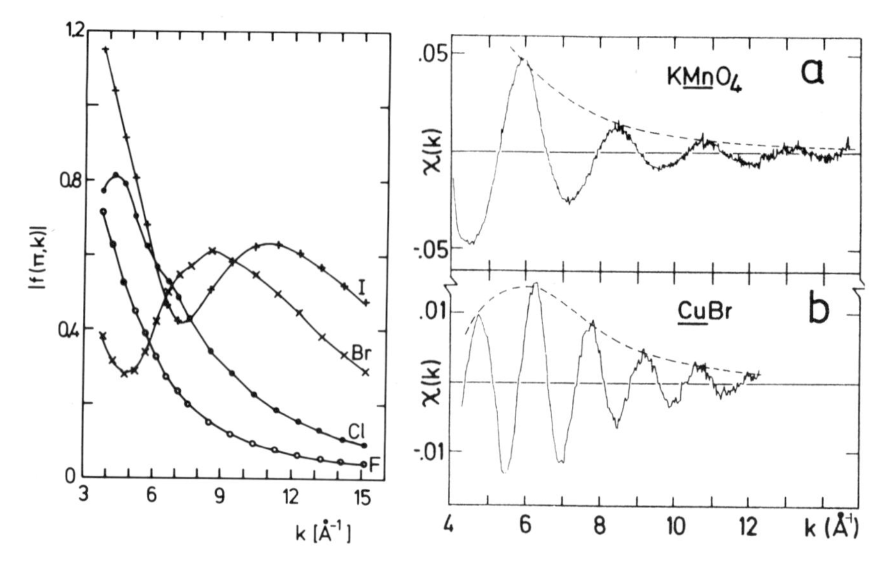

Fig.4 (left side): Scattering amplitudes obtained by Teo and Lee (10)

Fig.5 (right side): EXAFS at the K-edges of a) Mn in $KMnO_4$ (18) and b) Cu in CuBr (13)

For heavier atoms like Br $|f(\pi,k)|$ shows a maximum around $k \sim 8$ Å^{-1}. For even heavier elements the form of the backscattering amplitude becomes more complicated. A rough inspection of the EXAFS spectrum immediately can give some information about the type of the scattering atom. To illustrate this we compare in Fig.5 the EXAFS spectra of $KMnO_4$ and CuBr. The envelope functions show the expected k-dependent behaviour for light (O in $KMnO_4$(a)) and heavy (Br in CuBr(b)) scatterers respectively.

<u>The exponential $\exp(-2\sigma^2 k^2)$</u> describes the damping of the EXAFS spectrum towards high k values. This simple term is valid for the case of crystalline systems at low temperature, where the R_j-values of the discrete coordination shells are only smeared out by a small Gaussian broadening. Generally σ depends on the temperature. It increases with interatomic distances. Consequently the contributions from the second and higher coordination shells in the EXAFS spectrum decay much faster at higher k values (19,20). For large Gaussian broadenings or for asymmetric pair distribution functions, the simple exponential has to be modified by a more complicated term (21). This will be

discussed in section 4. The mean free path λ of the photoelectron wave in the second exponential of eq. (5b) depends on the lifetime of the core hole and the elastic and inelastic scattering of the photoelectrons in the environment of the central atom. Thus the range from which informations are carried in the EXAFS is confined to the immediate neighbourhood of the central atom, typically in the order of 6 Å.

The absolute determination of the coordination number N_j requires the precise knowledge of all other terms influencing the amplitudes of EXAFS. Even if all these values are known, a comparison of theoretical and experimental EXAFS spectra leads to inconsistencies. A reason for this is the assumption in the eq. (5), that all excited electrons coherently contribute to the interference effect. Other processes such as multielectron excitation are neglected in this model (22,23). Using the transferability of amplitude functions $A_j(k)$ (eq.6b) between experimental spectra in the sense as discussed for phases above typical accuracies of 10% can be achieved.

3. DATA EVALUATION

In simple cases with only one scattering shell the data reduction can be directly started from the EXAFS spectra in k-space. A fit with parametrized envelope functions $A_j(k)$ and phases $\alpha_j(k)$ can be applied to the experimental spectra. The increasing number of parameters in systems with more than one scattering shell limits the applicability of this technique. The most widely used method to separate the different contributions is a Fourier transform of $\chi(k)$ to real space (3). Each term in eq.(5) attributed to scatterers in a distance R_j shows up as a peak in the magnitude of the Fourier transform $F(r)$ (see Fig.6a). The magnitudes of these peaks are characterized by $A_j(k)$. The positions are determined by the argument of the sine function. According to the phase shift $\alpha_j(k)$ the positions of the peaks in $|F(r)|$ differ from the interatomic distances R_j by 0.2 to 0.5 Å.

There are several problems connected with the Fourier transform technique: The fact that only a limited k-range can be used (overlap of density of states below $k \sim 3Å^{-1}$ the damping of $\chi(k)$ at higher k-values due to the shape of $|f_j(\pi,k)|$ and the $\exp(-2\sigma^2 k^2)$ term) causes a broadening of the structu-

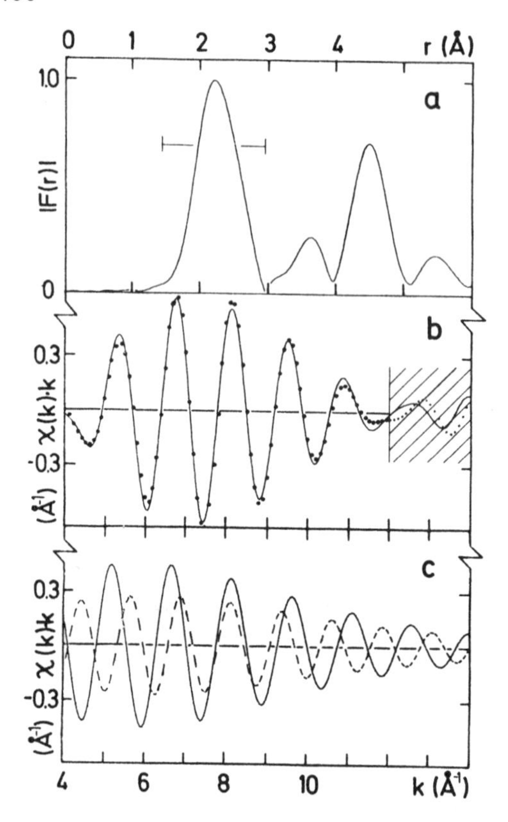

Fig.6:

a) Magnitude of the Fourier transform of EXAFS of Fe at the K-edge (80K)

b) Solid line: back transform of the first peak in a) and dotted line the best parameter fit

c) From parameter fit obtained contribution for the first (solid line) and the second (dashed line) scattering shell

res in $|F(r)|$; cutoff effects lead to lobes which can interfere with the real structures. Nevertheless, the Fourier transform yields a separation of the contributions to $\chi(k)$ with different frequencies and can serve as a starting point for a more refined data analysis.

We demonstrate this with the EXAFS spectrum of Fe (Fig.2). The next neighbours (bcc lattice) of the central Fe atom are two close lying Fe-shells. The first pronounced peak in $|F(r)|$ at 2.1 Å (Fig.6a) is a superposition of contributions from both shells. Fig.6b (solid curve) shows the result of an inverse Fourier transform to k-space over the range indicated by a bar in $|F(r)|$. A parameter fit (Fig.6b dotted line) for two shells using calculated phase shifts and backscattering amplitudes has been included (dots). The single contributions of the first two shells extracted from this fit are displayed in Fig.8c.

Table I: Structural parameters of crystalline Fe

Shell	R_E(Å)	N_E	R_C(Å)	N_C
1	2.44	8.16	2.48	8
2	2.84	6.61	2.87	6

E values: EXAFS data
D values: X-ray diffraction data (24)

The discrepancies between the absolute distances determined by EXAFS (R_E) and by X-ray diffraction (R_D) are larger than the accuracies stated above. They are due to the limited accuracies of the theoretical phases. Note, however, that the differences between these shells are in excellent agreement between both methods. The influence of the phases is cancelled in this case.

A special advantage that EXAFS has to offer is the freedom to select the central atom by measuring EXAFS above the absorption edge of this element. This simplifies the structure analysis from EXAFS for multicomponent systems as compared to X-ray diffraction studies. In the latter case in a system with N different types of scattering atoms the structure factor is caused by a superposition of N(N+1)/2 partial pair distribution functions. For EXAFS the corresponding number reduces to N.

4. EXAFS RESULTS ON METALLIC GLASSES

Before we summarize applications of EXAFS to metallic glasses we briefly discuss the consequences of the degree of disorder on experimental spectra of amorphous Ge. Recent measurements of the EXAFS of a-Ge showed contributions from only the nearest neighbours (15,25). In contrast electron diffraction experiments (26) showed evidence for the first and second coordination shells. Although in apparent discrepancy both results are compatible with calculations based on the continuous random network model (27).

A careful analysis of the EXAFS data, especially of σ^2 of the first two shells of c-Ge and of the first shell of a-Ge showed, that the reason for the suppression of the second shell of a-Ge is the larger sensitivity of EXAFS to static and dynamic disorder compared to other techniques (15). This effect leads to a complete decay of the amplitude of these contributions beyond $k \sim 3$ Å^{-1}. Since the range below $k \sim 3$ Å^{-1} cannot be used for the evaluation of the EXAFS data, the information about the second neighbour shell in a-Ge is permanently lost. This finding clearly demonstrates the high sensitivity of the EXAFS method to disorder.

Up to now we have described the pair distribution by a sum of δ-functions (static part $p_j(r)$) localized at R_j which are convoluted by Gaussians due to thermal broadening. For metallic glasses the static part of the distribution has to be replaced by a continuous distribution, asymmetric in the range of nearest neighbours.
For an arbitrary pair distribution $p_j(r)$ the EXAFS has to be described by (21,28)

$$\chi(k) = \sum_j A_j(k)|V_j(k)| \sin\left\{2k\left(R_j \frac{\arg V_j(k)}{2k}\right) + \alpha_j(k)\right\} \tag{6}$$

where $V_j(k)$ is the Fourier transform of

$$v_j(r) = \frac{p_j(r)}{r^2} \exp(-2r/\lambda) \tag{7}$$

The influence of magnitude and phase of $V_j(k)$ on the analysis of the local order can easily be demonstrated for an exponential distribution

$$v_j(r) = \exp\left(\frac{r-R_j}{a}\right) \quad \text{for } r > R_j$$

$$\text{and} \qquad = 0 \qquad \text{for } r \text{ elsewhere} \tag{8}$$

This form of $v_j(r)$ serves in the following as an approximation of the pair distribution of metallic glassed (2,28). The most severe consequence for the determination of distances is the additional k-dependent term $(1/2k) \arg V_j(k)$. It decreases monotonically from a at k=0 to zero for $k \to \infty$. This means that a different weighting of EXAFS in different k-ranges pretends different interatomic distances, if eq.(5) is

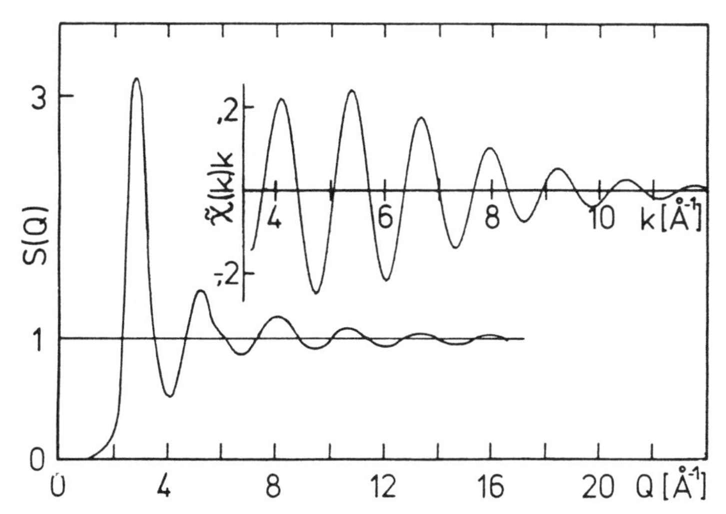

Fig.7: S(Q) structure factor for $Fe_{80}B_{20}$ obtained from X-ray diffraction; χ(k) EXAFS of $Fe_{80}B_{20}$ obtained from a two shell parameter fit to experimental data (Fig.9a) and modified to α(k) = 0

used. The same argument holds in a comparison of EXAFS with X-ray diffraction. This is demonstrated in Fig.7. It shows the diffraction structure factor S(Q) (29) together with the EXAFS (28) of $Fe_{80}B_{20}$. Whereas S(Q) has its main contribution at low Q-values the range where EXAFS is observed and can bei interpreted is confined to $k > 3$ $Å^{-1}$ (note that Q = 2k). Therefore, neglecting the assymmetry of $p_j(r)$ for metallic glasses yields for EXAFS distances which are typically 0.1 Å smaller than those from X-ray scattering.

The experimental results for $Fe_{80}B_{20}$ are shown in Fig.8 and in Table II. The Fourier transform (Fig.8b) is dominated by a maximum at 2 Å, which is a superposition of contributions from the nearest Fe- and B-neighbours. All higher coordination shells are substantially suppressed due to higher disorder through the effect described previously for Ge. Isolating the contributions contained in the first maximum by an inverse Fourier transformation of $|F(r)|$ for the range of 0.8 - 2.75 Å into k-space allows a separation of the partial Fe- and B-contributions with a two shell fit

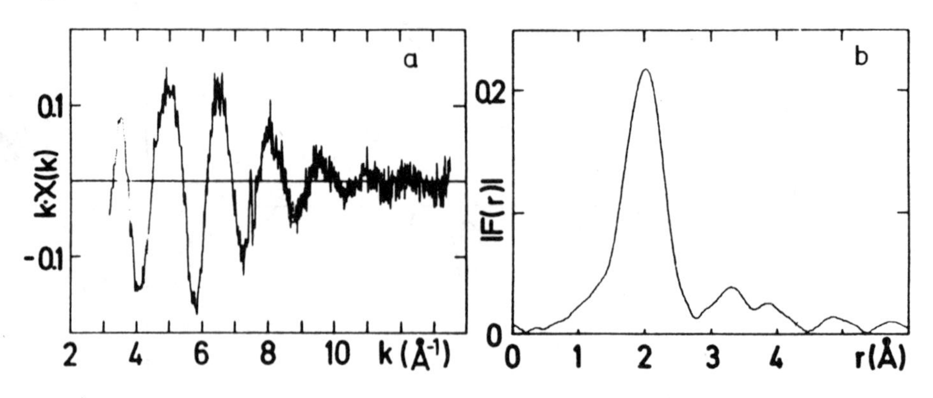

Fig.8: a) EXAFS of $Fe_{80}B_{20}$ above the Fe K-edge;
b) Magnitude of the Fourier transform of a.

Table II: Structural parameters for the nearest neighbours in $Fe_{80}B_{20}$

Shell	R_E	N_E	σ_E^2	R_E^c	N_E^c	R_D	N_D
B	1.96	1.2	0.009	2.06	2.2		
Fe	2.46	4.5	0.010	2.55	8.22	2.57	1.19

E values: after a parameter fit using eq.(5)
c values: after correcting for the asymmetric distribution (eq.6)
D values: X-ray diffraction data by Waseda and Chen (29)

(see Fig.9). The results using eq.(5) and experimental phase shifts and backscattering amplitudes of Fe are compiled in Table II (R_E, N_E and σ_E^2). A comparison with corresponding data from Waseda and Chen (29) obtained from X-ray scattering (R_D and N_D in Table II) shows deviations in R and even more dramatic in N. As has been mentioned above this disagreement is due to the fact that an inappropriate Gaussian pair distribution has been used to analyze the data. Assuming an asymmetry parameter of a = 0.17 Å which is consistent with DRP (dense random packing of hard spheres (30))

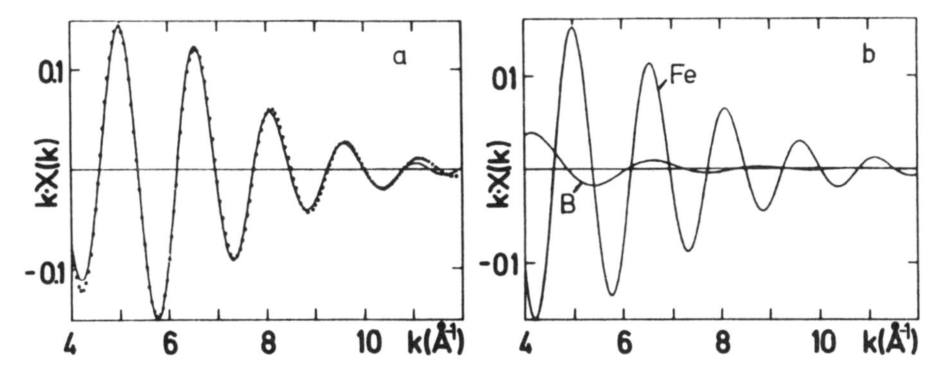

Fig.9: a) Solid line: back transform of the first peak in Fig.8b. Dotted line: Results of a paramter fit with contributions from the next Fe and B neighbours (part b)

distribution function for $Fe_{80}B_{20}$ the data of R_E and N_E can be corrected. The new values (R_E^C, N_E^C in Table II) are much closer to those of Waseda and Chen (29).

The first EXAFS measurements on metallic glasses have been performed by Hayes et al.(31) on $Pd_{80}Ge_{20}$ above the K-edge of the metalloid Ge. Their best fit of the data yields 8.6 nearest neighbours (Pd) around Ge with a distance of 2.49 ± 0.01 Å, slightly smaller than the value for crystalline PdGe (24). No evidence for Ge in the first coordination shell has been found in accordance with the Polk model (32) based on a DRPHS (dense random packing of single sized hard spheres) model. The small spread of ± 0.1 Å in the experimental data for the next neighbours distance, however, has been stated to be in contrast to this and to more refined model calculations (33). The latter model yielded a spread of 0.21 Å. Thus Boudreaux (33) concluded that Hayes et al.(31) used an inappropriate pair correlation. Calculations (28) show that exerimental data in fact can be approximated with eq.(5). As discussed above this yields N- and R-values smaller than those derived from X-ray diffraction. Furthermore one gets σ-values, which are smaller than the true widths of the asymmetric pair distribution. To conclude, differences in interatomic distances derived from EXAFS and X-ray scattering are an essential hint for the existence of a non-Gaussian pair distribution.

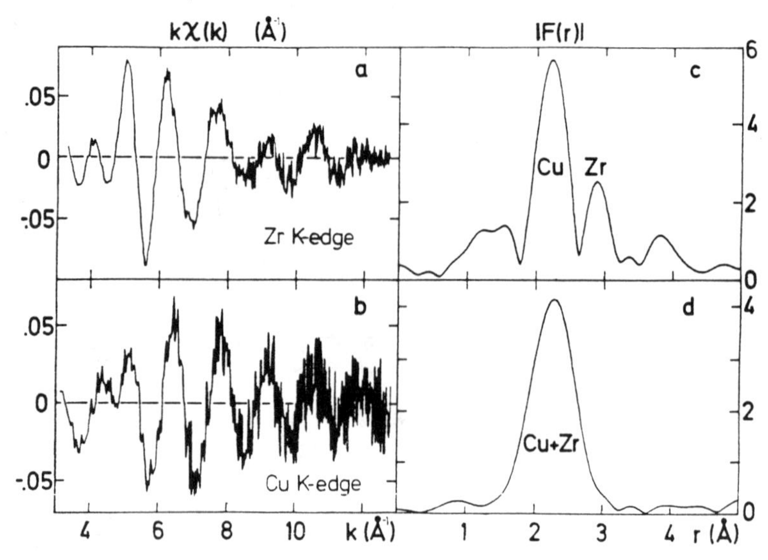

Fig.10: a) and b) EXAFS above the K-edges of Zr and Cu in the metallic glass $Zr_{54}Cu_{46}$; c) and d) the Fourier transform of a) and b) respectively

In contrast to the previous examples the K-edges EXAFS of both constituents have been investigated in $Zr_{54}Cu_{46}$ (2). The EXAFS above the K-edges of Zr and Cu are shown in Fig.10 a) and b) respectively. The parts c) and d) of Fig.10 show $|F(r)|$. Between 1.8 Å and 3.2 Å, two well separated maxima show up at 2.2 Å and 2.8 Å, which can be ascribed to the next Cu- and Zr-neighbours. The broader maximum at 2.3 Å in d) is attributed to the Cu- and Zr-neighbours of the central Cu-atom. The single Cu- and Zr-contributions in the Zr K-edge EXAFS have been isolated by inverse Fourier transform of each peak in $|F(r)|$ separately. Fits with the general form of $\chi(k)$ (eq.(6)) with $a = 0.12$ Å for Cu, but $a = 0$ for Zr and experimental phases and amplitudes using N, R and σ as free parameters yielded values which are in good agreement with X-ray diffraction data (see Table III). An inverse Fourier transform of the first peak in Fig.10 d) yields the Cu- and Zr-contributions simultaneously. This EXAFS has been fitted by eq.(5) using the R values of Zr-Cu determined above. The significantly larger Cu-Cu distance found in X-ray diffraction again points to an asymmetric distribution. To summarize we find coordination numbers in agreement with the stoechiometric composition of the alloy, but we

Table III: Structural parameters of $Zr_{54}Cu_{56}$

Central atom	Scatterer	R_E	N_E	R_D	N_D
Zr	Cu	2.74 ± 0.02^c	4.6 ± 1^c	2.75	5.0
	Zr	3.14 ± 0.02	5.1 ± 1	3.15	5.0
Cu	Cu	2.47 ± 0.03		2.53	5.8
	Zr	2.74 ± 0.03^c		2.75	5.6

E,c: see Table IV; D: X-diffraction data (34)

have different pair distribution characters, i.e. symmetric for Zr-Zr and asymmetric for Cu-Cu and Cu-Zr.

As a multicomponent system $Fe_{40}Ni_{40}B_{20}$ has been studied (2). Similar measurements have been performed by Wong et al.(35) on $Fe_{40}Ni_{40}B_{20-x}P_x$ (with x varying from 0 to 20 at %). Fig. 11 shows the data for $Fe_{40}Ni_{40}B_{20}$ (2) for both the EXAFS regions above the Fe and Ni K-edge. Similarily as for the $F_{80}B_{20}$ the $|F(r)|$ mainly contains one pronounced peak with a maximum at 2.08 Å, representing the nearest neighbour shell for the Fe- or Ni-atoms. We first consider the Fe K-edge spectrum. A comparison with the $F_{80}B_{20}$ data showed a good coincidence of the frequencies of the fine structure and hence the positions of the maxima in $|F(r)|$. The deviation is less than 0.02 Å. This leads to the statement, that replacement of 50% of the Fe-atoms by Ni-atoms does not alter the Fe vicinity significantly.

The amplitudes and phases of Fe and Ni are nearly identical and the Fe-Ni distances are comparable. This has the consequence that we cannot discriminate between Fe and Ni scatterers within our limits of accuracy. In the same way as previously done for $Fe_{80}B_{20}$ we perform an inverse transform of the first maximum back to k-space and by a parameter fit with eq.(5) using calculated backscattering amplitudes and phases we determine the R_E and N_E values for (Fe,Ni) and B neighbours (Table IV).

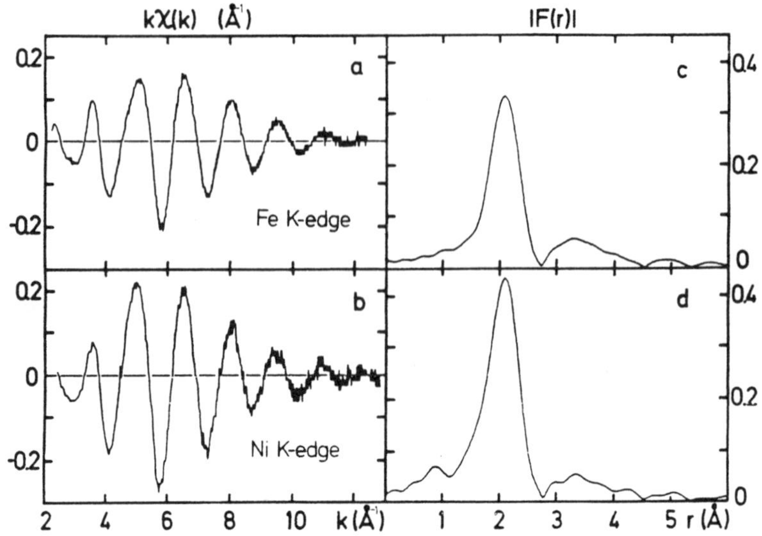

Fig.11: a) and b) EXAFS contributions of $Fe_{40}Ni_{40}B_{20}$ above the K-edges of Fe and Ni; c) and d) the structural functions $|F(r)|$ belonging to a) and b)

The Fe-(Fe,Ni) distances R_E and the coordination numbers N_E are in good agreement with the corresponding Fe-Fe values for $Fe_{80}B_{20}$. This means that the pair correlation functions are the same in both cases. With an asymmetry parameter of a = 0.17 Å found for $Fe_{80}B_{20}$ an R_E^c = 2.57 Å is obtained. Analyzing the Ni EXAFS with eq.(5) the Ni-(Fe,Ni) distance R_E turns out to be the same as the Fe-(Fe,Ni) value. However, a significantly different coordination number N_E is observed, resulting in N_E (Ni-EXAFS)/N_E (Fe-EXAFS) = 1.4.

There are two possible explanations for the behaviour: i) The number of (Fe,Ni) in the next neighbour shell of Ni is really higher by a factor of 1.4 as compared to the Fe-vicinity. ii) The coordination numbers are the same, but the asymmetry parameters for the distribution functions are different. The assumed asymmetry factor a = 0.17 Å for the Fe vicinity already enlarges N_E = 4.9 (from eq.(5)) to N_E^c = 12 (using eq.(6)). The same a-value applied to the Ni vicinity would yield a $N_E^c \sim 17$ atoms, certainly an un-

Table IV: Structural parameter of $Fe_{40}Ni_{40}B_{20}$

Central Atom	Scatterer	R_E (Å)	N_E	R_E^c	N_E^c
Fe	B	1.97			
	(Fe,Ni)	2.48	4.9	2.57	12
Ni	B	1.95			
	(Fe,Ni)	2.45	6.9	2.51	12

E values: after a parameter fit using eq.(5)

c values: after correcting for the asymmetric distribution (eq.6)

realistic number. Therefore, the solution ii) seems to be more appropriate. Assuming $N_E^c = 12$ for the Ni vicinity as well leads to an asymmetry factor $a = 0.11$ Å.

Independent of a more detailed analysis the data show distinct differences in the short range order around the Fe- and Ni-atoms in $Fe_{40}Ni_{40}B_{20}$. This necessitates the inclusion of chemical bonding effects in the description of the local structure.

Similar conclusions can also be drawn from the measurements on $Fe_{40}Ni_{40}B_{20-x}P_x$ by Wong et al.(35), who derived the fractional change $(\Delta\sigma^2/\sigma^2)$ in disorder around the Fe and Ni atoms as a function of x and of the annealing temperature from the peak heights in $|F(r)|$. For $x = 0$, i.e. $Fe_{40}Ni_{40}B_{20}$ the disorder around the Fe atom is larger than that around the Ni atoms; Wong et al.(35) stated that the boron preferentially coordinates with Ni. On the other hand for $Fe_{40}Ni_{40}P_{20}$ phosphorous coordinates preferentially with Fe.

ACKNOWLEDGEMENTS

This work was supported by the Bundesministerium für Forschung und Technologie and the Deutsches Elektronen-Synchrotron DESY. The support by the members of the synchrotron radiation group at DESY is gratefully acknowledged. We thank Prof. Güntherodt (Basel) for providing samples used in our work. Finally thanks are due to M. Höfelmeyer for a careful typing of the manuscript.

REFERENCES

1. R. de L. Kronig, Z.Phys. 70, 317 (1931); 75, 191 and 468 (1932)
2. A. Werner, Dr.rer.nat. thesis, Universität Kiel 1979
3. F.W. Lytle, in Adv. X-ray Analysis 9, 398 (1966) (Plenum Press)
   D.E. Sayers, E.A. Stern and F.W. Lytle, Phys.Rev. Letters 27, 1204 (1971)
   F.W. Lytle, D.E. Sayers and E.A. Stern, Phys.Rev. B 11, 4825 (1975)
   E.A. Stern, D.E. Sayers and F.W. Lytle, Phys.Rev. B 11, 4836 (1975)
4. E.A. Stern, Phys.Rev.B 10, 3027 (1974)
5. C.A. Ashley and S. Doniach, Phys.Rev.B 11, 1279 (1975)
6. P.A. Lee and J.P. Pendry, Phys.Rev. 11, 2795 (1975)
7. P.A. Lee and G. Beni, Phys.Rev.B 15, 2862 (1977)
8. B.K. Teo, P.A. Lee, A.L. Simons and P. Eisenberger, J.Am.Chem.Soc. 99, 3854 (1977)
9. P.A. Lee, B.K. Teo and A.L. Simons, J.Am.Chem.Soc. 99, 3856 (1977)
10. B.K. Teo and P.A. Lee, J.Am.Chem.Soc. 101, 2815 (1979)
11. C. Kunz, ed., Synchrotron Radiation, Topics in Current Physics, Vol.10 (Springer Verlag Berlin, Heidelberg, New York, 1979)
12. P. Rabe, G. Tolkiehn and A. Werner, Nucl.Instr. Meth. (to be published, preprint DESY SR-79/23)
13. G. Martens, P. Rabe, N. Schwentner and A. Werner, Phys.Rev.B 17, 1481 (1978)

14. P.H. Citrin, P. Eisenberger and B.M. Kincaid, Phys.Rev.Letters 36, 1346 (1976)
15. P. Rabe, G. Tolkiehn and A. Werner, J.Phys.C 12 L545 (1979)
16. R. Ingalls, G.A. Garcia and E.A. Stern, Phys.Rev. Letters 40, 334 (1978)
17. G. Martens, P. Rabe, N. Schwentner and A. Werner, Phys.Rev.Letters 39, 1411 (1977)
18. P. Rabe, G. Tolkiehn and A. Werner, J.Phys.C 12, 1173 (1979)
19. G. Beni and P.M. Platzman, Phys.Rev.B 14, 1514 (1976)
20. W. Böhmer and P. Rabe, J.Phys.C 12, 2465 (1979)
21. P. Eisenberger and G.S. Brown, Sol.State Comm. 29, 481 (1979)
22. J.J. Rehr, E.A. Stern, R.L. Martin and E.R. Davidson, Phys.Rev.B 17, 560 (1978)
23. E.A. Stern, S.M. Heald and B. Bunker, Phys.Rev. Letters 42, 1372 (1979)
24. R.W. Wyckhoff, Crystal Structures, Vol.I (Interscience Publishers, 1963)
25. Unpublished results of S. Hunter, see T.M. Hayes, J.Non-Cryst.Sol. 31, 57 (1979)
26. J.F. Grazcyk and P. Chaudhari, phys.stat.sol.(b) 58, 163 (1973)
27. P. Chaudhari and D. Trunbull, Science 199, 11 (1978
See also: Electronic Processes in Non-Crystalline Materials, 2nd edition by N.F. Mott and E.A. Davis (Clarendon Press, Oxford 1979) chapter 7
28. R. Haensel, P. Rabe, G. Tolkiehn and A. Werner (to be published)

29. Y. Waseda and H.S. Chen, phys.stat.sol.(a) 49, 387 (1978)

30. G.S. Cargill III, Solid State Physics Vol.30, 227 (1975) (Academic Press)

31. T.M. Hayes, J.W. Allen, J. Tauc, B.C. Giessen and J.J. Hauser, Phys.Rev.Letters 40, 1282 (1978)

32. D. Polk, Acta Metall. 20, 485 (1972)

33. D.S. Boudreaux, Phys.Rev. 18, 4039 (1978)

34. H.S. Chen and Y. Waseda, phys.stat.sol.(a) 51, 593 (1979)

35. J. Wong, F.W. Lytle, R.B. Greegor, H.H. Liebermann J.L. Walter and F.E. Luborsky, in Rapidly Quenched Metals III, ed. B. Cantor (The Metals Society, London 1978) Vol.II, p.345

POSITRON ANNIHILATION IN AMORPHOUS MATERIALS AND LIQUID METALS

W. Triftshäuser

Hochschule der Bundeswehr München
8014 Neubiberg, Federal Republic of Germany

ABSTRACT. The basic theory for the positron annihilation method is presented and the various experimental techniques - lifetime, angular correlation and Doppler broadening measurements - are reviewed. With these techniques information is obtained about electron states in perfect crystals and about the electronic configuration of defects and their agglomerates in real metals and of liquid and amorphous metals. The present status of positron annihilation experiments in liquid metals and alloys is assessed. Recently, the positron annihilation method has been applied to investigate amorphous metals and glassy alloys. New results in this field have been obtained and will be discussed.

1. INTRODUCTION

The positron was discovered by Anderson in 1932 in his studies of cosmic radiation /1/. Some years later it was confirmed that the positron is the antiparticle to the electron and that the positron was the particle predicted in Dirac's relativistic theory of electrons /2, 3/. As antiparticles to the electrons the positrons annihilate with electrons. When energetic positrons enter a metal they rapidly lose almost all their energy by collisions with the electrons. Theoretical calculations of Lee-Whiting /4/ show that the time required for the positron to reach thermal energy is of the order of 10^{-12} sec, a time that is short compared with the lifetime of a positron in a metal /5, 6/. It has been experimentally verified that a positron is in thermal equilibrium with the lattice in a metal down to about 20 K /5/. The annihilation is announced by the emergence of energetic photons whose energies, momenta and time of emission can be measured with high precision with modern detector systems. The utility of positron annihilation studies of metals relies on the fact that these characteristics of the annihilation process, which in principle involve the sophisticated considerations of quantum electrodynamics, nevertheless depend almost entirely on the initial state of the positron-many-electron system. Positron studies of the electronic structure have provided a valuable confirmation of many of the established concepts of solid state physics and, in many cases, information not obtainable from other techniques. The annihilation process will supply an energy of $E = 2m_oc^2$, the total rest mass energy of the annihilating pair. For slowly moving pairs, conservation of momentum demands that at least two other bodies (particles or quanta) be involved in the process /6, 7/. From the requirements of conservation of angular momentum and parity it follows that the electron-positron pair must decay under emission of three photons, if the spins of the two particles are parallel, and under emission of two phonons, if the spins are antiparallel. The probability of two-photon annihilation is considerably (372 times) greater than that for three photons, and therefore for almost all investigations this decay mode is used in positron annihilation experiments. A schematic representation of the positron annihilation process is shown in fig. 1. A relatively high energy ($\lesssim 0.5$ MeV) positron (e^+) is injected into a condensed medium from a radioactive positron source (e.g. ^{22}Na) and becomes rapidly thermalized by a succession of electron-hole excitations and phonon interactions resulting in a Bloch state in the lattice. Within a few picoseconds after the positron emission the ^{22}Ne nucleus emits a 1.28 MeV photon which serves as a birth signal. After a time δt the positron will annihilate with an electron in the material yielding two annihilation γ-rays. These γ-rays, in the laboratory reference frame, each have an energy equal to the rest-mass of an electron or positron (m_oc^2 = 511 keV) plus or minus some energy increment δE, and travel in opposite direction plus or minus some angular increment $\delta\theta$.

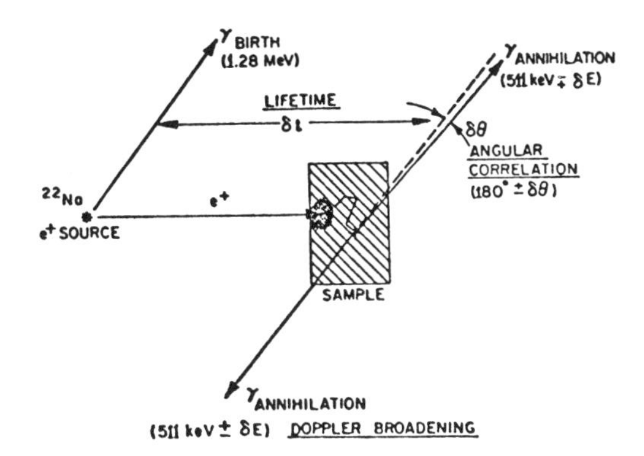

Fig. 1. The positron annihilation process. Positrons from a radioactive isotope (^{22}Na) annihilate in the sample material. The positron lifetime is determined from the time interval between the birth (1.28 MeV γ-ray) and the annihilation of the positron (0.511 MeV annihilation quanta). The momentum of the electron-positron pair is measured by the angular deviation $\delta\theta$ (angular correlation) or by the energy shift δE (Doppler broadening) /8/.

2. THEORY

The attempt to determine experimentally the momentum-distribution functions of metals and alloys from the angular correlation of the annihilating photons dominated the application of positron annihilation for a long time /9/. Fairly early it was recognized that the annihilation of a positron with the electrons in a metal is a many-body effect /10/. Many-body theory shows that correlation effects are indeed essential in explaining positron lifetimes and positron thermalization, but that the momentum dependence of these effects is relatively weak. Consequently, independent-particle theory forms a very good first approximation, to which many-body corrections can be applied at a later stage. Ferrell /11/ was able to account for the striking observation that the short lifetime of positrons in metals required electron densities at the positron exceeding the average density of conduction electrons by about one order of magnitude, while the momentum distribution was changed very little from that of a free electron model. In order to interpret the results of lifetime, angular correlation or Doppler broadening experiments, we require an expression for the probability that a positron in an arbitrary system of electrons and external fields will annihilate under emission of two photons having total momentum $\underline{p}$. The most general treatment of this problem was supplied by Chang Lee /12/ and Ferrell /11/. This is in particular a many-body problem and cannot be solved exactly. However, in the independent-particle model the probability $\Gamma(\underline{p})$ can be given by the following expression /13/

$$\Gamma(\underline{p}) = \frac{\pi r_o^2 c}{(2\pi)^3} \; \varrho(\underline{p}) \, d\underline{p} \tag{1}$$

with

$$\varrho(\underline{p}) = \sum_k \left| \int \psi_k(\underline{r}) \; \psi_+(\underline{r}) \, e^{-i\underline{p}\cdot\underline{r}/\hbar} \, d\underline{r} \right|^2 \tag{2}$$

where $r_o = \frac{e^2}{m_o c^2}$ is the classical electron radius, ψ_+ is the positron wave function and ψ_k is the one-electron wave function. The summation extends over all occupied electron and positron states. At room temperature the thermalized positron is near the bottom of the positron conduction band (the positron lifetimes and normally used source intensities are such that the number of positrons present in the sample at any given time is of the order of one). In a metal the positrons move in an array of repulsive ion potentials, and since Coulomb repulsion will tend to exclude the positrons from regions well inside the ion cores, positrons will annihilate mainly with conduction electrons. However, there exists a certain probability that the positron annihilates with the outer bound electrons of the ion cores, and the relative amount of these annihilations depends on the radial distribution of the core-electron-wave functions. Because the positron is in its lowest energy state its contribution to the total momentum of the electron-positron system is negligible and therefore $\varrho(\underline{p})$ of eq. (2) gives in very good approximation the momentum distribution of the electrons in a metal. More detailed theory is covered in a recent review by Mijnarends /14/. The total annihilation probability or the annihilation rate λ is obtained from eq. (1) by integrating over all photon-pair momenta

$$\lambda = \pi r_o^2 c n_e \qquad (3)$$

where n_e is the electron density at the site of the positron. By measuring the annihilation rate λ, inverse of which is the mean lifetime τ, one directly obtains the electron density n_e encountered by the positron. Thus a positron can serve as a test particle for the electron density of the medium.

3. EXPERIMENTAL TECHNIQUES

3.1 Lifetime measurements

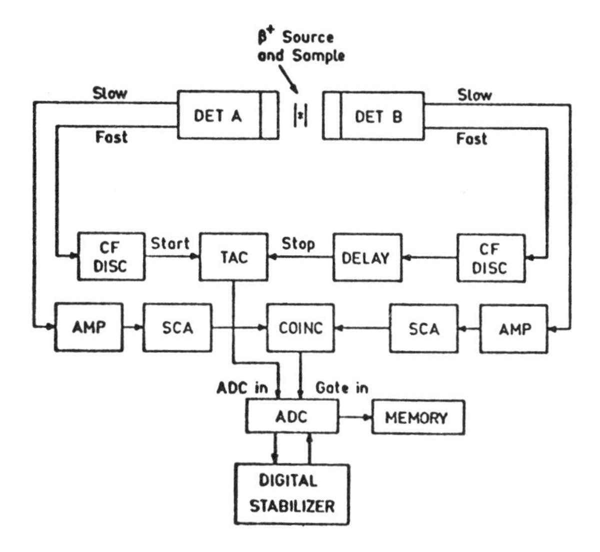

Fig. 2. Schematic circuit of a fast-slow coincidence system for positron lifetime measurements. The anode pulses of the photomultipliers are fed through fast constant-fraction discriminators into a time-to-amplitude (TAC) converter. The output of the TAC is connected to a stabilized analog-digital converter (ADC), which is gated by the slow circuit.

In positron lifetime experiments the favorable fact is used that isotopes like ^{22}Na (as indicated in fig. 1) and ^{44}Sc provide a zero of time signal of the entry of a positron into the sample by emitting a simultaneous (< 10 psec) γ-ray. The lifetime spectrometer is shown schematically in fig. 2. The positron source is sandwiched by two pieces of the sample material. The detectors consist of fast plastic scintillators (cone shaped) coupled to fast photo-multiplier tubes. The energy windows in the slow channel are adjusted in the way that the detectors A and B register the birth and annihilation γ-rays of individual positrons, respectively. The fast anode signals are fed through constant-fraction-timing discriminators into a time-to-amplitude converter (TAC). The output pulses of the TAC are transferred to an analog-digital-converter, which is digitally stabilized, and are finally stored in the memory. A prompt time-resolution of about 165 ps can be obtained under optimal conditions /15/. Fig. 3 shows a lifetime spectrum for sodium together with the resolution curve measured with ^{60}Co. The evaluation of lifetimes of the order

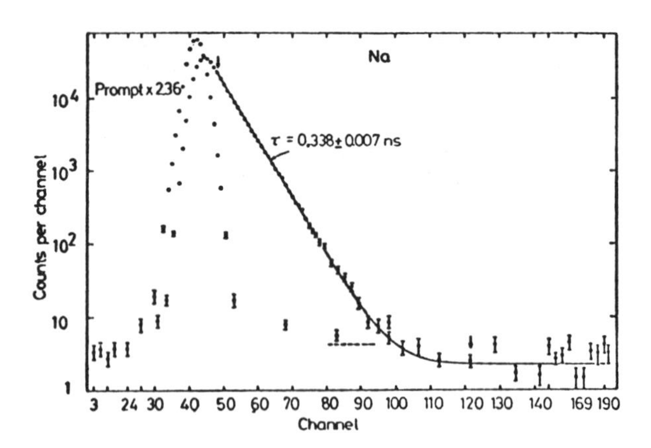

Fig. 3. Typical lifetime spectrum for positrons in sodium /16/. The prompt curve represents the time resolution of the spectrometer.

of a few hundred picoseconds is very close to the limit of present measurement techniques and the ultimate success of such experiments depends greatly on source and sample preparation (surface effects), electronic and temperature stability and realistic statistical and mathematical analysis, especially if more than one lifetime has to be extracted.

3.2 Angular correlation measurements

A typical experimental set-up for the measurement of the angular correlation of annihilation photons is shown in fig. 4. A radioactive source is placed close to the sample but shielded by lead from a direct view by the two detectors. Additional lead collimators in front of the scintillation counters define the instrumental angular resolution. The coincidence counting rate from the two detectors is measured as a function of the displacement of one detector. To achieve adequate counting rates the detectors and slits are made in the x-direc-

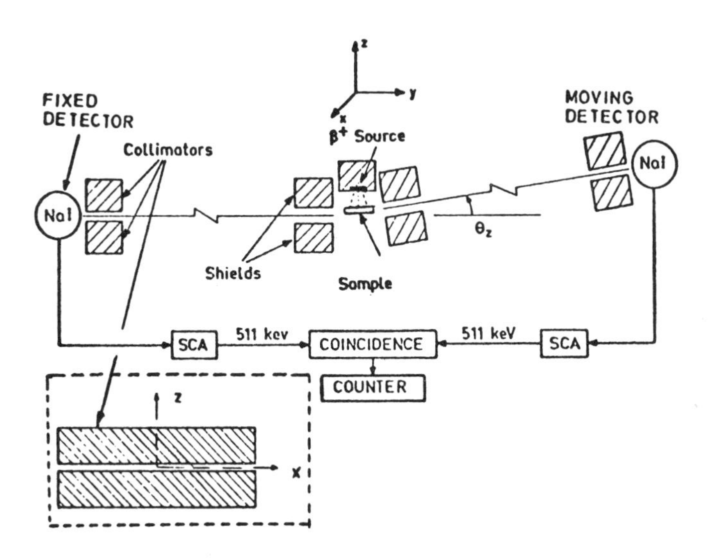

Fig. 4. Schematic arrangement of an angular correlation apparatus with long-slit geometry. The coincidence count rate is measured as function of θ_Z.

tion as long as possible. In order to minimize errors resulting from source decay and drift of the electronics an angular correlation curve is usually obtained from the results of many cycles of the movable detector through the necessary angular region. The arrangement shown in fig. 4 is an apparatus with a long slit geometry, a compromise between good angular resolution (0.25 mrad) and reasonable counting rates (activity of about 1 Curie for ^{64}Cu). With this geometry only one component, p_z, of the transverse momentum of the photon pair can be determined, whereas p_x and p_y remain undefined.

$$N(p_z) = \int_{-\infty}^{+\infty} \int_{-\infty}^{+\infty} \rho(\underline{p})\, d p_x\, d p_y \qquad (4)$$

A positron angular correlation curve for an aluminum single crystal is shown in fig. 5. Indicated is also a difference curve

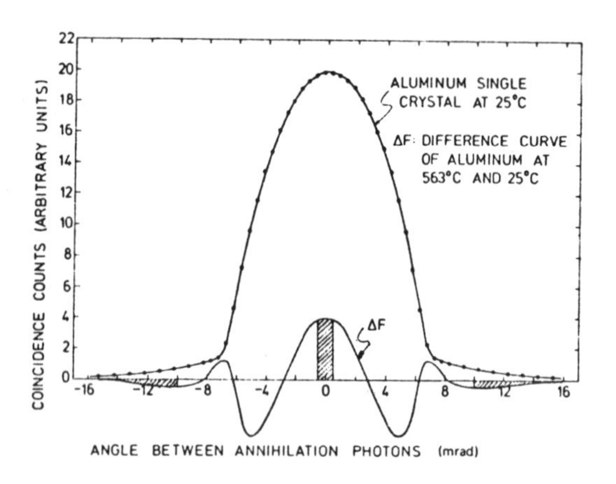

Fig. 5. Positron angular-correlation curve of aluminum single crystal at 25°C and difference curve of between the angular-correlation curve at 563 and 25°C /17/. The vertical scale for F is expanded by a factor of two. Indication (cross hatched area) is the region of $F^{(peak)}$ and $F^{(core)}$ for calculating R of eq. 6.

obtained from measurements at high and low temperature /17/. Reducing the resolution function in x-direction to the same width as along the z-direction results in a point-detector geometry. Fig. 6 shows the result obtained for aluminum /18/.

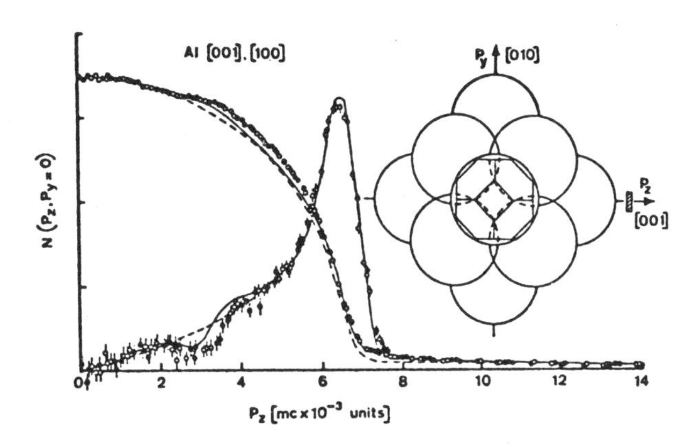

Fig. 6. Angular correlation in aluminum with p_x, p_z along the directions [100, 001] and p_y = 0 together with its first derivative /18/. Full curve is an orthogonal-plane-wave prediction, dashed curve corresponds to a simple sphere model shown in insert. The shaded rectangle indicates the resolution along p_y and p_z.

3.3 Doppler broadening measurements

The finite momentum of the electron-positron pair causes a Doppler shift on the energy of the annihilation radiation. The energy of the two annihilation quanta is

$$E_{1,2} = m_o c^2 \pm \frac{p_y c}{2} \tag{5}$$

with p_y the momentum component of the electron-positron pair in the direction of the photon emission. For an electron of 10 eV and a thermalized positron this results in an energy shift of about 1.6 keV. A high performance intrinsic-Ge detector has an energy resolution of 1.2 keV at 514 keV corresponding to an equivalent angular resolution of about 4 mrad. A typical arrangement is shown in fig. 7. Because the intrin-

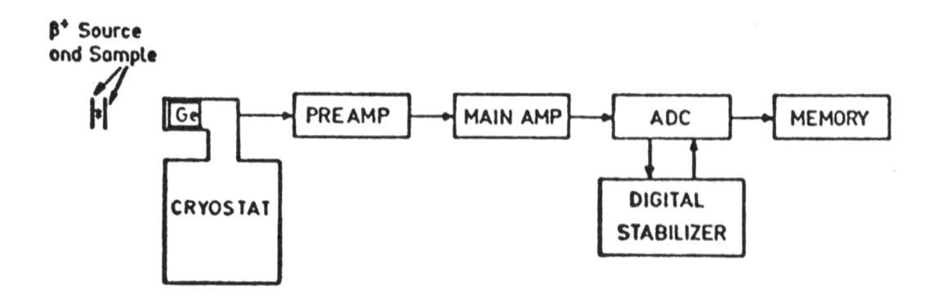

Fig. 7. Schematic set-up for Doppler-broadening measurements of the annihilation line.

sic energy resolution is almost of the same magnitude as the energy shift, sophisticated stabilization techniques are required if subtle changes in the width of the energy line are to be measured. However, the necessity for only one detector results in a rapid accumulation of data even with weak sources and allows for the investigations of samples in difficult environmental conditions (e.g. cryostat). Because of the similar source-sample geometry it is advantageous to combine both Doppler broadening and lifetime experiments for more precise information. Fig. 8 shows the effect of Doppler broadening on the energy distribution of the annihilation line in annealed and deformed copper. The 514-keV peak of ^{85}Sr represents the response of the detector system for monoenergetic radiation. The line shape of the curves is different for various materials and sensitive to the positron trapping effect by lattice defects (e.g. vacancies and/or dislocations in deformed copper). A defect specific parameter R can be deduced as /17/

$$R = \frac{\Delta F^{(peak)}}{\Delta F^{(core)}} = \frac{F_V^{(peak)} - F_f^{(peak)}}{F_V^{(core)} - F_f^{(core)}} \qquad (6)$$

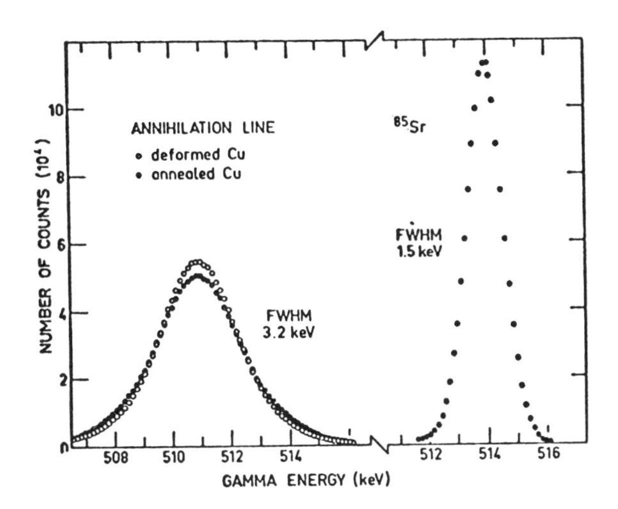

Fig. 8. Doppler-broadened spectrum of annealed and deformed copper. The ^{85}Sr line represents the energy resolution of the system. All curves have been normalized to equal areas /19/.

where F_v and F_f are the characteristic values for positrons annihilating from a defect site and in the defect free region, respectively. This applies for Doppler broadening as well as for angular correlation spectra (indicated by the cross-hatched areas in fig. 5).

4. RESULTS AND DISCUSSION

4.1 Liquid metals and alloys

Investigations of liquid metals and studies of the effect of melting involved mainly angular correlation /20-22/ and life-time measurements /16, 22-26/. Before it was recognized that positrons can be trapped by temperature induced defects below melting, the observed changes in the angular distribution of the annihilation photons were to a large extend attributed to electronic structure changes. The changes at the melting point and in the liquid phase were considered to be due to the effect of disorder on the conduction electron states /27/. Significant changes in angular correlation experiments between the solid and the liquid phase were found in a variety of metals including tin, gallium, mercury and bismuth. The changes which are very pronounced in mercury (fig. 9) are

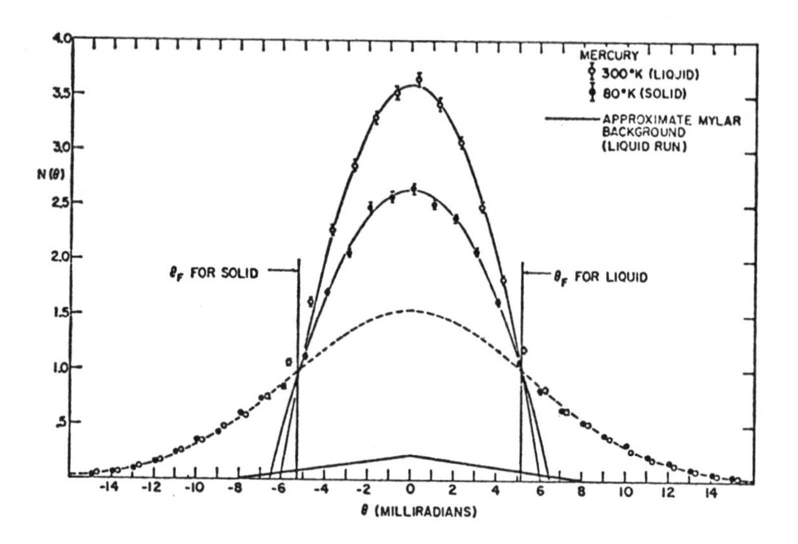

Fig. 9. Positron angular-correlation curve for solid and liquid mercury /20/. θ_F is the angle due to the Fermi momentum.

qualitatively similar to those arising from vacancy trapping in solid metals. This was supported by accurate measurements of the coincidence counting rate at the peak of the angular correlation curve as a function of temperature in solid and liquid aluminum, indium and lead /17/. At the melting point the coincidence count rate rises abruptly. The transition from the solid to the liquid is also accompanied by a volume increase of 6%, 3.4% and 2.7% for Al, Pb and In, respectively /27/. Because of this volume increase and the nonregular arrangements of the atoms in the liquid, the structure of the trapping sites for the positrons in the liquid is different from the one in the solid. The Coulomb repulsion between the positron and the positively charged ions in the liquid causes an increase in the effective volume of the trapping site and this leads to a stronger localization of the positron wave function, and hence to a change in the annihilation characteristics. The defect specific R-parameter (eq. 6) determined for single vacancies and trapping centers in the liquid metal supports the vacancy-like regions of the disordered liquids /17/. Similar results are obtained from positron lifetime measurements. Lifetime studies in liquid metals and particularly comparisons of solid and liquid spectra are made difficult by source-sample interface problems. Wetting or meniscus effects and the high reactivity of most liquid metals demand special care in source-sample assembly design /29/. Large changes in positron lifetimes on melting are observed in gallium /24/ (fig. 10). In sodium only a small increase of the lifetime takes place, which is typical for all alkali metals. Large melting effects occur only in metals where vacancy trapping might be expected but is either missing or very weak in the solid phase. When large changes occur they are essentially similar to those exhibited by vacancy trapping effects. The smearing and narrowing of the conduction electron part of the angular distribution is almost the same as observed in vacancy trapping and may be similarly interpreted in terms of positron localization and local density effects. The nature of the effective trapping sites in the liquid phase is of considerable interest. In spite of a possible size distribution of potential positron traps and their transient nature because of density fluctuations in the liquid it is most likely that a positron creates its own hole and trapping site. This is supported by a temperature independent R-parameter in liquid aluminum /17/ and by lifetime studies in liquid indium, where only one lifetime is observed /29/. Because of this "hole-creating" effect in the disordered liquid phase and the therefore trapped positron state, the measured angular distribution of the annihilation photons is not representative for the electron momentum distribution in the liquid itself, but rather for the one inside the trapping sites. Therefore particular care should be taken for the interpretation of angular correlation data between solid and liquid metals, especially if the observed changes are attributed to electronic structure effects due to the disordered phase in liquids. Most likely the desired information about the disordered and amorphous structure of

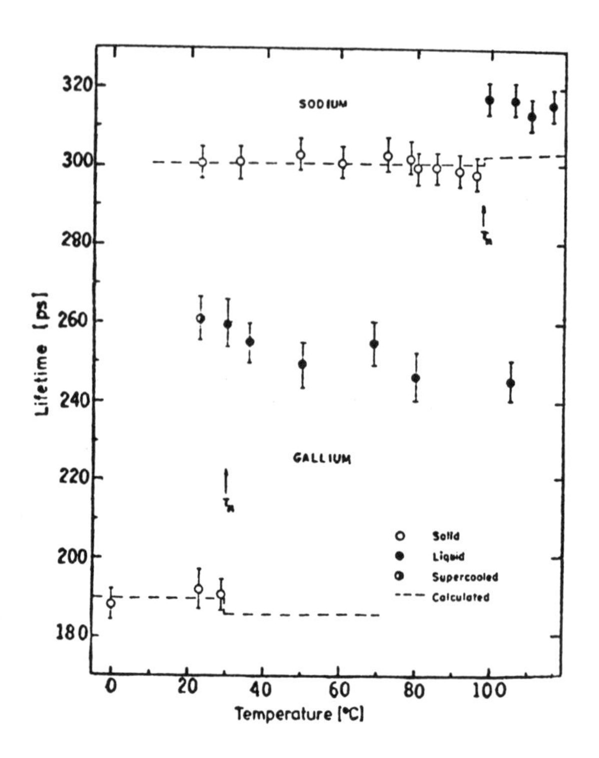

Fig. 10. Positron lifetimes in gallium and sodium as function of temperature. T_M is the melting temperature. The dotted lines indicate the change in lifetime on melting as predicted for the density change by the free electron gas model /24/.

liquid metals and liquid alloys will not be possible to obtain. This applies for positron studies in liquid alloys as well. Investigations in a bismuth-mercury alloy /21/ provide an angular distribution of a form very close to that of a composition weighted mixture of the pure metal curves. Thus no particular positron affinity for one or the other of the two components was found. Later studies of liquid Cu-Bi and Cu-Sn alloys (fig. 11) are interpreted as large positron affinities for bismuth and tin attributable to the atomic arrangement in these alloys /30/. Recent studies of alloys of germanium with

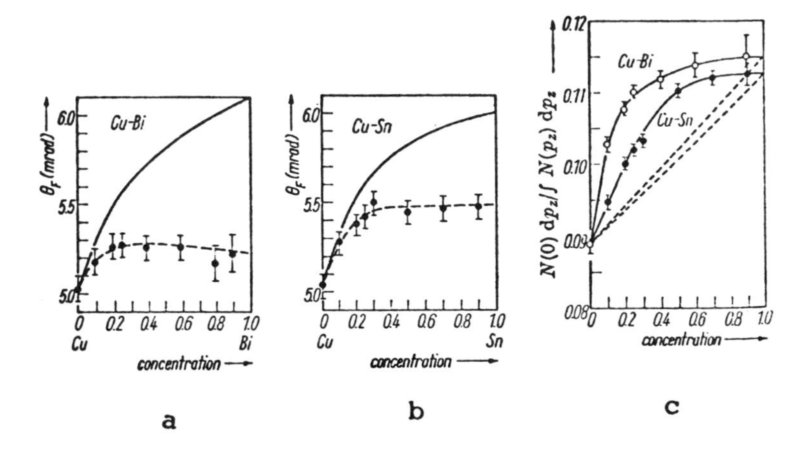

Fig. 11. The concentration dependence of the Fermi momentum in Cu-Bi alloys (a) and Cu-Sn alloys (b) in the liquid state /30/. (c) The normalized coincidence count rate at the peak of angular correlation curve as a function of Bi and Sn concentration, respectively /30/. Positrons are localized to Bi or Sn in these alloys. Dashed straight lines are the linear extrapolation between pure elements.

noble and transition metals show complex trends in angular distribution more suggestive of electronic structure influence than positron affinity effects /31/.

4.2 Amorphous and glassy alloys

Metallic glasses have been studied recently by positron annihilation techniques /32-34/. For the alloy $Pd_{77.5}Cu_6Si_{16.5}$ a slight decrease of the positron lifetime was found on crystallization /35/. The angular correlation curves of Ni-P and Co-P alloys in the glassy and crystalline state do not differ much /32/. The effect of cold rolling and electron irradiation of glassy alloys was also studied and it was concluded that no vacancy-like defects are introduced additionally /36/. All these positron annihilation experiments were carried out at room temperature or above. However, first Doppler-broadening measurements as a function of temperature down to 5 K on various alloys in the amorphous and crystalline state, respectively, show differences for the two phases /37/. The temperature dependence of the line shape parameter I_V (fraction of counts in the annihilation peak region from −1.5 mrad to +1.5 mrad) suggests the existence of vacancy-like defects. A similar behavior has been observed for simple metals containing vacancy and interstitial loops /38, 39/. More recent and extended studies of various glassy alloys by positron lifetime and Doppler-broadening measurements /40, 41/ confirm the trapping of positrons at large empty Bernal holes /42/. Fig. 12 shows Doppler-broadening results

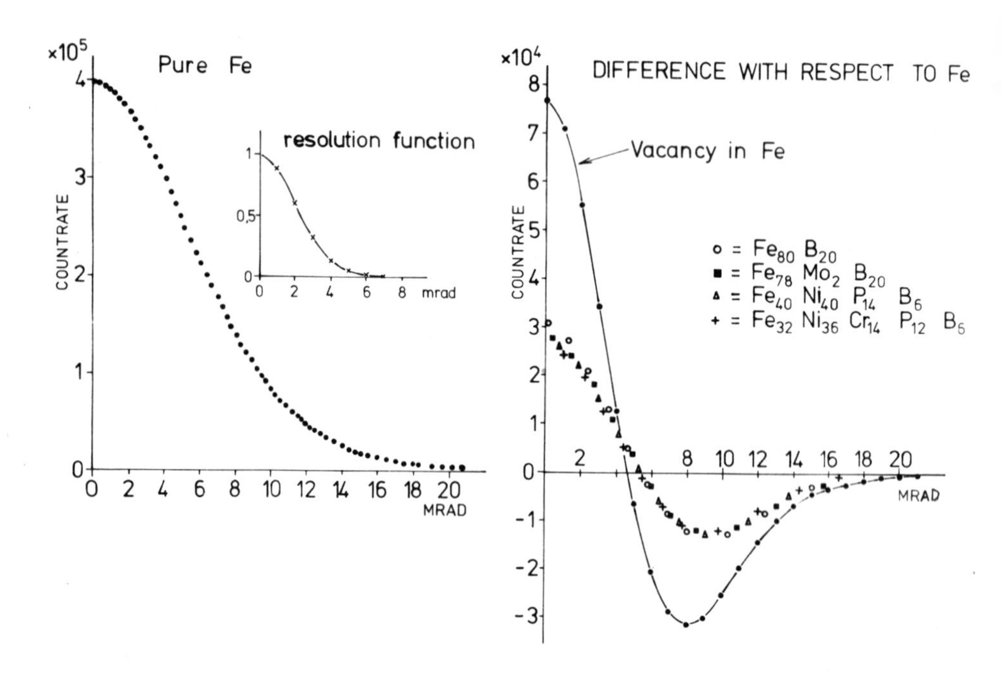

Fig. 12. Doppler-broadening curve for pure iron (left) with resolution function of the system. Comparison of difference curves for iron containing vacancies and various alloys of metallic glasses (right).

for iron based glassy alloys compared to pure iron with and without vacancies. Before taking the difference curve all annihilation lines were normalized to equal areas. The largest effect is observed for vacancies in iron. However, a vacancy-like effect is observed as well in the glassy alloys. This trapping effect is the same for all the alloys investigated, independent of the individual composition. This indicates that the local electronic structure seen by the positron at the trapping sites is the same for these various alloys. Since iron is the major constituent in these glassy alloys, it is concluded that the trapping sites for the positrons are associated with the iron atoms. The smaller trapping effect compared to vacancies in iron indicates that the "free-volume" of the trapping center in the alloys is slightly smaller than that of a vacancy. This is supported and verified by positron lifetime measurements /40/. The lifetime spectra for the glassy alloys show an exponential decay involving only one single lifetime indicating that all positrons annihilate from the same state. This lifetime ranges from 143 psec for $Fe_{80}B_{20}$ and $Fe_{78}Mo_2B_{20}$ to 160 psec for $Fe_{32}Ni_{36}Cr_{14}P_{12}B_6$ and is always less than the 170 psec for vacancies in iron /40/. The bulk lifetime in iron (defect free lattice) is 110 psec compared to 116 psec for crystalline $Fe_{78}Mo_2B_{20}$ /40/. Both lifetime and Doppler-broadening results clearly show that the positron annihilates from a definite state, the nature of which lies between the pure bulk and the pure vacancy state. First calculations and comparisons with experimental data indicate that the positrons in these glassy alloys are trapped at trigonal prisms and at Archimedian antiprisms /42/. Since the "free-volume" of the trapping sites is smaller than for a vacancy, the binding energy for a positron is smaller as well. This may lead to a temperature dependent trapping since at higher temperatures detrapping from a weak bound state becomes possible. Fig. 13 shows the temperature dependence of the positron lifetime τ and the lineshape parameter I_v as a function of temperature for $Fe_{78}Mo_2B_{20}$ in the glassy and crystalline state. I_v and τ show the same basic tendency in both states. However, lineshape parameter and lifetime are larger in the amorphous state. Below 50 K in the crystalline state there is an increase in I_v and τ indicating positron trapping at shallow trapping sites, e.g., grain boundaries. Similar effects are observed at other alloys. Further experiments are in progress to investigate and perhaps clarify the structure of metallic glasses.

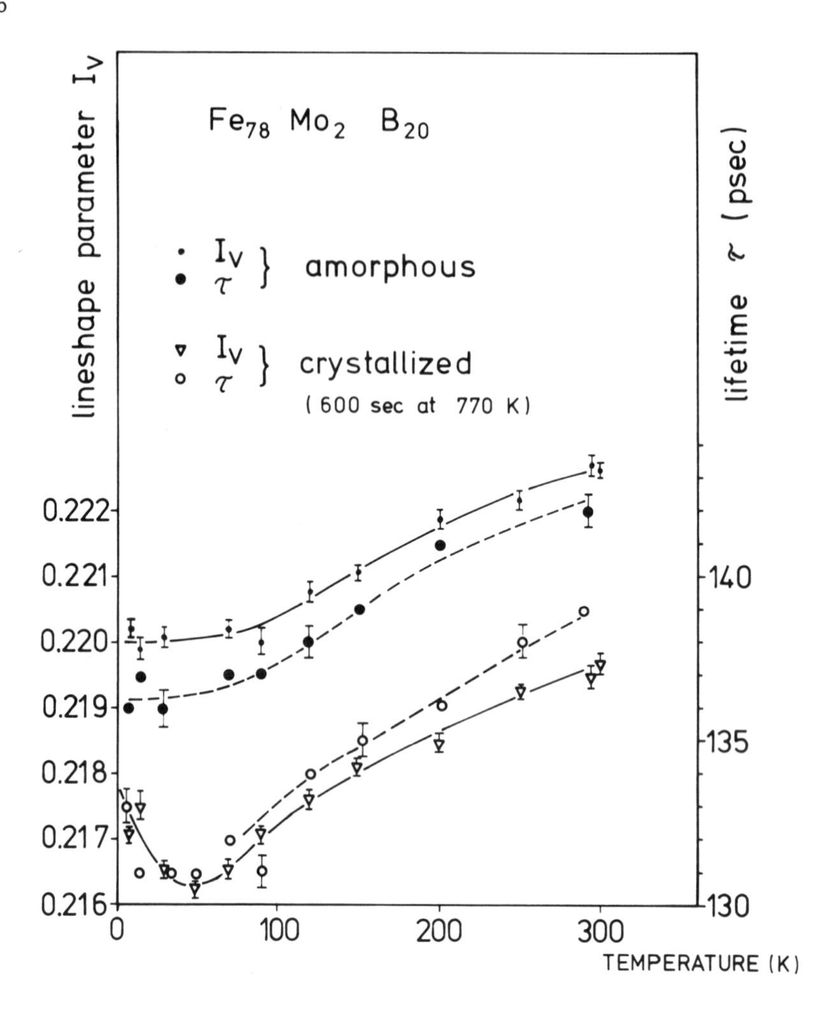

Fig. 13. Doppler-broadening line shape parameter I_V and positron lifetime τ as a function of temperature in the amorphous and crystalline state.

ACKNOWLEDGEMENTS. The author gratefully acknowledges the financial support of the Bundesministerium für Forschung und Technologie and helpful discussions with G. Kögel, J. Winter and Z. Kajcsos.

REFERENCES

1. C.D. Anderson, Science 76, 238 (1932).
2. P.M.S. Blackett and G.P.S. Occhialini, Proc. Phys. Soc. (London) A 139, 699 (1933).
3. P. Dirac, Proc. Phys. Soc. (London) A 126, 360 (1930); ibid A 133, 60 (1931).
4. G.E. Lee-Whiting, Phys. Rev. 97, 1557 (1955).
5. P. Kubica and A.T. Stewart, Phys. Rev. Lett. 34, 852 (1975).
6. V.I. Goldanskii, Atom. Energy Rev. 6, 3 (1968).
7. P.R. Wallace, Solid State Physics 10, 1 (1960).
8. R.W. Siegel, J. Nucl. Mat. 69-70, 117 (1978).
9. A.T. Stewart, Positron Annihilation, p. 17, A.T. Stewart and L.O. Roellig (eds.), Academic Press, New York (1967).
10. S. de Benedetti, C.E. Cowen, W.R. Konnecker, and H. Primakoff, Phys. Rev. 77, 205 (1950).
11. R.A. Ferrell, Rev. Mod. Phys. 28, 308 (1956).
12. Chang Lee, Soviet Phys. JETP 6, 281 (1958).
13. W. Triftshäuser, Positron Studies of Metals in Festkörperprobleme Vol. XV, p. 381, H.J. Queisser (ed.), Pergamon/Vieweg, Braunschweig (1975).
14. P.E. Mijnarends, Positrons in Solids, p. 25, P. Hautojärvi (ed.), Springer Verlag, Heidelberg (1978).
15. G. Kögel, Proc. 5th Int. Conf. on Positron Annihilation, p. 383, R.R. Hasiguti and K. Fujiwara (eds.), Lake Yamanaka, Japan (1979).
16. H. Weisberg and S. Berko, Phys. Rev. 154, 249 (1967).
17. W. Triftshäuser, Phys. Rev. B 12, 4634 (1975).
18. J. Mader, S. Berko, H. Krakauer and A. Bransil, Phys. Rev. Lett. 37, 1232 (1976).
19. P. Hautojärvi, Positrons in Solids, p. 1, P. Hautojärvi (ed.), Springer Verlag, Heidelberg (1978).
20. D.R. Gustafson, A.R. Mackintosh, D.J. Zaffarano, Phys. Rev. 130, 1455 (1963).
21. R.N. West, R.E. Borland, J.R.A. Cooper, N.E. Cusack, Proc. Phys. Soc. 92, 195 (1967).
22. O.E. Mogensen and G. Trumpy, Phys. Rev. 188, 639 (1969).
23. J.D. McGervey, Positron Annihilation, p. 305, A.T. Stewart and L.O. Roellig (eds.), Academic Press, New York (1967).

24. W. Brandt and H.F. Waung, Phys. Lett. A 27, 700(1968).
25. M.V. Chu, C.J. Jan, P.K. Tseng and W.F. Huang, Phys. Lett. A 43, 423 (1973).
26. R.N. West, V.H.C. Crisp, G. DeBlonde and B.G. Hogg, Phys. Lett. A 45, 441 (1973).
27. L.E. Ballentine, Can. J. Phys. 44, 2533 (1966).
28. T. Gorecki, Z. Metallkunde 65, 426 (1974).
29. R.N. West, Positron in Solids, p. 89, P. Hautojärvi (ed.), Springer Verlag, Heidelberg (1978).
30. F. Itoh, K. Kai, M. Kuroka and S. Takeuchi, Phys. stat. sol. (b) 75, 559 (1976).
31. K. Tsuji, H. Eudo, Y. Kita, M. Ueda and Z. Morita, Liquid Metals, p. 367, R. Evans and D.A. Greenwood (eds.), Inst. of Physics, Bristol and London (1977).
32. M. Doyama, S. Tanigawa, K. Kuribayashi, H. Fukushima, K. Hinode and F. Saito, J. Phys. F 5, L 230 (1975).
33. H.S. Chen and S.Y. Chuang, Phys. stat. sol. (a) 25, 581 (1974).
34. S.Y. Chuang, S.J. Tao and H.S. Chen, J. Phys. F 5, 1681 (1975).
35. H.S. Chen and S.Y. Chuang, J. Electronic Materials 4, 783 (1975).
36. H.S. Chen and S.Y. Chuang, Appl. Phys. Lett. 27, 316 (1975).
37. Zs. Kajcsos, J. Winter, S. Mantl and W. Triftshäuser, Phys. stat. sol. (in press).
38. S. Mantl, W. Kesternich and W. Triftshäuser, J. Nucl. Materials 69–70, 593 (1978).
39. R.M. Nieminen, J. Laakkonen, P. Hautojärvi, and A. Vahanen, Phys. Rev. B 19, 1397 (1979).
40. G. Kögel, this volume.
41. G. Kögel and W. Triftshäuser, to be published.
42. J.D. Bernal, Nature 183, 141 (1959).

3.3 Magnetic properties of disordered systems

FERROMAGNETISM IN AMORPHOUS METALS

S. Methfessel

Ruhr-Universität, D-4630 Bochum,
W-Germany

ABSTRACT:

1. INTRODUCTION

The existence of amorphous ferromagnetism is known as a curiosity for about 30 years, but the possibility of technical applications stimulated very active research during the last 4-5 years. What can be learned from studying amorphous magnets beside of improving their technical usefulness?

We know from crystalline alloys that the strength of their magnetization and their critical temperature Θ for spin ordering depend strongly on the overlapping of the partially filled electron orbitals between neighboring atoms i.e. on the chemical bonding mechanism. When it is large compared to electron-electron interactions at the same atom, as is nearly always the case for s- and p-type electrons, then we obtain nearly free band electrons. An applied magnetic field H produces then the temperature independent Pauli susceptibility $\chi_P = M/H = 2\mu_B N(E_F)$. In case of extremely weak interatomic overlapping, such as for 4f-electrons in rare earth materials, the electrons remain in well quantized $4f^n$ states as in free atoms and their effective magnetic moment p can be derived from Hund's rules. The susceptibility follows the Curie-Weiss law $\chi = C/(T-\Theta)$ with $C \sim 0.125\ p^2$ and $\theta = 2/3k \Sigma z_i J_i S(S+1)$ as sum of exchange interactions J_i with all neighboring atomic spins S_i.

In the ferromagnetic transition metals Fe, Co, Ni the partially filled 3d-orbitals overlap in covalent bonds so strongly that a discussion lasts now for more than 40 years whether they should be better treated as "localized atomic states" with strong perturbation from neighbor atoms or as "itinerant band states" with strong scattering by exchange forces near the ion cores. In any way, the strong interatomic overlap brings the ferromagnetic Curie temperatures up to about 1000 K and the effective atomic moments follow, of course, not Hund's rules. One finds for Fe, Co, Ni p-values of 3.2; 3.15 and 1.61 which correspond to nonintegral Bohr magneton numbers of about 2.2, 1.7 and 0.6 μ_B/atom, respectively.

The directional distribution and exponential decay of the outer part of the d-wave functions make interatomic overlap, chemical bonding and magnetic properties strongly dependent on crystal structure, atomic coordination and kind of neighbors in alloys. However, the situation of the 3d-electrons influences also crystal structures, melting points, cohesive energies etc. The melting points of ferromagnetic 3d-elements are lower than for corresponding 4d and 5d elements because the electron-electron interaction which produces ferromagnetism is comparable in size to the band width which gives the strength of the metallic bond. A particularly strong interaction between magnetism and mechanical properties which is observable as large pressure dependence of magnetization and Curie temperature and large magnetostriction, occurs in Fe-Ni alloys of the socalled invar-region, where the crystal structure changes with increasing d-electron number from bcc to fcc.

We derive from this that the magnetic properties of amorphous materials should be useful for our understanding of

1. the dependence of magnetic properties on atomic coordinations, average number and kind of nearest neighbors,

2. the dependence of magnetic and mechanical properties on d-electron numbers, because the chemical composition of alloys can be varied continuously disregarding crystal structures with specific stoichiometries,

3. the dependence of atomic neighbor correlations on preparation conditions of amorphous alloys. The magnetic investigations offer a complimentary method to X-ray and neutron scattering which

give average radial distribution functions only.

In addition, we may be rewarded by finding new materials with interesting magnetic and elastic properties for technical applications.

In recent years, such a large number of new results have been published about amorphous magnets that only a very general survey can be given here. Our list of literature contains many survey articles where more details and references to the original work can be found.

2. MATERIALS OF INTEREST

Nearly all amorphous magnets, investigated so far, are metals. We know very little about magnetic insulators such as ferrites, garnets etc, in the amorphous state. Probably, they will not become really amorphous but polycrystalline with very small crystallites, because the charge transfer and compensation required in heteropolar bonding is not compatible with the fluctuations in chemical composition, which are considered to be characteristic for metallic glasses. We shall also not discuss here the so called "ferromagnetic glasses" which are conventional SiO_2, B_2O_3 or P_2O_5 glasses with small additions of Fe or Fe_2O_2 which, in most cases, precipitate to superparamagnetic particles. The "spin glasses" where diluted magnetic atoms are randomly distributed in a nonmagnetic matrix, which is crystalline in most cases, are also excluded here, although "spinglass like" properties have been observed in some concentrated amorphous magnets, when the magnetization is strongly reduced by suitable alloying.

The concentrated magnet glasses to be discussed here can be subdivided in three main groups:

1. Pure Fe, Co, Ni and their solid solutions can be made amorphous only by vapor deposition on very cold substrates in high vacuum. The impurities trapped from the residual gas (O_2, N_2, CO, CO_2, H_2O etc) are very helpful. For impurity levels below 0.04 %, the Ni is always crystalline, Fe amorph at 4K for thickness less than 50 Å, while Co, Cr and Mn remain amorphous up to 38, 50 and 400 K, respectively [9]. Ni with about 1 % occluded gas is amorphous up to 77 K. Materials with such low crystallization temperatures are not of interest for technical applications and, therefore, less intensively investigated, although the fundamental understanding of amorphous ferromagnetism could be easier for pure materials.
2. Alloys of Mn, Fe, Co, Ni with other transition metals (e.g. Sc, Ti, Zr, V, Nb or rare earth elements) or with normal metals (e.g. Zn, Al) over a wide concentration range. Both components should not form solid solutions but rather have, in their phase diagrams, deep eutectics between high melting intermetallic compounds. This means, that for the eutectic composition the atoms cannot find a crystalline space group with a local point symmetry,

which allows easy hybridization of electron orbitals near the Fermi energy. Such amorphous alloys have sufficiently high crystallization temperatures, but the large concentration of nonmagnetic atoms reduces the magnetization too much for many technical applications. Special attention have found amorphous GdCo films which develop a strong perpendicular magnetic anisotropy when produced by cathode sputtering in Ar under special conditions. They may replace the thin single crystalline garnet layers used in bubble devices for electronic data storage.

3. Metglasses are alloys of ferromagnetic metals with additions of 15 to 25 % of "glass forming" metalloids (B, C, Si, Ge, P). The strong chemical interaction between metal and metalloid atoms lowers the eutectic temperature much deeper than expected by theory and stabilizes the amorphous state up to about 450 ^{o}C; they are usually produced by spinning ribbons from fast rotating cold drums, but other methods can be used also. At present, the ribbons are limited in thickness to about 50 μm and in width to a few cm. Samples can be obtained from Allied Chemical, Morristown, New Jersey, USA and Vacuumschmelze, Hanau. Metglasses of various compositions have a good potential for several technical applications, their properties have been investigated very intensively during the last four years and they can be considered to be prototypes of amorphous magnets.

3. SATURATION MAGNETIZATION AT T = OK

3.1 Itinerant Magnetization

The magnetization of Fe, Co, Ni is usually explained as "itinerant electron magnetism" or "band magnetism". The exchange energy splits the d-band for up and down spin electrons by an energy $\Delta \sim IM$ which Stoner suggested to be taken proportional to the magnetization $M = N\uparrow - N\downarrow$ times an exchange parameter I. The Pauli susceptibility χ_o becomes exchange enhanced $\chi = \chi_o . F$ by the Stoner factor $F = (1-I\chi_o)^{-1}$. The socalled "Stoner criterion" expects the occurence of ferromagnetism for $F \to \infty$, i.e. $I\chi_o \geq 1$. Because of $\chi_o \sim N(E_F)$ it is usually written as $IN(E_F) \geq 1$, saying that large density of states at the Fermi surfaces favors ferromagnetism. When the function N(E) is known from band calculations the magnetization at T = OK can be calculated by integration over all up and downspin electrons in the conduction band. It depends for different elements on the position of E_F in good agreement with experiments. Particularly, the maximum of magnetic moments between Fe and Co is well explained by complete filling of the spin up band. Additional electrons must go in the spin down band and reduce magnetization.

The problem with the Stoner model is that in reality the exchange parameter I and the magnetic moments do not disappear for $M \to 0$ at the Curie temperature but become only disordered. The reason is that the symmetry of wave functions in transition metals is rather complicated and results in a nearly localized character of the magnetic electrons, as observed by neutron diffraction. However, the geometrical size of the local moment is not related to the atomic size but larger and includes wave functions of neighbor atoms. The origin of the magnetic splitting is not the Weiss field, as Stoner assumed, but the Coulomb repulsion $U \sim 0.6$ eV between electrons at the same lattice site. In some crystalline solid solutions, the average magnetic moments per atom behave as if only E_F varies while the band structure remains unchanged by alloying. Slater introduced this as the "rigid band approximation" for itinerant electrons. Pauling used a local description where molecular orbitals are formed by hybridization of s, p, d orbitals of the center and surrounding atoms which are then filled according to Hund's rule to give the magnetic moment. However, the famous Slater-Pauling curve for the concentration dependence of average μ_B-numbers per atom (shown by dashed line in Fig. 2) is only applicable to alloys between partners which are not too far separated from one another in the periodic table. Otherwise, a too large difference in ionic charge will prevent the mixing of orbitals at E_F or their hybridization into molecular orbitals, depending on the model. There is a change in crystalline structure from bcc to fcc around $N \sim 8.5$ to 8.7 and two Slater-Pauling curves for different atomic coordinations and with different maxima overlap one another in the invar region with very strong magneto-elastic effects.

Since the size of magnetic moments is so closely related to the orbital overlap, i.e. chemical bonding, in wich the magnetic d-electrons are involved, it is interesting to ask about the changes which shall occur, when the crystal symmetry is destroyed in amorphous materials:

1. The magnetic moment will change radically, when atomic coordination and overlap of wave functions are very different from the situation in crystals. Fe, as an example, has in small particles with fcc structure only 0.7 μ_B/atom instead of the 2.2 μ_B/atom in bcc crystals.

2. The moment will change mildly, when the local atomic symmetry is approximately maintained, but the interatomic distances and number of neighbors fluctuate locally. The sharp peaks in N(E) curves of transition metals average out and the resulting in $N(E_F)$ can increase or decrease the magnetic moment [23].

3. The metalloid or other additions, which are necessary for stabilization of the amorphous state, must be chemically activ, i.e. they must form local chemical bonds with the transition atoms and, therefore, change the average magnetic moment. This effect is often called "charge transfer" but is rather a change in local covalent bonds than in valency of single atoms. The latter effect should result in changes of ionic sizes which is usually not observed in metallic alloys.

3.2 Amorphous transition elements

For amorphous pure elements the experimental situation is complicated due to the low stability mentioned earlier. There is a large risk of having stabilizing impurities or small crystalline regions intermixed. The experimental results show a clear trend of reducing the magnetization in amorphous Fe, Co, Ni films compared to the crystalline state, but the numbers depend very much on sample preparation. Some typical numbers for the ratio of amorphous to crystalline magnetization are [9]: Fe: 0.2 - 0.6, Co: 0.95 - 1.0, Ni: 0.3 - 0.6. Bennemann [23] has predicted a ratio of 0.8, 0.7 and 0.6, respectively, from the reduction of $N(E_F)$ by increased interatomic distances in the amorphous state. The d-band becomes a little narrower and electrons can be transferred to the s-band [15]. The vision, that the amorphous state is like a crystal which is blown up by increased atomic volumes is, for most cases, much too simple. How important changes in number and coordination of neighbor atoms are, is demonstrated by Fe where the amorphous magnetization is comparable with that of small particles with fcc structure and deviates strongly from Bennemann's prediction.

3.3 Amorphous Alloys

The magnetic moment per transition metal ion is here in most cases lower than in the pure crystalline elements. However, this has nothing to do with disorder, but with "electron transfer" between the glassforming and 3d-ions. The changes in magnetic moments are related to the chemical bonding mechanisms responsible for the formation of the very deep eutectics which are so useful for the stabilization of the amorphous state. In the simplest case, which applies to most Co-alloys, the effective average number of outer electrons per 3d ion controlls the magnetization according to the Slater-Pauling curve. The magnetization has been observed to decrease linearly with the concentration of glassforming ions (Fig. 1). The magnetic moments per Co- or Ni-atom are surprisingly identical in amorphous and crystalline samples at the same chemical composition [18]. The electron transfer is in metals not really a change in valence but consists rather of contributions of s, p wave functions of metalloids to the electronic environment of the 3d atoms where the magnetic moment is generated. The number n of electrons transferred per metalloid atom can be calculated easily from the reduction $\Delta\mu$ of the magnetic moment per Co-atom observed at the metalloid concentration x by using the relation $\Delta\mu = n.x\ (1-x)^{-1}$, which is often called rigid band approximation. The magnetization of the new material is simply shifted along the Slater-Pauling curve by a corresponding amount of tranfered electrons.

This model is, of course, not very convincing for amorphous materials where no regular bands exist and it should not surprise when the derived numbers of transferred electrons do not agree with chemical valencies. The better model is here Pauling's molecular interpretation of the Slater-Pauling curve. In this model, one has to ask first for the probability to find metalloid atoms as next neighbors around a transition metal ion. This depends not only on

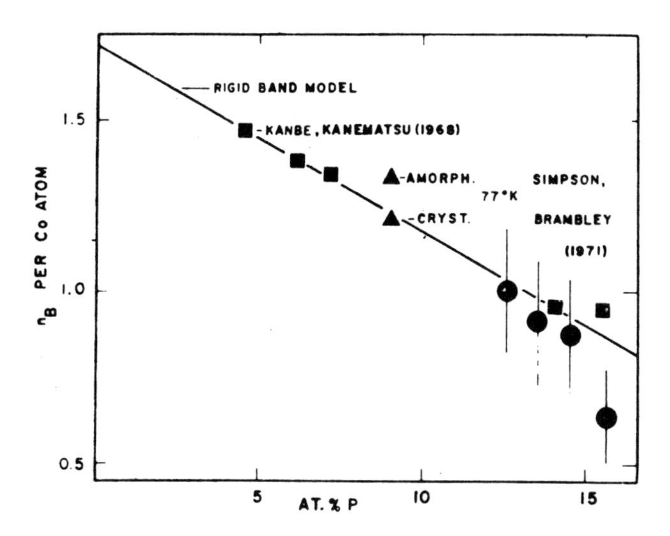

Fig. 1: Magnetic moment per Co-atom in amorphous Co-P alloys of different origin [24]. Solid line: rigid band model

concentration x but also on the available sites. The variation of magnetic moment with the number of metalloid neighbors due to local orbital overlapping can be derived from the experiments. Kouvel [25] has worked out the statistics for disordered crystalline alloys which also apply to amorphous systems. This analysis gives Δn values for transferred electrons which are closer to what one might expect from p-electron configuration of the metalloid atoms (e.g. $\Delta n = 3$ for each P-atom, $\Delta n = 2$ for C, Si and $\Delta n = 1$ for B). The observed transfered electron numbers are in Fe-glasses usually lower (e.g. B: 0.3 - 1.0; P: 1,0 - 2.1) and decrease in the wrong sequence B - P - Si.Deviations from the linear approximation have been observed in some Fe alloys with G, Si, as examples [18]. They are indications for more complicated chemical bonding conditions in neighboring intermetallic compounds or to other irregularitis in the phase diagram.

The situation is particularly simple in alloys where transition ions are replaced by one another at constant metalloid concentrations. Fig. 2 shows, as an example, the results of Mizoguchi [26] for amorphous $(Fe_{1-x}M_x)_{0.8}B_{0.1}P_{0.1}$. The Slater-Pauling curve has the same character as in corresponding crystalline binary solid solutions but is apparently shifted to lower numbers by about 0.4 electrons transfered from the metalloids. Therefore, Ni has no moment at the new Slater-Pauling curve whilst Co, Fe have their moments reduced to about 2 or 1 μ_B per atom.

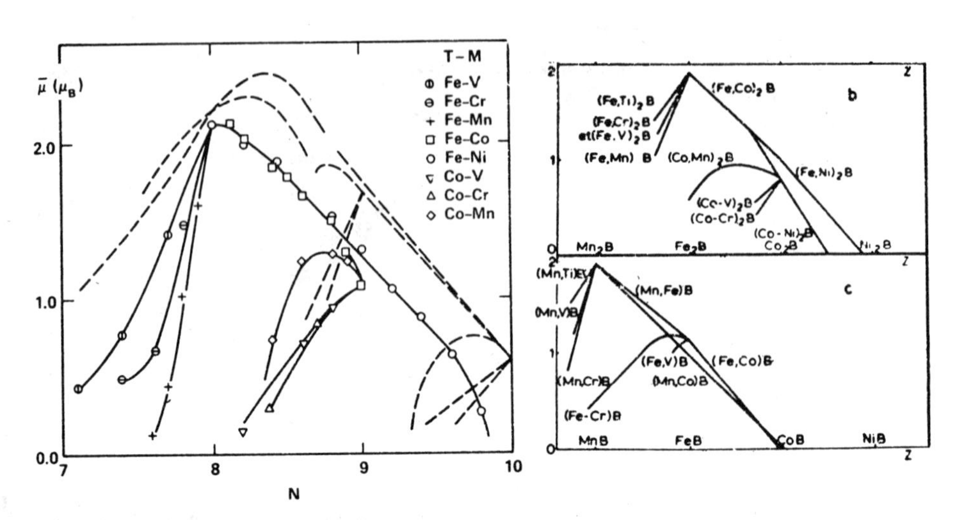

Fig. 2: Variation of saturation moment with outer electron number N = 3d + 4s (Slater-Pauling curve) for:
a) crystalline binary alloys: --------
amorphous $(Fe_{1-x}M_x)_{0.8}B_{0.1}P_{0.1}$ [8,26]: ————
b) in crystalline $(Fe_{1-x}M_x)_2B$ and $(Fe_{1-x}M_x)B$ alloys [from ref. 21]

3.4 Amorphous alloys with rare earth ions (RE)

The phase diagrams of 3d-elements with RE have a large number of crystalline compounds which have certain structural similarities. The RE-ions transfer their 3 outer s and d electrons into the 3d orbitals and reduce their magnetization (e.g. Co has in $GdCo_2$ 1.0 instead of 1.7 μ_B). Ni has no moment in most compounds, the atomic moment of Co disappears for RE-ion concentrations larger than 25 %. Crystalline pseudo binary solid solutions between different 3d-ions of constant RE concentration follow with their magnetization the Slater-Pauling curve so that the rigid band model seems to be applicable, although the electronic band structures are very complicated and very different from the free electron case. The heavy RE align their spins so that $M = M_{RE} - M_{3d}$ varies irregularly with temperature and can have compensation points with M = 0, because the magnetic RE-RE exchange is very weak and the RE-3d exchange weak compared to the 3d-3d exchange.

All those crystalline properties are qualitatively maintained in the amorphous state. The Re-ions take here obviously the role of the glass forming atoms which have strong chemical interaction with the 3d-ions and produce even lower eutectic temperatures than the metalloids. The RE-3d electron transfer is a little smaller in the amorphes than in the crystalline state because the smaller density reduces orbital hybridization. Therefore, the 3d-moments are higher (e.g. Co in $GdCo_2$:1,4 μ_B) and a RE-concentration of about 50 % is now needed to cancel the Co-moment.

The random distribution of magnetic anisotropy directions due to random fluctuations in atomic order has interesting effects for RE-ions with $L \neq 0$, e.g. in amorphous Tb Fe, DyFe alloys. The strong anisotropies compete with the RE-Fe exchange and the RE-spins are not simply antiparallel to the Fe-spins, as in GdFe, but are distributed over cones whose angle ($\sim 140^o$) is determined by the relative directions and strength of anisotropy and exchange fields [28]. This random ferrimagnetic spin order has been called "sperimagnetic". In TbNi random ferromagnetism or "speromagnetism" occurs because Ni has no atomic moment.

4. TEMPERATURE DEPENDENCE OF MAGNETIZATION

Spin wave theory for Heisenberg magnets is proven to be a very good approximation for crystalline materials with short range exchange interactions and at low temperatures relative to the Curie temperature. The spin waves have for small wave vectors q the energy

$$E = E_o + Dq^2 .$$

The so called "spin wave stiffness constant" $D \sim 2JSr^2$ depends on exchange constant J, magnetic moment S and interatomic distance r. The magnetization decreases with increasing T by spin wave excitation

$$M(T) = M(0)\ (1 - BT^{3/2} - CT^{5/2}\ . . .)$$

Comparison with experimental $M(T^{3/2})$ curves gives B which is related to the stiffnes constant D:

$$B = 0.0587 \frac{g\ \mu_B}{M_o} (k_B/D)^{3/2} ,$$

which can be determined independently by inelastic neutron scattering. D is expected to vary in Heisenberg magnets as $T^{5/2}$. Magnetization measurements and neutron diffraction show unanimously that amorphous magnets follow the spin wave approximation even to higher relative temperatures $T/T_c \sim 0.4$ than crystalline materials (~ 0.2). However, the parameter B is by a factor 2 to 4 larger in amorphous than in crystals indicating that spin waves with low q and large wave length are easier excited in the amorphous state and the magnetization falls off faster with increasing temperature. Moreover, the value of B depends strongly on chemical composition as seen in Fig. 3. In some cases, discrepancies have been found between the B and the D values, which were obtained by magnetization and neutron measurements, respectively. It is not yet clear whether the effect is real and what its origin could be.

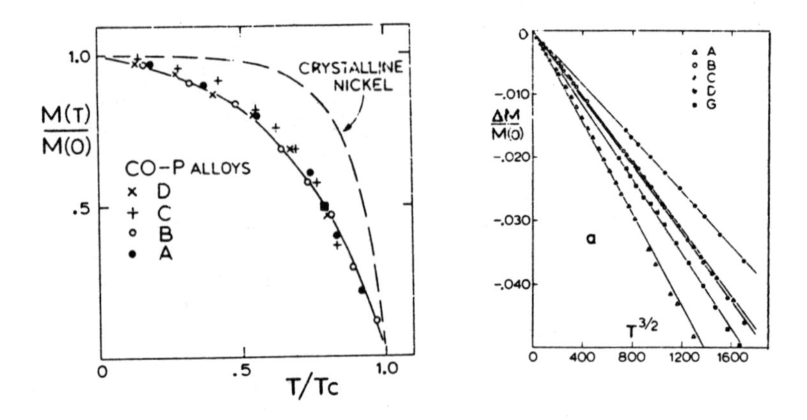

Fig. 3: Temperature dependence of relative magnetization for amorphous CoP-alloys with P-concentrations between 21 and 24 % [29].

The good agreement of experiments with Heisenberg spin wave theory for long wave lengths indicates that the observed deviations of the M(T) curves from Brillouin functions near T = 0 are not caused by local fluctuations in anisotropy or exchange forces but by the smaller spin wave stiffness D due to long range averaging over local exchange fluctuations. The local fluctuations become visible for shorter wave length magnons in neutron scattering as a steep minimum in spin wave energy (Fig. 4) which can be assigned to excitations of localized spin waves [31].

The magnetization at higher temperatures can be described by introducing exchange fluctuations $\bar{J} + \Delta J$ into the molecular Weiss field, which lead to corrections in the argument of the Brillouin function [32]. The standard deviations δ of the nearest neighbor exchange integral can be derived from the experimental M(T) curve. In rapidly quenched amorphous $Fe_{75}P_{15}C_{10}$, as an example, a standard deviations of $\delta = 0.3$ was found for temperatures above 0.6 T_c [33]. Such large local fluctuations are also obtained by NMR and Moessbauer measurements. For $\delta \sim 1$ the ferromagnetic phase will break up in local fluctuations and spin glass like peaks in the initial susceptibility will appear.

We have already mentioned earlier that in 3d-RE alloys ferri- and sperimagnetic spin orderings occur which have much mor complicated M(T) curves not to be discussed here.

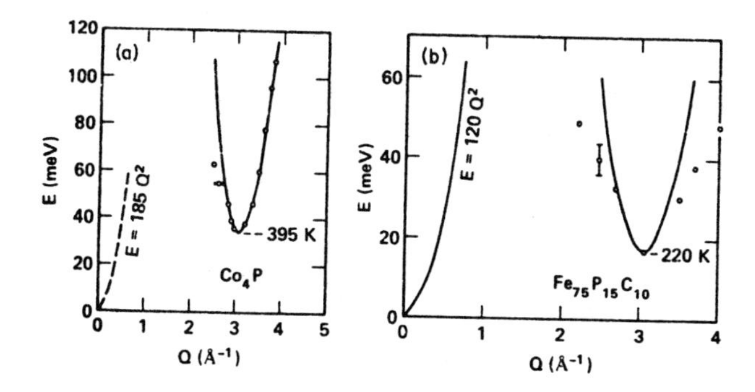

Fig. 4: Spin wave dispersion curves for amorphous Co and Fe alloys [30].

5. MAGNETIC CURIE TEMPERATURE T_C

At T_C the thermal motion of the spins freezes in the molecular field $W = 1/2 \Sigma J_i S_i (S_i+1)$ because $3k_B T_C < W$. The number of spins S_i and their exchange interactions J_i with the central spin depend on the nature of chemical bonds involving magnetic electron orbitals. Three exchange mechanisms are usually under consideration:

1. direct or Heisenberg exchange by direct overlap of d-orbitals between neighbor atoms in a molecule. Since the overlap depends on the ratio of interatomic distance to orbital radius, the T_C values are often plotted as functions of this parameter in so called Slater-Bethe curves.

2. superexchange by overlap of 3d-orbitals of several magnetic atoms with p-orbitals of the same anion in the center. This is related to the covalent bonding part in mostly ionic compounds, e.g. MnO, and depends on the ratio of overlap integral to Coulomb repulsion.

3. indirect or RKKY exchange by band electrons which have become spin polarized by intraatomic exchange with the local moments of magnetic ions. The spin polarization of the band electrons decays as r^{-3} and oscillates with a wave length which depends on the Fermi momentum of the conduction electrons. The long range RKKY interactions applies, in particular, to the magnetism of rare earth with localized 4f moments but can be used also for 3d metals and insulators when band electrons are replaced by other suitable wave functions.

In amorphous metals the Heisenberg exchange has to be modified by the radial distribution function of the neighbor atoms [32].

The statistics of distribution of B atoms around an A-atom in an alloy $A_{1-x}B_x$ and its influence on Tc has been worked out by Kouvel [25]. The specific interaction temperatures between pairs can be determined from the best fit of experimental $T_c(x)$ curves.

The disorder will have two effects on the RKKY interaction. Since the main free path of the conduction electrons will become of comparable size to the interatomic distances, the interaction will be cut off exponentially at the same short distance. Furthermore, it is important whether the local fluctuations of interatomic distances are small or large compared to the RKKY oscillations. In the latter case the spacial fluctuations can change the sign of the exchange interactions.

The critical behavior around T_c is not very much influenced by disorder. The specific heat anomaly at T_c remains very sharp, the critical exponents change their values less than 30 %. Sharp transitions into the ferromagnetic state at T_c are also observed in magnetization and resonance measurements.

The difference between amorphous and crystalline is much larger for T_c than for the magnetization. The effect of metalloid concentration on T_c depends not only on the kind of metalloid but also on the d-electron concentration of the transition metal ions and the interatomic distances. Experiments show that

1. T_c varies linear with the concentration of nonmagnetic metalloid atoms such as B, C, Si, P. This is understandable because molecular field and T_c are expected to be proportional to the ion spins S which vary proportional to the metalloid concentration because of electron transfer.

2. T_c varies in amorphous Fe, Co, Ni alloys with the number of outer d electrons in a similar way as in crystalline alloys but the maximum of T_c is shifted to lower n-values due to electron transfer (Fig. 5a).

3. T_c values of different glasses follow a Slater-Bethe curve when plotted in function of R/r, where the d-orbital radii r are taken from normal metals and the interatomic distances R from radial distribution functions (Fig.5b) The corresponding curve for crystals has its maximum around R/r = 1.75 and goes through zero near 1.5. Most Ni rich glasses increase Tc when their density increases after annealing at about 300° C.

4. The decrease of T_c with decreasing interatomic distance R, which occurs in the Bethe curve for Fe - rich amorphous alloys, suggests similar invar properties as well known for fcc crystalline Fe-Ni alloys in the region $8.5<n<8.7$. Fe has here in the fcc structure low magnetic moments and T values which decrease further with increasing Fe concentration and under hydrostatic pressure.

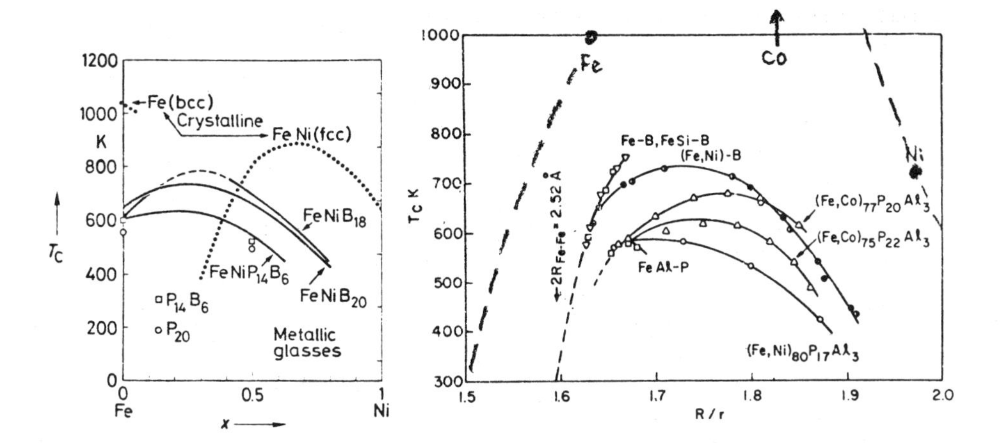

Fig. 5: Curie temperatures in amorphous and crystalline FeNi alloys as function of concentration [16] and Slater-Bethe curves [35].

6. MAGNETIC PROPERTIES AND CHEMICAL STRUCTURE

We have learned that the magnetic moment and ferromagnetic Curie temperature behave in amorphous and crystalline materials of comperable composition nearly identical. This suggest a search for similarities in transition atom coordinations for both cases and one finds immediately regular tetrahedra of closely packed Fe-atoms for tetragonal Fe_2B, which is formed by crystallization from the eutectic (Fig. 6). Some of the tetrahedra have B-atoms in their center. Such Fe- or Co-tetrahedra are also found in $GdCo_2$ and related structures, where the Slater-Pauling curve applies (Fig. 6b). In pure bcc or fcc transition metals the same tetrahedra are packed closely together, because they are not surrounded by atoms of another kind. Ganzhorn has shown that for electron numbers between N = 4 and 8 the tetrahedron is the basic unit which can be stabilized by s-d hybridization.

Computer models for amorphous materials prefer also regular tetrahedra as seed unit and let it grow by placing new atoms at the tetrahedra faces. The result is a completely disordered structure. The radial density functions have best agreement with experiments, and show, in particular, the right splitting of the second peak, when the Fe-atoms within the tetrahedra touch one another [9]. The formation of octahedra from tetrahedra is hard in a disordered material because the intermediate step is the regular pyramid which is a statistically very improbable configuration. Therefore, the tetrahedra have already by themself a rather large resistance to crystallization which is further increased by the surrounding

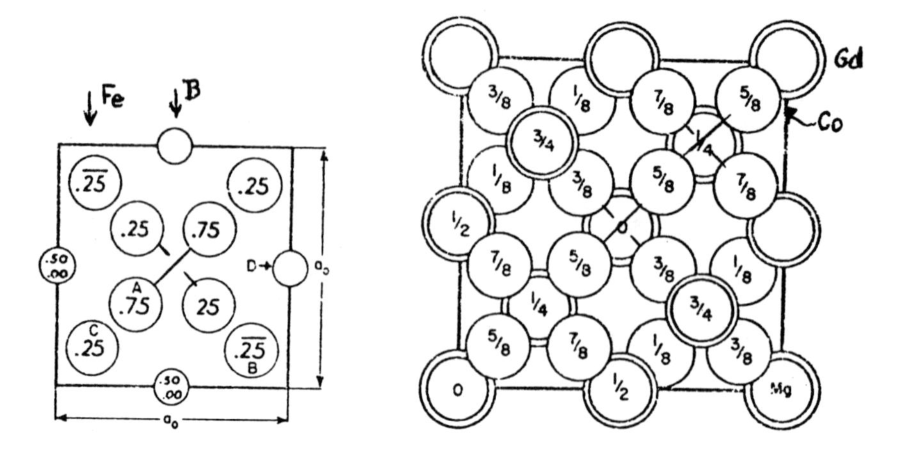

Fig. 6: Atom positions in $Fe_2B(CuAl_2$: C16) and $GdCo_2(MgCu_2$: C15)

metalloid atoms which stick strongly to them. The tetrahedra interact not sufficiently strong with one another to form a random network, like SiO_2 glass, but leave many voids and defects in packing, which can be decreased by careful annealing at about 300° C. More annealing makes the material brittle, because the metalloid atoms precipiate when the tetrahedra rearrange to a more improved order.

The importance of the local coordination in amorphous as well as crystalline compounds has been demonstrated nicely for $Fe_3(P_{1-x}B_x)$. A crystal structure change at $x \sim 0.4$ which moves Fe-atoms closer, which form bridges between the tetraeder, influences magnetization and Curie temperature for the crystalline and amorphous phase in the same way [45]. The observation, that the magnetization and Curie temperature are mainly determined by arrangement and chemical nature of the nearest neighbors in the first coordination sphere corresponds to the so called Joffe principle for liquid semiconductors. Anomalies in the viscosity of liquid Fe-Ni alloys indicate, that tetrahedra are formed already in the melt about 40 - 50° C above the melting point [47]. They could be even more stable in alloys with metalloids. The strong adherence of surrounding metalloid atoms blocks the growth of tetrahedra into long range crystalline order so efficiently that the melting point of the eutectic becomes very low, amorphous solids can be formed.

7. MAGNETIC ANISOTROPIES

All deviations from spherical symmetry produce preferred directions for the magnetization, i.e. magnetic anisotropy. Samples of extreme shape such as thin ferromagnetic ribbons or sheets have, first of all, a large "shape anisotropy", which tries to keep the magnetization in plane.

7.1 Crystalline Anisotropy

The symmetry of the local atomic coordination is reflected by the "crystalline anisotropy". The"single ion" anisotropy of local moments is caused by the non spheric electrostatic field produced by the molecular environment which lifts the degeneration of the magnetic orbital and hinders or even quenches the rotation of magnetic orbitals in external fields. In itinerant electron systems the symmetry of d-wave functions at the Fermi energy and their spin-orbit interaction are important. The anisotropy will change with alloying when E_F is shifted into wave functions with other symmetry. Crystal anisotropy is the larger, the more local symmetry deviates from spherical.

In some space groups the non cubic local site symmetries are combined to an overall cubic symmetry and the magnetic anisotropy is reduced from second to fourth order in angle of magnetization to crystal axes. In solid solutions with several kinds of atoms, these atoms must be distributed also at random over all equivalent lattice sites in order to keep cubic symmetry. When this randomness is perturbed, e.g. because the different atoms prefer different positions during crystal growth, the high local anisotropy breaks through the overall cubic anisotropy and establishes an uniaxial "growth-induced" anisotropy, which recalls the direction of crystal growth. This effect occurs in garnet single crystals and, possibly, also in sputtered and evaporated amorphous RE-3d films which develop a strong uniaxial anisotropy relative to the plane of the film [36]. In amorphous materials with random distribution of local symmetry axes and alloy components the crystalline anisotropy should be averaged out to zero. In practice, the crystalline anisotropies are about 2 orders of magnitude smaller in amorphous metals than in crystals.

7.2 Magnetostriction

The "magnetostriction" couples magnetization to distortions of local symmetry and mechanical stress and is related to the variation of crystalline anisotropy with interatomic distances. Since magnetostriction is not much smaller in amorphous materials than in crystals, it is more conveniently used than the crystal anisotropy for studying the orbital character of wave function near E_F. The sign and magnitude of magnetostriction depends sensitively on the chemical composition. The saturation magnetostriction has for Co-rich glasses the same sign as in polycrystalline materials, but the opposite in Fe-rich materials. Its temperature dependence as $M(T)^2$ indicates that the single ion model is still a valid approximation [37]. The change in sign of magnetostriction with 3d-composition can be related to the shift of Fermi energy from Ni or Co 3d-orbitals to Fe-orbitals. This change of sign is important for technical applications because it allows to produce magnets insensitive to

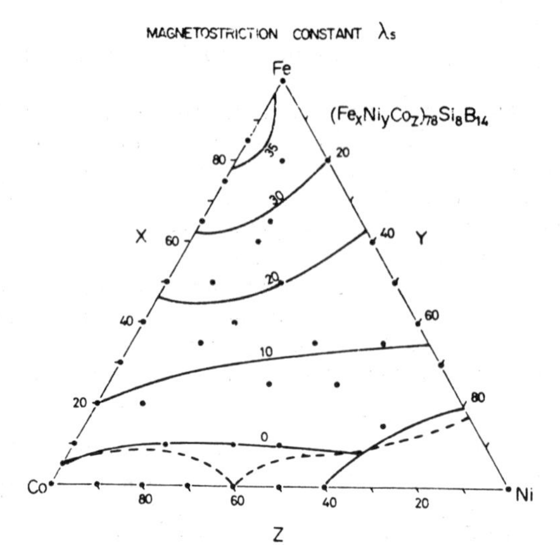

Fig. 7: Saturation magnetostriction (in units 10^{-6}) as function of transition metal concentration in
——— $T_{78}Si_8B_{14}$ glasses,
------- $\lambda = 0$ in the crystalline state

mechanical stress and deformation. However, there are also Fe rich glasses, such as $Fe_{80}P_{13}C_7$, which have very large magnetostriction and ΔE effect, i.e. change of elastic modulus in applied field. They can be used for electro acustical devices [38].

7.3 Directional Order Anisotropy

Pairs of magnetic atoms, impurities, defects and any other entities which define a direction can produce also magnetic anisotropies when their intrinsic directions are not distributed at random. Deviation from randomness can be produced by annealing in a magnetic field which saturates the magnetization, or under stress. The pairs diffuse into positions, where the coupling energy with magnetization is reduced, and freeze after cooling. The deviations from random distribution establish now an uniaxial anisotropy in the direction of magnetization during annealing. The magnitude of anisotropy is proportional to $M(T)^2$ and to the impurity concentration n^2. The pair ordering can be enhanced by neutron or electron radiation, crystalline disorder or anything else that improves diffusion at low temperature, where the magnetization is high.

The existance of induced pair anisotropy is well established for 3d glasses [39] and has the properties predicted by theory(Fig.8

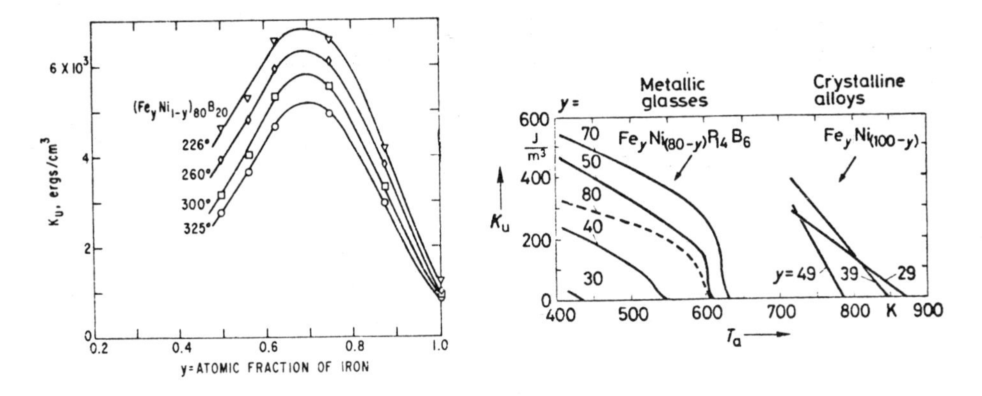

Fig. 8: Induced anisotropy as function of Fe concentration and annealing temperature in amorphous and crystalline Fe-Ni-B alloys [39], [16].

However, the annealing temperatures required for changes in anisotropy are much smaller than in crystals, because the short range diffusion is much faster in the disordered state. The activation energies for diffusion are comparable with values obtained in very fast quenched crystalline alloys. They reduce with annealing time [18].

There was a long discussion whether the large perpendicular anisotropies in sputtered GdCo films for bubble devices are caused by pair ordering also. Surprisingly, they appear only under certain sputter conditions, which made it plausible that gas atoms are built into the film. Films evaporated under higher vacuum show a perpendicular anisotropy only when at least 3.10^{-8} Torr oxygen are present as residual gas [40].It has been proposed that oxydation along walls of columnar structures is the origin of this anisotropy [41].This is mentioned as an example, how many complicated processes, such as fine precipitations, alignments, short range order and many other things can produce anisotropies in amorphous materials.

8. HYSTERESIS PROPERTIES AND TECHNICAL APPLICATIONS

The technical use of ferromagnetic materials depends on the way they reverse their magnetization in external fields. This can occur by rotation of the magnetization in larger areas, what brings relatively low energy losses, or by domain wall motion. The domain walls are pinned by local structural or magnetic perturbations which have comparable size with domain wall thickness $d \sim (D/K)^{1/2}$, with D as exchange stiffness and K as anisotropy. The walls break free and make Barkhausen jumps when the applied field reaches the coercive force H_c. When the anisotropy K averages out to small values in amorphous materials a large domain wall width, d ∿ 1000 Å, and low H_c values and hysteresis losses are expected and observed (Fig. 9). H_c values of below 10 A/cm are easy to obtain [8]. The hysteresis depends in the same principle way as in crystals on inhomogenities, impurities, inclusions, surface roughness etc. Theory of micromagnetism can be applied without special considerations of

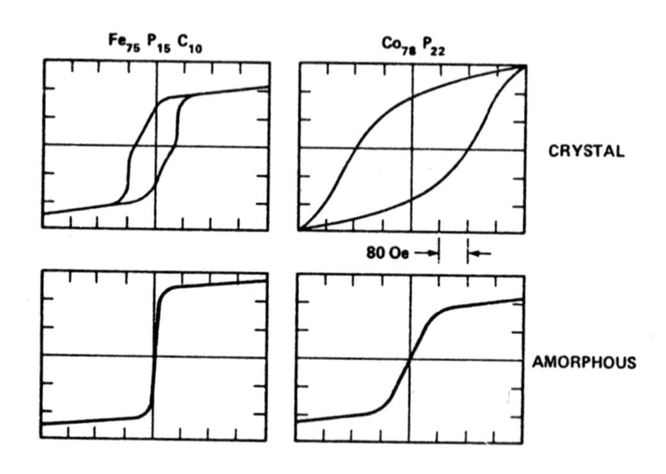

Fig. 9: Hysteresis loops in crystalline and amorphous films of Fe and Co

fluctuations in magnetization or exchange. In freshly prepared Metglass ribbons the stress induced by the thermal shock during preparation increases H_c and reduces the remanent magnetization and must be relieved by annealing at around 300° C for 1 hour. The domain patterns observed by magnetooptical Kerr effect change from narrow complicated structures to large domains similar to those in single crystals with very few perturbations [43]. However. the hysteresis properties still depend mostly on magnetostriction and internal stress. Values of H_c ∿ 5 - 10 mOe can be obtained for low magnetostrictiv alloys which can compete with 80 - 20 permalloy.

The role of amorphous materials in the magnetics industry of the future has been discussed recently in detail [20, 13, 16]. An attractive feature are the low hysteresis losses for the construction of distribution transformers. A reduction of core losses by a factor of 3 in all transformers used now in the USA would save annually 200 million Dollars. Disadventages are the small dimensions of ribbons and the relatively low saturation magnetization. They are less important for electronic applications in tape heads etc. For such special applications the larger flexibility in chemical composition compared to crystalline materials is a major adventage. However, the scientist has to realize that the technical interest for amorphous magnets is so large, because they are so similar to conventional crystalline materials and will bring some technical progress, but not a revolution.

9. CONCLUSION

The fundamental ferromagnetic properties of amorphous alloys are reasonably well explored and show no sensational deviations from principles known from crystalline materials. The similarity of intrinsic magnetic properties in the amorphous and crystalline state indicates, that the situation in the first coordination shell is more important than long range order. The directional d-d bonds and the screening by attached metalloid atoms stabilize the atomic structure in the first coordination shell so strongly that it survives melting and the usual preparation methods for amorphous materials. The most probable fundamental cluster is a closely packed tetrahedron of transition atoms. The random distribution of the tetrahedra and the metalloids sticking on them produces local fluctuations, which are, however, averaged out often by long range exchange so that surprisingly "normal" magnetic properties result in most cases. Some particularities of the disordered state, such as the larger homogenity and miscibility of alloys, result in improved magnetic properties which are of interest for technical applications.

LITERATURE

Introductory Articles:

[1] Amorphous Magnetism (Ed.: H.O. Hooper and A.M. de Graaf) Plenum Press, New York-London 1973
[2] G.S.Cargill: AIP Conf.Proc. 24, 138 (1975)
[3] T.Egami, P.J.Flanders and C.D.Graham: AIP Conf.Proc. 24, 697 (1975)
[4] P.J.Flanders, C.D.Graham and T.Egami: IEEE Trans.Magn.Mag. 11, 1323 (1975
[5] J.J.Gilman: Physics Today 28, 46 (1975)
[6] E.M.Gyorgy, H.L.Leamy, R.C.Sherwood and H.S.Chen: AIP Conf.Proc. 29, 198 (1976)
[7] R.Hasegawa, R.C.O'Handley and L.I.Mendelsohn: AIP Conf.Proc. 34, 298 (1976)
[8] T.Mizoguchi: AIP Conf.Proc. 34, 286 (1976)
[9] J.G.Wright: IEEE Trans.Magn. MAG 12, 95 (1976)
[10] Rapidly Quenched Metals (Ed.: N.J.Grant and B.C.Giessen) MIT-Press, Cambridge 1976
[11] Amorphous Magnetism II (Ed.: R.A.Levy and R.Hasegawa) Plenum Press, New York-London 1977
[12] G.Dietz: J. Magn.Magn.Mat. 6, 47 (1977)
[13] A.Hubert: J. Magn.Magn.Mat. 6, 38 (1977)
[14] S.Kobe and A.R.Ferchmin: J. Mat.Sci 12, 1713 (1977)
[15] U.Krey: J. Magn.Magn.Mat. 6, 27 (1977)
[16] F.Assmuss: Siemens Forsch. u. Entw. Ber. 7, 118 (1978)
[17] U.Krey: J. Magn.Magn.Mat. 7, 150 (1978)
[18] F.E.Luborsky: J. Magn.Magn.Mat. 7, 143 (1978)
[19] F.E.Luborsky, J.J.Becker, P.G.Frischmann and L.A.Johnson: J. Appl.Phys. 49, 142 (1978)
[20] F.E.Luborsky, P.G.Frischmann and L.A.Johnson: J. Magn.Magn.Mat. 8, 318 (1978)
[21] E.P.Wohlfahrt: J. Magn.Magn.Mat. 7, 113 (1978) IEEE Magn.MAG 14, 933 (1978)
[22] F.E.Luborsky, J.J.Becker, J.L.Walter and H.H.Liebermann: IEEE Trans.Mag.MAG 15, 1146 (1979)

Other interesting papers:

[23] K.H. Bennemann: J. de Physique, Coll C 4, 305 (1974)
[24] G.S.Cargill and R.W.Cochrane: Amorphous Magnetism I, 313 (1973)
[25] J.S.Kouvel: Magnetism and Metallurgy (Ed.: A.E.Berkowitz and E.Kneller) Academic Press N.Y. 1969
[26] T.Mizoguchi, K.Yamauchi and H.Miyajima in [1], p. 325
[27] J.A.Tarvin, Shirane, Birgeneau and Chen: Phys.Rev. B 17, 241 (1978)
[28] A.K.Bhattacharjee, R.Jullien and M.J.Zuckermann: J. Phys. F; Metal Phys. 7, 393 (1977
[29] R.W.Cochrane and G.S.Cargill: Phys.Rev.Letters 32, 476 (1974)

[30] H.A.Mook et al.: Phys.Rev.Letters 34, 1029 (1975);
Phys.Rev. B 16, 2184 (1977)
[31] Y.Takahashi and M.Shimizu: Phys.Lett. A 58, 419 (1976)
[32] K.Handrich: Phys.Stat.Sol. 32, K 55 (1969)
[33] C.C.Tsuei and H.Lilienthal: Phys.Rev. B 13, 4899 (1976)
[34] J.Durand: IEEE Magn.MAG 12, 945 (1976)
[35] H.S.Chen, R.C.Sherwood and E.M.Gyorgy:
IEEE Magn.MAG 13, 1538 (1977)
[36] J.J.Becker: IEEE Magn.MAG 14, 938 (1978)
[37] R.C. O'Handley: Phys.Rev. B 18, 930 (1978)
[38] N.Tsuya and K.I.Arai: IEEE Magn.MAG 13, 1547 (1977);
MAG 14, 946 (1978)
[39] F.E.Luborsky and J.L.Walter: IEEE Magn.MAC 13, 953 (1977)
[40] A.Brunsch and J.Schneider: IEEE Magn.MAG 14, 731 (1978)
[41] H.Hoffmann, A.J.Owen and F.Schröpf: Phys.Stat.Sol.52,161(1979)
[42] H.Kronmüller and J.Ullner: J. Magn.Magn.Mat. 6, 52 (1977)
[43] B.Gröger and H.Kronmüller: J. Magn.Magn.Mat. 9, 203 (1978)
[44] H.Kirchmayr and C.Poldy: J. Magn.Magn.Mat. 8, 1 (1978)
[45] J.Durand: IEEE Magn.MAG 12, 945 (1976)
[46] K.Ganzhorn: Dr. Thesis Stuttgart 1952 and Z.S.f. Nat.Forsch.
7a, 291 (1952)
[47] A.Adachi, Morita, Ogino and Ueda:
Prop. of Liquid Metals (Ed.: S.Takeuchi)
Taylor and Francis, London 1973.

Papers about amorphous magnetism are also accumulated in:

Physica 86-88 B, 745-813 (1977): ICM 1976 in Amsterdam
IEEE Trans.Magn.MAG 11, 1323-1338 (1975): INTERMAG London
dito MAG 12, 921-953 (1976): INTERMAG Pittsburgh
dito MAG 13, 1532-1558, 1598-1620 (1977):
INTERMAG Los Angeles
dito MAG 14, 719-733, 933-957 (1978): INTERMAG Florence
J. Magn.Magn.Mat. 6, 27-86 (1977): AG Magn. Münster
J. Magn.Magn.Mat. 7, 143-206 (1978): Haifa

AN INTRODUCTION TO SPIN GLASSES

J.A. Mydosh

Kamerlingh Onnes Laboratorium
der Rijks-Universiteit Leiden
Leiden, The Netherlands.

ABSTRACT. We present a working definition and a general description of a spin glass. A number of different systems are discussed and a connection is made whenever possible with the amorphous metals. We then review the various concentration and temperature ranges for spin glass behavior. A series of recent experiments are discussed and a percolation type of model is gradually developed. The two current approaches to the theory of spin glasses are briefly outlined. Finally we conclude with some speculations as to open questions and future work.

1. DEFINITION AND GENERAL DESCRIPTION OF SPIN GLASSES

The term spin glass has been broadly applied to the freezing behavior and low temperature magnetic properties of many different classes of random magnetic systems. Presently this concept is the subject of intensive study, from the experimental side in order to systematically characterize the exact nature of the spin glass behavior, and theoretically to determine the best and simplest model with which to describe these properties. The experimental results encompassing a large variety of different experiments are somewhat muddled by problems of sample preparation and metallurgy. While among the existing theories, there is a strong rivalry between the phase transition approach which mainly focuses upon the freezing process and the model of superparamagnetic blocking which treats primarily the low temperature effects.

A general, working definition of a spin glass may be given as a random, mixed-interacting, magnetic system characterized by a random freezing of spins at a well-defined temperature T_f below which distinctly different magnetic behavior appears without the presence of long range order. The two most essential ingredients of a spin glass are therefore the randomness and the mixed interactions. Normally there are two ways of creating a random magnetic system. The first method consists of simply substituting a good magnetic element for a nonmagnetic one in a multicomponent system. The best-known examples are the archetypal spin glass alloys <u>Cu</u>Mn and <u>Au</u>Fe [1]. A second and more modern technique takes a magnetic intermetallic compound and crystallographically disorders it, i.e. makes it amorphous usually through sputtering. Some typical examples utilizing this technique are $GdAl_2$ crystalline Curie temperature T_c = 170 K, amorphous T_f = 16 K [2]; YFe_2 crystalline T_c = 540 K, amorphous T_f = 58 K [3], and MnSi crystalline T_c = 30 K amorphous T_f = 22 K [4]. It should be mentioned here that the general techniques for producing amorphous metals have been usefully employed to fabricate novel or more homogeneous spin glasses. For, many nonsoluble or nonexisting alloys may be formed over a large concentration region in the amorphous state, e.g. LaGaAu [5]. In addition some of the well-studied Metglass-like systems naturally exhibit spin glass behavior [6]. The second essential ingredient of a spin glass is the mixed or competing ferro-antiferromagnetic interactions. Experimentally there does not seem to exist today a spin glass where only antiferromagnetic exchange is present. Yet two attempts with well-defined and randomly distributed, antiferromagnetically coupled impurities, In doped into CdS [7] and P doped into Si [8], have failed to find spin glass-like freezing down to temperatures of about 3 mK. However, solely antiferromagnetic spin coupling will certainly lead to "frustrated" spins and according to Toulouse [9] these frustration effects are highly important in the formation of the spin glass state. So at least for the present we must rely on the spatially oscillating spin-spin interactions such as the RKKY [10], the dipolar and mixtures of ferro- and antiferromagnetic superexchange. We illustrate the RKKY polarization of conduction electrons and the coupling between two spins in Fig. 1. Note that the spin-spin interaction is strong, oscillatory and of long range so that spin glass behavior is commonly found in metallic systems.

Now what is this random freezing of spins at T_f, the freezing temperature? This question is not so easily answered, for, the essence of the freezing process is at present unresolved and in dispute. Nevertheless, we may offer the following naive description. First, as the temperature is lowered from $T \gg T_f$ many of the randomly positioned and freely rotating(paramagnetic) spins build themselves into locally correlated clusters or domains which can then rotate as a whole. This development of interacting spins is nucleated by the always present concentration fluctuations. At a given temperature, a portion of the total spin system will remain loose or free from correlations with other spins. Thus, the cluster growth process is strongly dependent upon the

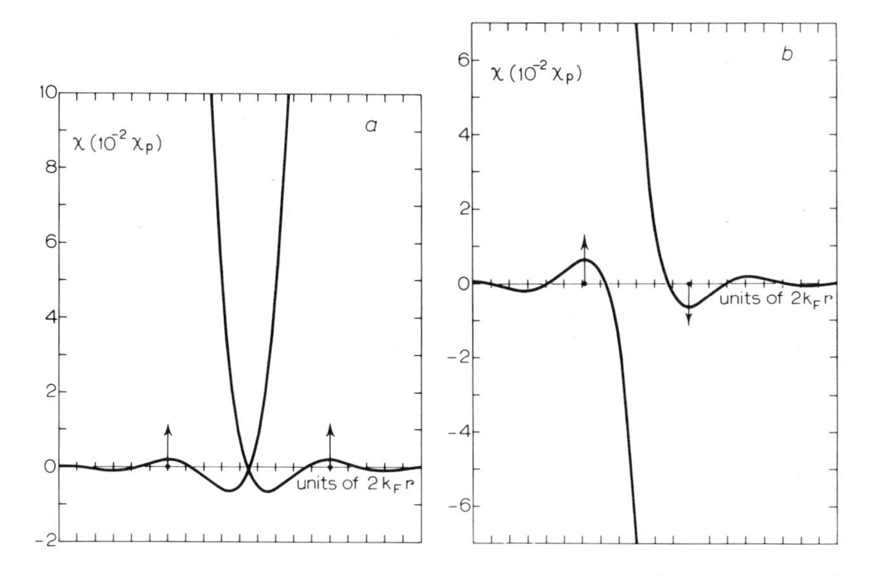

Fig. 1 The RKKY interaction between two local moments different distances apart. χ_p is the Pauli paramagnetic susceptibility.

alloy randomness and the temperature. As $T \to T_f$ the various spin components begin to interact with each other over a longer range and the different size clusters become viscously frozen with random orientations. Thereby the effective cluster spin is locked into a favorable alignment axis which is caused by the geometric anisotropy of the cluster. Certainly not all the spins or clusters participate in the freezing at T_f since there are always at finite T some loose spins. Now it would seem appealing to make a direct comparison with the solidification process of a real glass (or with the crystallization procedure in an amorphous metal). Here there is no sharp phase transition but a gradual change in the behavior of the viscosity, density, volume, etc. as the molten liquid is cooled into an amorphous solid. Most important are the irreversibilities and dependences of the experimental properties upon the rates and directions of temperature variation. Although such considerations for the spin glasses have been discussed there have been no systematic investigations to elucidate the connection. Thus, the value of treating the spin glass in terms of vitrification theory remains an open question.

The change of magnetic behavior below T_f concerns a transition from a (super)paramagnetic state into a highly irreversible, metastable, frozen state. Here the application of an external field induces many ferromagnetic-like properties, e.g. remanence, hysteresis, etc., and in addition a long time relaxation which results from any change of field. Although the zero-field cooled and measured spin glass has no net magnetization and the truely reversible susceptibility can only be measured in the limit $H_{ext} \to 0$ [11], there is a definite reaction of the spin glass to fields ranging from ≈ 10 μT [12] up to 40 T [13]. Even in such large (pulsed) fields complete saturation is not reached in a frozen spin glass at low temperatures. This is perhaps the most difficult feature of the spin glass state - - its response to such a wide spectrum of fields or energies. An early suggestion was originally made to treat these low temperature spin glass properties together with those of an amorphous solid (including now amorphous metals [14]) in terms of a two level system model [15]. Very recently this suggestion has been reemphasized by the inclusion of a reaction field appearing at T_f which generates the two level system in the spin glasses [16]. While a sophisticated understanding exists for the two level system and its detailed properties in amorphous materials, very little has been done to directly investigate this model for the spin glasses.

The question of long range magnetic order involves a spatial order or periodicity which is put into an order parameter. For even the most complicated magnetic structures, one can always predict or calculate the orientation of a certain spin a given distance from a fixed spin. There is no long range order of this type in a spin glass. The usual ordering is prevented by a randomness of spin directions coupled with a multidegenerate ground state. The latter is a result of frustrated spins which have no energy difference between pointing in several orientations. For the spin glass we must use a time dependent correlation function $\langle \overline{S(t)}\ S(o)\rangle \neq 0$ which may be transformed into an order parameter $q=\overline{\langle S\rangle^2}$ where the $\langle\ \ \rangle$ represents a thermal average and $\overline{\quad\quad}$ a configurational one [17].

In the above we hope that a number of connections have been made to the area of amorphous metals even though the spin glass behavior can be observed in perfect single crystal alloys. The next Section will consider how the spin glass fit into the general scheme of magnetic alloys - - from the single impurity Kondo effect to the nearest neighbor percolation for long range magnetic order. Then in Section 3 we shall discuss some salient and recent experimental features of the spin glasses. In Section 4 an introduction to the theory of spin glasses is presented by sketching the two current approaches - - the Edwards-Anderson model and the Néel superparamagnetism. Section 5 outlines a combined model which accounts for many of the experimental properties and furnishes, at least for the experimentalist, a modus operandi. Finally Section 6 poses a few of the open questions and offers some suggestions for future work on spin glasses.

2. CONCENTRATION AND TEMPERATURE REGIMES FOR A SPIN GLASS

If we begin with a single magnetic impurity substituted within the lattice of a nonmagnetic metal, the idealized Kondo effect arises. This is an antiferromagnetic coupling between the conduction electrons of the metal and the localized d (or f) electrons of the impurity, $- J \vec{s} \cdot \vec{S}$. At high temperatures the impurity behaves like a strong, paramagnetic moment. However, below a characteristic temperature T_K, specific for each alloy system, the impurity becomes nonmagnetic due to its interaction with the conduction electrons. This temperature is known as the Kondo temperature and it signifies the beginning of a broad temperature transition to a quasi-bound, singlet, state which is unable to response to a magnetic field (nonmagnetic) and possesses an enhanced electron scattering cross-section (resistivity minimum). It should be mentioned here that similar low temperature resistivity minima are commonly found in amorphous metals. For such materials one must be very careful to distinguish between a resistivity minimum caused by the Kondo effect of a magnetic component and that resulting from the amorphous structure of the metal.

Of course it is unrealistic to talk about a single impurity and we must consider a finite concentration alloy with the resulting inter-impurity RKKY interaction. Now at low temperatures we have two opposing effects: the Kondo moment weakening and the interactions which tend to strengthen the moments via a magnetic environment. Upon sufficiently increasing the concentration, the impurity-impurity couplings will reduce the average T_K, for, a strongly interacting pair of impurities possesses a lower T_K than a single impurity and for a triplet, etc., T_K is even further decreased. In Fig. 2 we show schematically how the average T_K might vary with the concentration. Note the gradual emergence of the spin glass state (T_f) for $c>c_o$. As a typical example <u>Au</u>Fe has a $T_K = 0.2$ K and a $c_o \simeq 50$ ppm. Available evidence on many different alloys indicates that a spin glass phase always appears in a certain concentration regime once "good" moments have been formed. We would call attention to very weak moment systems such as VFe ($c_o \approx 25$ at.%) and CuNi ($c_o \approx 40$ at.%)[18] which nicely show this spin glass phase.

On the high concentration side there exists a percolation limit (or critical concentration) c_p for the onset of long range, albeit inhomogeneous, magnetic order. This is most simply illustrated for the case of nearest neighbor direct ferromagnetic exchange. Here an infinite cluster of correlated spins is created with a minimum site occupancy concentration, c_p, of 16 at.% for a fcc lattice e.g. <u>Au</u>Fe [19] and 19 at.% for a bcc lattice e.g. <u>Cr</u>Fe [20]. Modified forms of this percolation concept may be used for weak moment-cluster systems and for the ordering in the giant moment alloys, such as Pd with Fe, Mn or Co [21]. The long range magnetism at T_c for $c>c_p$ (see Fig. 2) is of interest itself

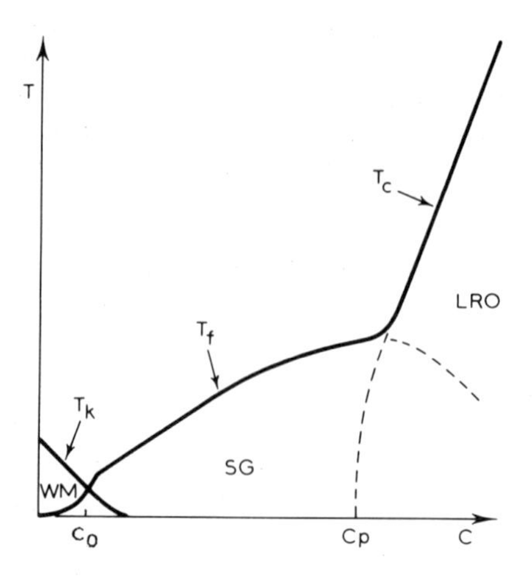

Fig. 2 Schematic T-c phase diagram for a typical magnetic alloys illustrating the weak moment (Kondo), spin glass, and long range ordered regimes.

since the intrinsic randomness and the nonuniformity of the ordering cause definite changes in the critical phenomena and critical exponents. Also there are background or fluctuation effects due to the nonconnected smaller clusters. Such type of phase transitions in random systems can be treated with the renormalized group methods [22].

In between the low-c Kondo regime and the high-c, percolated (long range order) phase lies the spin glass region. It is customary to subdivide this spin glass portion of the phase diagram into three parts: a) a low c<0.5 at.% scaling regime formed by strictly the RKKY interaction where $T_f \propto c$ and the different experimental quantities are universal functions of T/c and H_{ext}/c, b) an intermediate c-regime where the scaling relations break down and where, with increasing probability for direct exchange, ferromagnetic clusters begin to appear and to affect the magnetic properties, and c) a higher concentration region $c \approx 10$ at.% extending up to the percolation limit. In this latter regime very large ferromagnetic clusters 20-20,000 μ_B exist and cause the dominance of metastable ferromagnetic effects. Here the terms mictomagnetism or cluster glass are usually applied to describe the magnetic behavior.

Recently there has been considerable attention devoted to possibilities and difficulties of observing a double or mixed transition for c just above c_p [23]. In Fig. 2 we illustrate this paramagnetic→(T_c)→ferromagnetic→(T_f)→spin glass transformation with dotted lines. Compelling experimental evidence is available on a number of systems [24] that the ferromagnetic state is broken up into a distribution of randomly aligned clusters or domains. Neutron scattering measurements indicate that the average extent of these regions is 30-50 Å. Most probably a random anisotropy is present which destroys the ferromagnetic infinite-cluster by reducing the temperature. Since this effect is closely associated with the increased presence of antiferromagnetic exchange which offsets the predominance of ferromagnetic exchange, we used the concepts of a spin glass to describe the behavior at temperatures below the second transition. For $T<T_f$ many of the experimental effects are similar to a pure spin glass.

Therefore in Fig. 2 we suggest a general T-c phase diagram for most magnetic-nonmagnetic substitutional alloys. The importance of the spin glass state is clearly seen and we must now try to fully understand this phase.

3. SOME SELECTED EXPERIMENTS ON SPIN GLASSES

3.1 Susceptibility

One of the most effective methods with which to investigate the onset of the spin glass ordering remains the low field susceptibility χ[25]. For, the sharp peaks or cusps in the $\chi(T)$ behavior determining T_f represent a fundamental property of spin glasses. An ac mutual inductance technique is commonly used to measure $\chi(T)$ with field and frequency as external parameters. The great advantage of the ac susceptibility is that the driving field is usually very small <1mT and its rotation-frequency (normally in the low audio frequency range) is large enough to prevent the appearance of long-time metastable effects. Thus the true reversible susceptibility is measured, i.e. $\chi(T) = \lim_{H\to 0}[M(T)/H]$. Two current approaches to susceptibility experiments have been improved dc sensitivity, especially due to SQUID techniques, in small external dc fields (<10μT) and the determination of the ac $\chi(T)$ with frequency variations of the ac driving field from 1 Hz to 10 GHz.

Let us begin by considering the ultra low field, dc, χ-measurements. One such study has been recently carried out by Krusin-Elbaum et al [12] on the $CoO\cdot Al_2O_3\cdot SiO_2$ (14 at.% Co) insulating, real glass, "spin glass". Their data for χ_{dc} versus T with $3\mu T<H_{ext}<80\mu T$ is shown in Fig. 3. Note that the broad leveling off (or at best a knee) in $\chi(T)$ as the temperature is lowered developes into rather sharp peak as H_{ext} is reduced below 15 μT. So a peak which is always present in χ_{ac} finally emerges in χ_{dc} for low enough H_{ext}. Furthermore, for $H<15\mu T$ thermal hysteresis or relaxation effects are not observed. $\chi(T)$ is quite reversible with temperature below this 15μT threshold field. These results clearly demonstrate the $H\to 0$ limit for the reversible susceptibility and the surprising sensitivity of the spin glass to external fields. In addition, such properties may be taken as

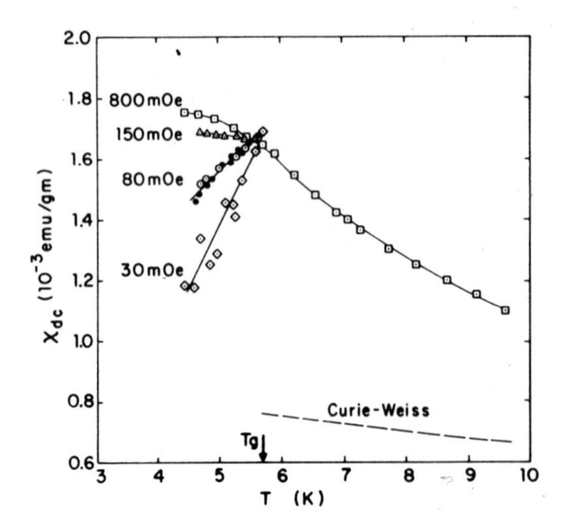

Fig. 3 The very low external field dc susceptibility of a Co aluminosilicate glass versus the temperature. After Krusin-Elbaum, Raghavan and Williamson [12].

evidence for a cooperative transition at T_f below which correlations extend over many spins [12].

The frequency dependence of $\chi(T)$, especially at T_f, is a highly controversial situation. A number of experiments have been performed on similar systems with seemingly contradictory results. Initially we must offer a word of caution. The susceptibility of metallic materials is a complicated quantity to measure at high frequencies. For, $\chi(\nu) = \chi'$ (dispersion) + $i\chi''$ (absorption) and in conducting systems χ'' can become very large thereby affecting the magnetic susceptibility χ'. This occurs not only through skin depth and eddy current (heating and magnetic field) effects but also through a possible decoupling of the local spin-conduction electron system from the lattice. So extreme care must be taken with simultaneous measuring the absolute values of χ' and χ'', and additionally working with powdered or fine grain samples.

A high-frequency, ac-susceptibility (χ'-only) study of AgMn (0.1 to 0.5 at.% Mn) in powder (100 μm) and thin foil (18 μm) form has been carried out by Dahlberg et al [26]. With frequencies ranging between 16 Hz and 2.8 MHz, the freezing temperature T_f (peak in χ') was frequency independent within experimental error. Mulder et al [27] have performed $\chi'(\nu, H_{ext}, T)$ and $\chi''(\nu, H_{ext}, T)$ measurements on powdered CuMn (0.23 to 1.5 at.%). The frequency ν was varied from 1 Hz until χ'' became approximately 10 % of χ' at T_f which was about 10 kHz for the ≈100 μm grain size. Within an experimental error, of <2% there was no frequency dependence of the χ-peak temperature. A rapid increase in χ'' (T_f) along with a weak decrease in $\chi'(T_f)$ was observed at frequencies above 5 kHz even in these powdered samples. The shape of the χ-peak, but not its temperature, was strongly dependent on sample preparation and heat treatment.

In contrast to these results a number of authors [28] have reported a frequency shift (always positive) of T_f for a number of metallic spin glasses including CuMn. The alloys were in bulk form and there was a concentration dependence of ΔT_f, i.e. the larger the concentration the larger the $\Delta T_f(\nu)$-shift. In an attempt to resolve these differences along physical grounds Tholence [29] has suggested a general curve of $\Delta T_f/\Delta\log\nu$ versus c, which fits all of the experiments and is based upon the superparamagnetic blocking model. For the very low concentration of Dahlberg et al. [26], $\Delta T_f/\Delta\log\nu$ is very small and might not have been seen in their experiments. On the other hand, the work of Mulder et al. [27] on CuMn (1.5 at.%) should show some discernible frequency shift. Another possibility which is related to the concentration dependence of the frequency shift concerns the amount of ferromagnetic short range order. The systems which clearly show a large $\Delta T_f(\nu)$ have ferromagnetic clusters present; the same is true for the insulating "spin glasses" [30] which are dominated by ferromagnetic cluster effects. So it would seem that low concentration AgMn and CuMn are unlikely candidates to exhibit a significant frequency dependence.

3.2 Specific Heat and Energy Flux.

The specific heat of the spin glasses remains a puzzlement with no anomalous behavior having been observed at T_f. The most accurate measurements to date reveal a broad maximum at $T>T_f$. Yet, the phase transition theories of spin glasses predict some type of anomaly at T_f. This may be only a broad maximum or a small peak superimposed upon a Schotty-type or X-Y linear-chain broad specific heat maximum. An increased sensitivity, than that presently available via calorimetry experiments, is probably required to detect such a fine temperature-interval peak as predicted by the cluster (modified Edwards-Anderson) theory [31]. We would also like to recall the suggestion that T_f is related to a maximum or knee in C_m/T or dS_m/dT [32]. Here C_m is the magnetic contribution to the specific heat and S_m is the magnetic entropy equal to cRln (2S+1). At low temperature clear evidence has been found for deviations from the famous linear initial specific heat with the addition of a T^2-term [33].

One recent attempt to determine the fine structure of the specific heat near T_f has been to measure the ultrasonic velocity $\Delta v/v$ and to thermodynamically relate this quantity to the specific heat [34]. These authors derive

$\frac{\Delta v}{v} = A \int_0^T C_m dT + BTC_m$ where A and B are temperature independent constants. The results of their experiments on Cu + 5 at.% Mn are given in Fig. 4. Here a fit to the high temperature back-

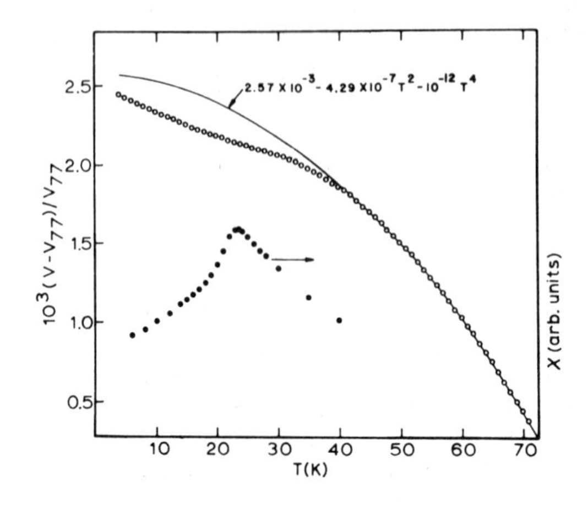

Fig. 4 Change in the ultrasonic velocity versus temperature for a CuMn spin glass. The ac susceptibility is shown for comparison. After Hawkins, Thomas and de Graaf [34].

ground is made and then subtracted to obtain the "critical" part of $\Delta v/v$ which in turn is compared to the specific heat written in critical exponent form $C_m = P-Q\varepsilon^{-\alpha}/$, $(\varepsilon = (T-T_f)/T_f)$. Also shown in Fig. 4 are the ac susceptibility data for the same sample. Notice that there is no sharp anomaly in $\Delta v/v$. By analyzing these $\Delta v/v$ data, the critical exponent α for the $T<T_f$ is found to be -1.9. This large negative value for α means that the specific heat C_m will only exhibit a broad maximum at T_f. However, the best specific heat measurements to date all show a broad maximum, but at a temperature ≃20% larger than T_f.

Another related experiment concerns the measurement of the energy flux for $T<T_f$ as the external field is changed. For both possibilities (field cooling and then setting $H_{ext}= 0$ at $T<<T_f$ or zero field cooling and then turning on a field at $T<<T_f$) heat flows out of the sample with a long-time relaxation (see Fig. 5). At very low temperatures the energy flux $\dot{Q}\propto\frac{1}{t}$ and the energy $Q(H_{ext})$ = (const.) log t [35,36]. Recently such measurements have been extended to higher temperatures and interpreted in terms of the two level system model [36]. At temperatures approaching T_f and for long times, $\dot{Q}^{-1}$ is not simply proportional to t. This behavior is similar to the remanent magnetization which is usually described with the Néel superparamagnetic blocking model. However, more is required for the existence of an energy flux. The latter

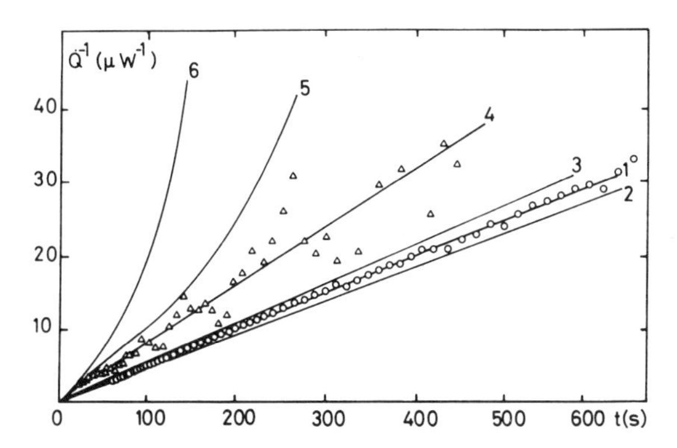

Fig. 5 The inverse energy flux $\dot{Q}^{-1}$ out of a AuFe spin glass at various temperatures (1:T=1.65 K to 6:T=4.15 K) for field cooling (2.5 T) and reducing this field to zero. After Berton et al. [36].

authors [36] argue that, in addition to a distribution of energy barrier heights P(V) which is sufficient for describing the remanent magnetization, a distribution of splittings of the double well (nonequal depths) P(W) is further required for an energy flux. A P(V) and a P(W) are the two prerequisites for the application of the two level system model. Thus there is once again a direct connection between the low temperature phenomenon in the spin glass and that in amorphous materials. Nevertheless, the quantum mechanical tunneling which is essential for the amorphous materials has not yet been verified in the spin glasses - - thermal activated processes alone can explain the data. It would be very interesting to further search for tunneling processes at very low temperatures in the spin glasses, for, then, a direct correspondence with the real glasses could be made.

3.3 Neutron Scattering and μ^+ Spin Relaxation

Since the phenomena in a spin glass are very complicated we must consider a variety of experimental techniques including, in particular, microscopic measurements. Two highly sophisticated techniques which directly probe the spin correlation at the atomic level, namely neutron scattering and positive muon spin relaxation, have recently been brought to bare on the spin glass problem. Murani and Heidemann [37] have studied a CuMn 8 at.% alloy via neutron scattering and determined the elastic scattering cross section as function of the temperature and the energy resolution or window, ΔE, of the spectrometer. The energy resolution is related to longest relaxation time of the spin system, $\tau_o \approx \hbar/\Delta E_o$. For spin relaxation times longer than τ_o the scattering appears

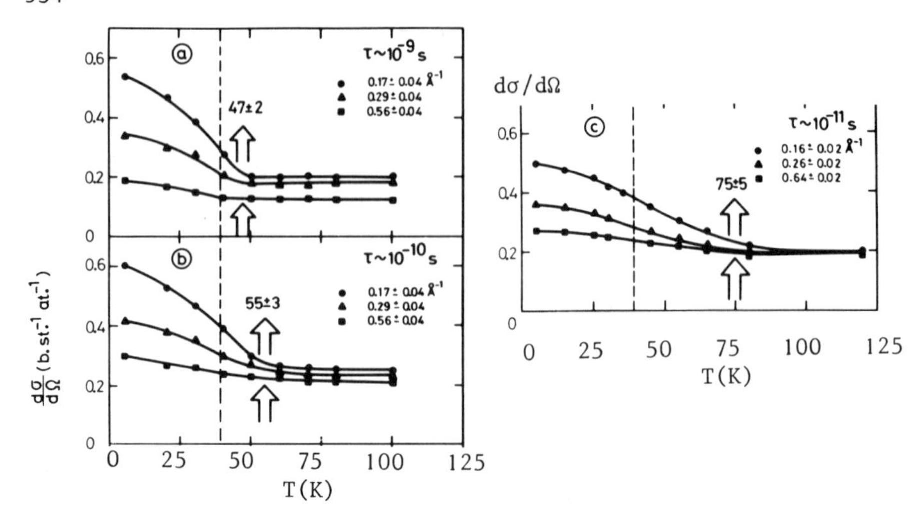

Fig. 6 Elastic scattering cross section for a CuMn alloy versus temperature with different time constants of measurements. The various shaped points represent different scattering vectors. After Murani and Heidemann [37] .

elastic. Inelastic or quasielastic processes can only be resolved for $\tau<\tau_o$, they lie outside the elastic energy window of the spectrometer. In addition the Edwards-Anderson order parameter (see Sections 1 and 4) is describable in terms of the elastic scattering at any given temperature. So by employing different spectrometer techniques Murani and Heidemann [37] have measured the temperature dependence of the elastic scattering cross section as τ_o (or ΔE) is varied from 10^{-9} to 10^{-11} sec. The onset of spin freezing (slowing down of the spin relaxation processes) is indicated by an increase in the elastic scattering cross section. In Fig. 6 we have reproduced the data of Murani and Heidemann. Note that the temperature for this scattering increase decreases as the time constant of the measurement τ is reduced. These results indicate that a wide spectrum of spin relaxation time exists at each temperature around T_f in a spin glass and further that depending upon the time constant of the measurement, a particular freezing temperature may be defined. We illustrate this schematically in Fig. 7.

Most recently Yamazaki et al. [38] have studied the time correlations of random moments in two CuMn (0.8 and 0.5 at.%) alloys via zero field muon spin relaxation technique (in contrast to the field dependent μ^+spin rotation (μ^+S R) employed previously in the study of spin glasses [39]). The zero field μ^+ method probes the atomic dipolar fields from the randomly distributed Mn moments and their relaxation function as a function of temperature. If the Mn moments are completely frozen $T<T_f$, the static

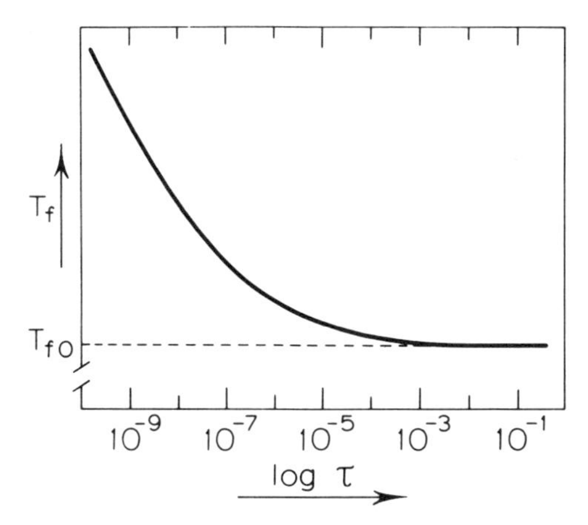

Fig. 7 Schematic representation of the freezing temperature as determined by different measurement techniques having different "time constants" of measurement.

spin relaxation function should be a highly damped exponential decaying to a constant value of 1/3. However, for fluctuations in the spin system, this constant 1/3 also decays exponentially to zero. Experimentally the latter is observed indicating that even for $T \approx 0.5\ T_f$ the Mn spins are slowly fluctuations with an average time constant $\approx 10^{-5}$ sec. In the high temperature region $T \gg T_f$ the μ^+ relaxation function shows a fast decay which can be halted with the application of a small external field. Here the average Mn relaxation time is of order 10^{-10} sec. As $T \to T_f$ from above the experiments which are rather preliminary indicate a smooth variation or a continuous slowing down. This persists through T_f down into the low temperature region where the fluctuations are relatively slow $\approx 10^{-5}$ sec. It must be emphasized that this single Mn fluctuation time represents a rough average over all the Mn spins for the particular temperature. Thus, as suggested by Yamazaki et al. [38], a distribution of fast relaxing regions may coexist with other regions which relax slowly. These results are consistent with the aforementioned neutron scattering experiment and also with the concept which was introduced in Section 1 of an infinite cluster of randomly frozen spin in addition to many small clusters and loose spins.

4. THEORY OF SPIN GLASSES

In 1975 Edwards and Anderson (EA) [17] proposed a simple model for describing spin glass behavior. They replaced the randomly distributed magnetic impurities, interacting via the RKKY interaction (see Fig. 1), with a uniform lattice of spins (every site occupied) which have a random distribution of exchange interactions $P(J_{ij})$. Thus,

$$P(J_{ij}) = [(2\pi)^{\frac{1}{2}}J]^{-1}\exp[-(J_{ij}-J_o)^2/2J^2]$$

where $J_o \neq 0$ in the case of a ferromagnetic-spin glass [40]. EA proceed to treat this model via the standard Hamiltonian and the methods of statistical mechanics.

$$\mathcal{H} = \tfrac{1}{2}\sum_{ij} J_{ij}\vec{S}_i\vec{S}_j + g\mu_B H_o^z \sum_i S_i^z ,$$ defining

$$\left.\begin{aligned} \langle\vec{S}_i\rangle &= 0 \text{ for } T>T_f \\ \langle\vec{S}_i\rangle &\neq 0 \text{ for } T<T_f \end{aligned}\right\} \text{ thermal averages.}$$

But $M \equiv \overline{\langle\vec{S}_i\rangle} = 0$ always via a configuration (bar) average. This states that the net magnetization is always zero which is the spin glass condition of random freezing. Using the fluctuation-dissipation susceptibility,

$$\chi = \chi_{cw} \sum_{ij} [\overline{\langle\vec{S}_i\cdot\vec{S}_j\rangle} - \overline{\langle\vec{S}_i\rangle\langle\vec{S}_j\rangle}] \frac{1}{NS(S+1)}$$

where $\chi_{cw} = \dfrac{NS(S+1)(g\mu_B)^2}{3\ kT}$

For the case of no short range order (no long range order is instrinsic to the spin glass problem) each i-spin has a j-spin neighbor with up or down spin directions of the same probability. Therefore,

$$\overline{\langle\vec{S}_i\cdot\vec{S}_j\rangle} = S(S+1)\,\delta_{ij}$$

$$\overline{\langle\vec{S}_i\rangle\langle\vec{S}_j\rangle} = \overline{\langle\vec{S}_i\rangle^2}\,\delta_{ij} \quad \text{and}$$

$$\chi = \chi_{cw}[1-q]$$

where q is the EA spin glass order parameter, $q \equiv \overline{\langle\vec{S}_i\rangle^2}$. Now one must look for a free energy F such that for $\partial F(q)/\partial q = 0$, a self-consistent equation for q(T) is obtained below a "critical temperature" T_f. EA performed these difficult calculations in the mean field approximation via the replica method. The free energy equation to be solved has the form

$$\begin{aligned} F = \overline{F\{J_{ij}\}} &= \int P(J_{ij})F(J_{ij})dJ_{ij} \\ &= \int P(J_{ij})[-k_B T\ln(\mathrm{Tr}\ \exp[-\mathcal{H}/k_B T])]\,dJ_{ij}. \end{aligned}$$

A specific q(T) dependence was found for $T<T_f$, and by taking temperature and field derivatives of the free energy, the specific

heat $C_m(T)$ and susceptibility $\chi(T)$ are available for comparison with the experimental results. A cusp is predicted for $\chi(T)$ at T_f with a low temperature behavior also in accord with experiment. The EA theory further predicts a semi-cusp in $C_m(T)$ at T_f which is not found in the measurements. The EA model has stimulated an enormous amount of interest, for, many properties of a spin glass may be calculated using the statistical mechanical techniques of the model. In addition the basic model permits a variety of calculations beyond the mean field approximation [41] and is easily simulated via the Monte Carlo method [42]. There is presently much discussion over the existence of a phase transition in the EA model going beyond the mean field approximation. Most theoretical indications point towards the lack of a phase transition in 3-dimensions and currently a number of attempts are being made to retain the basic EA model while improving the calculations.

The other approach to describing spin glass behavior is the phenomenological Néel model of superparamagnetism [43]. This model has been widely used for small particlesand rock magnetism. Here superparamagnetic (large effective moment) clusters, m, exist at high temperatures and become "blocked" at low temperatures by an energy barrier associated with an anisotropy energy E_A and an external field energy $H_{ext}m$. Spin flip transitions require thermal activation over the energy barrier E which for a spin glass has a certain probability distribution P(E). These transitions occur at a rate $1/\tau$ = const $\exp(-E/k_BT)$ where the const is typically of order $10^9 s^{-1}$. If the relaxation time τ becomes larger than a typical time of measurement τ_m, then the particle magnetization is frozen with a blocking temperature T_B given by $E(T_B) = kT_B \ln(\text{const}\ \tau_m)$. Thus, there is also a distribution of blocking temperatures $P(T_B)$. The origin of this anisotropy which must be distributed in random directions over the different clusters is not certain. Dipolar interactions which have an intrinsic anisotropy and the shape anisotropy associated with the nonspherical cluster geometry both have been proposed. The random crystalline anisotropy which is a special property of the amorphous 4f magnetic system is not as useful in the mainly 3d impurity spin glasses. Additional work is required to better understand the crucial anisotropy in a spin glass. The freezing occurs when a significant number of the clusters are blocked and thereby unable to contribute to the susceptibility; thus a χ-peak appears at T_f. Recently Wohlfarth [44] has developed a procedure for determining the distribution of blocking temperatures $P(T_B)$ from $\chi(T)$: $P(T_B) = \frac{d(\chi T)/dT}{\lim_{T\to\infty}(\chi T)}$. For a sharp spin glass χ-cusp, $P(T_B)$ exhibits a very sharp step at T_f which then goes through a maximum and slowly tapers off at low T_B [27]. This rather unique distribution function indicates a cooperative behavior of the clusters at T_f. On the basis of random statistics and the random molecular field model the distribution functions for P(H) should be Gaussian or Lorentzian. At low temperature, this randomly frozen chain of clusters still possesses sufficient thermal excitations with which to activate over certain small energy barriers. These processes are well-describable within the Néel model and lead to the remanence, irreversibilities and long-time relaxation effects characteristic of the frozen spin glass state.

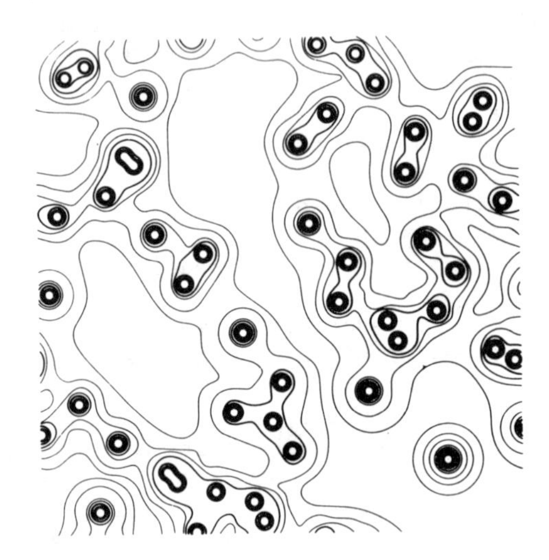

Fig. 8 Computer simulation for a random lattice of points (spins) with the contours and their overlapping representing the different ranges of spin-spin interactions. After Verbeek [45].

5. SCHEMATIC MODEL FOR SPIN GLASS FREEZING

Based upon these two approaches for describing the salient features of a spin glass, we would propose the following compromise. We rely upon our experience with the giant moment, dilute ferromagnetic alloy [21] which exhibit a similar type of transition, but into a long range ordered ferromagnetic state. Now we expand upon the description given in Section 1. The gradual growth of correlated spins (evolution of clusters) is certainly different than the well-defined chemical particles of the original Néel superparamagnetism. This difference is illustrated in Fig. 8 by the computer simulation of temperature dependent regions of spin correlations [45]. Here the initially spherical-symmetric correlation contours extend further as the temperature or thermal disorder is reduced. A large k_BT prevents the correlation from overlapping and forming large clusters. This represent the high temperature regime of a spin glass, i.e. a collection of dynamic independent supermagnetic moments. As the temperature is lower towards T_f the growing clusters of various sizes and strengths begin to make contact with each other. There is a viscous rubbing or a competition of the loose spins at the periphery of two touching cluster. Now with this diminished freedom to rotate or to response to a magnetic field, the random anisotropy can play the deciding role. Note the irregular pattern which develops in Fig. 8 for the randomly distributed lattice of central spins. This elucidates our concept of shape anisotropy. Various groups of spins have a preferred axis with which to align; the orientations of the axes are random throughout the sample. At a certain temperature T_f an "infinite" (macroscopic) region of viscously touching small clusters form, the moments are no longer free to respond as a superparamagnet. The friction at the periphery combined with the internal anisotropy causes a locking in to the perferred axis and the system is frozen. The manifestation of this process on a macroscopic scale represents a percolation, i.e. an infinite cluster of randomly oriented smaller clusters. As is well-known percolation phenomenon is a cooperative effect and can be fully treated within the theory of phase transitions. There are also many free small clusters which exist outside the infinite cluster. For a giant moment ferromagnet, the predominance of ferromagnetic exchange produces mainly parallel couplings as two small clusters begin touch. The role of the random anisotropy is overcome by this exchange. So for a spin glass antiferromagnetic exchange is necessary which then weakens or frustrates the coupling at the periphery and allows the anisotropy to fix the orientation. We argue in general that a competition exists between the ferromagnetic exchange on the one side and the antiferromagnetic exchange together with random anisotropy on the other. The winner determines whether the alloy is a percolated ferromagnet or a frozen spin glass. Once the temperature is reduced through T_f, there are still some weak links in the infinite cluster and the free noninfinite clusters. Here the full Néel theory may be brought to bear with thermally activated processes over various magnitude energy barrier determining the low temperature behavior. Indeed this concept does nicely describe the remanence, irreversibilities and long time relaxation. Since for $T \ll T_f$ we are dealing with rather large magnetic entities possessing a great deal of friction, the relaxation times will be very long, even approaching temporal periods beyond what is normally observable in a laboratory. As mentioned previously there is still the open question of quantum mechanical tunneling processes in a spin glass at very low temperatures.

6. OPEN QUESTIONS -- FUTURE WORK

The spin glass problem is a difficult one and is far from being solved. The experimental situation is moving into a second generation of highly sophisticated and accurate measurements. In addition the importance of metallurgy is being emphasized and more care is now devoted to sample preparation and treatment.

It is recognized that a combination of systematic experimentation and careful metallurgy can produce new insights into the spin glass behavior. Present attention is centered upon dynamic and time dependent effects. The question of the frequency dependence of the susceptibility remains unclear. Also the relaxation and critical slowing down of a spin glass as studies via NMR, μ^+SR and neutron scattering have not been fully correlated. At low temperatures there is renewed interest in the field cooled displaced hysteresis loop and the sharp switching from positive to negative magnetizations with the application of a small negative field. Here a phenomenological model is evolving. The significance of a ferromagnetic to spin glass transition is unresolved as to the exact nature of the more disordered ground state. Further an exact comparison between the metallic spin glass and their insulating counterparts have shown certain differences. Are these differences severe enough to warrent to use of another term to distinguish then? Would superparamagnetism be sufficient to describe the insulators?

More measurements are needed to quantify the anisotropy which is present in different forms in both the 3d and 4f moment spin glasses. Additionally with the rare earth materials, the 4f wave functions are highly localized, the RKKY interaction much weaker and crystal field effects play a significant role. Finally there is still the interesting problem of the coexistence of superconductivity and magnetic ordering. We know that a frozen spin glass can coexist with a superconductor but the details of how these two different states involving the conduction electrons affect each other are not known. Here help could be gotten from the recent investigations of the ferromagnetic-superconductivity coexistence in certain rare earth intermetallic compounds.

On the theoretical side we can name three possible areas for greater study. Firstly, how useful is the EA model for describing spin glass behavior in real systems? Can the statistical mechanical difficulties be eliminated and a phase transition or even a continuous phase transition as in the X-Y model be obtained? How should the short range order which certainly exists in the real systems be incorporated into this model? And what about a theoretical calculation involving the modified percolation ideas sketched in the previous Section? Secondly more fundamental theory, i.e. first principle calculations are required for the Néel model. This is especially needed to come more realistically to grips with the sharp cooperative-like freezing of a spin glass. Finally the theory and experiment must develop hand in hand beginning with simple theoretical calculations for the given experiment. This should then lead to a stronger connection with the more general theory of spin glass and allow the particular experiment to be better understood. With so many sophisticated methods being used in the spin glasses it is very difficult to see the overall association. The theory in many cases is simply for the theorist with little or no connection to experiment. The experiments are sometimes so complicated that only a few experts usually workin on the same experiment can appreciate the significance and the weaknesses of the particular technique. Here the theorist is excluded by the technical details and reverts to the position of a (non)believer.

In conclusion we are confident that many of these obstacles will be removed and a more effective understanding of the spin glasses shall soon appear. Nevertheless the spin glass topic will certainly remain a fruitful area of magnetic research in the coming years with an intimate relationship to the general problem of amorphous magnetism and materials.

REFERENCES

1. J.A. Mydosh, A.I.P. Conf. Proc. 24,131 (1975).
2. T. Mizoguchi, T.R. McGuire, S. Kirkpatrick and R.J. Gambino, Phys. Rev. Letters 38, 89 (1977).
3. D.W. Forester, N.C. Koon, J.H. Schelleng and J.J. Rhyne, Solid State Commun. 30, 177 (1979).
4. J.J. Hauser, Solid State Commun. 30, 201 (1979); and R.W. Cochrane, J.O. Strom-Olsen and J.-P. Rebouillat, to be published in J. Appl. Phys. 1980.
5. S.J. Poon and J. Durand in Amorphous Magnetism II (R.A. Levy and R. Hasegawa eds.) (Plenum Press, New York, 1977) pg. 245, Solid State Commun. 21, 743 (1977), and ibid 21, 999 (1977).
6. J. Durand and S.J. Poon, J. Physique 39 C6-953 (1978), and C.L. Chien et al., G. Dublon et al., and K.V. Rao et al., all to be published in J. Appl. Phys. 1980.
7. R.B. Kummer, R.E. Walstedt, S. Geschwind, V. Narayanamurti and G.E. Devlin, Phys. Rev. Letters 40, 1098 (1978).
8. K. Andres, Bull. Amer. Phys. Soc. 24, 262 (1979).
9. G. Toulouse, Commun. Phys. 2, 115 (1977).
10. M.A. Ruderman and C. Kittel, Phys. Rev. 96, 99 (1954), T. Kasuya, Prog. Theo. Phys. 16, 45 (1956) and K. Yosida, Phys. Rev. 106,893 (1957).
11. J.A. Mydosh, J. Mag. and Mag. Mat. 7, 237 (1978).
12. L. Krusin-Elbaum, R. Raghavan and S.J. Williamson, Phys. Rev. Letters 42, 1762 (1979).
13. J.J. Smit, G.J. Nieuwenhuys and L.J. de Jongh, Solid State Commun. 30, 243 (1979).
14. J.L. Black and B.L. Gyorffy, Phys. Rev. Letters 41, 1595 (1978).
15. P.W. Anderson, B.I. Halperin and C.M. Varma, Philos. Mag. 25, 1(1972), and W.A. Philips, J. Low Temp. Phys. 7, 351 (1972).
16. M. Cyrot, Phys. Rev. Letters 43, 173 (1979).
17. S.F. Edwards and P.W. Anderson, J. Phys. F5, 965 (1975).
18. H. Claus, to be published in J. Appl. Phys. 1980.
19. B.R. Coles, B.V.B. Sarkissian and R.H. Taylor, Philos. Mag. B37, 489 (1978).
20. S.K. Burke and B.D. Rainford, J. Phys. F8, L239 (1978).
21. J.A. Mydosh and G.J. Nieuwenhuys in Handbook of Ferromagnetic Materials (E.P. Wohlfarth ed.) (North Holland, Amsterdam, 1979) Chap.II.
22. A. Aharony, J. Mag. and Mag. Mat. 7, 198 (1978).
23. B.H. Verbeek, G.J. Nieuwenhuys, H. Stocker and J.A. Mydosh, Phys. Rev. Letters 40, 589 (1978), B.H. Verbeek and J.A. Mydosh, J. Phys. F8, L109 (1978), and H. Maletta and P. Convert, Phys. Rev. Letters 42, 108 (1979).
24. For a review see G.J. Nieuwenhuys, B.H. Verbeek and J.A. Mydosh, J. Appl. Phys. 50, 1685 (1979).
25. V. Cannella and J.A. Mydosh, Phys. Rev. B6, 4220 (1972).

26. E.H. Dahlberg, M. Hardiman, R. Orbach and J. Souletie, Phys. Rev. Letters 42, 401 (1979).
27. C.A.M. Mulder, A.J. van Duyneveldt and J.A. Mydosh to be published in J. Mag. and Mag. Mat. 1980.
28. H. von Löhneysen, J.L. Tholence and R. Tournier, J. Physique 39 C6-992 (1978), G. Zibold, J. Phys. F8, L229 (1978) and J. Physique 39 C6-896 (1978), and F.S. Huang, L.H. Bieman, A.M. de Graaf and H.R. Rechenberg, J. Phys. C11, L271 (1978).
29. For a review of the frequency dependence of the susceptibility see J.L. Tholence to be published in J. Appl. Phys. 1980.
30. H. Maletta and W. Felsch, to be published in Phys. Rev. B 1979.
31. C.M. Soukoulis and K. Levin, Phys. Rev. Letters 39, 581 (1977).
32. H.A. Zweers, W. Pelt, G.J. Nieuwenhuys and J.A. Mydosh, Physica 86-88B, 837 (1977).
33. W.H. Fogle, J.C. Ho and N.E. Phillips, J. Physique 39 C6-901 (1978) and D.L. Martin, J. Physique 39 C6-903 (1978) and to be published in Phys. Rev. B 1979.
34. G.F. Hawkins, R.L. Thomas and A.M. de Graaf, J. Appl. Phys. 50, 1709 (1979).
35. G.J. Nieuwenhuys and J.A. Mydosh, Physica 86-88B, 880 (1977).
36. A. Berton, J. Chaussy, J. Odin, R. Rammal, J. Souletie and R. Tournier, J. Physique 40, L-391 (1979).
37. A.P. Murani and A. Heidemann, Phys. Rev. Letters 41, 1402 (1978).
38. T. Yamazaki, Y.J. Vemura, M. Takagi and C.Y. Huang, to be published in J. Mag. and Mag. Mat. 1980.
39. D.E. Murnick, A.T. Fiory and W.J. Kossler, Phys. Rev. Letters 36, 100 (1976).
40. D. Sherrington and S. Kirkpatrick, Phys. Rev. Letters 35, 1792 (1975).
41. See for example A.P. Young, J. Appl. Phys. 50, 1691 (1979) and M.A. Moore and A.J. Bray to be published in J. Mag. and Mag. Mat. 1980.
42. K. Binder in Festkörperprobleme, Advances in Solid State Physics Vol XVII (J. Treusch ed.) (Vieweg, Braunschweig, 1977) pg. 55.
43. J.L. Tholence and R. Tournier, J. Physique 35 C4-229 (1974), and C.N. Guy, J. Phys. F7, 1505 (1977) and ibid 8, 1309 (1978).
44. E.P. Wohlfarth, Phys. Letters 70A,489 (1979).
45. B.H. Verbeek, Thesis, Rijksuniversiteit, Leiden 1979 (unpublished).

CHARACTERISTIC PROPERTIES OF MICTOMAGNETS (SPINGLASSES)

Paul A. Beck

University of Illinois, Urbana, IL, USA
at present Physik-Department E13
Technische Universität München
8046 Garching, Germany

ABSTRACT. It became recognized during the last several years that there is a large class of magnetic materials, in which the spins "freeze" on cooling (i.e. the spin directions become essentially fixed with respect to the lattice) but, unlike in ferromagnets or in anti-ferromagnets, without long range spin order /1a/. In the following, the characteristic magnetic properties of this group of materials will be briefly described, taking into consideration a selection of very recent developments, which appear particularly significant. A more detailed review of the experimental situation, as of about the end of 1977, was published recently /1b/.

1. THERMOMAGNETIC HISTORY EFFECTS; TIME DEPENDENCE

As seen in Fig. 1, for $Cu_{83}Mn_{17}$ at higher temperatures the magnetization in a fixed field increases with decreasing temperature, corresponding approximately to the modified Curie-Weiss law:

$$\chi = \chi_o + \frac{C}{T - \Theta} \tag{1.1}$$

where χ_o is a temperature-independent susceptibility. However, at lower temperatures deviations from this kind of temperature-dependence occur. (a) If the specimen is measured in the course of cooling in a magnetic field, at low temperatures the magnetization becomes essentially temperature-independent (data points: o in Fig. 1), (b) If the specimen is cooled

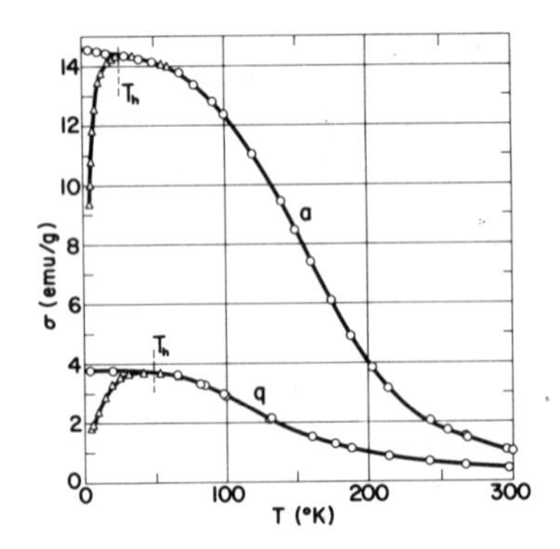

Fig. 1. Magnetization vs. temperature for quenched, q, and aged, a, $Cu_{83}Mn_{17}$. Thermomagnetic history effect below T_h.

to 4.2 K in zero field, after which a field of 12.6 kOe is applied and measurements are made in the course of increasing the temperature (data points: Δ), then at low temperatures the magnetization is considerably lower than that measured in the course of cooling in the field. Clearly, at low temperatures the magnetization is not uniquely determined by the applied field and the temperature. Rather, it depends also on the previous thermo-magnetic history. In the zero-field cooled condition it is observed that, after the application of the field, the magnetization also varies with time at a constant temperature. The rate of this time-dependence increases very much with increasing temperature and also with increasing field. An analogous time-dependence at constant temperature is found in the field-cooled condition, if the cooling field is removed (or decreased). The thermo-magnetic history effect, as well as the gradual approach to equilibrium magnetization through a relatively slow relaxation process, are phenomena typical of mictomagnetic materials.

The curves in Fig. 1 marked a and q represent data for $Cu_{83}Mn_{17}$ in two different metallurgical conditions. Curves q give the magnetization in a field of 12.6 kOe after quenching (fast cooling) the alloy from 850 °C to room temperature, while curves a give similar data obtained after the quenched alloy was subsequently aged for a month at 100 °C. The great increase in the magnetization due to the aging treatment is particularly pronounced for Cu-Mn alloys containing 10 to 25 %Mn.

2. DEVIATION FROM THE CURIE-WEISS LAW; GIANT MOMENTS

The freezing of the spin orientations on cooling and the resultant essentially temperature-independent magnetization in the field-cooled condition at low temperatures have also been observed in very much more dilute alloys. As shown by Hirschkoff, Symko and Wheatley /2/, Fig. 2, even for alloys with less than 100 ppm Mn the magnetization becomes essentially temperature-independent at sufficiently low temperatures in the field-cooled state. In Fig. 2 the magnetization divided by the field is plotted on a reciprocal scale, so that the Curie-Weiss relationship is shown very clearly by a straight line as a function of temperature. Fig. 2 also shows that the deviation from the Curie-Weiss relationship and the approach to a temperature-independent magnetization value in the field-cooled state takes place gradually (not at a sharply defined temperature), even for these very dilute alloys.

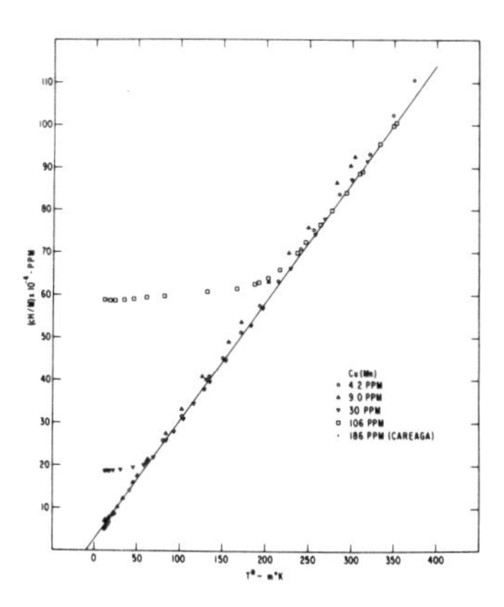

Fig. 2. Reciprocal susceptibility vs. temperature for dilute Cu-Mn alloys, from /2/.

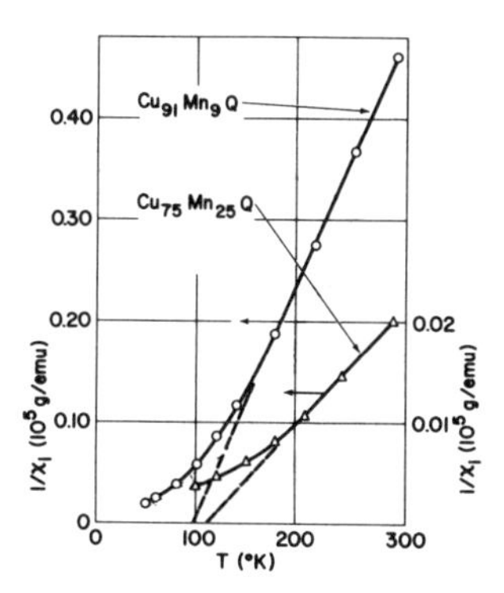

Fig. 3. Reciprocal initial susceptibility for quenched $Cu_{91}Mn_9$ and $Cu_{75}Mn_{25}$ vs. temperature, from /3/.

Fig. 3 shows a similar gradual deviation from the Curie-Weiss law for alloys with 9 % and 25 % Mn in the quenched condition. In the case of these alloys with high Mn content, the nature of the magnetic changes that give rise to the gradual deviation from the Curie-Weiss relationship at low temperatures was studied by means of magnetization, σ, vs. field, H, isotherms, such as those shown in Fig. 4.

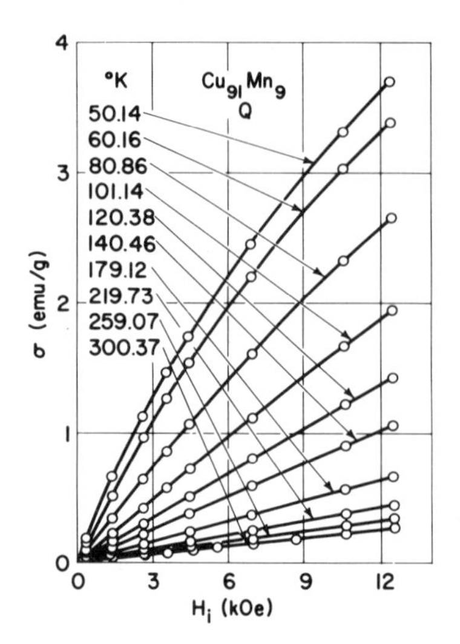

Fig. 4. Magnetization vs. field for quenched $Cu_{91}Mn_9$, from /3/.

At the higher temperatures where the Curie-Weiss law is valid, the isotherms are straight lines up to fields of at least 12.5 kOe, i.e. under such conditions the susceptibility is field-independent. However, at lower temperatures the isotherms are curved. These curved isotherms can be empirically described as the sum of a susceptibility term and of a Brillouin function term, equation (2.1)

$$\sigma = \chi H + \mu \cdot c \cdot B[\mu, \frac{H+\lambda(\sigma-\chi H)}{T}] =$$

$$= \chi H + c\{(\mu+1)\coth[(\mu+1)\frac{\mu_B}{k} \cdot \frac{H+\lambda(\sigma-\chi H)}{T}] -$$

$$- \coth[\frac{\mu_B}{k} \cdot \frac{H+\lambda(\sigma-\chi H)}{T}]\} \qquad (2.1)$$

All magnetic dipoles are here represented by some kind of an "average moment" value μ; c is the concentration of this "average" dipole moment. If one assumes that g = 2, i.e. the orbital moment of all dipoles is quenched, then $\mu = gS$, where the spin quantum number, S, is not required to be an integral multiple of 1/2. λ is the Weiss molecular field coefficient, taking account of the interaction between dipoles of an "average moment" μ in the molecular field approximation. The parameters in Equ. (2.1) can be least squares fitted quite well to data /3/ such as those shown in Fig. 4, provided that one uses data only from not very different temperatures, e.g. from two adjacent isotherms. This requirement arises from the fact that the parameters here are in fact dependent on the temperature, but this is neglected in the fitting. By least squares fitting consecutive pairs of isotherms one can obtain approximate average values for μ and for the other parameters as a function of temperature, Fig. 5.

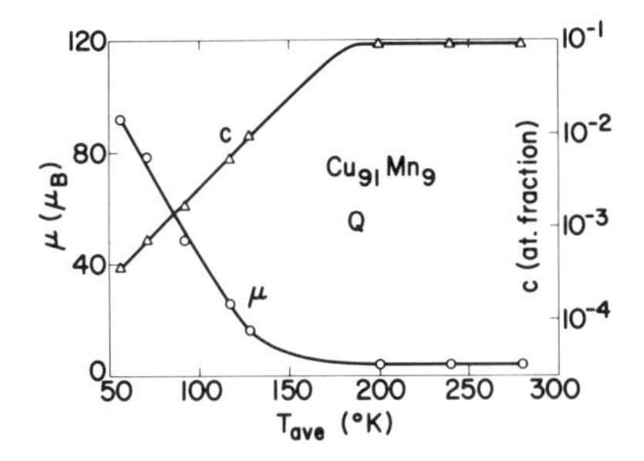

Fig. 5. Average moment, μ, and its concentration, c, vs. T for $Cu_{91}Mn_9$, from /3/.

In the temperature range of the Curie-Weiss law, Equ. (1.1) was used to determine the average dipole moment and the value of $\mu = 4(\mu_B)$ was arrived at for the 9 % Mn alloy in the quenched condition. This value is equal to that derivable from the data by Hurd /4/ for much more dilute alloys, indicating that in the 9 % Mn quenched alloy one still deals with independent Mn atomic moments and that these account for the entire Langevin paramagnetism in the Curie-Weiss temperature range. At lower temperatures, where the reciprocal susceptibility vs. temperature graph deviates from the Curie-Weiss straight line, one obtains much higher values by fitting pairs of isotherms to Equ. (2.1), as mentioned above. The giant moments represent magnetic clusters, but the "average dipole moments" determined by fitting include, in addition to the giant moments, also individual Mn atomic moments that may still be present in a given temperature range. As seen from Fig. 5, the dipole moment concentration descreases by more than two orders of magnitude on decreasing the temperature. It reaches values of less than 0.1 at. % at the lowest temperatures used. This very low concentration indicates that, at these lowest temperatures, the individual Mn atomic moments play practically no role so that here the "average dipole moment" includes essentially only the various giant moments present.

The very strong apparent temperature dependence of μ, indicated in Fig. 5, comes about in large part as a result of the increasing admixture to μ of the single Mn atomic moments as the temperature increases.

3. DISPLACEMENT AND UNIDIRECTIONAL REMANENCE

Another characteristic feature of mictomagnetic alloys is the displacement of the magnetization curve on cooling in a magnetic field to very low temperatures /5/, as shown in Fig. 6. A result of this displacement is the unidirectional remanence. At such low temperatures hysteresis is almost completely absent. If the specimen is cooled to a low temperature in the absence of a magnetic field, then the magnetization vs. field data are represented by a straight line, which passes through the zero of the field at zero magnetization, i.e. there is no remanence. When the specimen is cooled in a magnetic field to the same low temperature and then the field is removed, the resultant magnetization in zero field is designated as the unidirectional thermo-remanent magnetization (TRM). As seen in Fig. 6, on reversal of the field the magnetization continues to decrease along the same line to a relatively small value of the reversed field. When the reversed field is increased further, a steep decrease in the magnetization occurs, followed by its reversal symmetrically with respect to the straight line measured after zero-

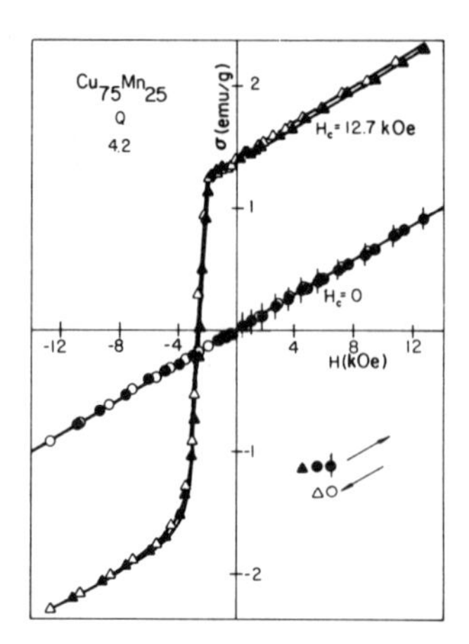

Fig. 6. Magnetization vs. field at 4.2 K for $Cu_{75}Mn_{25}$ after zero-field cooling, H_c = 0, and after field-cooling, H_c = 12.7 kOe.

field cooling. This reversal indicates that the aligned magnetic moments, giving rise to the unidirectional remanence, react very differently to an applied field, as compared with their reaction after zero-field cooling, when they are not aligned but, presumably, randomly oriented. It seems that any explanation of this behaviour would require the assumption that there is a magnetic interaction between the dipoles present. Fig. 6 shows that after all the aligned moments have been reversed by a sufficiently large reversed field, further increase in the reversed magnetization occurs only to a relatively minor extent, corresponding to the slope of the straight line for the zero-field cooled specimen. The entire magnetization curve is displaced with respect to the zero of the field in the direction opposite to that of the cooling field. The extent of the displacement, H_d, is the field at the intersection of the magnetization curves after field-cooling and after zero-field cooling, Fig. 6.

A very interesting feature is the following: when the reversed field is decreased, then the reversed dipoles return to their original orientation (which they arrived at during field-cooling) <u>in opposition to the still remaining reversed field</u>. The simplest and most straightforward model for this phenomenon (originated in a somewhat different form by Kouvel /5a/) involves interaction between the magnetic moments that are aligned by field-cooling and the surrounding moments, which are not.

It is noteworthy that the collective reversal of the aligned moments by a reversed field can take place at a very low temperature in a relatively low field, without any noticeable delay and that, if the temperature is sufficiently low, no hysteresis occurs when the aligned moments revert to their original orientation as the applied reversed field is decreased, see Fig. 6. At somewhat higher temperatures hysteresis appears, but the hysteresis loop is still displaced, as shown in the original detailed studies of these "exchange anisotropy" effects by Kouvel /5b/. With increasing temperature the displacement decreases, while the hysteresis at first increases. The remanence due to the hysteresis can be separated from that due to displacement if data are available for a complete cycle and the field range is large enough to obtain closure at the loop, but not so large as to give rise to appreciable isothermal remanent magnetization (see Section 4). The <u>average</u> of the magnetization values on the two branches of the loop at zero field is the unidirectional remanence due to displacement, while the half of their difference is the uniaxial remanence due to hysteresis /1b/. That differentiating between remanence due to displacement and that due to hysteresis is meaningful is shown not only by their different temperature-dependence, but also by the fact that torque measurements /5c/ can indicate the unidirectionality of the corresponding anisotropy for one and its uniaxial nature for the other, as well as their presence side-by-side in the same specimen in a certain temperature range. At somewhat higher temperatures the displacement disappears altogether and the hysteresis loop is then symmetrical. In some instances the presence of mictomagnetism can still be recognized at these temperatures, in spite of the absence of displacement after field-cooling, by the fact that the first magnetization curve measured after zero-field cooling is partially outside of the hysteresis loop measured at the same temperature in the field-cooled state /6/.

Recent very interesting experiments by Monod and Préjean /7/ show that, at least in fairly dilute Cu-Mn alloys (2 % Mn or less), displaced hysteresis loops after field-cooling may consist of very large thermally (or mechanically) activated steps. Under the conditions of those experiments a considerable amount of hysteresis also occurs. The very abrupt partial, or even nearly complete, reversal of the magnetization in a single step apparently initiated by a single activation event, as observed by Monod and Préjean, contrasts sharply the seemingly continuous hysteresis loops previously observed at higher Mn concentrations /5b, 3/.

4. IRREVERSIBLE CHANGES IN MAGNETIZATION; "VISCOUS" MAGNETIC BEHAVIOUR

It has been known since the earliest studies on exchange anisotropy in Cu-Mn alloys that, in addition to field-cooling, unidirectional remanence can be also produced isothermally, even at quite low temperatures, by the application of sufficiently large fields to a zero-field cooled specimen. The slope of the straight line in Fig. 6, for the zero-field cooled specimen, represents its susceptibility at 4.2 K. Increasing or decreasing the field (within the limits shown), or reversing it, has no influence on the susceptibility; it is, therefore, referred to as reversible. As seen in

Fig. 6, the reversibility extends to at least 12.6 kOe at 4.2 K for quenched $Cu_{75}Mn_{25}$. As may be seen in Fig. 7, line q, the reversibility for quenched $Cu_{83}Mn_{17}$ at 4.2 K does not extend to as high fields. Starting from the zero-field cooled state, the straight line extends at least somewhat above 5 kOe. However, at higher fields the measured magnetization values deviate from the extrapolated straight line and this deviation is irreversible. When the field is subsequently decreased, the magnetization decreases approximately along a straight line, parallel to the original one through the zero of the field and of the magnetization, but it is now displaced to higher magnetization values. Correspondingly, when the field is decreased to zero, there is a finite remanent magnetization, approximately equal to the deviation from the original straight line reached at the maximum field applied. The appearance of remanence upon isothermal increase of the field is a result of the lack of reversibility. The isothermal remanence (IRM) values, produced by consecutive excursions to fields exceeding 5 kOe, are cumulative. Recently, the isothermal remanence in Cu-Mn alloys at high fields was studied in significant detail by Claus and Kouvel /8/. They found that (at a given rate of measurement) this isothermal irreversible increase in magnetization (and the remanence) start at a certain field H_1, just as shown in Fig. 7, but that at much higher fields the $\sigma(H)$ isotherm has an inflection point. At still higher fields the remanence approaches saturation. At reversed fields, larger than a value H_2, the displacement itself is isothermally reversed /1b/.

From the comparison of Figs. 6 and 7, it is clear that the limiting field for reversibility increases with increasing Mn content. This limit is found to decrease with increasing measuring temperature for a given alloy. Another very significant fact, which must be kept in mind when considering these questions, is the time dependence of the deviation from reversibility. In fact, the limiting field for reversibility is defined only for a given rate of making the successive measurements at increasing fields. If more time is allowed at any one value of the field, the irreversible increase of the magnetization becomes larger. The rate at which the increase of the irreversible magnetization takes place becomes larger with increasing applied field, as well as with increasing temperature. The temperature-dependence of the rate of irreversible change in the magnetization indicates that the corresponding relaxation process is thermally activated.

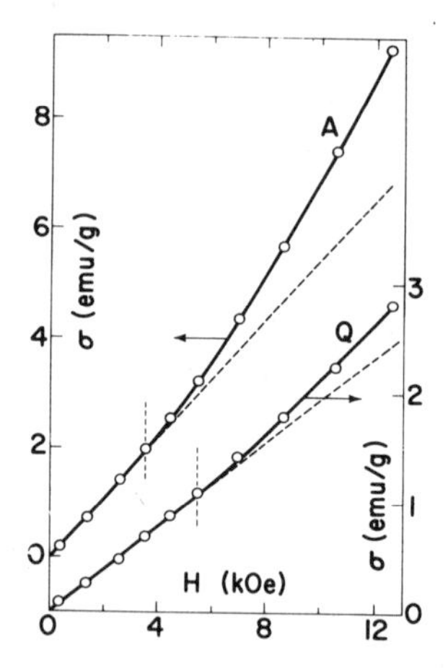

Fig. 7. Magnetization vs. field at 4.2 K for quenched, Q, and aged, A, $Cu_{83}Mn_{17}$ after zero-field cooling. Vertical dashed lines indicate H_1, where irreversibility starts.

The effect of the factors T and H can be schematically expressed /1a/ by stating that the rate of relaxation (i.e. the rate of irreversible change in the magnetization) is proportional to expression (4.1). Here E is the potential barrier that must be overcome to initiate an increase in the magnetization by a small rotation of the dipoles under the influence of the field, H, acting on them. The magnetic energy available is $\mu \cdot H$ and, if this is not sufficient to overcome the energy barrier, thermal activation must be relied upon to furnish the difference. Thus, schematically, the rate at which thermal activation brings about the sequence of small consecutive irreversible changes in magnetization might be considered to be proportional to the Arrhenius expression:

$$\exp\left[-(E-\mu \cdot H)/kT\right] \tag{4.1}$$

Other effects, e.g. the decrease in the rate of change as saturation is approached must be also taken into consideration in a more detailed account. One way of doing this is by considering the distribution of the parameters E and μ, as recently suggested by Schwink et al. for E /9/.

The very interesting and significant question as to

the identity and the magnitude of the dipole moments contributing to the unidirectional remanence and affected by the described moment reversal phenomena at low temperatures, has been answered with a reasonable degree of assurance for the relatively concentrated alloys of copper with approximately 25 % Mn. The neutron scattering work of Werner, Sato and Yessik /10/ has clearly shown the presence of a $TiAl_3$-type short-range atomic order in aged $Cu_{75}Mn_{25}$. The width of the diffuse peaks, at (1, 1/2, 0) and other equivalent locations in the reciprocal lattice, allowed the conclusion that the atomically ordered regions in aged $Cu_{75}Mn_{25}$ at room temperature have an average diameter of approximately three f.c.c. cell edges. Earlier work on the Cu_2AlMn Heusler alloy and on Au_4Mn /6/ established that in such ordered alloys, where the Mn atoms are not nearest neighbors, their moments are ferromagnetically aligned. Since in the $TiAl_3$-type ordered regions in $Cu_{75}Mn_{25}$ the Mn atoms are 2nd-nearest, or more distant, neighbors, it is quite reasonable to assume that, here too, the Mn moments are ferromagnetically aligned within each atomically ordered region. This assumption leads to an estimate of the average giant moment of the magnetic clusters, corresponding to the individual atomically ordered regions in aged $Cu_{75}Mn_{25}$ at room temperature, of approximately 108 μ_B. This estimate was found /3,11/ to be in agreement well within a factor of 2 with the average giant moment value obtained by least squares fitting of the magnetization data for aged $Cu_{75}Mn_{25}$ near room temperature. It is, therefore, very likely that the giant moments found in aged $Cu_{75}Mn_{25}$ in the super-paramagnetic state by the analysis of the magnetization vs. field isotherms do correspond to the atomically ordered regions.

Furthermore, it seems reasonable to assume that the giant moments can be relatively easily aligned by field-cooling and that, thus they may also be responsible for the unidirectional remanence and the moment reversal phenomena at low temperatures. In fact, it was found /11a/ that the unidirectional remanence in aged $Cu_{75}Mn_{25}$ corresponds, within a factor of 2, to the total sum of the giant moments, $\mu \cdot c$, derived by least squares fitting of the parameters in Equ. (2.1) to the magnetization data at temperatures just above T_f (in the superparamagnetic state), as described in section 2.

This assumption leads, furthermore, to a very reasonable explanation of the effect of aging and of plastic deformation on the magnitude of the unidirectional remanence in $Cu_{75}Mn_{25}$ /1b/ in terms of changes in the magnitude and of the concentration of the giant moments, resulting from the effect of the metallurgical treatments on the atomic short range order. It turns out that, for the three metallurgical conditions (aged, quenched and deformed), the unidirectional remanence values after field cooling to 4.2 K, which differ from each other by more than an order of magnitude, are equal to 0.6 times the values of $\mu \cdot c$, derived from the analysis of the corresponding superparamagnetic data /1b/.

Although neutron scattering information is not available for $Cu_{91}Mn_9$, it is very likely that the observed effect of metallurgical treatment on the unidirectional remanence of this alloy (an effect that is qualitatively very similar to, but in magnitude somewhat smaller than, that for $Cu_{75}Mn_{25}$) is also a result of the effect of the metallurgical treatments on the short range atomic order and, hence, on the magnitude and concentration of the giant moments associated with small atomically ordered regions.

The recent brilliant NMR work by Alloul /12/ resulted in a very different model for remanence in more dilute Cu-Mn alloys, e.g. 1 % Mn. In this model nearly all Mn atomic moments contribute to the remanence and the fact that the magnitude of the remanence in these alloys, too, corresponds only to a small fraction of the sum of all Mn-moments present, is considered to be a result of the rather limited degree of alignment of the moments by field-cooling. The remanent magnetization is assumed to favor coherent motion of the spins in domains, which nearly fill the sample volume. However, it is at present not yet clear how this model can explain the displacement of the hysteresis loop, which does occur at these low Mn-contents as well /7/. It would be particularly interesting to see whether or not the hysteresis will at such low Mn contents also approach zero at sufficiently low temperatures, as it does for far more Mn-rich alloys. If it will, then the model must also confront the problem of finding a framework for explaining the return of the moments to their orientation in the field-cooled state in opposition to a reversed field, as described in section 3.

5. THE CUSP OF THE REVERSIBLE SUSCEPTIBILITY

Because the irreversible increase in the magnetization starts at lower and lower fields as the temperature increases, the range of static fields, within which the reversible susceptibility can be measured at a given rate, becomes more and more restricted. Thus, by very careful steady field measurements, Tustison /12c/ succeeded in determining the reversible susceptibility for quenched $Cu_{75}Mn_{25}$ only up to 40 K. Within the experimental error he found a linear increase from 3.7 to 9.5 (10^{-4} emu/g) in the temperature range from 4.2 to 40 K.

A very convenient and simple method can be used to measure the reversible susceptibility up to much higher temperatures. As expression (4.1) suggests, the approach to equilibrium magnetization by relaxation processes can be minimized by making the field small and by allowing only a very short time for the field to act in any one direction on the giant moments responsible for unidirectional remanence. These conditions can be realized in a simple way by using an alternating low field method. With fields of the order of one Oersted and frequencies of the order of 100 Hertz irreversible relaxation processes can be effectively suppressed in Cu-Mn alloys. Canella, Mydosh and Budnick /13/ discovered that the rise of the reversible susceptibility with increasing temperature in Au-Fe spinglass alloys is terminated by a fairly sharp cusp, above which the reversible susceptibility decreases with increasing temperature, Fig. 8. The temperature of the cusp has been considered the "freezing temperature" of the spins, T_f.

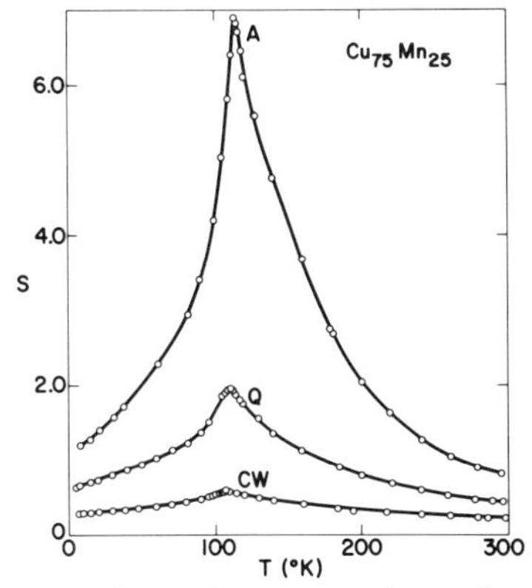

Fig. 8. Alternating low field susceptibility index, S, in arbitrary units vs. temperature for $Cu_{75}Mn_{25}$ in the quenched, aged and deformed condition, from /11b/.

6. THE TRANSITION FROM PARAMAGNETISM TO MICTOMAGNETISM

The sharpness of the reversible susceptibility cusp has, interestingly, led some investigators to consider the change from the paramagnetic to the "spinglass" state as an equilibrium phase transition, even though the reversible susceptibility certainly does not correspond to any equilibrium state, as mentioned in section 5. In fact, as shown in Figs. 1 and 2, the magnetization values obtained in the course of cooling in a steady field, where equilibrium states for the magnetic field values used are more nearly approached, are never found to show a cusp and in most cases not even a broad maximum. This characteristic absence of a sharp transition is by no means limited to high alloy concentration or to high fields. The absence of any maximum was found by Symko for Ag-Mn alloys with about 1 ppm Mn and for measuring and cooling fields down to 1 Oersted /14/. To summarize: in mictomagnetic alloys from very dilute to concentrated, when equilibrium is approached by field cooling, nothing like a sharp phase transition is observed between the paramagnetic and the mictomágnetic states, but instead a very gradual transition even at very low fields.

The Edwards-Anderson theory /15/ which considers the change from paramagnetism to the spinglass state as an equilibrium phase transition, calls for a cusp in the specific heat at T_f, the temperature of the reversible susceptibility cusp. The fact that the postulated specific heat cusp has not been found experimentally /16, 17, 18/ is an important indication that this interesting and very ingeneous theory is not relevant to the real physics of the paramagnetism to mictomagnetism transition.

One of the arguments advanced /15/ in favor of an equilibrium phase transition was the supposed agreement between T_f, the alternating low field susceptibility cusp temperature and the temperature where the Mössbauer spectrum has the sharp onset of hyperfine splitting. Indeed, for Cu-Mn alloys these two temperatures /19/ and also the transition temperature determined by static low field susceptibility measurements are in good agreement with each other /1b/. However, for Au-Fe alloys the three methods mentioned give different results. At a concentration of 0.01 % Fe the Mössbauer hyperfine splitting temperature is 28 times higher than the transition temperature defined by steady field susceptibility measurements /1b/. Similar differences were previously noted by Murani for micto-magnetic Rh-Fe alloys /20/.

Another argument that has been given in favor of an equilibrium phase transition is based on the absence of any frequency-dependence of T_f within the experimental accuracy available with the alternating low field susceptibility measurements on Ag-Mn alloys /21/. Similar results were previously obtained also for Cu-Mn alloys /3/. However, the frequency dependence of T_f was recently well established for amorphous Au-Fe alloys /22/, for the metallic Laves phase mictomagnet $(La_{0.4}Gd_{0.6})Al_2$ /23/ and for the insulating mictomagnet (Eu,Sr)S, which has been very thoroughly studied by Maletta and Felsch /24/. The seemingly anomalous lack of (or very low /25/) frequency-dependence of T_f for Ag-Mn and Cu-Mn alloys may be due to the well defined freezing temperature of the matrix surrounding the magnetic clusters, where the short range antiferromagnetic coupling between Mn-moments in the matrix /1b, 11; see sections 7 and 8 below/ leads to a coupling of these moments to the lattice as well at the cusp temperature (the thermal energy becomes smaller than their anisotropy energy). The presence of massive short range antiferromagnetism at T_f in these Mn alloys may well account also for the agreement between the freezing temperatures determined by various methods, as discussed in the previous paragraph. In dilute Au-Fe or Rh-Fe alloys where the agreement is not good, or in (Eu,Sr)S where there is a strong frequency-dependence of T_f, no massive short range antiferromagnetism is expected to occur.

The nature of the transition from the paramagnetic to the mictomagnetic state can be inferred from Fig. 9, which shows the magnetization divided by the applied steady field as a function of temperature for a quenched $Cu_{91}Mn_9$ alloy specimen warmed up and measured in various fields, after being cooled to 4.2 K in zero field from above T_f. The figure shows that the maximum becomes sharper and sharper as the steady field becomes smaller. This increasing sharpness with decreasing steady field has been explained /3/ as a very elementary result of the "viscous" magnetic behaviour of the mictomagnetic alloys at temperatures below T_f and of their Langevin paramagnetic behaviour as room temperature is approached. It follows from expression (4.1) that, after zero-field cooling to a low temperature, a high applied field in the "viscous" temperature range will give a larger increase in the magnetization within a given period of time than a lower field. Thus, de-

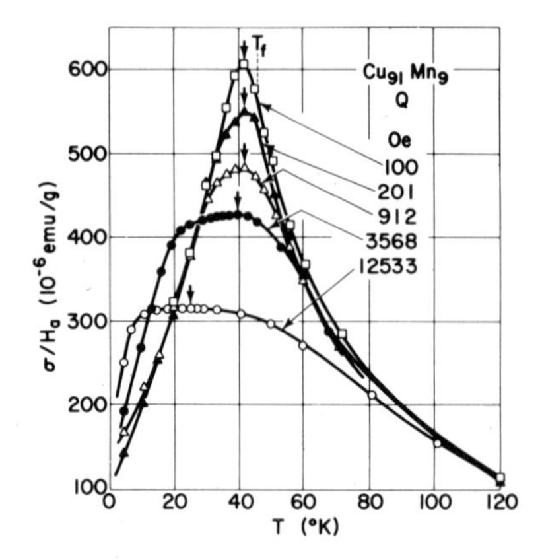

Fig. 9. Magnetization divided by applied field vs. temperature at fields indicated, for quenched $Cu_{91}Mn_9$ after zero-field cooling. As the field decreases the maximum of the curves (↓) approaches T_f /3/.

creasing the field will tend to push the low temperature (ascending) part of the curves in Fig. 9 toward higher temperatures. On the other hand, the high temperature (descending) part of all curves tend to coincide, since at high temperatures the susceptibility is not field-dependent. At intermediate temperatures, the σ/H value, which is an average of the field-dependent differential susceptibility corresponding to the lower temperature curved isotherms in Fig. 4, increases with decreasing field, approaching the initial slope of these curves as $H \rightarrow 0$. Accordingly, the maxima in Fig. 9 become not only narrower but also higher as the field decreases. It is seen that these maxima, marked by arrows in Fig. 9, occur at temperatures approaching the alternating low field cusp temperature, T_f, as the field decreases. Even at a field as high as 100 Oe (the lowest steady field used in obtaining the results shown in Fig. 9), the maximum is indeed not far from T_f. A typical feature of the σ/H vs. T curves for different values of the field is that they intersect one another at relatively low temperatures, as seen in Fig. 9. In this interpretation the occurrence of the alternating low field susceptibility cusp is a direct result of the presence of giant moments in mictomagnetic alloys; the more than an order of magnitude variation of the value of the reversible susceptibility at the cusp for $Cu_{75}Mn_{25}$, as a result of different metallurgical treatments (see Fig. 8) is a straightforward consequence of the change in short-range atomic order and of the corresponding changes in the magnitude and in the concentration of the giant moments of magnetic clusters. The "shoulder", apparent in the curves for the highest fields in Fig. 9 (12.5, 3.6 and 0.9 kOe) at temperatures below their maxima, is a result of the relatively fast initial increase in the magnetization at high fields, followed by an approach to saturation for the particular field concerned. The temperature of the "shoulder" clearly decreases with increasing field, as expected from the present interpretation.

7. MASSIVE SHORT RANGE ANTIFERROMAGNETISM IN Cu-Mn ALLOYS

The straight line relationship between the reciprocal susceptibility and the temperature for quenched $Cu_{91}Mn_9$ in Fig. 3 above about 180 K indicates the validity of the Curie-Weiss law in this temperature range. By least squares fitting the parameters of Equ. 1 one obtains a value for the Curie constant, $\underline{C}$. Assuming that $g = 2$ (i.e. that the orbital momentum is quenched for all dipoles), the Curie constant per solute atom:

$$C = P^2_{eff}/3k \qquad (7.1)$$

allows the calculation of $\mu = g.S$. For the 9 % Mn alloy one obtains the value of $\mu = 4$ (μ_B), i.e. the same value that fits the data of Hurd for Cu-Mn alloys with much lower Mn concentrations. One may conclude that, in the temperature range between 180 and 300 K, this alloy is a simple Langevin paramagnet with each Mn atomic moment an independent dipole. The straight line portion of the graph in Fig. 3 for quenched $Cu_{75}Mn_{25}$ can be similarly analyzed. The result in this case is an average dipole moment per Mn atom of $\bar{\mu}=g.\bar{S}.=3.3$ (μ_B), which is markedly lower than that for alloys very dilute and up to 9 % Mn. Such a lower μ-value was previously obtained for a 24 % Mn alloy by Kouvel /5b/, who interpreted this result in terms of antiferromagnetic coupling between some of the Mn moments at temperatures below approximately 320 K. Above that temperature, he found $\mu = 4.5$ (μ_B). Consequently, Kouvel identified 320 K, where the reciprocal susceptibility vs. T line changes its slope, as the temperature where the antiferromagnetic interaction, present at lower temperatures, breaks down. On the basis of Kouvel's interpretation, one finds that in quenched $Cu_{75}Mn_{25}$ approximately 26 % of the Mn moments present are not contributing to the Curie constant below 300 K since they are coupled to one another antiferromagnetically. Tustison /11b/ found that for quenched $Cu_{75}Mn_{25}$, the Curie constant per Mn atom decreased upon severe plastic deformation from 39.5 to 29.6 (10^{-5} μ_B deg/Oe), indicating that in his deformed specimen 45 % of the Mn moments present were compensated by antiferromagnetic coupling. These observations can be interpreted on the basis of short range atomic order. It is reasonable to assume that severe plastic deformation partially disrupts the short-range atomic order present in the quenched specimen, making the distribution of Mn atoms more random, and that it thereby increases the concentration of Mn nearest neighbor pairs, the moments of which are coupled to each other antiferromagnetically. The above interpretation is given further support by the observation /11a/ that an increase of the Mn concentration above 25 % also drastically decreases the Curie constant, Fig. 10. For quenched $Cu_{50}Mn_{50}$, the Curie constant is about 10 times smaller than that for quenched $Cu_{75}Mn_{25}$. In quenched $Cu_{34}Mn_{66}$ approximately 98 % of the Mn moments are compensated by antiferromagnetic coupling at temperatures between 160 and 300 K, so that only 2 % of the Mn moments present contribute to the Curie constant /11a/.

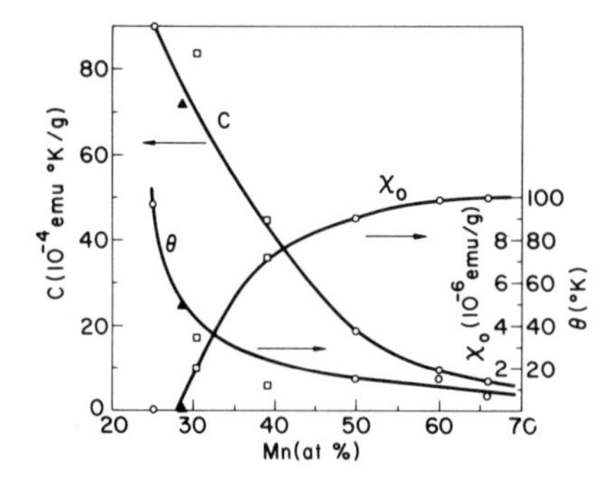

Fig. 10. The Curie constant, C, the Weiss temperature, Θ, and the temperature-independent susceptibility, χ_o, vs. Mn-content for Cu-Mn alloys, from /11a/.

It is noteworthy that, in spite of the very extensive antiferromagnetic coupling between Mn moments in these alloys, no long range antiferromagnetism is present even at temperatures well below T_f. As indicated by neutron diffraction results /26, 27/, in Cu-Mn alloys long range antiferromagnetism sets in only at Mn concentrations larger than 68 %. In the concentration range between 25 and 66 % Mn, we are dealing with a very massive short range antiferromagnetism between T_f and room temperature /11a/. Although at the higher Mn concentrations within this range most of the Mn moments are coupled to each other in an essentially antiferromagnetic manner, they are clearly free in the temperature range between T_f and 300 K to change their directions with respect to the lattice in periods of time shorter than 10^{-7} sec. This is shown by the absence of hyperfine splitting in the Mössbauer spectrum above T_f in the entire concentration range studied (up to 50 at. % Mn) /19/. Vance, Smith and Sabine /28/ gave very convincing arguments in support of their interpretation of the strong diffuse neutron scattering observed by them in $Cu_{32}Mn_{68}$ near the (100) position up to temperatures of at least 550 K as largely, or perhaps entirely, attributable to "quasistatic" magnetic short range order (which appears static to the thermal neutrons during their time of passage of the order of 10^{-11} seconds). Those authors found similar neutron diffuse scattering near the (100) position for f.c.c. alloys with still higher Mn contents, above their Néel temperatures, extending up to at least 550 - 600 K. These neutron scattering observations strongly support the interpretation of the measured Curie constant values in terms of massive short range antiferromagnetism. The presence of massive short range antiferromagnetism in the high Mn alloys, and according to the neutron scattering evidence /28/ also in $Mn_{50}Fe_{50}$, may well explain the temperature-independence /29/ or even increase /26/ of the magnetic susceptibility with increasing temperature above the Néel point. These anomalous temperature-dependences were attributed to "band antiferromagnetism". In case of $Fe_{50}Mn_{50}$ large magnetic specific heat contributions were observed at temperatures up to about 200 K above T_N /30/. These observations are quite consistent with a gradual decrease of the massive short range antiferromagnetism with increasing temperature as indicated by the gradual decrease of the neutron diffuse scattering intensity near (100) /28/.

The quenched alloys with 25 to 66 % Mn obey Equ. (1) from room temperature down to approximately 140 - 210 K, depending on their composition, but the temperature-independent susceptibility χ_o in Equ. (1), for these concentrated alloys no longer corresponds to the diamagnetism of copper as it does for dilute alloys /4/. Instead, it is a temperature-independent paramagnetism, which increases with increasing concentration, c_a, of the antiferromagnetically coupled Mn moments not contributing to the Curie constant /11a/. As shown in Fig. 11, when χ_o is plotted against c_a, the point for plastically deformed $Cu_{75}Mn_{25}$ fits on the curve drawn through the points for quenched alloys with different Mn contents. It has been suggested by J. Bardeen /31/ that this temperature-independent paramagnetism may be of the Van Vleck type and that the changes in the Mn atomic energy levels required to give rise to Van Vleck paramagnetism may be a result of the presence of nearest-neighbor Mn atoms. This would explain the correlation between χ_o and c_a. It would be interesting to see whether the required changes in atomic energy levels could be accounted for theoretically on this basis.

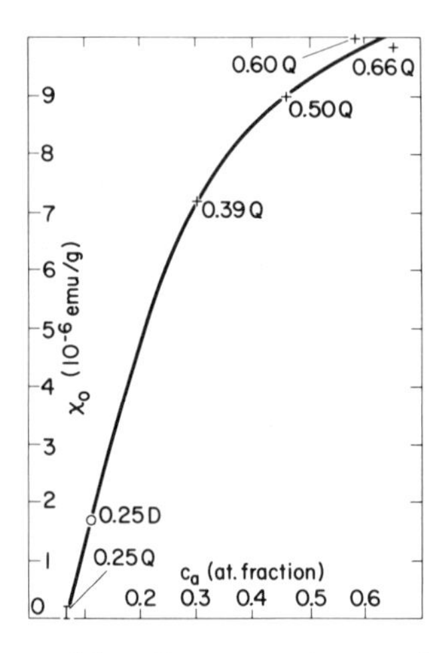

Fig. 11. Temperature-independent susceptibility, χ_o, vs. concentration of antiferromagnetically coupled Mn moments for quenched, Q, Mn alloys at Mn-contents indicated and for a deformed alloy with 25 % Mn, D, from /11a/.

8. SHORT RANGE ANTIFERROMAGNETISM JUST ABOVE T_f IN Cu-Mn AND IN Ag-Mn

Electron spin resonance (ESR) measurements on Cu-Mn alloys by various investigators /32-35/ indicate that, above a certain measuring temperature depending on the composition, the resonance line width increases nearly linearly with the temperature and that, in this temperature range, the resonance line occurs at a fixed value of the field, corresponding approximately to g=2. It is clear that in this temperature range the electron spin resonance is due to independent dipoles corresponding to individual Mn atomic moments. This behavior was also observed for quenched $Cu_{75}Mn_{25}$, in agreement with the conclusion from the magnetic susceptibility measurements on this alloy, that in the temperature range between 210 and 280 K only approximately 26 % of the Mn atomic moments present are antiferromagnetically coupled. The majority of 74 % of the Mn atomic moments account for the Langevin paramagnetism as well as for the ESR. At lower temperatures the ESR studies referred to above show considerable deviations from the described single moment behavior found at higher temperatures. Within a relatively narrow temperature range the slope of the resonance line width vs. temperature graph is reversed; below this temperature range the line width increases with decreasing temperature. In the same temperature range where this change in slope takes place, the resonance line position starts to shift, corresponding to increasing g values with decreasing temperature. One may take the temperature of the minimum of the line width, T_{ESR}, as an approximate indication of the center of the temperature range where the described transition in the ESR behavior takes place. Fig. 12 shows T_{ESR}, along with T_f, for Cu-Mn alloys as a function of the Mn concentration. It is seen that, up to 25 % Mn, T_{ESR} is approximately 60 to 100 % higher than T_f. The resonance line width increases with decreasing temperature und eventually the ESR line disappears altogether, as the freezing temperature, T_f, is approached.

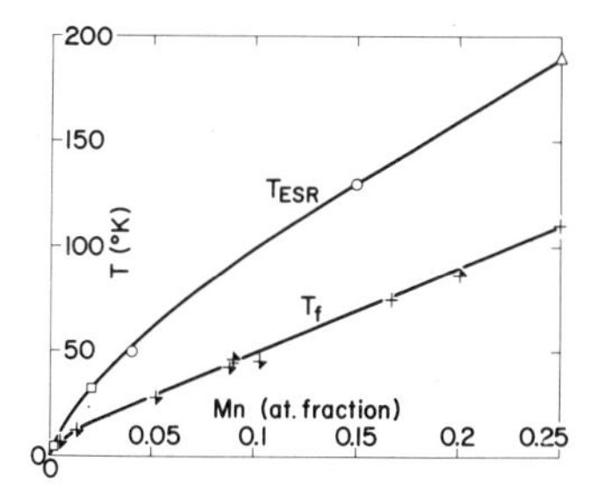

Fig. 12. T_{ESR} and T_f for Cu-Mn alloys vs. Mn-content, from /1b/.

Referring to the temperature range below T_{ESR}, the observed ESR behavior has been described as an "unusual form of antiferromagnetic resonance absorption" /26/. It is now known that there is no long range antiferromagnetism at any temperature in Cu-Mn alloys in the composition range up to 25 % Mn. However, it seems reasonable to attribute the observed ESR behavior below T_{ESR} to increasing short range antiferromagnetic coupling between the Mn atomic moments and to increasing prevalence of the anisotropy energy over the thermal energy with decreasing temperature. It appears that all of the Mn moments, which at higher temperatures give rise to Langevin paramagnetism and to free moment ESR behavior, become antiferromagnetically coupled with one another giving rise to massive short range antiferromagnetism as the temperature is lowered. The only Mn moments which are not compensated by antiferromagnetic coupling in the vicinity of T_f appear to be those relatively few, which are ferromagnetically coupled and, thus, participate in the magnetic clusters. This explains why essentially all temperature-dependent magnetic phenomena slightly above, at and below T_f, including the cusp of the alternating low field susceptibility, are dominated by the magnetic clusters, as described in section 2. It would be desirable to confirm the prescence of short range antiferromagnetism near T_f in Cu-Mn alloys with less than 25 % Mn by neutron diffuse scattering.

9. THE TRANSITION FROM FERROMAGNETISM TO MICTOMAGNETISM

Although for years both experimental and theoretical work centered largely on the transition from the paramagnetic to the mictomagnetic state and on the attendant alternating low field susceptibility cusp, it was found recently that $Fe_{75-x}Al_{25+x}$ alloys in the range of x=2 to 5, ordered in the Fe_3Al structure, are ferromagnetic at room temperature, but become mictomagnetic on cooling to low temperatures/36/. Fig. 13 shows the magnetization in a steady field of 50 Oe as a function of temperature. Although the magnetization decreases at low temperatures even on field cooling (in contrast to Cu-Mn alloys, Fig. 1), the decrease after zero-field cooling is distinctly larger, so that the data clearly indicate a thermomagnetic history effect typical for mictomagnetism. The vertical arrows for the alloys with 30 to 35 % Al show T_f from the cusps of the alternating low field susceptibility graphs in Fig. 14. Clearly, in the alloys with 30 % and more Al, the mictomagnetic state is formed on cooling from the paramagnetic state. The fact that $Fe_{70}Al_{30}$ is ferromagnetic at room temperature, implies then that somewhere between 300 K and T_f = 92 K this alloy undergoes first a transition from the ferromagnetic to the paramagnetic state on cooling, before changing from para-

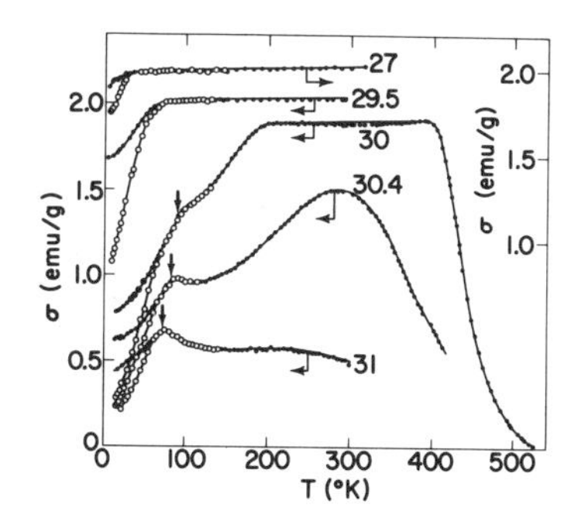

Fig. 13. Magnetization in a field of 50 Oe vs. temperature for Fe-Al alloys of Fe_3Al structure at Al-contents indicated, after field-cooling (filled symbols) and after zero-field cooling (empty symbols), from /36/

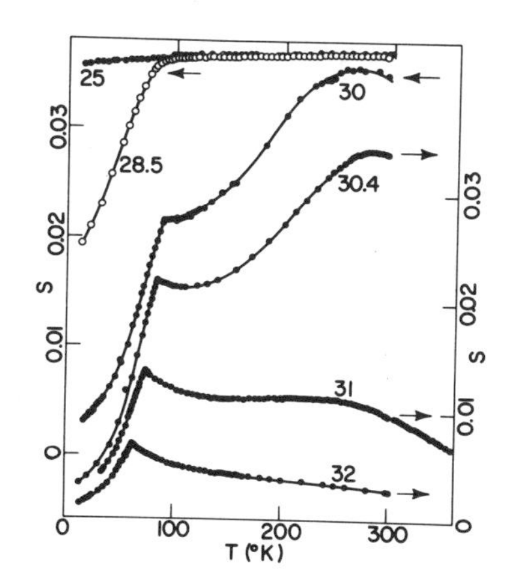

Fig. 14. Alternating low field susceptibility index, S, vs. temperature for Fe-Al alloys from /36/

magnetism to mictomagnetism on further cooling. This conclusion is supported by the temperature-variation of the remanence for field-cooled $Fe_{70}Al_{30}$, as shown in Fig. 15. Such double transition from ferro- to para- to mictomagnetism on cooling was first observed in partially ordered Au_4Mn /6/. The alloys with 27 to 29.5 % Al change from the high temperature ferromagnetic to the low temperature mictomagnetic state directly, i.e. without going through an intermediate paramagnetic state, as shown for $Fe_{70.5}Al_{29.5}$ in Fig. 15. In these cases the transition temperature is not marked by a cusp in the alternating low field susceptibility, as seen in Fig. 14 for $Fe_{71.5}Al_{28.5}$.

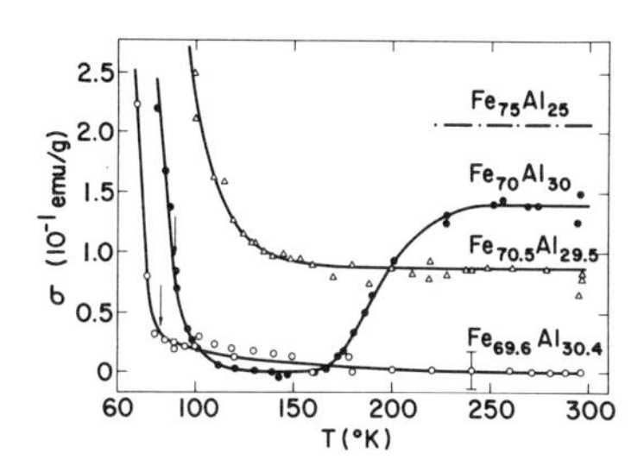

Fig. 15. Remanent magnetization vs. temperature for Fe-Al alloys, after field-cooling, from /36/.

It is worth noting that the alternating field (Fig.14) data show the cusp for alloys with 30 to 32 % Al to be superimposed on a general decrease of the susceptibility with decreasing temperature, starting well above T_f. This tendency is clearly visible also in Fig. 13 as a general decrease in the low steady field magnetization over a wide temperature range, even in the course of field cooling. The large increase of the temperature-linear magnetic contribution to the low temperature specific heat /37/ of the negative pressure coefficient of the saturation magnetization /38/ and of the high-field differential susceptibility /39/ with the Al-content between 25 and 30 % can all be interpreted as indication of the presence in these alloys of parasitic antiferromagnetism, increasing with the Al content and with the density (through compression). This picture is entirely consistent with the observed decrease in magnetization and in susceptibility with decreasing temperature (increasing density) over a wide temperature range.

The question as to the temperature limit where increasing parasitic antiferromagnetism will eventually destroy the overall ferromagnetic spin alignment and leads to mictomagnetism on cooling (or in the case of $Fe_{70}Al_{30}$ to superparamagnetism) can not be answered on the basis of the available data. Neither can it be decided for the alloys with less than 30 % Al whether or not a sharp temperature limit, that might be considered indicative of a phase transition, actually occurs. Certainly, in general, there is not justification whatsoever for regarding the temperature, at which the measured low field susceptibility or magnetization begins on cooling to decrease from its temperature-independent value, as a phase transition temperature. As is well known, the measured temperature-independent "susceptibility value" is, in fact, determined by the demagnetization factor, which depends only on the geometry of the specimen and not on its magnetic properties. The temperature at which the true susceptibility, decreasing on cooling because of the increase in parasitic antiferromagnetism, happens to intersect the reciprocal demagnetization line is, therefore, co-determined by geometric factors. An example for the insignificance of the intersection can be seen in Fig. 14, where the alternating low field susceptibility of $Fe_{70}Al_{30}$ drops gradually from about 260 K to T_f=92 K, without any anomaly indicating a phase transition between these temperatures. The data approximately reflect the temperature-dependence of the true reversible susceptibility, since the latter never quite reaches the value of the reciprocal demagnetization factor. However, in Fig. 13 the magnetization for $Fe_{70}Al_{30}$ shows a well defined "shoulder" or "knee" at about 190 K which, by the criteria currently in vogue, might be mistakenly considered as the temperature of the phase transition from ferromagnetism to mictomagnetism. Fig. 14, by contrast, shows that the transition is, in fact, quite gradual and that no phase-transition in any generally accepted sense can be found near 190K. Since in both measurements the same spherical specimen was used, the difference must be the result of an increased true susceptibility when the measurement is made with a steady field of 50 Oe, causing the limitation by the reciprocal demagnetization factor to become effective. In view of the lack of evidence for a sharp equilibrium phase transition for the change from either paramagnetism or ferromagnetism into mictomagnetism, the now fashionable use of "magnetic phase diagrams" for such systems must be viewed with grave reservations.

Recently, in the course of a broadly based detailed investigation by Maletta and associates of the insulating mictomagnetic sulfides, $(Eu_xSr_{1-x})S$, using various types of measurements, Maletta and Convert /40/ presented particularly thorough documentation for the change from ferromagnetism to mictomagnetism on cooling, quite similar to that in $Fe_{3-x}Al_{1+x}$. Their neutron diffraction spectra clearly show the presence and/or increasing concentration of magnetic clusters with decreasing temperature as ferromagnetism changes into mictomagnetism, as well as their presence in the superparamagnetic state, above the Curie temperature. (Eu,Sr)S has emerged within the last year, or so, as one of the best known mictomagnetic materials, as a result of the dual efforts of Maletta on the experimental side and of Binder in providing theoretical interpretation.

A remarkably careful and discerning study of the transition from ferromagnetism to mictomagnetism has been very recently completed by H. Claus and associates /41/ on V-Fe and Cu-Ni alloys as a function of composition and of metallurgical treatment (i.e. of short range atomic order). In this work the sharpness of the alternating low field susceptibility cusp was used as an indicator of the homogeneity of both the chemical composition and the short range atomic order of the alloys. By using thin sheet specimens and measuring them with the sheet parallel to the field direction, the demagnetization factor was made very small, so that the alternating low field susceptibility had a cusp at the Curie temperature as well as at T_f. Both of these temperatures were found to vary to a surprising extent with the degree of short range order.

REFERENCES

1a. P.A. Beck, Metall.Trans. 2 (1971) 2015:
P.A. Beck, Magnetism in Alloys (edited by P.A. Beck and J.T. Waber), Met.Soc.AIME, New York, 1972, p.211
1b. P.A. Beck, Properties of Mictomagnets (Spinglasses) Progress in Materials Science 23 (1978) 1-49
2. E.C. Hirschkoff, O.G. Symko and J.C. Wheatley, J.Low Temp.Phys. 5 (1971) 155
3. A.K. Mukopadhyay, R.D. Shull and P.A. Beck, J. less-common Metals 43 (1975) 69
4. C.M. Hurd, J.Phys.Chem.Solids 28 (1967) 1345
5a. J.S. Kouvel, J.Phys.Chem.Solids 24 (1963) 795
5b. J.S. Kouvel, J.Appl.Phys. 31 (1960) 107 and other papers listed in /1b/
5c. J.S. Kouvel, J.Appl.Phys. 30 (1959) 313S
6. D.J. Chakrabarti and P.A. Beck, Int.J.Mag. 3 (1972) 319
7. P. Monod and J.J. Préjean, J.de Physique 39 (1968) C6-910
8. H. Claus and J.S. Kouvel, Sol.State Comm. 17 (1975) 1553-1555
R.W. Knitter, J.S. Kouvel and H. Claus, J.Magn.M. Materials 5 (1977) 356
9. Ch. Schwink, K. Emmerich and U. Schulze, Zeitschr. f.Physik B31 (1978) 385-389
10. S.A. Werner, H. Sato and M. Yessik, AIP Conf.Proc. 10 (1) (1972) 679
11a. R.W. Tustison and P.A. Beck, Sol.State Comm. 20 (1976) 841
11b. R.W. Tustison, Sol.State Comm. 19 (1976) 1075
11c. R.W. Tustison, University of Illinois, Urbana, Thesis, 1976
12. H. Alloul, Phys.Rev.Lett. 42 (1979) 603
13. V. Cannella, J.A. Mydosh and J.I. Budnick, J.Appl.Phys. 42 (1971) 1689
14. O. Symko, personal communication
15a. S.F. Edwards and P.W. Anderson, J.Phys. F5 (1975) 965 and J.Phys. F6 (1976)1927
15b. P.W. Anderson, Amorphous Magnetism II(edited by R.A. Levy and R. Hasegawa) Plenum Press, New York, 1977, p.1

16a. L.E. Wenger and P.H. Keesom, Phys.Rev. B11 (1975) 3497
16b. L.E. Wenger and P.H. Keesom, AIP Conf.Proc. 29 (1976) 233
17. W.H. Fogle, J.C. Ho and N.E. Phillips, J.de Physique 39 (1978) C6-901
18. D.L. Martin, Phys.Rev. B20 (1979) July 1
19. B. Window, J.Phys. C3 (1970) 922 and B. Window, J.Phys. C2 (1969) 2380
20. A.P. Murani, Phys.Rev.Lett. 37 (1976) 450 and J.Magn. and Magn.Mat. 5 (1977) 95
21. E.D. Dahlberg, M. Hardiman, R. Orbach and J. Souletie, Phys.Rev.Lett. 42 (1979) 401
22. G. Zibold, J.Phys. F8 (1978) L229
23. H.V. Löhneisen, J.L. Tholence and R. Tournier, J.de Physique 39 (1978) C6-922
24. H. Maletta and W. Felsch, Phys.Rev. B20 (1979) July 1
25. T.A. Meert and L.E. Wenger, Bulletin APS 24, No.3 (1979) E13, p.333
26. G.E. Bacon, I.W. Dunmur, J.H. Smith and R. Street, Proc.Roy.Soc. A241 (1957) 223
27. P. Wells and J.H. Smith, J.Phys. F1 (1971) 763
28. E.R. Vance, J.H. Smith and T.M. Sabine, J.Phys. C3, Metal Phys.Suppl. 1 (1970) S34
29. V.L. Sedov, Soviet Phys. JETP 15 (1962) 88
30. T. Hashimoto and Y. Ishikawa, J.Phys.Soc. Japan 23 (1967) 213
31. J. Bardeen, personal communication
32. J. Owens, M.E. Browne, V. Arp and A.F. Kip, J.Phys.Chem.Solids 2 (1975) 85
33. D. Griffiths, Proc.Phys.Soc. 90 (1967) 707
34. A. Nakamura and N. Kinoshita, J.Phys.Soc. Japan 22 (1967) 335; A. Nakamura and N. Kinoshita, J.Phys.Soc. Japan 26 (1969) 48
35. M. Salamon and R. Herman, unpublished results
36. R.D. Shull, H. Okamoto and P.A. Beck, Sol.State Comm. 20 (1976) 863
37. C.H. Cheng, K.P. Gupta and P.A. Beck, J.Phys. Chem.Solids 25 (1964) 759
38. J.E. Noakes, H. Sato and A. Arrott, J.A.P. 42 (1971) 1608 and M. Shiga, R. Minakata, S. Fujimoto and Y. Nakamura, Phys.Stat.Solidi (a) 37 (1976) K33
39. T. Wakiyama, J.Phys.Soc. Japan 32 (1972) 1222
40. H. Maletta and P. Convert, Phys.Rev.Lett. 42 (1979) 108
41. D.W. Carnegie, Jr., C.J. Tranchita and H. Claus, to be published in Proc. of the Conf.Magn. M. Materials, 1979

NMR EXPERIMENTS ON LIQUID AND AMORPHOUS METALS

J. Goebbels, H.R. Khan*, K. Lüders, and D. Ploumbidis

Institut für Atom- und Festkörperphysik,
Freie Universität Berlin, D-1000 Berlin 33, Germany (F.R.)

ABSTRACT. Some NMR experiments on metals and alloys in the liquid and the amorphous state are reported. Whereas the NMR method for liquid metals is well established, its application to amorphous metals has recently started. In the case of liquid metals some few selected experiments are described: 1. Simple metals: Ga, 2. Transition metals: Co, Mn, Zn, and 3. Alloys: In-Hg, -Zn and Bi-Mn, -Ni, -Co, -Fe. In the case of amorphous metals nearly all up to date experiments are considered: Ni-Pt-P, Ni-Pd-P, Ni-P, Ni-P-B, and the superconducting alloys Mo-Ru-P and Nb-Be-Zr.

1. INTRODUCTION

The nuclear magnetic resonance (NMR) technique is a well-known method for studying metals in both the solid and the liquid state [1-3]. Its application to metallurgy dates back to the discovery of the resonance frequency shift in metals by Knight [4] in 1948. This Knight shift K expresses the enhancement of the local magnetic field at the metallic nucleus relative to the field which would exist in a non-metallic non-magnetic compound of the same nucleus. Unlike other methods the NMR technique can not be used for each metallic system. For the NMR measurements, it is necessary that the samples contain at least one component with nuclei favourable to NMR. Fig. 1 shows NMR intensities for some selected nuclei. As can be seen this quantity varies within a wide range. The white marked nuclei have a spin larger than 1/2 which can lead to quadrupolar

* Also at Forschungsinstitut für Edelmetalle und Metallchemie, D-7070 Schwäbisch Gmünd, Germany (F.R.)

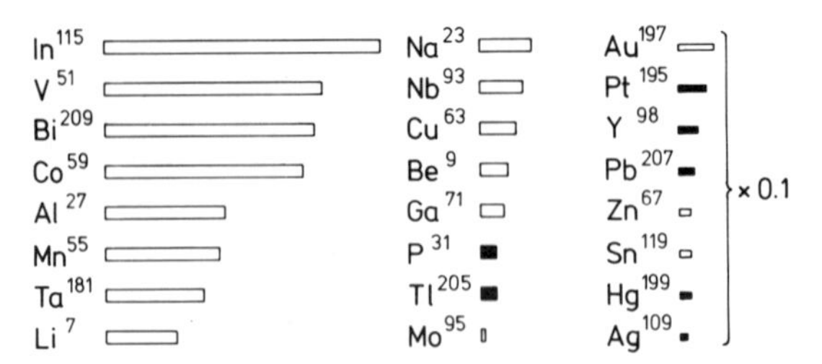

Fig. 1. Comparison of NMR intensities for some selected nuclei (arbitary units).

broadening of the NMR line connected with a considerable intensity loss under certain conditions. This quadrupolar broadening results from the quadrupole interaction with electric field gradients. It is especially troublesome in the investigation of amorphous metals.

The most important NMR parameters are 1. Resonance frequency f (Knight shift K), 2. Line shape (width δ, moments M_i, intensity I), and 3. Spin lattice relaxation time T_1. In most cases K is positive and ranges from 0.03 to about 1.5%. The reason for this shift is the hyperfine interaction between nuclei and electrons. For the simple non-transition metals K is rather well described by

$$K = \frac{8\pi}{3} \chi_s \langle |\Psi(0)|^2 \rangle_F \quad , \tag{1}$$

where only the contact hyperfine interaction is used. χ_s and $\Psi(0)$ are the Pauli spin susceptibility and the electron wave function at the nucleus. In case of the transition metals further terms must be added to take into account the d-electrons:

$$K = K_s + K_d + K_{orb} \quad . \tag{2}$$

Line shape calculations are rather difficult and were performed only in some special cases. Besides the line width the moments M_i are used for line shape descriptions. The dipole-dipole interaction of the nuclei leads to a second moment given by

$$M_2 = \frac{3}{4} I(I+1)a \sum_j (\gamma^2 h r_{ij}^{-3})^2 (3\cos^2 \theta_{ij} - 1)^2 \tag{3}$$

with a = natural abundance, r_{ij} = vector to the neighbour nuclei and θ_{ij} = angle between r_{ij} and the magnetic field B_o, which holds for the light metals. In other cases indirect interactions via the conduction electrons or quadrupole interactions must be taken into account.

T_1 characterizes the energy exchange of the spin system with the environment. The T_1 values for metals range from 10^{-1} to 10^{-3}s at 300 K. With the assumption that only the contact interaction is effective one obtains

$$1/T_1 = \pi k h\, T(\frac{8\pi}{3}\mu_B\gamma\, N(E_F)\langle|\Psi(0)|^2\rangle_F)^2 \quad (4)$$

(μ_B = Bohr magneton, $N(E_F)$ = density of states at the Fermi surface) and the Korringa relation

$$K^2 T_1 T = \mu_B^2/(\pi k h \gamma^2) = \text{const.} \quad (5)$$

In the following chapters some special features of NMR experiments on metals and alloys in the liquid and the amorphous state will be discussed. Whereas the NMR method for liquid metals is well established, its application to amorphous metals has recently started. Therefore only some few selected liquid metal experiments will be described. For the case of amorphous metals nearly all up to date experiments will be considered.

2. LIQUID METALS

For NMR experiments on metals in the liquid state high temperatures are required. Thus, special spectrometers with suitable furnaces are necessary. Detailed descriptions of NMR furnace design have been given by Kerlin [5], El-Hanany et al. [6] and Ploumbidis [7]. The latter developed a furnace suitable for working in an 8T field of a superconducting magnet system at temperatures up to 2200 K. The high field application allows to compensate partly the loss of signal intensity (signal-to-noise ratio) due to the high temperature. Instead of bifilar wires the heating element consists of two coaxial cylinders. By this design disturbing magnetic fields at the NMR sample caused by the heating current or by temperature dependent magnetization effects are avoided. Furthermore, the temperature gradients over the sample volume are kept small and a high mechanical stability is achieved.

2.1 Simple metals: Gallium

The temperature dependence of the Ga^{69} and Ga^{71} Knight shift in liquid gallium was measured in detail by many authors [8-14]. K decreases from 0.455% in the supercooled region to 0.43% at 1250 K. A detailed analysis including T_1 measurements and using the Korringa relation led Kerlin and Clark [12] to the conclusion, that in liquid gallium only the direct contact interaction is responsible for K and T_1. The decrease of K and $(T_1T)^{-1}$ at higher temperatures is caused by the reduction of χ_s and $\langle|\Psi(0)|^2\rangle_F$.

It is interesting to compare the Knight shift of pure liquid gallium with that of gallium antimonide and of pure liquid antimony, also measured by Kerlin and Clark [12]. The Knight shift of Ga in liquid GaSb is 7% larger than in liquid Ga, whereas the shift of Sb in liquid GaSb is 6% smaller than in pure liquid Sb. Although a quantitative treatment is difficult, this result is reasonable since the average number of electrons per atom in GaSb is in between the valences of Ga and Sb.

Nuclear acoustic resonance (NAR) in liquid metals. When talking about liquid Ga the first successful NAR experiments in a liquid metal should be mentioned. They were performed with liquid Ga by Reusche and Unterhorst [13] and Unterhorst et al. [14]. Fig. 2

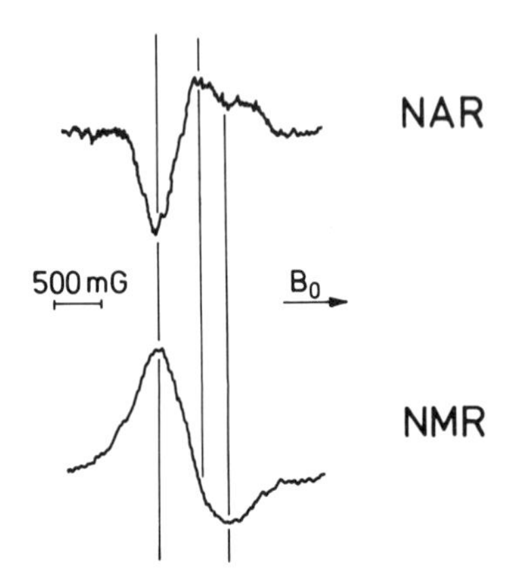

Fig. 2. Derivative of NAR and NMR absorption lines at f = 19.73 MHz and T = 30°C.

shows a NAR signal compared with a NMR signal (derivatives of the resonance lines). The NAR line width is about 25% smaller than the NMR line width. Both signals are asymmetric. In the case of NMR this is due to the skin effect and in the case of NAR it is due to the fact that the absorption line is a mixture of NMR absorption and dispersion [14]. Nuclear acoustic spin transitions can be induced by two coupling mechanisms between sound and spin system: the electric quadrupole interaction and the magnetic dipole interaction. The latter is restricted to metals only but is also applicable to liquid metals. In the presence of a magnetic field sound waves produce coherent electromagnetic waves in the interior of a metal which can couple directly to the nuclear spin system. In a certain sense this kind of NAR is an indirect NMR. Measurements of the temperature dependence of the Ga NAR Knight shift [14] agree well with the NMR measurements mentioned above.

2.2 Transition metals: Co, Mn, Zn

Liquid transition metals mean a hard challenge to NMR experimentalists. The very few existing data in this field are doubtless connected with the experimental difficulties caused by the high melting points and chemical reactivities of these metals.

The highest temperatures up to now were reached in Knight shift measurements of liquid Co by Shaham et al. [15]: K was found to be negative and to vary from -6% at 1675 K to -1.5% at 1875 K. The discontinuity at the melting point is about 1%. A value of the d-spin hyperfine field was obtained using susceptibility data of Müller and Güntherodt [16] for liquid Co: $H_{hf}^{(d)} = (-116.1 \pm 6.3) kG/\mu_B$.

El-Hanany and Warren have performed NMR investigations on liquid Mn [6]. The temperature range of the liquid state investigated is relatively narrow (~ 60 degrees). In this region K decreases slightly and T_1 is nearly constant. An analysis based on the comparison with susceptibility data and including several assumptions (e.g. only χ_d is temperature dependent) leads to the conclusion that K and χ of liquid Mn are dominated by a d-electron contribution.

First NMR measurements on metallic Zn were published last year. Bucklisch and Ploumbidis [17] as well as Kerlin and McNeil [18] reported Knight shift data of liquid Zn in the temperature range from the melting point (693 K) to about 1000 K. In the meantime Bucklisch and Ploumbidis have extended their measurements up to 20 degrees below the boiling point (Fig. 3). The sign of K is

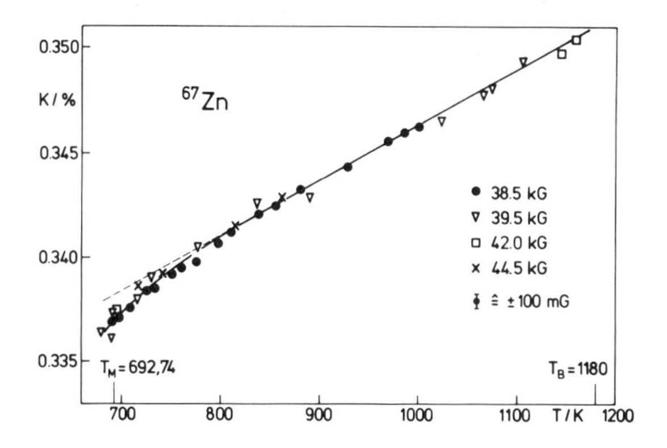

Fig. 3. Knight shift K for ^{67}Zn in liquid Zn.

positive and at the melting point its value was determined to $(0.3368 \pm 0.0004)\%$. For the linear part of the function K(T) the temperature coefficient amounts to $2.7 \times 10^{-5} \% K^{-1}$. Towards lower temperatures the slope increases. No NMR signal could be detected for solid Zn between room temperature and the melting point. This may be mainly due to quadrupolar broadening in the non-cubic structure of solid Zn. The interpretation of the temperature dependence of K is not yet clear. The main change is probably caused by the thermal expansion but an increase of the density of states may also contribute. For a further analysis one needs the pressure dependence of the Knight shift and the temperature dependence of the susceptibility $\chi(T)$. In the liquid $\chi(T)$ was so far only measured up to 823 K [19].

Measurements of T_1 [18] in liquid Zn show that the Korringa relation is fulfilled if an enhancement factor of about 1.25 is added on the right-hand side. (This seems to be a general rule for liquid metals.). It seems, that a quadrupolar contribution to the relaxation mechanism is negligible in liquid Zn.

2.3 Alloys: In-Hg, -Zn; Bi-Mn, -Ni, -Co, -Fe

The electronic properties of liquid Hg and its alloys have been investigated by means of different methods, e.g. magnetic susceptibility, Hall coefficient, and electrical conductivity measurements by Güntherodt et al. [20] and NMR studies by Seymour and Styles [21]. Detailed measurements of the temperature and the concentration dependence of the ^{115}In Knight shift in liquid In-Hg were carried out recently by Ploumbidis and Sutter [22]. Their results are shown in Fig. 4. Earlier measurements in relatively restricted ranges of temperature or concentration [21] are in very good agreement. The K variation is caused by both the intrinsic temperature variation and the corresponding volume variation. Their experimental separation is of considerable interest and can be done using the following equation:

$$(\partial K/\partial T)_V = (\partial K/\partial T)_p + \alpha/\beta\ (\partial K/\partial p)_T \quad . \qquad (6)$$

α and β are the thermal expansion coefficient and the compressibility. In the case of In-Hg, all data are available: $(\partial K/\partial T)_p$ and $(\partial K/\partial p)_T$ from measurements of Oshima et al. [23], and α and β from measurements of Vol and Kagan [24] and Lee and Stephens [25], respectively. The resulting concentration dependence of the intrinsic temperature coefficients is shown in fig. 5 (curve b). It increases rapidly with increasing Hg concentration at the Hg-rich side of the system. Such a behaviour cannot be explained in terms of the free electron model. A possible interpretation can be given using Mott's pseudogap model [26]. With increasing Hg concentration the ratio $g = N(E_F)/N(E_F)_{free}$ decreases. ($N(E_F)_{free}$ is the density of states corresponding to the free electron model.) The results can be explained with the additional assumption that the temperature coefficient $(\partial g/\partial T)_V$ (being proportional to $(\partial K/\partial T)_V$) increases with increasing pseudogap depth.

As second example ^{115}In Knight shift measurements in liquid In-Zn may be mentioned [27]. In the temperature range investigated (650 - 900 K) K increases rapidly with increasing Zn concentration: from 0.78% at 100 at%In to 0.86% at 5 at%In (650 K). Possibly this is due to the increase of the probability density of electrons at the In sites (charge transfer).

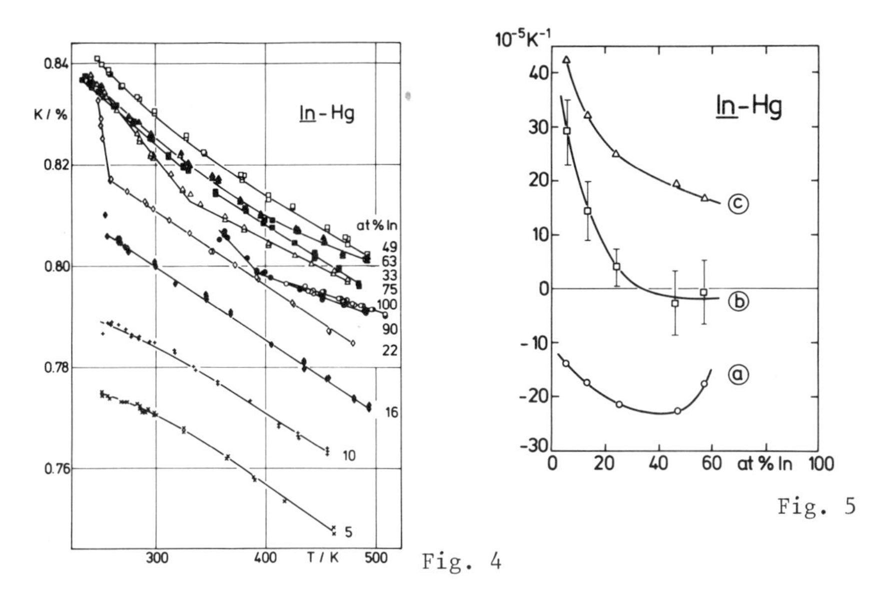

Fig. 4. Temperature dependence of the ^{115}In Knight shift in liquid In-Hg alloys. □49, ▲63, ■33, △75, ○100, ●90, ◊22, ♦16, +10, ×5at%In.

Fig. 5. Concentration dependence of the temperature coefficient of the ^{115}In Knight shift at room temperature. Curves a and c represents the terms on the right-hand side of equ. (6).

A further interesting class of liquid binary alloys are those containing 3d-transition metals. Very recently NMR investigations were performed on the systems Bi-Mn [28,29], Bi-Ni, Bi-Co and Bi-Fe [29]. Fig. 6 shows the ^{209}Bi Knight shift for various systems as a

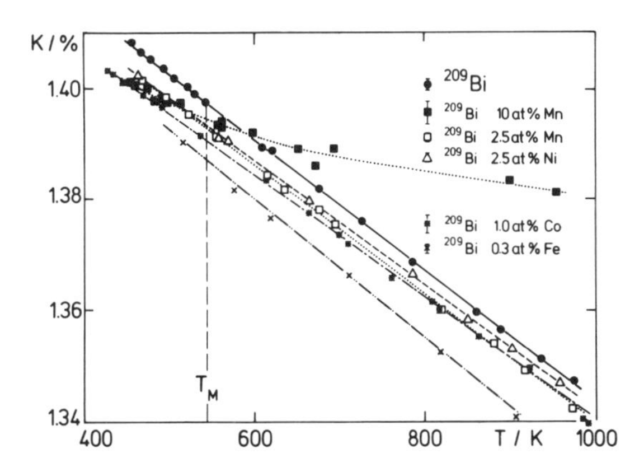

Fig. 6. Temperature dependence of the ^{209}Bi Knight shift in liquid Bi with d-element impurities.

function of temperature. As can be seen K is strongly affected by the presence of d-element impurities. Ni has the smallest influence compared with the others, although its concentration is higher in these measurements. Probably the almost filled d shell does not have a strong magnetic moment and therefore there is only a week polarization of the conduction electrons. A special tendency was found for 10% Mn. In this curve the temperature coefficient changes which may be due to the strong paramagnetism of the Mn ions. But to clarify this behaviour further detailed experiments are necessary.

3. AMORPHOUS METALS

The main difference between NMR experiments on liquid metals and those on amorphous metals becomes evident in the different line shapes. In liquids due to atomic motion, the dipolar and quadrupolar interactions can be averaged out leading to comparatively narrow NMR lines (motional narrowing). In the amorphous state this mechanism does not work and furthermore due to the random structure additional line broadening can occur. Thus, line shape analysis is very important and may provide information about the interactions involved. On the other hand, however, extreme broadening also reduces the signal intensity considerably and in the worst case no signal can be observed.

3.1 Ni-Pt-P, Ni-Pd-P, Ni-P, Ni-P-B

In order to avoid difficulties with quadrupolar line broadening, the use of nuclei with spin 1/2 is advantageous. Hasegawa et al. [30] reported ^{31}P and ^{195}Pt NMR measurements in the metallic glass Ni-Pt-P for the first time. Later, these authors also investigated the system Ni-Pd-P using the ^{31}P resonance [31]. Fig. 7 shows their Knight shift results of ^{31}P as a function of P and Ni concentration. K decreases with increasing P concentration whereas it remains constant over the entire range of Ni concentration. Such a behaviour might be expected since a change in P concentration would vary the average number of electrons per atom while a change of Ni concentration with respect to Pt or Pd does not effect the electron concentration. Assuming two principal Knight shift contributions, $K=K_s+K_d$, the authors attribute the strong decrease in K with P concentration to filling up of the transition metal d-states because of a charge transfer from the P atoms.

The most striking feature of the ^{195}Pt Knight shift is its reduction to -0.1% compared with -3.5% in pure Pt metal*. This also can provide evidence that the transition metal d-states are filled.

---

* A comparable ^{195}Pt Knight shift reduction was found in the A15- and σ-phase of Nb-Pt [32].

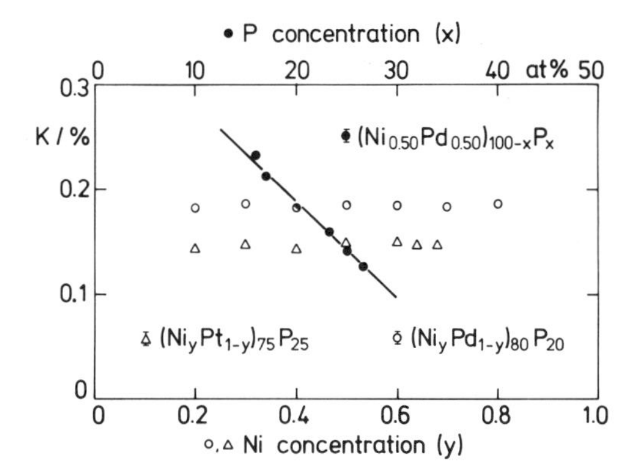

Fig. 7. ^{31}P Knight shift as a function of P and Ni concentration in amorphous Ni-Pd-P and Ni-Pt-P. (Ref. [31]).

As already mentioned, line shape analysis is important for the interpretation of NMR investigations in amorphous metals. In the case of the ^{31}P resonance line in amorphous Ni-Pt-P and Ni-Pd-P a characteristic feature is the frequency dependence of the line width δ [30,31]. For every composition it increases continuously with frequency. This indicates a broadening mechanism resulting from a distribution of the Knight shift, which means that the P atoms have a variety of environments in the amorphous structure. The line width data were analyzed by considering two contributions: a frequency dependent part ($\delta_1 = cf$) and a frequency independent part (δ_2), e.g. due to dipolar broadening, with the total line width given by

$$\delta = [(cf)^2 + \delta_2^2]^{1/2} \quad . \tag{7}$$

For the system $(Ni_xPt_{(1-x)})_{75}P_{25}$, the resulting c and δ_2 values are given in table 1. c and δ_2 remain constant for $0.20 \leq x \leq 0.50$

Tab. 1. $(Ni_xPt_{(1-x)})_{75}P_{25}$

x	0.20	0.30	0.40	0.50	0.60	0.64	0.68
c(G/MHz)	0.40	0.40	0.40	0.40	0.35	0.31	0.28
δ_2(G)	2.0	1.9	1.8	1.8	1.4	1.4	1.3

and then decrease continuously for $0.50 \leq x \leq 0.68$. This shows that the K distribution and correspondingly the distribution in P sites is reduced for $x \geq 0.50$. When Pt is replaced by Pd then the situation is reversed, i.e., c increases above $c = 0.50$. This different behaviour could result from different structures caused by atomic size effects.

Bennett et al. [33] have studied the ^{31}P NMR in electroplated and chemically deposited films of Ni-P with P contents between 15% and 25%. They found different Knight shifts for the samples prepared by different techniques: 0.16% for the electroplated and 0.20% for the chemically deposited samples (For comparison: for crystalline Ni_3P and Ni_5P_2 the K values are 0.17% and 0.10%, respectively.) which is obviously due to different local structures.

The line width increases with increasing frequency which corresponds to the behaviour mentioned above. The extrapolated zero frequency line width of $Ni_{0.82}P_{0.18}$ is about 3 kHz. If one calculates the line width using the expression for the second moment (Equ. 3) for a random distribution of Ni and P atoms, a line width of 6 kHz is obtained. Considering another distribution where P has only Ni neighbours [34], the calculated line width results, however, to about 2 kHz, which is in reasonable agreement with the experimental value. For crystalline Ni_3P, P also has only Ni neighbours and a narrow line width of about 2 kHz is observed. This indicates that the short-range bonding and the nearest neighbours of P are similar in crystalline and amorphous materials.

^{31}P and ^{11}B NMR measurements in amorphous $Ni_{78-80}\,P_{14-12}\,B_8$ were performed by Guerra et al. [35]. Their Knight shift values are 0.10% and 0.27% for ^{11}B and ^{31}P, respectively. The value for ^{31}P compares fairly well with that of Ni-Pd-P [31] and Ni-P [33]. T_1 measurements and comparison with the Korringa relation lead to a charge transfer picture, filling up the Ni d-band.

3.2 Superconducting alloys: Mo-Ru-P, Nb-Be-Zr

NMR measurements on superconducting $(Mo_{0.5}Ru_{0.5})_{80}P_{20}$ metallic glass in the normal and the mixed superconducting state are reported by Guerra et al. [36]. $(Mo_{0.5}Ru_{0.5})_{80}P_{20}$ has T_{co} and dH_{c2}/dT values of 5.4 K and 25 kG/K, respectively. The ^{31}P line width shows a frequency dependence like Ni-Pd-P and suggests a distribution of Knight shifts. Knight shift and relaxation rate are temperature independent in the normal state within the limits of experimental accuracy: $K_n = 0.50\%$, $R_n = 0.122\ \text{sec}^{-1}K^{-1}$. The resulting Korringa ratio is 1.27, which is slightly higher than the ideal value of unity. The temperature dependence of T_1 as a function of T_{co}/T is shown in fig. 8. It follows the general behaviour as predicted by the theory of Hebel and Slichter for the type I superconductors [37]. For the range $1.2 < T_{co}/T < 2.5$, T_1 follows an exponential law yielding a value of 3.52 for the ratio $2\Delta_o/\,k_BT_c$ (Δ_o = gap parameter) in excellent agreement with the value predicted by BCS Theory for weakly coupled superconductors. This shows that strong coupling effects do not necessarily result from a disordered amorphous structure.

Another superconducting amorphous alloy was investigated by Freyhardt et al. [38]: $Nb_{2.5}Be_{32.5}Zr_{65}$. The transition temperature T_c was found to be 2.92 K (measured inductively), which is a little bit higher than the value measured by Hasegawa and Tanner [39] (2.62 K, measured resistively). The ^9Be Knight shift could be detected in the temperature range from 4.2 K to room temperature. It varies from K = -0.016% at 4.2 K to K=-0.010% at room temperature.

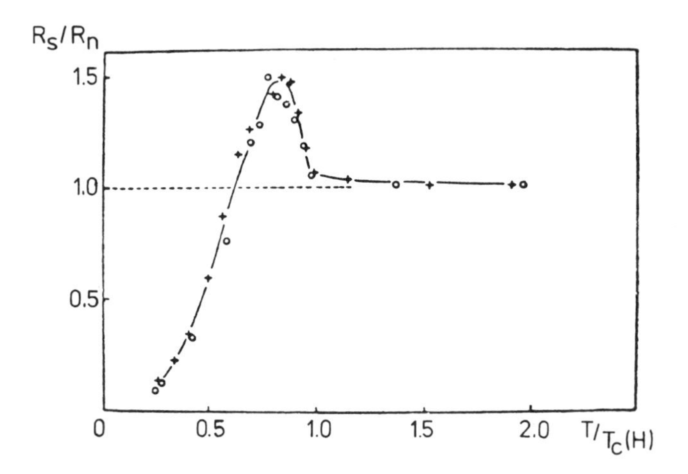

Fig. 8. Reduced relaxation rate R_s/R_n versus reduced temperature $T/T_c(H)$. $R_n = (0.122 \pm 0.002) s^{-1}K^{-1}$, H = 5.2 and 8.7 kG. (Ref. [36]).

All values are more negative than the Knight shift of pure Be. This may indicate the dominating influence of the d-electrons. At room temperature this factor amounts to a value of about 4. The line width is comparatively small (~ 3.2 G) and is not temperature dependent.

REFERENCES

1. L.E. Drain, Metals and Materials 1, 195 (1967).
2. J. Winter, Magnetic Resonance in Metals, Oxford, At the Clarendon Press, 1971.
3. I. Ebert, G. Seifert, Kernresonanz im Festkörper, Leipzig, 1966.
4. W.D. Knight, Phys. Rev. 76, 1259 (1949).
5. A.L. Kerlin, Ph.D. thesis (UCLA, 1972) (unpublished).
6. U. El-Hanany, W.W. Warren, Jr., Phys. Rev. B12, 861 (1975).
7. D. Ploumbidis, Rev. Sci. Instrum. (1979), to be published.
8. D.A. Cornell, Phys. Rev. 153, 208 (1967).
9. K. Suzuki, O. Uemura, J. Phys. Chem. Solids 32, 1801 (1971).
10. B.R. McGarvey, H.S. Gutowsky, J. Chem. Phys. 21, 2114 (1953).
11. D. Hechtfischer, R. Karcher, K. Lüders, J. Phys. F3, 2021 (1973).
12. A.L. Kerlin, W.G. Clark, Phys. Rev. B12, 3533 (1975).
13. M.J. Reusche, E.J. Unterhorst, Physics Letters 51A, 275 (1975).
14. E.J. Unterhorst, V. Müller, G. Schanz, phys. stat. sol. (b) 84, K53 (1977).
15. M. Shaham, J. Barak, U. El-Hanany, W.W. Warren, Jr., Solid State Commun. 29, 835 (1979).
16. Ref. given in [15].
17. R. Bucklisch, D. Ploumbidis, Phys. Rev. B17, 4160 (1978).
18. A.L. Kerlin, J.A. McNeil, Solid State Commun. 27, 757 (1978).

19. E. Wachtel, E. Übelacker, Z. Metallkd. 56, 349 (1965).
20. H.-J. Güntherodt, A. Menth, Y. Tiéche, Phys. kondens. Mat. 5, 392 (1966).
21. E.F.W. Seymour, G.A. Styles, Proc. Phys. Soc. 87, 473 (1966).
22. D. Ploumbidis, T. Sutter, phys. stat. sol. (b) 91, 185 (1979).
23. Ref. given in [22].
24. Ref. given in [22].
25. Ref. given in [22].
26. N.F. Mott, Phil. Mag. 26, 505 (1972).
27. D. Ploumbidis, R. Bucklisch, to be published.
28. R. Dupree, R.E. Walstedt, F.J. Di Salvo, Phys. Rev. B19, 4444 (1979).
29. R. Bucklisch, D. Ploumbidis, to be published
30. R. Hasegawa, W.A. Hines, L.E. Kabacoff, P. Duwez, Solid State Commun. 20, 1035 (1976).
31. W.A. Hines, L.T. Kabacoff, R. Hasegawa, P. Duwez, J. Appl. Phys. 49, 1724 (1978).
32. H.R. Khan, K. Lüders, Ch.J. Raub, Z. Szücs, phys. stat. sol. (b) 84, K33 (1977).
33. L.H. Bennett, H.E. Schone, P. Gustafson, Phys. Rev. B18, 2027 (1978).
34. J.F. Sadoc, J. Dixmier, A. Guinier, J. Non-Cryst. Solids 12, 46 (1973).
35. D.A. Guerra, P. Panissod, J. Durand, Solid State Commun. 28, 745 (1978).
36. D.A. Guerra. J. Durand, W.L. Johnson, P. Panissod, Solid State Commun. 31, 487 (1979).
37. L.C. Hebel, C.P. Slichter, Phys. Rev. 113, 1504 (1959).
38. H.C. Freyhardt, J. Goebbels, K. Lüders, J. Reichelt, to be published.
39. R. Hasegawa, L.E. Tanner, J. Appl. Phys. 49, 1196 (1978).

4. CONTRIBUTED PAPERS

4.1 Liquid metals

STRUCTURE FACTORS AND EFFECTIVE PAIR POTENTIALS OF LIQUID METALS

O.J. Eder, B. Kunsch and M. Suda

Forschungszentrum Seibersdorf, A-2444 Seibersdorf, Austria

ABSTRACT. We report on the attempt to calculate effective pair potentials for liquid Ni, Cu and Al from measured structure factors via the BG, HNC, PY, RPA and STLS approximations.

Recently we have published structure measurements on liquid nickel at 1873K (2) and on liquid copper at 1393K and 1833K (3). We have also performed structure measurements on liquid aluminum at several temperatures in the range between 953K and 1128K (1).

In this communication we want to report on our attempts to calculate pair potentials from the measured structure factors. There has been a large number of approaches proposed to achieve this goal. So far we have used the approximations commonly known as Born Green (BG), Hypernetted Chain (HNC), Percus Yevick (PY), Random Phase (RPA) and Ailawadi's and Naghizadek's (AN) modification of the Singwi, Tosi, Land, Sjölander (STLS) theory. The related references and formulae may be found in our paper on liquid copper (3). The applicability of these approximations to liquid metals has been considered over the last years by several authors with different conclusions (4-7).

Before we discuss our results in some detail, we want to summarize our experiences with the calculation of pair potentials from our structure factors:

i) All pair potentials depend strongly on the statistical accuracy of the data and on the fact, whether smoothed or unsmoothed data are used in the calculations. The results are particularly sensitive to smoothing in the range $0,5 \lesssim Q \lesssim 1,5$.

ii) The pair potentials are to a less, but for quantitative considerations still important degree dependent on the particular way the data are extrapolated to $Q \rightarrow 0$.

These points indicate that the interpretation of pair potentials obtained by the above approximations has to be performed with great care even disregarding the approximate nature of the approaches themselves.

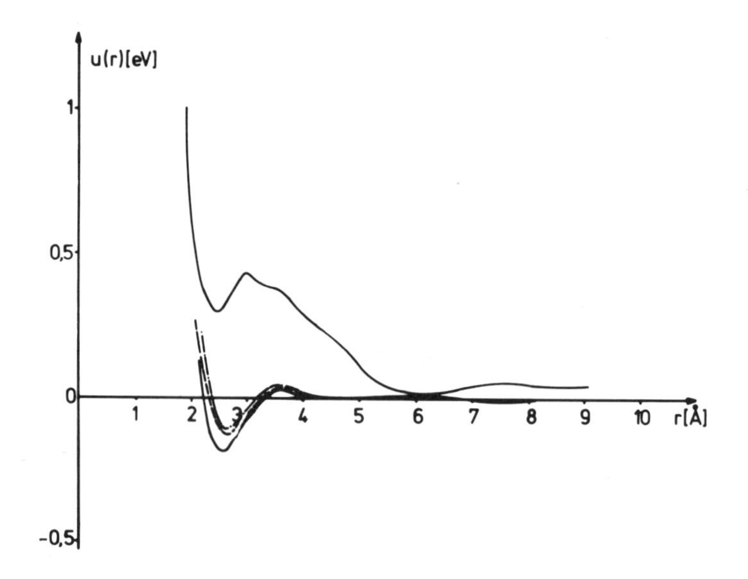

Fig.1: Effective pair potentials of liquid nickel at 1837K, as obtained using the BG (upper full curve), HNC (dashed dotted curve), PY (dashed curve) and AN (solid lower curve).

Fig.1 shows the pair potentials for nickel at 1873K. The potentials except BG look qualitatively similar (RPA is not shown and lies between PY and AN). When the effective spin quantum number S of liquid nickel is chosen to be 0, the minimum of the potentials is enhanced by about 15%. The agreement of our PY potential with the potentials obtained by Johnson et al. 1976 and Mitra 1976 is shown in Fig.2.

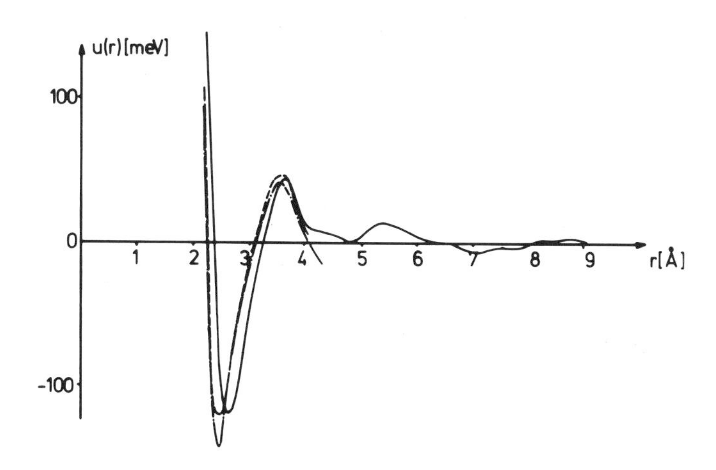

Fig.2: Effective pair potentials for liquid nickel at 1873K as obtained in this work using the PY approximation (full curve) and by Johnson et al. 1976 (dashed curve) and Mitra 1976 (dashed dotted curve).

It is surprisingly good, bearing in mind that two different experimental structure factors are involved and that the other authors have both used molecular dynamics to refine their final result.

As we have shown (3), the systematical behaviour of the potentials observed for nickel remains the same in the case of copper. Again, the depth of the first minimum is increasing from HNC to AN and BG is positive over the whole r-range. The temperature increase of about 440K results in a deeper first minimum for the HNC, PY, RPA and AN potentials and a shift of the BG potential towards smaller energies without affecting its shape.

The availability of structure factors for Aluminum at several temperatures gives us the opportunity to check out the consistency of the dependence of the potentials on temperature in more detail. However since our investigations are not finished yet, we will limit ourselves here to a short preliminary account.

The BG potential changes its shape monotonically with temperature. It does not alter its form when the low Q data, which still contain after smoothing some spurious structure, due to statistical and systematical errors, are replaced by an even polynomial up to Q^4, the coefficients of which are obtained from a least square fit. In contrast, the other four potentials are affected considerably by this measure. The results obtained this way are in qualitative agreement with the potentials found by Ruppersberg and Wehr 1972, in particular the small relative minimum is reproduced on the repulsive part of the PY potential at about r = 3Å. While its position remains unchanged, this minimum goes negative when one uses a quadratic parabola to fit the data at low Q. The HNC, PY, RPA and AN potentials do not change their shape monotonically with temperature, that means that e.g. the depth of the minimum does not systematically decrease with temperature. Further, this change depends on the assumptions for the low Q behaviour of S(Q). We conclude therefore that theoretical efforts to calculate the low Q behaviour of S(Q) would be very helpful for extracting pair potentials in a selfconsistent way.

REFERENCES

1. Eder O.J. and Kunsch B. 1976, Proc. Conf. Neutr. Scatt. Gatlinburg, Vol.II, 1028-33
2. Eder O.J., Erdpresser E., Kunsch B., Stiller H., Suda M. and Weinzierl P. 1979a, J.Phys.F: Metal Phys. 9 1215-22
3. Eder O.J., Erdpresser E., Kunsch B., Stiller H. and Suda M. 1979b, J.Phys.F: Metal Phys. in print
4. Howells S.W. 1973 in The Properties of Liquid Metals Proc. 2nd Int. Conf., Tokyo, 43-49
5. Johnson M.D., Hutchinson P. and March N.H. 1964, Proc. Roy. Soc. A282, 283-302
6. Mitra S.K. 1976 in Liquid Metals, Proc. 3rd Conf. on Liquid Metals, Bristol, 146-55
7. Ruppersberg H. and Wehr H. 1972, Phys. Lett. 40A 31-32
8. Johnson M.W., March N.H., McCoy B. and Mitra S.K., Page D.I. and Perrin R.C. 1976, Phil. Mag. 33 203-06

HOW TO GET STRUCTURAL INFORMATIONS FROM PAIR DISTRIBUTION FUNCTIONS

B.Steffen

Fritz-Haber-Institut, Berlin 33, Germany

It has been shown, that the determination of a most preferred threedimensional structure is possible for a melt, if the instantaneous atomic positions are known [1]. The most preferred nearest neighbour arrangement of the ions in a KCl melt, simulated by molecular dynamics can be described by a suitable deformation of the corresponding solid state ionic - arrangement. This result stimulates the formulation of a lattice approximation for static liquid and glassy structures, which is based on the following assumption:

It is possible, to label the atoms within a certain distance from an arbitrarily chosen atom μ according to a threedimensional lattice whose origin is atom μ. The lattice type can be chosen to be the same for all atoms, merely the orientation may depend on μ.

Then after parallel alignment an average distribution $W_{pqr}(\vec{x})$ for all atoms labeled p,q,r can be defined. The pair distribution function g(r) is given by

$$g(r) = \frac{1}{N \cdot \rho_o} \int \sum_{pqr} W_{pqr}(\vec{x})\, d\Theta d\psi \qquad (1)$$

(Θ,ψ polar angles, N number of atoms, ρ_o number density)

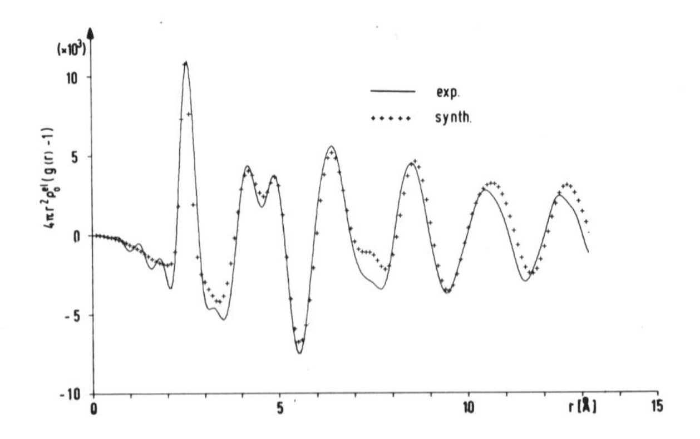

Fig.1 Atomic distribution in glassy $Fe_{40}Ni_{40}B_{20}$ compared to paracrystalline fit.

Nearest neighbours, their number, their average distance and their standard deviations from the ideal lattice positions can then be defined exactly as in solid state physics, namely by looking only at those W_{pqr}, which fulfill

$$p^2 + q^2 + r^2 = \min$$

A first approximation for a description of the average threedimensional density distribution $\sum_{pqr} W_{pqr}$ is given by the paracrystalline model of R.Hosemann [2], based on the assumption, that the average nearest neighbour distribution is the same for all atoms. Then the threedimensional density distribution can be calculated by means of a convolution polynom. After performing the rotational average (1) g(r) can be compared to the experiment, as has been done for the simple liquid metals Li, Na, K, Rb, Cs, Pb [3,4].

The application of this model to glassy metals is aleo successfull, as can be seen in fig.1 for glassy $Fe_{40}Ni_{40}B_{20}$. The distribution of the metal atoms can in this case be described, starting from a bcc lattice with a special vacancy distribution. Contrary to the simple liquids for the metallic glasses additionally to paracrystalline distortions, strain-like distortions had to be introduced [5].

This means, that there exist in the metal relatively large regions with an average "lattice constant" which deviates from the overall average. Such distortions can also be caused by chemical inhomogeneities. Generally this way of describing the atomic structure in the noncrystalline state by giving the centers of mass and the standard deviations of the atomic positions, represents an alternative to those approximations which are exact only in the gaseous state, i.e. at low densities.

REFERENCES

1. B.Steffen, R.Hosemann, K.Heinzinger, L.Schäfer Z.Naturforsch. 32a, 1426 (1977)
2. R.Hosemann, S.N.Bagchi *Direct Analysis of Diffraction by Matter* North Holland Publ.Comp., Amsterdam 1962
3. M.S.Zei, R.Hosemann Phys.Rev. B18, 6560 (1978)
4. B.Steffen, R.Hosemann Phys.Rev. B13, 3232 (1976)
5. B.Steffen, to be published

METHODS TO CALCULATE THE COMPRESSIBILITY OF A LIQUID METAL

I. Ebbsjö

The Studsvik Science Research Laboratory
s-611 82 Nyköping, Sweden

ABSTRACT. The isothermal compressibility evaluated from the long-wave limit of the static structure factor is in the context of computer simulation of liquid aluminium compared with the results from the pressures at neighbouring densities as well as from the pressure fluctuations at the equilibrium density. The total interaction potential is assumed to consist of one part depending only on the electron density and another part consisting of density dependent ionic pair potentials. With a pair potential based on the Ashcroft pseudopotential and by adjusting a structure independent term in order to make the theoretical pressure equal to zero at the equilibrium density for real aluminium the results for the isothermal compressibility from the methods above are consistent with each other and also show satisfactory agreement with experiments.

1. INTRODUCTION

Molecular dynamics calculations(MD) have been made for 'liquid aluminium' at the density for real liquid aluminium at atmospheric pressure and at 975 K[1]. The principal aim of the work reported in [1] was to calculate the dynamical structure factor but there as well as in ref[2] we have also compared the results of alternative methods to calculate the isothermal bulk modulus $B_T=-V(\partial p/\partial V)_T=\kappa_T^{-1}$ with the value on the compressibility κ_T obtained from the long-wave limit of the static structure factor. The three pair potentials $\phi(r;r_s)$ used in the MD-simulation are all calculated from pseudopotential formalism. The one used in the present work is deduced from Ashcroft's pseudopotential with Geldart-Vosko screening(see [1] for details). The dependence on the electron density is indicated by r_s: we have $V=NZ4\pi r_s^3/3$.

2. B_T FROM THE LONG-WAVE LIMIT OF THE STATIC STRUCTURE FACTOR

In order to obtain the isothermal bulk modulus from the relation $S(k=0)=\rho k_B T \kappa_T$ we have fitted our MD-data for the static structure factor in the small k-region to

$$S(k)=s_o + s_2 k^2.$$

The S(k)-values have been taken from the t=0 values of the intermediate scattering function and the fit has been done for four values of $|k|$ from 0.30 to 1.22 $Å^{-1}$(the main peak in S(k) is at about 2.7 $Å^{-1}$). We obtain S(k=0)=0.01791 and, as seen in table 1 (method I), the value on B_T agrees well with the experimental value[3].

Table 1. The bulk moduli calculated by different methods in units of 10^{10}dyn/cm for aluminium at the density corresponding to r_s/a_o=2.167 and at T=988 K. a_o is the Bohr radius

Method	B_T		B_s
I 'Long waves'	40.2		
II Pressure difference	41.6		
III Pressure differentiation	42.5		
IV Pressure fluctuations	43.4	←	50.8
Exp(Seemann and Klein[3])	41.2		51.1

3. B_T FROM THE DIFFERENTIATION OF THE PRESSURE

As shown before(e.g. Waller and Ebbsjö[2]) the pressure for a system with the total interaction potential given by

$$\Phi=\Phi_o(r_s) + \sum_{j>i}\phi(r_{ij};r_s) \tag{1}$$

is

$$p=\rho k_B T - d\Phi_o/dV - \frac{1}{3V}\left\langle \sum_{j>i} \left\{ r_{ij}\frac{\partial\phi(r_{ij};r_s)}{\partial r_{ij}} + r_s\frac{\partial\phi(r_{ij};r_s)}{\partial r_s}\right\}\right\rangle. \tag{2}$$

Evaluation of (2) from molecular dynamics gives a pressure far from zero! But if we now, as done for example by Ashcroft and Langreth[4], adjust the term Φ_o in order to get agreement between the theoretical and the experimental pressures then the difference between the pressures at r_s/a_o=2.167± 0.023 gives a value on B_T which compares well with the experimental value, see table 1(II). We also note that this adjustment brings the internal energy into agreement with experiment.

The pressure given in (2) may be expressed in terms of the pair distribution function $g(r;r_s)$ and scaling the coordinates as done in the derivation of the pressure gives a formula for B_T given in [1], the result is here given in table 1(III). In principle this is equivalent to the method described above: instead of the pressure difference we now need information on $r\partial g/\partial r + r_s \partial g/\partial r_s$ which is obtained from the molecular dynamics calculations.

4. B_s FROM THE PRESSURE FLUCTUATIONS AND CONVERSION TO B_T

As shown by Waller and Ebbsjö [2] the adiabatic bulk modulus B_s for systems with the total interaction potential given by (1) is

$$B_s = \frac{5}{3}\rho k_B T + Vd^2\Phi_o/dV^2 - \frac{N}{\rho k_B T}\langle(\delta P_i)^2\rangle_m + B_{I2}.$$

Here $\delta P_i = P_i - p$ where P_i is the pressure at time point t_i and the microcanonical average is taken over the time points t_i in the simulation with N=500 particles. The term B_{I2} is specified in [1]. As seen in table 1(IV) the result on B_s agrees well with the experimental value. This method has the advantage to give the bulk moduli at each density considered.

The isothermal bulk modulus is calculated from $B_T = B_s - T\gamma_v^2/\rho c_v$ where the thermal pressure coefficient $\gamma_v = (\partial p/\partial T)_v$ is calculated from the correlation coefficient between the temperature and the pressure in the simulation and the specific heat from the temperature fluctuations [1]. The specific heat is about 10% too large and the thermal pressure coefficient too small with the same amount in comparison with experiments. We consider, however, the value on B_T given in table 1(IV) to be in satisfactory agreement with the other methods.

5. DISCUSSION

The value on the isothermal compressibility κ_T obtained from the long-wave limit of the static structure factor as described in section 2 agrees well with both the other methods considered here and with experiments. We note that the long-wave limit result does not depend on Φ_o (the structure independent part of the total interaction potential) and also that the density dependence of the pair potential does not enter.

The relation $S(k=0) = \rho k_B T \kappa_T$, valid for real experiments, should in the context of pseudopotential theory not give the same value on κ_T as that obtained from differentiation of the pressure (see Evans [5] and references therein). One possible reason why the present results do not conform to this expectation is that the extra contribution to S(k) is seen only for very small magnitudes of the wave vector. Also because the parameter in the Ashcroft pseudopotential is determined from Fermi surface data terms beyond second-order are taken into account in a way which is hard to see.

REFERENCES

1. Ebbsjö I, Kinell T and Waller I 1979 Accepted by J.Phys.C
2. Waller I and Ebbsjö I 1979 Accepted as a Letter by J.Phys.C
3. Seemann HJ and Klein FK 1965 Z.angew.Phys.19 368
4. Ashcroft NM and Langreth DC 1967 Phys.Rev.155 682
5. Evans R 1977 Microscopic Structure and Dynamics of Liquids ASI serie B 33(New York: Plenum) p 153-220

INVESTIGATION OF THE STRUCTURE FACTOR OF LIQUID RUBIDIUM UP TO TEMPERATURES OF 2000 K AND PRESSURES OF 140 BAR

E. Schneider, G. Franz* and W. Freyland*
Physik-Department, Technische Universität München
*Physikalische Chemie, Universität Marburg

ABSTRACT. A series of neutron elastic scattering experiments on liquid rubidium have been performed to determine the structure factor S(Q) over a range of temperatures and pressures along the vapor pressure curve up to the critical point. The evaluation procedure to extract S(Q) from the scattered intensity will be outlined in some detail. The results will be discussed qualitatively

1. Experimental

Liquid metals show strong changes of many of their macroscopic properties, e.g. their electrical resistivity, when expanded between their melting point and conditions near their critical point. To understand this behaviour on a microscopic scale is the objective of these structure determinations.
A special high-temperature, high-pressure furnace has been developed which is described elsewhere in more detail /1/.

The neutron scattering experiment has been performed at the D1B-two-axis spectrometer at the high-flux reactor in Grenoble. The main advantage of this spectrometer is, that the scattered intensity is accumulated for 400 angular positions at a time by a multidetector covering an angular range of 80 degrees in steps of 0.2 degrees, so a good angular resolution is achieved besides good counting statistics within relatively small measuring times. As several materials additional to liquid rubidium were in the beam careful attention was necessary to the various corrections which must be applied to the experimental data.

2. Data Analysis

In order to extract the structure factor from the measured intensity five different runs have to be performed /2/.

a) Sample, container and furnace
b) Sample-container and furnace
c) Vanadium and furnace
d) Furnace
e) Cadmium rod and furnace

Vanadium plus furnace being the calibration run and a cadmium plus furnace run is necessary for the correction at the small Q-values /3/.

1) First from each spectrum a)-e) the parasitic Bragg peaks originating from the molybdenum-cell, the tungsten-heater and the autoclave consisting of an aluminium-alloy have to be removed.
2) To perform the background subtractions several attenuation-factors according to Paalman and Pings /4/ have to be computed.
3) Placzek-corrections, which take into account the deviation from the static approximation.
4) Multiple scattering events in both sample and container plus furnace have to be computed and subtracted.

3. Results and discussion

Four experimental curves (preliminary results) are shown in the figure for the following temperatures, pressures and the corresponding densities:

Temperature Kelvin	Pressure Bar	Density g/cm^3
350	1	1.46
900	2	1.22
1700	100	0.79
2000	139	0.54

Three effects on S(Q) by changing the temperature and pressure along the vapor pressure curve are apparent (see the figure):

1) The intensity of the first peak of S(Q) is strongly reduced with increasing temperature indicating a significant reduction of the number of nearest neighbours.
2) The position of the first maximum is shifted to smaller Q-values by heating from 350 K to 900 K and remains essentially

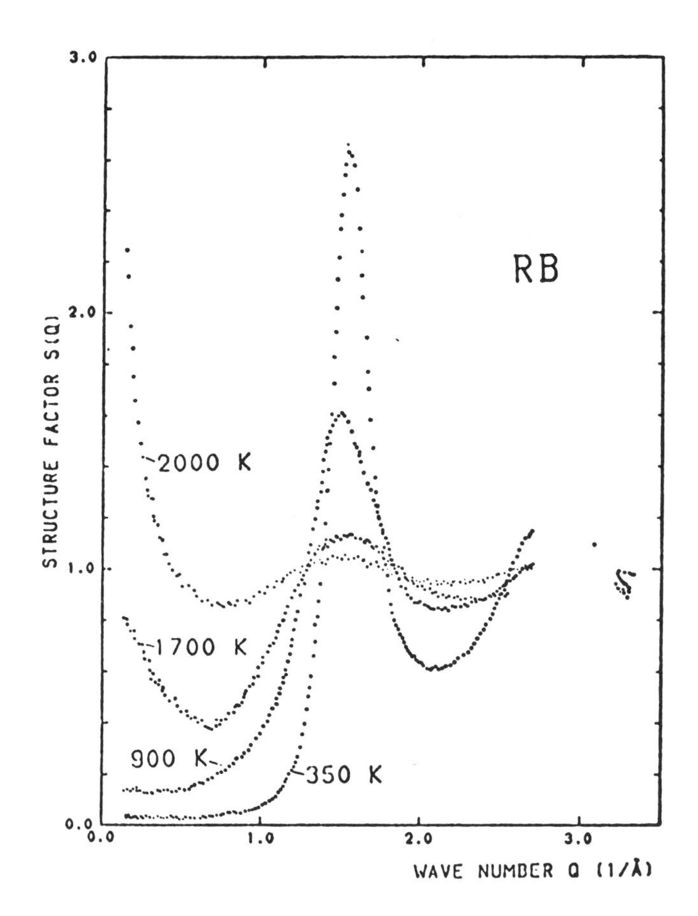

constant for the higher temperatures. The distance between the nearest neighbours does not change much.

3) A strong increase of the small angle scattering at high temperatures.

REFERENCES

/1/ W. Freyland, F. Hensel and W. Gläser, Ber. Bunsenges. Phys. Chem., 83, 884-889 (1979)

/2/ W.S. Howells, ILL-internal technical report, 77HO46T (1977)

/3/ P.F.J. Poncet, PHD Thesis, J.J. Thomson Physical Laboratory, University of Reading (1976)

/4/ H.H. Paalman and C.J. Pings, J. Appl. Phys. 33, 2635 (1962)

APPLICATION OF THE FABER-ZIMAN THEORY OF THE RESISTIVITY OF LIQUID BINARY ALLOYS ON THE RESISTIVITY OF PURE LIQUID LEAD, TIN AND GALLIUM

M. Pokorny

Department of Solid State Physics, Royal Institute of Technology, S-100 44 Stockholm 70, Sweden

ABSTRACT. Faber-Ziman theory expression for the resistivity of a liquid binary alloy as a function of concentration is modified to allow the treatment of the temperature dependent resistivity of a monoatomic and, by assumption, two-component (two-state) liquid metal. This "binary model" perfectly reproduces the observed temperature derivatives of resistivity of liquid Pb, Sn and Ga.

The resistivities of liquid Pb, Sn and Ga were measured with emphasis on a high relative accuracy and found to be nonlinear functions of the temperature [1-3]. The curvature of the resistivity vs. temperature curves is such that the temperature derivatives of resistivity (TDR) of liquid Pb and Sn are exponentially decreasing functions of temperature. The TDR of liquid Ga exhibits first a maximum (cca 60 degrees above the freezing point) and afterwards decreases exponentially in the same manner as that observed for liquid Pb and Sn, Fig.1. These results are described very accurately by a simple expression

$$\mathrm{TDR}(T) = a_1 + a_2 \cdot a_3 \cdot \exp(-a_3 \cdot x) - b_2 \cdot b_3 \exp(-b_3 \cdot x) \qquad (1)$$

where $x=T-T_{FP}$ (FP: freezing point) and a_1, a_2, a_3, b_1 and b_3 are constants determined by the least-squares fit. b_1 and b_3 are necessary only for the description of liquid Ga. Expression (1) is, however, only an analytical device and gives no suggestion about the mechanism causing the observed phenomenon.

Appearance of maximum on the TDR vs. T curve of liquid Ga can be seen as an influence of a saturation process. When this was realized the Faber-Ziman theory of the resistivity of liquid binary alloys has been suggested as a suitable tool [4]. It is assumed that the liquid metal can be treated as a two-component or two-state

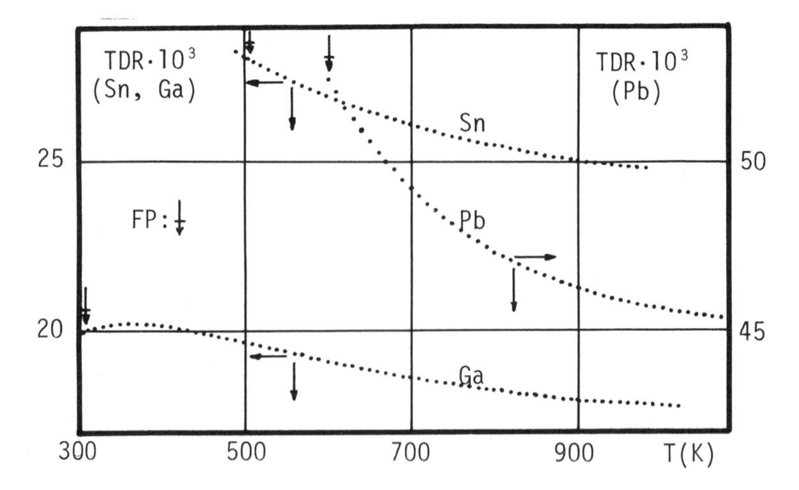

Fig. 1. Temperature derivatives of resistivity of pure liquid lead, tin and gallium

medium. The role of solvent is played by those ions which form a disordered matrix, that of solute by the ions forming groupings or clusters. The concentration of ions bound in clusters, C, is temperature dependent and, in the simplest approximation, proportional to the Boltzmann factor

$$C \sim 1-\exp(-Q/kT) \qquad (2)$$

where Q is interpreted as an energy necessary to break the bond between the ion and its cluster. Faber-Ziman theory [5] gives for the concentration dependence of the resistivity, R,

$$R = a' + b'\cdot C + c'\cdot C^2 \qquad (3)$$

In the original theory a', b', c' are weakly temperature dependent integrals over the partial structure factors and suitable pseudopotentials. For both b' and c' this T-dependence is neglected with respect to that of C given by expression (2). a', describing the solvent's contribution to the resistivity, is assumed to depend linearly on the temperature. This leads to the following expression for the TDR as a function of temperature

$$\mathrm{TDR}(T) = a + b\cdot\frac{Q}{kT^2}\cdot\exp\left(-\frac{Q}{kT}\right) + c\cdot\frac{2Q}{kT^2}\cdot\exp\left(\frac{2Q}{kT}\right) \qquad (4)$$

Expression (4) gives a perfect fit to the data generated via expression (1) both for liquid Ga and for liquid Pb and Sn. Values of a, b, c and Q as well as those of other relevant quantities are given in Table I.

Provided the signs of parameters b and c are opposite the analytical form of expression (1) quarantees the appearance of a maxi-

TABLE I

Quantity	Pb	Sn	α	Ga	β
a(μΩcm/K)	+0.0467	+0.0236		+0.0164	
b(μΩcm/K)	+60.661	+9.8205		-36.399	
c(μΩcm/K)	-47.257	-5.7306		0.0085	
Q (eV)	0.0325	0.0492			
T_{FP} (K)	600.60	505.13	302.91	[11]	256.8
Q(fusion) (eV/atom)	0.0491	0.0724	0.0579		0.0276
T_{max} (K)	157.50	247.82		346.00	
ρ(solid, FP) (μΩcm)	49 [9]	22.8 [9]		15.4 [10]	
% cl. at FP	2.56	5.85		23.46	
% cl. at FP+500	0.03	0.99		8.82	

mum on the TDR vs. T curve. T_{max} values in Table I give the positions of maxima in all three cases investigated. If one interprets T_{max} as the temperature at which the liquid is saturated with clusters one can understand why it is so easy to subcool liquid Ga. The structure of liquid Ga seems to be closely connected with that of solid Gaβ [6]. The liquid is already 60 degrees above the FP of Gaα saturated with Gaβ-like clusters. These cannot act as crystallization nuclei for Gaα and in fact their presence effectively prevents crystallization. A similar mechanism acts even in liquid Pb and Sn. There, however, the saturation temperatures are deep below the FP's and the cluster concentrations are considerably lower. Moreover the structure of clusters seems in both cases to resemble more closely that of the solid phase at the FP [7,8]. Very crude estimate of the cluster concentrations can be made if one assumes their resistivity to be that of the solid phase at the FP. The last two rows of Table I show the cluster concentrations calculated with this assumption. The concentration is indeed highest for liquid Ga, nearly 25% at the FP.

REFERENCES

1. M.Pokorny and H.U.Åström, Phys.Chem.Liquids 3, 115 (1972)
2. M.Pokorny and H.U.Åström, J.Phys.F: Metal Phys.5, 1327 (1975)
3. M.Pokorny and H.U.Åström, J.Phys.F: Metal Phys.6, 599 (1976)
4. N.H.March, private communication
5. T.E.Faber, An Introduction to the Theory of Liquid Metals, Cambridge Univ.Press, 1972
6. A.Bizid, A.Defrain, R.Bellisent, G.Tourand, J.Phys.39, 54 (1978)
7. H.Krebs, H.Hermsdorf, H.Thurn, H.Welte, L.Winkler, Z. Naturf. 23a, 491 (1968)
8. B.Steffen, R.Hosemann, Phys.Rev.B13, 3232 (1976)
9. A.Roll, H.Motz, Z. Metallk.48, 272 (1957)
10. R.W. Powell, Proc.Roy.Soc. (London) Ser.A209, 525 (1951)
11. L.Bosio, R.Cortes, A.Defrain, G.Folcher, J.Chim.Phys.70, 357 (1973)

MAGNETIC FLUCTUATIONS IN Fe AND Co ABOVE CURIE-TEMPERATURE IN THE SOLID AND LIQUID STATE

G. Rainer-Harbach, S. Steeb

Max-Planck-Institut für Metallforschung
Institut für Werkstoffwissenschaften
D-7000 Stuttgart 1, Western-Germany

1. INTRODUCTION

Magnetic fluctuations with iron and cobalt in solid and also in molten state were investigated by means of neutron-small-angle-scattering. The temperature of investigation reached from 813 °C to 1606 °C for iron and from 1192 °C to 1630 °C for cobalt. The two-circle apparatus D4 at the HFR Grenoble with neutrons of wavelength 0.7 Å was used.

2. EXPERIMENT AND RESULTS

From the measured intensities the differential cross sections $d\sigma/d\Omega$ were obtained within the quasistatic approximation, by applying the usual correction and normalization procedures /1/.

The run of $d\sigma/d\Omega$ versus the scattering vector q shows a rise toward smaller q-values due to spin fluctuations for Fe as well as for Co at all temperatures under investigation even in the molten state. Fig. 1 shows examples for iron and cobalt at various temperatures. In the case of iron the four temperatures are due to the four different phases, namely the bcc α-phase (up to 911 °C), the fcc γ-phase (911 °C - 1395 °C), the bcc δ-phase (1395 °C - 1535 °C) and the melt (above 1535 °C).
For Co two temperatures above melting point (1494 °C) are shown.

Thus spin fluctuations exist in Fe at least up to 837 °C above Curie- or 71 °C above melting-temperature. In Co at least up to 509 °C above Curie- or 136 °C above melting-temperature.

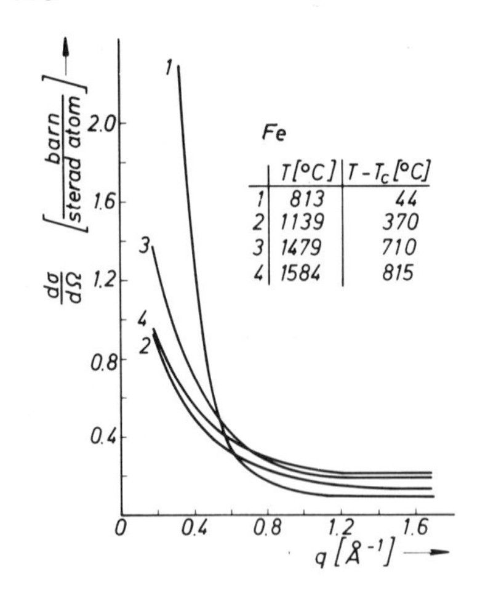

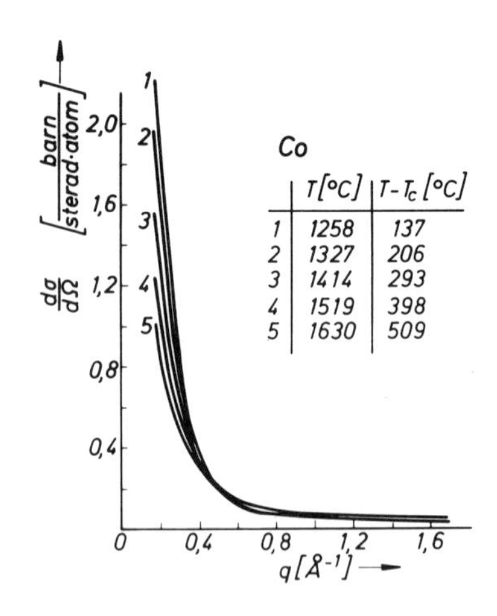

Fig. 1. Fe and Co: Differential cross sections $d\sigma/d\Omega = d\sigma/d\Omega/_{koh} + d\sigma/d\Omega/_{mag}$ at different temperatures

According to Van Hove /2/ the differential magnetic cross section at small q-values may be expressed as

$$\frac{d\sigma}{d\Omega}\Big/_{mag} = (r_o\gamma)^2 \; \frac{2}{3} \; S(S+1) \; f^2(q) \cdot \frac{1}{r_1^2(q^2+K_1^2)}$$

with K_1 the inverse of the correlation length ξ. The correlation length ξ were thus determined from the measured $d\sigma/d\Omega(q)$-function after subtraction of the very small contribution of nuclear scattering.

Fig. 2. shows the depencende of ξ from reduced temperature

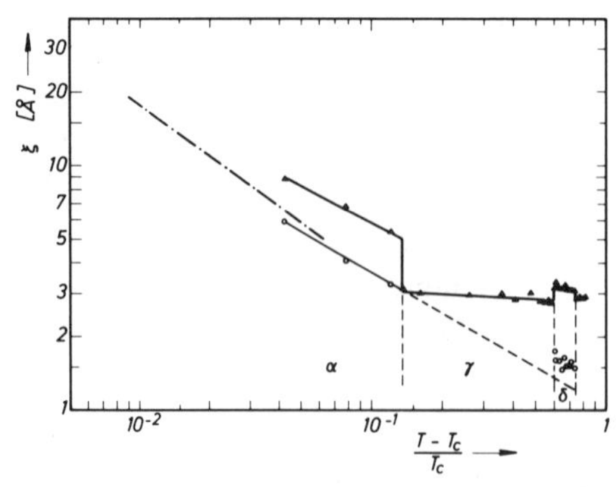

Fig. 2. Fe: Correlation length ξ. ΔΔΔ without corrections for inelasticity, ooo with correction, -.-.- measured data according to Als-Nielsen /3/

$(T-T_c)/T_c$ for Fe. ξ shows for the bcc δ-phase a similar temperature dependence as for the bcc α-phase. The fcc γ-phase which is known to exhibit antiferromagnetic behaviour shows within the whole region from 911 °C up to 1395 °C no variation of the correlation length, which is smaller than that of the δ-phase and that of the melt. ξ in the melt lies between the correlation length of the γ- and the δ-phase and shows no change with temperature.

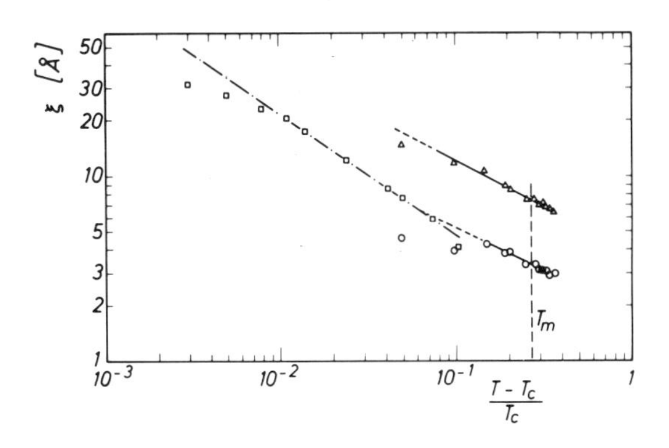

Fig. 3. Co: Correlation length ξ. ΔΔΔ without correction for inelasticity, ooo with correction, □□□ measured data according to Glinka et al. /4/

In Fig. 3 ξ for Co is plotted. It shows a continous decrease with rising temperature. No discontinuity can be observed at the solid-liquid phase transition, indicating that the spin fluctuations are not influenced by the melting process.

As the critical magnetic neutron scattering with Fe and Co near the critical point shows high inelasticity /5,4/ corrections for inelasticity due to the theoretical assumptions of Dynamical Scaling according to /5/ were applied. The values of ξ thus obtained are also shown in Fig. 2 and 3.

REFERENCES

1. G. Rainer-Harbach, Thesis work, University of Stuttgart Germany (1979).
2. L. van Hove, Phys. Rev. 95 (1954) 1374.
3. J. Als-Nielsen, in: Phase Transitions and Critical Phenomena, C. Domb and M.S. Green, ed. (Acad. Press, London, N.Y. 1975) 87.
4. C.J. Glinka, V.J. Minkievicz, and L. Passell, AIP Conf. Proceedings 24 (1975) 283.
5. J. Als-Nielsen, Phys. Rev. Letters 25 (1970) 730.

SELF- AND IMPURITY DIFFUSION IN CU AND SN IN THE LIQUID PHASE

A. BRUSON, M. GERL

Laboratoire de Physique des Solides - Université de NANCY - I, C. O. 140 - 54037 - NANCY-CEDEX (France)

1. INTRODUCTION

The dynamics of simple or model liquids has been extensively studied over the last decade, theoretically, experimentally or by computer simulation [1, 2, 3, 4]. All calculations show that (i) the actual diffusion coefficient D differs markedly from the binary collision diffusion coefficient D_E calculated using ENSKOG's theory ; (ii) the velocity autocorrelation function $\psi(t)$ exhibits a long time tail which can be traced to collective motions in the fluid, because the motion of a given atom builds up hydrodynamic modes which decay slowly [5]. This interpretation is supported by the dependence of the diffusion coefficient on the number of particles used in molecular dynamics (MD) calculations and by the fact that $\psi(t)$ asymptotically varies as $t^{-d/2}$, where d is the dimensionality of the system. Detailed MD calculations on hard sphere (HS) systems show that two regimes of velocity correlation can be defined : (i) at large densities, the backscattering of a given atom on its first neighbour shell leads to an actual D which is smaller than D_E and $\psi(t)$ takes large negative values ; (ii) at intermediate densities, long wavelength modes are excited by the initial motion of a given atom and give rise to an increase of D with respect to D_E . The behaviour of a test particle of reduced mass $M = m_i/m_s$ and reduced size $\Sigma = \sigma_i/\sigma_s$ (where m_i and m_s denote the mass of the solute and solvent respectively, and σ is their diameter), can be easily understood in this framework : (i) as M increases (M > 1) backscattering effect decreases or hydrodynamic modes are more easily excited ; therefore in both regimes, the ratio D/D_E increases when the mass of the solute atom increases ; (ii) as Σ increases, the backscattering effect increases and therefore D/D_E decreases when the backscattering is predominant but, as hydrodynamic modes are more easily excited by a large particle than by a small one, D/D_E increases in the persistance regime.

2. EXPERIMENTAL RESULTS

In order to investigate these effects, systematic measurements of the diffusion coefficient of a number of impurities has been undertaken in liquid copper and liquid Sn, using a shear cell which provides accurate data and which has been described elsewhere [6]. The comparison of the diffusion coefficient of ^{64}Cu, ^{110} mAg and ^{195}Au gives an experimental determination of the mass effect and measurements on very dilute alloys of Cu or Sn with ^{110}mAg, ^{111}In, ^{113}Sn and ^{124}Sb provide an estimation of the size effect.

2.2 Self diffusion

As there is no calculation of the binary diffusion coefficient in Sn and Cu, we use the prescription of PROTOPAPAS [7] who defines the equivalent hard sphere diameter

$$\sigma(T) = \sigma_o \left(1 - a \left(\frac{T}{T}\right)^{1/2}\right) \qquad (1)$$

where σ_o is calculated from the density and packing fraction at the melting temperature T_m. The $T^{1/2}$ dependence of $\sigma(T)$ accounts for the softness of the actual repulsive potential. Using $\sigma(T)$ given by Eq. 1, we calculate the ENSKOG HS diffusion coefficient

$$D_E = \frac{3}{8} \frac{1}{n\, g\,(\sigma)\, \sigma^2} \left(\frac{kT}{\pi m}\right)^{1/2} \qquad (2)$$

where g (σ) is estimated using CARNAHAN and STARLING's formula. The experimental data in Sn are recorded in fig. 1 as a function of the packing fraction y, and compared with the results obtained by ALDER [2].

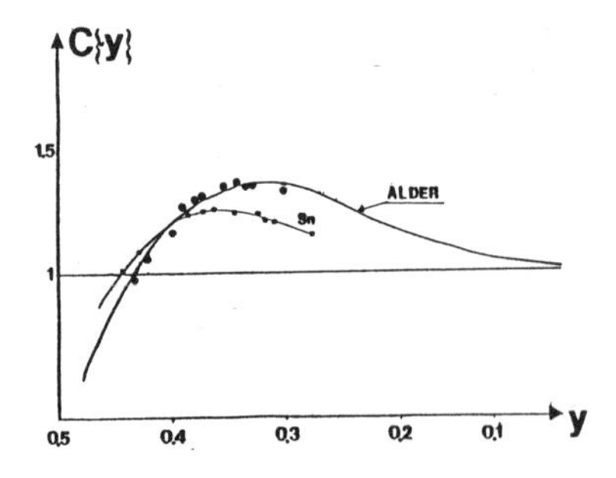

Fig. 1. Experimental values of the ratio D/D_E in Sn (squares) compared with ALDER's data.

Figure 1 shows that the experimental variation of C (y) = D/D_E is correctly predicted by MD calculations. Moreover, it is possible to calculate, at each density, the value $\sigma'(T)$ of the HS diameter which, when used in formula 2, would lead to the value of D/D_E predicted by ALDER. Using this procedure, we get

$$\sigma' (T) = 3.16 \left[1 - 0.082 \left(\frac{T}{T_m}\right)^{1/2}\right]$$

instead of

$$\sigma (T) = 3.30 \left[1 - 0.112 \left(\frac{T}{T_m}\right)^{1/2}\right]$$

when the prescription given by PROTOPAPAS is used.

2.2 Impurity diffusion

In the case of impurity diffusion, we calculate the ENSKOG diffusion coefficient given by :

$$D_E^i = \frac{3}{8} \frac{1}{n\, g\, (\sigma_{is})\ \sigma_{is}^2} \left(\frac{kT}{\pi\mu}\right)^{1/2} \tag{3}$$

where $\sigma_{is} = \frac{1}{2} (\sigma_i + \sigma_s)$ where i and s stand for impurity and solvent respectively and μ is the reduced mass ; $g\ (\sigma_{is})$ is calculated using the prescription given by MANSOORI [8].

$$g_{is} = (\sigma_i\, g_{ss}\, (\sigma_s) + \sigma_s\, g_{ii}\, (\sigma_i))/\ (2\, \sigma_{is})$$

and

$$g_{\alpha\alpha} = (1 - y)^{-2} \left[1 + \frac{y}{2} + \frac{3\, y_\beta}{2\, \sigma_\beta} (\sigma_\alpha - \sigma_\beta)\right]$$

The measured diffusion coefficient D_i can be written

$$D_i = C_1\ (y)\ C_2\ (M, \Sigma)\ D_E^i \tag{4}$$

and we assume that the correction factor C_1 (y) accounting for density effects is the same as the factor D_S/D_E^S obtained from self-diffusion. The values of C_2 (M, Σ) calculated in this way are plotted against temperature in figure 2.

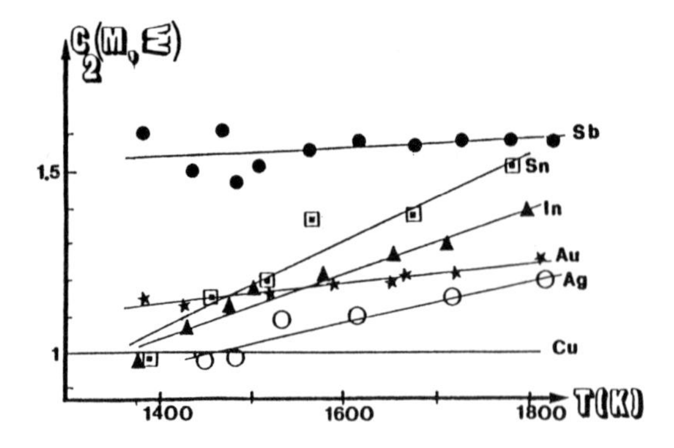

Fig. 2. Experimental values of the correction factor C_2 (M, Σ) for Ag, Au, In, Sn and Sb in liquid copper.

Mass effect : in figure 3, the experimental dependence of C_2 (M, Σ) on M is given at a reduced volume V/V_o = 1.6 and 1.75. The results are compared with those calculated by ALDER (or extrapolated from ALDER's data) at V/V_o = 1.6. In Cu, the experimental values are in good agreement with MD calculations. In Sn, although both experimental data and calculations show that C_2 increases with M, a discrepancy of about 20 % between the absolute values of C_2 is observed.

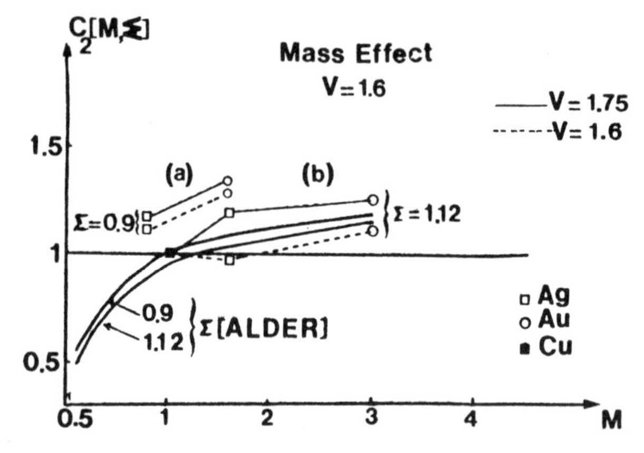

Fig. 3. Mass dependence of C_2 (M, Σ)
(a) Ag and Au in liquid Sn
(b) Ag and Au in liquid Cu

Size effect : In high density fluids one would expect that C_2 decreases as the size Σ of the impurity increases. Our experimental results, shown in fig. 2 for Cu, seem to indicate the opposite trend when one goes from Ag to In, Sn and Sb. This can be interpreted as the effect of the valence of the impurity on its diffusion coefficient. Because of the (screened) electrostatic potential around the impurity, the effective HS diameter used in ENSKOG's formula should be different from its value in the pure solute. Some calculations are currently in progress to evaluate this effect.

REFERENCES

1. L. SJÖGREN, A. SJÖLANDER, Ann. Phys., 110, 122 (1978)
2. B. J. ALDER, T. E. WAINWRIGHT, Phys. Rev. Lett., 18, 988, (1967)
3. A. RAHMAN, M. J. MANDELL, J. P. McTAGUE, J. Chem. Phys., 64, 1564 (1976)
4. K. TOUKUBO, K. NAKANISHI, N. WATANABE, J. Chem. Phys., 67, 4162 (1977)
5. B. J. ALDER, T. E. WAINWRIGHT, Phys. Rev. A, 1, 18 (1970)
6. A. BRUSON, M. GERL, Phys. Rev. B, 19, 6123 (1979)
7. P. PROTOPAPAS, N. A. D. PARLEE, Chem. Phys., 11, 201 (1975)
8. G. A. MANSOORI, N. F. CARNAHAN, K. E. STARLING, T. W. LELAND, J. Chem. Phys., 54, 1523 (1971).

4.2 Liquid alloys

STRUCTURE OF MOLTEN Au-Cs ALLOYS BY MEANS OF NEUTRON DIFFRACTION*

W. Martin, P. Lamparter, S. Steeb
Max-Planck-Institut für Metallforschung
Institut für Werkstoffwissenschaften,
D-7000 Stuttgart 1, Western Germany

W. Freyland
Inst. f. Physikalische Chemie, Universität Marburg,
D-3550 Marburg-Lahn, Western Germany

ABSTRACT. Structural studies on molten Au-Cs alloys by neutron diffraction [1] led to the conclusion, that the alloy $Au_{0.5}Cs_{0.5}$ with the stoichiometric composition is a molten salt.
$Au_{(1-x)}Cs_x$ alloys with x > 0.5 show the effect of microsegregation.

The Au-Cs system in the liquid state exhibits a metal-nonmetal transition, where the electrical conductivity drops more than three orders of magnitude below typical values of liquid metals within a very narrow range of composition about $Au_{0.5}Cs_{0.5}$ [2]. The measured molar volume shows strong contraction with a maximum deviation from the ideal behaviour of 46 % at $Au_{0.5}Cs_{0.5}$ [1]. These effects indicate a considerable change of the liquid structure caused by charge transfer from Cs to Au when the elements are alloyed.

1. STOICHIOMETRIC COMPOSITION AuCs

The structure factor of $Au_{0.5}Cs_{0.5}$ (Fig. 1) exhibits a premaximum at $q_p = 1.2$ Å^{-1} which arises from short range order of the Au^- - and the Cs^+-ions. According to the Bhatia-Thornton representation of S(q)

$$S(q) = \langle b \rangle^2 . S_{NN}(q) + (\Delta b)^2 . S_{CC}(q) + 2\Delta b \langle b \rangle . S_{NC}(q) \Big/ \langle b^2 \rangle$$

*This work has been supported by the Deutsche Forschungsgemeinschaft

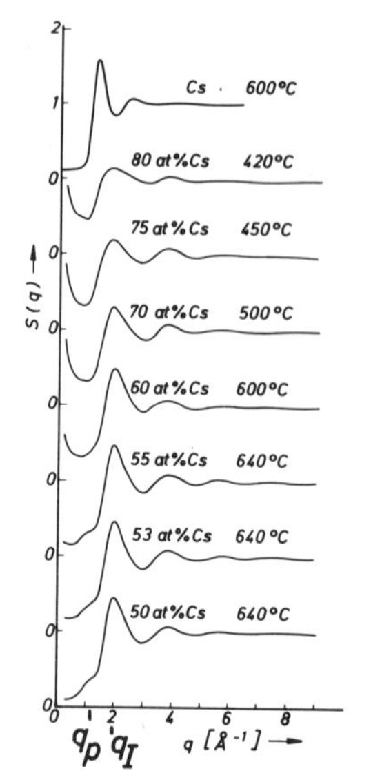

Fig. 1. Structure factors.

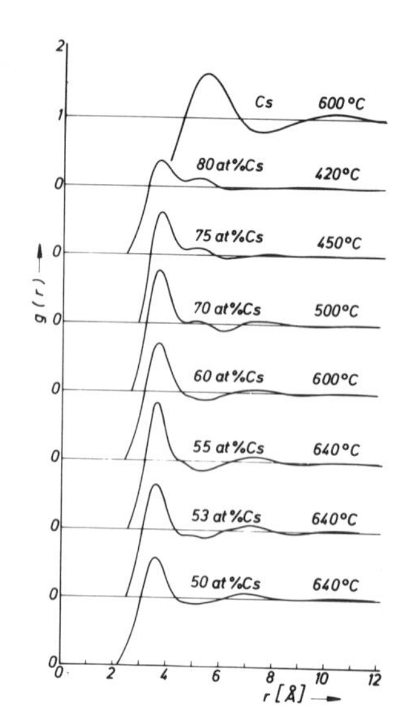

Fig. 2. Pair correlation functions.

this premaximum is attributed to the partial S_{CC} structure factor of the concentration fluctuations, whereas the principal maximum at q_I is mainly given by the partial structure factor S_{NN} of the density fluctuations. Neglecting the influence of the cross term S_{NC} the amplitudes of $S_{CC}(q_v)$ and $S_{NN}(q_I)$ have been estimated from the total S(q) curve and compared with the corresponding values of molten salts and LiPb and $Mg_{0.7}Bi_{0.3}$, respectively (see Table 1). From this follows that liquid $Au_{0.5}Cs_{0.5}$ belongs to the group of molten salts. Because of the salt-like structure the nearest neighbour peak in the pair correlation function at 3.58 Å (Fig. 2) is given by the distance between the Au^-- and the Cs^+-ions.

2. AuCs + Cs

$Au_{(1-x)}Cs_x$ alloys with x > 0.5 are not completely homogeneous but tend to microscopic phase segregation into the two components AuCs and Cs. This was concluded from two facts:

i The structure factors exhibit a pronounced small angle scattering effect which gives evidence for an inhomogenious structure in those melts.

ii From the pair correlation functions one can see that on alloying additional Cs to AuCs the position of the peak at 3.58 Å, given by the Au^--Cs^+ distance in AuCs, does not change, whereas an additional peak is growing up which corresponds well to the Cs-Cs distance in pure liquid Cs.

	LiPb	$Mg_{0.7}Bi_{0.3}$	CsCl	RbCl	KCl	AuCs
Ref.	/3/	/4/	/5/	/6/	/5/	This work
$\frac{S_{CC}(q_p)}{c_A \cdot c_B}$	1.7	2.7	3.2	3.7	4.8	4.1
$S_{NN}(q_I)$	2.2	1.76	1.47	1.49	1.33	1.44
elec.neg. difference	o.8	o.6	2.3	2.2	2.2	1.7

Table 1. Comparison of AuCs with molten salts and LiPb and $Bi_{0.3}Mg_{0.7}$.

The contribution of $S_{CC}(q)$ to $S(q)$ at small q-values has been evaluated according to the Ornstein-Zernike relation:

$$S_{CC}(q) = S_{CC}(q=0)/(1 + \xi^2 . q^2)$$

This yielded the correlation length ξ of the concentration fluctuations which has its maximum value at the composition $(AuCs)_{0.5}Cs_{0.5}$ (see Fig. 3). The temperature dependence of ξ is shown in Fig. 4. It can be well described by the relation from mean field theory:

$$\xi = \xi_o \cdot \left[(T-T_C)/T_c\right]^{-0.5}$$

Extrapolation to $\xi \rightarrow \infty$ yields the critical temperature $T_C \sim 570K$. This temperature is below the liquidus.

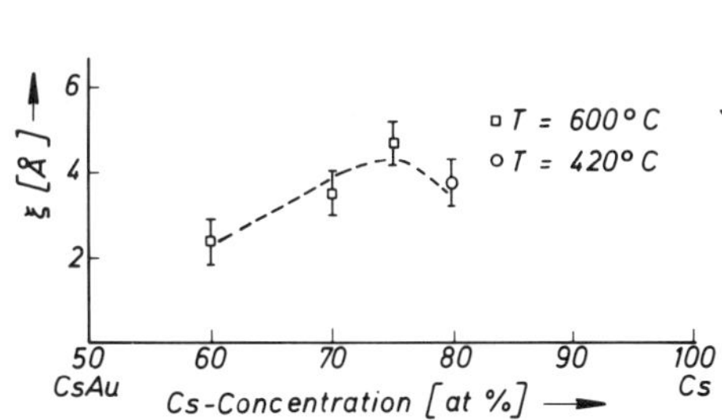

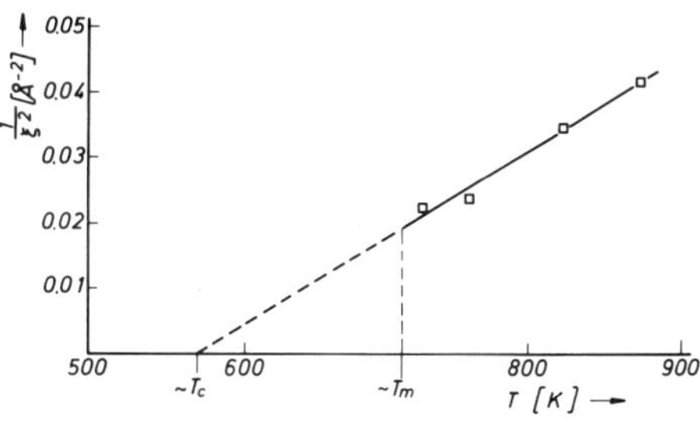

Fig.3. Concentration dependence of the correlation length.

Fig. 4. Temperature dependence of the correlation length.

REFERENCES

1. W. Martin, Neutronenstreuexperimente und Dichtemessungen an Cs-Au-Schmelzen, Dissertation, Universität Stuttgart (1979). To be published in Phys. Chem. Liq., 1980.
2. R.W. Schmutzler, H. Hoshino, R. Fischer, F. Hensel, Ber. Bunsenges.phys. Chem., 80, 107 (1976).
3. This data are based on measurements by H. Ruppersberg, H. Egger, J. Chem. Phys., 63, 4095 (1975) and have been taken from: J. Bletry, Z. Naturforsch., 33a, 327 (1978).
4. A. Boos, S. Steeb, Phys. Lett., 63A, 33 (1977).
5. J. Y. Derrien, Thèse, Université Lyon (1976).
6. P. F. J. Poncet, Thesis, University of Reading, UK (1976).

AN ACCESS TO SHORT RANGE ORDER IN LIQUID METALLIC ALLOYS

M. von Hartrott, D. Quitmann, J. Roßbach[+], E. Weihreter[++] and F. Willeke
Institut für Atom- und Festkörperphysik (WE 1A)
Freie Universität Berlin
D-1000 Berlin 33, Fed. Rep. Germany

In a liquid, the spatial arrangement and the motion are determined by the effective interaction between the particles. If a nucleus has electric quadrupole moment $Q \neq 0$, as a corollary of these interactions, a torque is exerted on the nuclear spin $\vec{I}$ due to random fluctuating electric field gradients. This leads to a spin relaxation rate $R_Q = 1/(T_1)_Q$ (for the present purpose, we may neglect the magnetic relaxation). A theory has been worked out for liquid metals by Sholl |1| and Warren |2| and Schirmacher |3|. Under several approximations

$$R_Q = Q^2 \cdot f(I) \int V^2(k)\ \hat{S}(k,\omega=0)\ d^3k \qquad (1)$$

$$V(k) = \int r^2\ V_2(r)\ g(r)\ j_2(kr)\ dr \qquad (2)$$

$V(k)$ may be viewed as the Fourier transform of the effective field gradient, when a neighbour atom at a distance r produces a field gradient $V_2(r)$. The hat on $\hat{S}(k,\omega)$ designates the modification of $S(k,\omega)$ which takes account of the fact that both, the atom which carries the spin I and the neighbour atoms, move. $\hat{S}(k,\omega)$ may be approximated by the convolution of $S(k,\omega)$ with $S_{self}(k,\omega)$; the resulting $\hat{S}(k,0)$ has a similar k-dependence as $S(k)$. The similarity between $S(k) = \int S(k,\omega)\, d\omega$ and $S(k,\omega=0)$ is discussed for Ar in |4|, for Ga and Rb in |5|, and also in |6|. f(I) is a known function of I.

The rate R_Q can be measured by classical NMR if favorable nuclear moments exist (Ga, Sb, Hg). Our group uses isomeric nuclei,

---

[+] Now at DESY, Hamburg
[++] Now at BESSY, Berlin

produced and aligned by pulsed irradiation at the Karlsruhe cyclotron, say by $(\alpha,2n)$ reactions, and observes the γ-decay radiation. The latter has an angular anisotropy which decreases at the rate R_Q (PAD-technique). Advantages of the method are the availability of $Q\neq 0$ nuclei in cases where the NMR nuclei (stable ground states) have $Q=0$, and temperature independent sensitivity. The added complication that the probe atom is chemically most often slightly different from the matrix has not shown up as a problem in the analysis of these experiments. Ref. |7| gives an extensive report on recent work.

If we are interested in short range order in liquid mixtures like alloys, we have to look for deviations of the Bhatia-Thornton structure factor S_{CC} from $c(1-c)$. The major experimental access is at present by neutron and/or X-ray scattering, see e.g. |8| . There is also an effect on electrical resistivity, see e.g. |9|. Furthermore, the thermodynamic properties, in particular the so called stability function $\partial^2 G/\partial c^2 = k_B T\, N/S_{CC}(k=0)$, depend on the degree of short range order; here G is Gibb's free enthalpy, N the number of particles. As an approximate model ("conformal solution model"), one introduces the energy change upon replacing an AA and a BB pair by two AB pairs, W; within the model

$$S_{CC}(0) = \frac{c(1-c)}{1-2c(1-c)W/k_B T} \tag{3}$$

and W can be evaluated from thermodynamic data of liquid alloys. Negative values of W, $S_{CC}(0)<c(1-c)$, correspond to an energetic preference for AB association and to an increased stability against long range concentration fluctuations.

The quadrupolar nuclear spin relaxation rate R_Q has been found to be especially large in such alloys which have a large stability. Within the Sholl-Warren theory this can be understood in the following way |10,7|. When the development of eqs.(1) and (2) is repeated for a probe sitting in an A-B alloy, and the structure factors $S_{NN}(k,\omega)$, $S_{CC}(k,\omega)$, and $S_{NC}(k,\omega)$ are introduced, one obtains three terms of which the NC-term may be neglected:

$$R_Q = Q^2 f(I) \cdot \int V_N^2(k)\, \hat{S}_{NN}(k,0)\, d^3k + \int V_C^2(k)\, \hat{S}_{CC}(k,0)\, d^3k = R_{QNN} + R_{QCC} \tag{4}$$

$$\begin{aligned} V_N(k) &= c\, V_A(k) + (1-c)\, V_B(k) \\ V_C(k) &= V_A(k) - V_B(k) \end{aligned} \tag{5}$$

where $V_A(k)$ is the effective field gradient produced by an A neighbour. V_N is thus an average field gradient just as S_{NN} is an average structure function, and R_{QNN} is R_Q for the "averaged liquid". When one goes to metallic alloys, large changes with respect to a pure liquid metal occur through the field gradient functions V_X because in the linear screening approximation they contain the ionic charge of neighbour X (+1 for Cu, Ag, Au, etc) as a factor. But the change of V_N with concentration is monotonous, the change of S_{NN} can also be expected to be gentle (that of $S_{NN}(0)$ is known to be small), so that the first term will change in a monotonous way, and in many cases not very much; this is born out by a study of Pb-alloys |12|.

For the second term, R_{QCC}, the V^2-factor will be comparable to that of the first term (but without c-dependence). The point is now that $S_{CC}(k)$ changes considerably between alloys, depending on the sign and magnitude at W, and that it develops maxima at finite k if there is a minimum of k=0, i.e. if W<0. Note that on the average the compositions of element A and B must be c and (1-c) which implies $\int [S_{CC}(k)-c(1-c)]d^3k = 0$. For experimental or model shapes see |8| or |13|. The first maximum of $S_{CC}(k)$ occurs (for W<0) at intermediate k, clearly before the first maximum of $S_{NN}(k)$; it appearently increases the overlap with $V_C^2(k)$, eq.(4), and causes the rise of R_Q observed. This interpretation |10| was derived from the correlation presented in fig. 1, between R_Q as observed near the melting temperature and the proportionality factors for the second term of eq.(4). The latter were assumed to be given for the field gradient by the ionic charges (linear screening), and for the strength of the $S_{CC}(k)$-maximum by its minimum at k=0, so that $R_{QCC} \propto (Z_A-Z_B)^2/S_{CC}(0)$. Note that the drastic changes of R_Q with X (and c) are very probably not explainable by dynamic effects: For $In_{1-c}Sb_c$ the diffusion constant changes only by 20% with c |11|.

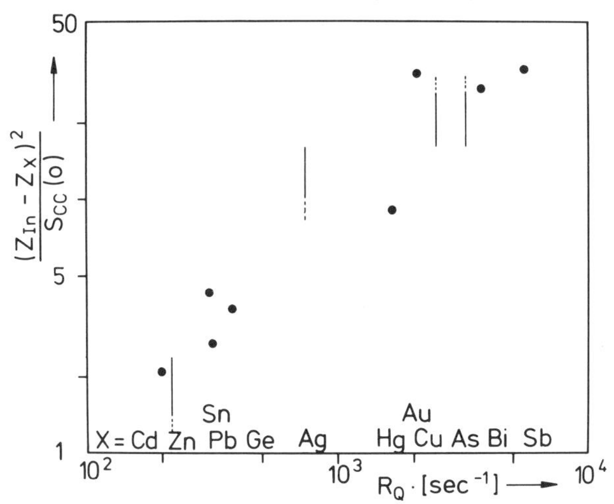

Fig.1. Comparison of quadrupolar relaxation rate R_Q of ^{117m}Sb in liquid $In_{0.5}X_{0.5}$ alloys with prediction derived from linear screening of ionic charges and conformal solution model; from |10|. For Zn, Ag, Cu, and As the information about W is uncertain.

Some of the $In_{0.5}X_{0.5}$ alloys with large values R_{QCC} also have a strong decrease of R_Q, i.e. of R_{QCC}, with rising temperature. In two cases where an unambiguous comparison is possible, the appearent activation energy for R_Q is not far from the value W of eq.(3) $In_{0.5}Sb_{0.5}$: $W/k_B \approx 1600^oK$; $In_{0.5}Bi_{0.5}$: $W/k_B \approx 900^oK$. This observation requires, however, a critical analysis of more cases especially in view of the difficulty to explain the temperature dependence of R_{QNN} (see e.g. |12|).

The present material seems to suggest that, at least in liquid metallic alloys and for favourable probe atoms, already a moderate deviation from ideal mixing makes the short range order observable by NMR methods.

This work was supported by Deutsche Forschungsgemeinschaft (Sfb 161) and by Kernforschungszentrum Karlsruhe. One of us (J.R.) is obliged to Hahn-Meitner Institut for financial support.

REFERENCES

1. C.A. Sholl: Proc. Phys. Soc. 91 (1967) 130; J. Phys. F4 (1974) 1556
2. W.W. Warren Jr.: Phys. Rev. A10 (1974) 657
3. W. Schirmacher: Ber. Bunsengesellschaft f. phys. Chemie 80 (1976) 736. Inst. Phys. Conf. Ser. 30 (1977) 610; J. Phys. F6 (1976) L 157
4. K. Sköld et al.: Phys. Rev. A6 (1972) 1107
5. M.I. Barker et al.: 2nd Int. Conf. Liquid Metals,Tokio, Taylor and Francis 1973, p. 99
6. A. Sjoelander: this volume
7. M. von Hartrott et al.: Phys. Rev. B19 (1979) 3449
8. H. Ruppersberg: this volume
9. D.N. Lee and B.D. Lichter in: Liquid Metals ed. S.Z. Beer, Marcel Dekker 1972
10. E. Weihreter et al.: Phys. Lett. 67A (1978) 394
11. D.K. Belashchenko et al.: Fiz. metal. metalloved. 32 (1971) 791
12. J. Rossbach et al.: J. Phys. F in print
13. J. Blétry: Z. Naturforsch. 33a (1978) 327

KNIGHT SHIFT AND ELECTRICAL RESISTIVITY OF SOME LIQUID LITHIUM ALLOYS

C. van der Marel

Solid State Physics Laboratory, Materials Science Center, Melkweg 1, 9718 EP Groningen, The Netherlands

ABSTRACT. Measurements of the Knight shift and the electrical resistivity on liquid alloys of Li with Cd, In and Pb have been performed as a function of concentration; in this series of alloys an increasing tendency to a MNM transition is observed.

INTRODUCTION

During the last ten years several liquid alloys of simple metals have been found which exhibit large deviations from metallic behaviour around some specific composition. Examples are the systems Mg-Bi, Cs-Au, Li-Bi and Li-Pb. In liquid Li-Bi alloys a transition occurs to non-metallic behaviour near the composition Li_3Bi; although less pronounced, the same kind of behaviour is observed in liquid Li-Pb alloys: in this system the electrical resistivity peaks up to about 500 μΩ cm at the composition Li_4Pb [1]. Also the density, enthalpy of mixing and activity exhibit large deviations from ideal behaviour in these systems [2]. These effects are ascribed to charge transfer from Li to the less electropositive Bi or Pb, resulting in a saltlike mixture near the composition Li_3Bi or Li_4Pb. If strong charge transfer occurs, one may expect that this will be reflected in the Knight shift. Therefore we have measured the Li^7 Knight shift in liquid Li-Pb alloys. Assuming that the difference in workfunction, $\Delta\phi^*$, of the pure components is a measure for the charge transfer, in the systems Li-Cd and Li-In the same amount of charge transfer is expected as in Li-Pb ($\Delta\phi^*$ = 1.20, 1.05, 1.20 and 1.30 in Li-Cd, -In, -Pb and -Bi respectively [3]). The results of our measurements of the Knight shift in these systems are presented in this paper. Furthermore we present resistivity measurements on liquid Li-Cd and Li-In alloys.

RESULTS

Li-Cd - In figure 1a both the experimental and the calculated electrical resistivities, ρ, at 550 ^{o}C, are plotted as a function of concentration. The theoretical curves are calculated by means of the well known Faber-Ziman formula, using Percus-Yevick hard sphere structurefactors, Shaw modelpotentials and different types of screening; for details we refer to [4]. $(d\rho/dT)_p$ was found to be slightly negative for Cd concentrations between 45 and 90 at%; this is in agreement with the diffraction model. In figure 1b the Li^7 Knight shift vs. concentration is plotted. The Knight shift decreases linearly when Cd is added to Li, whereas a change of slope occurs at the 50% Cd composition. From this behaviour it seems that the 50% composition is of special importance in this system. Unfortunately the only measured physical quantity of liquid Li-Cd alloys which could be found in the literature is the density of the 50% alloy: a volume contraction of 13% is reported. For a discussion we refer to [4].

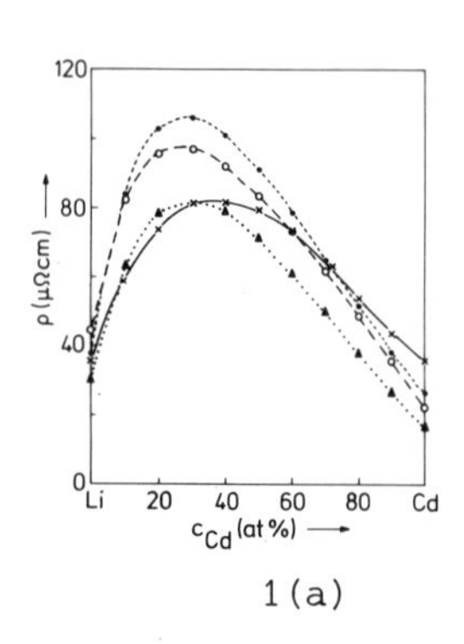

1(a)

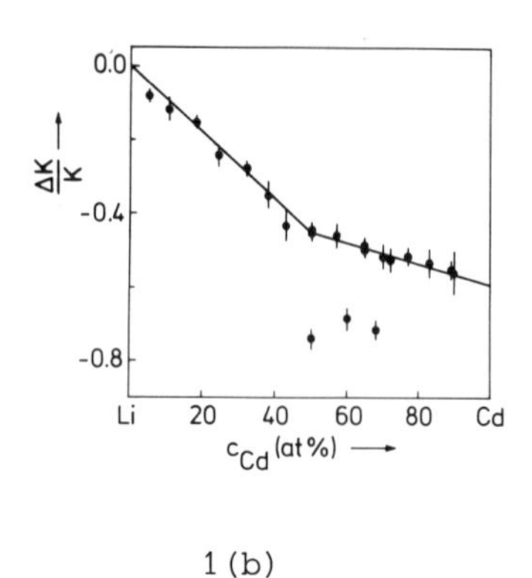

1(b)

FIG. 1. Liquid Li-Cd alloys at 550 ^{o}C. (a): electrical resistivities; ———: experiment; -----: theory. (b): Li^7 Knight shift; $\Delta K/K \stackrel{def}{=} (K_{Li}(c) - K_{Li}(0)/K_{Li}(0)$ where $K_{Li}(0)$ denotes the Knight shift of pure Li.

Li-In - In figure 2a the electrical resistivity, ρ, at 650 ^{o}C is plotted as a function of composition. The resistivity exhibits a distinct peak at c_{Li} = 75 at%, whereas the $(d\rho/dT)_p$ is strongly negative at this composition (fig. 2b). In this respect the Li-In system resembles the Li-Pb system [1], although the effects are much stronger in the latter one. In the Li-Pb system a large volume contraction is observed, with a maximum of 18% at the composition Li_4Pb [5]. It was therefore interesting to measure also the density of liquid Li-In alloys. The measurements were performed using a stainless steel pycnometer. The highest attainable temperature was 600 ^{o}C; therefore no measurements could be done between 40 and 65% In. The measured mean atomic volume, Ω, at 600 ^{o}C is plotted in

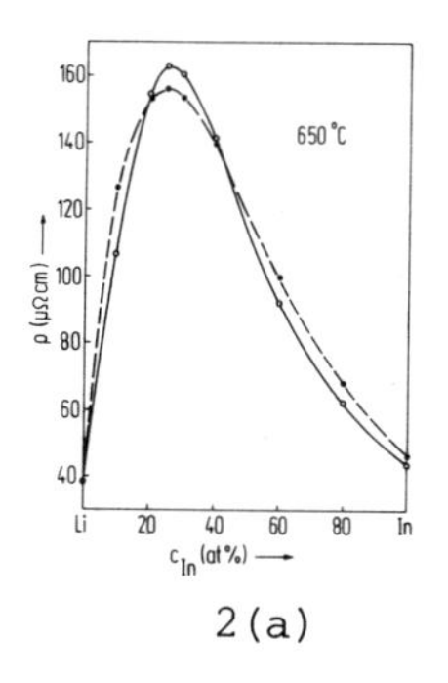

2(a)

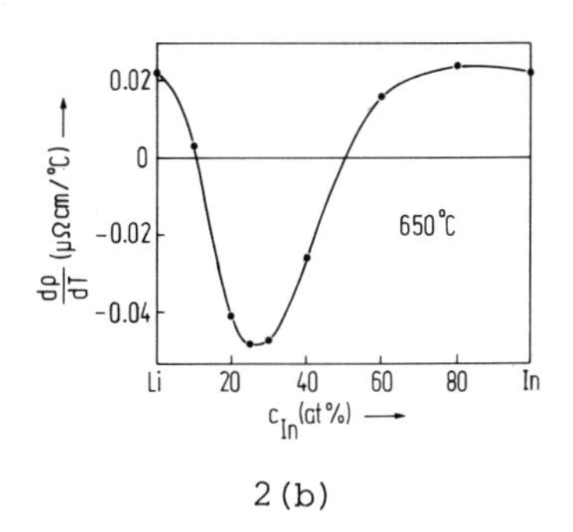

2(b)

FIG. 2. Liquid Li-In alloys at 650 °C. (a): electrical resistivities; ——: experiment; ----- theory. (b) temperature derivative of ρ.

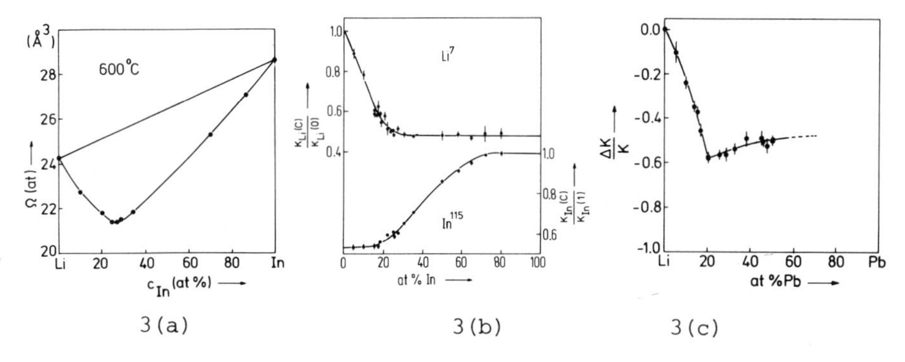

3(a) 3(b) 3(c)

FIG. 3. (a): mean atomic volume Ω in liquid Li-In at 600 °C. (b): Li^7 and In^{115} Knight shift vs. concentration in liquid Li-In. (c): Li^7 Knight shift in liquid Li-Pb; ΔK/K defined as in the caption of figure 1.

figure 3a. The volume contraction has a pronounced maximum of 15% at the 25% In composition. Both the strong volume contraction and the observation that $(d\Omega/dt)_{alloy} < (d\Omega/dt)_{ideal}$ for all compositions support the assumption of an ionic mixture. From a linear extrapolation of the right hand side of the Ω vs. concentration curve we obtain an estimate for the size of a Li^+ ion in these alloys. A value of about 18 Å^3 is found. This is the same value as is found in liquid Li-Pb for $c_{Pb} > 20$ at% [5], and in liquid Li-Cd alloys for $c_{Cd} > 50$ at% (assuming a linear dependence of Ω on concentration for $c_{Cd} > 50\%$). The results of our Knight shift measurements on both the Li^7 and the In^{115} nuclei are plotted in figure 3b and discussed in [6].

Li-Pb - The results of our Knight shift measurements are plotted in fig. 3c. The Li^7 Knight shift decreases rapidly when lead, upto 20 at% Pb, is added to pure lithium; when more Pb is added K increases somewhat. For Pb concentrations above 50% the Li resonance became very weak or even unobservable. This may be related to our qualitative observation that these alloys are - in both the solid and the liquid state - much more magnetic than pure components. In the alloys containing about 20% Pb the Knight shift was found to increase about 1 ppm per 10 $^{\circ}C$ with increasing temperature. The behaviour of the Knight shift fits qualitatively well in the model generally accepted: charge transfer occurs from Li to Pb resulting in a partially saltlike mixture near the composition Li_4Pb, whereas the electrical transport properties are determined by a strongly scattered electron gas. Finally we mention that in the alloys with Pb concentrations between 15 and 35% a second Li^7 resonance line is observed, with a Knight shift of 80 ± 10 ppm ($\Delta K/K = -0.70 \pm 0.04$), independent of temperature or composition. No explanation for this observation has been found yet.

ACKNOWLEDGEMENTS

This work is part of the research program of the "Stichting voor Fundamenteel Onderzoek der Materie" (Foundation for Fundamental Research on Matter - FOM) and was made possible by financial support from the "Nederlandse Organisatie voor Zuiver Wetenschappelijk Onderzoek" (Netherlands Organization for the Advancement of Pure Research - ZWO).

REFERENCES

1. J.E. Enderby, J. Physique C4, (1974), 309.
2. M.L. Saboungi, J. Marr and M. Blander, J. Chem. Phys. 68, (1978), 1375.
3. R. Boom, F.R. de Boer and A.R. Miedema, J. Less-Common Metals, 46, (1976), 271.
4. C. van der Marel and W. van der Lugt, to be published in J. Phys. F: Metal Physics.
5. H. Ruppersberg and W. Speicher, Z. Naturforsch. 31a, (1976), 47.
6. C. van der Marel, E.P. Brandenburg and W. van der Lugt, J. Phys. F: Metal Physics 11, (1978), L273.

ELECTRICAL RESISTIVITY OF LIQUID Ge - Sb ALLOYS

J.G. GASSER J.D. MULLER

Laboratoire de physique des milieux condensés
Faculté des sciences Ile du Saulcy 57000 METZ FRANCE

ABSTRACT. The electrical resistivity of Ge-Sb liquid alloys has been measured over the whole phase diagram from liquidus to 1150°C. The composition has been varied by ten per cent steps. Special attention has been given to the region near the eutectic between twelve and twenty four atomic per cent, by the mean of two per cent steps.

I INTRODUCTION

The electronic properties of liquid Ge - Sb alloys have (to our knowledge) never been investigated very carefully. Some results exist on the Ge - Sb - Se ternary system in amorphous and liquid state (Haisty and Krebs 1969). These measurements were performed by an electrodeless method (falling sample method) but the estimated accuracy was about ten to twenty per cent.

II MEASUREMENTS

Resistivity measurements were performed by the four probes method, using a quartz cell fitted with tungsten electrodes. A pressure of argon or vacuum can be applied over the sample. The composition of the alloys could be modified by adding one component in a storage tank. The capillary is cleared out by making vacuum and filled by applying the pressure of argon. Bubbles can be detected by the resistivity change when a variation of pressure is applied over the sample. They are eliminated by the same procedure as above for composition modifications.
The accuracy of the DC resistivity measurement is estimated to

be .2% and its resolution is of about .02%.
A three zone furnace is required to obtain a small temperature gradient along the cell. The temperatures along the capillary tube are measured with four Chromel - Alumel thermocouples. The accuracy of the temperatures is estimated to be .5%. The homogeneity of the temperature is better than 3°C.

The main error arises from the knowledge of the composition which is estimated to .5 atomic per cent.

III EXPERIMENTAL RESULTS

The temperature dependance of the resistivities of $Ge_x\ Sb_{(1-x)}$ alloys are shown on fig. 1 ; x is varied from 0 to 1 by .1 steps. The resistivity of all the alloys increases with temperature and lies on equally spaced parallel curves.
The temperature coefficient of the resistivity increases with temperature. Typical values lie between 10 and30 $n\Omega.cm.K^{-1}$.
Near the eutectic composition, we observe a lowering of about 5 $n\Omega.cm.K^{-1}$ in the temperature coefficient versus concentration curve at 900, 1000, and 1100°C.
We have specially investigated the region 100°C over the eutectic temperature by varying the composition from two to two atomic per cent (Eutectic : 17 at % of germanium).
Very accurate measurements (fig. 2) reveal the existence of a small negative temperature coefficient (typically - 10 $n\Omega.cm.K^{-1}$ at 590°C) near the eutectic temperature and composition.
This negative coefficient cannot be confounded with a negative coefficient occuring because of segregation or solidification by attaining the liquidus curve.

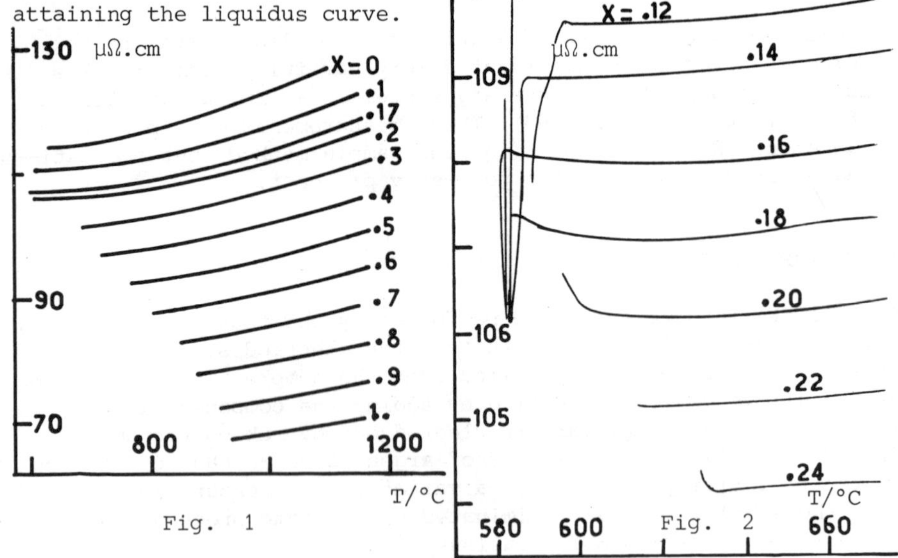

Fig. 1

Fig. 2

Resistivity of liquid $Ge_x - Sb_{(1-x)}$

Indeed we observed two types of curves above and below the eutectic composition. We have drawn them schematically :

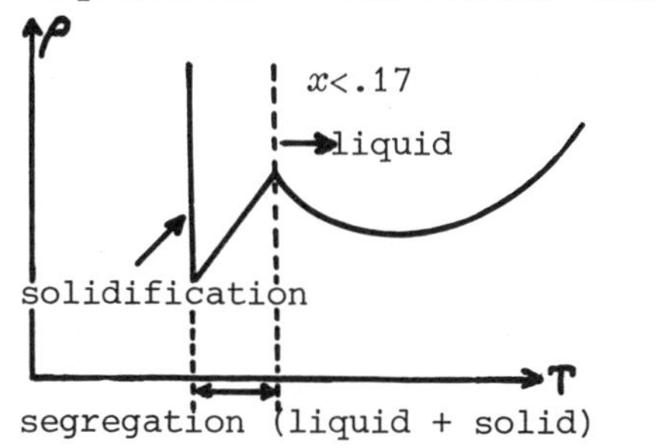

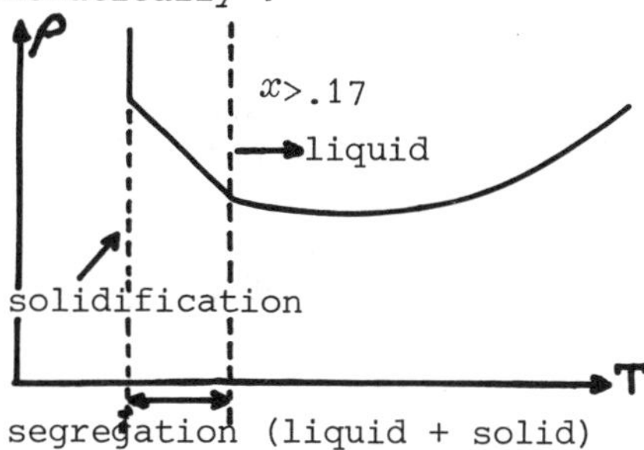

The resistivity versus concentration curves are reported on fig. 3. We must observe that they are quasi linear at the different temperatures.

IV DISCUSSION

Numerical calculation of the electrical resistivity of these alloys have been performed in the framework of the Faber Ziman (1965) theory. We calculated the partial structure factors following the prescriptions of Ashcroft and Langreth (1967) with a packing fraction of .38 for germanium and .40 for antimony (Waseda 1976).
The empty core model potential of Ashcroft (1966, 1968) is used. The parameter Rc has been fitted on the resistivity of the pure metal by using Ziman's formula (1961) (Rc = .6797 Å for germanium and Rc = .7790 Å for antimony).
The hard-sphere diameter is taken from pure metal and held constant over the wholecomposition range. The atomic volume of the alloy is linearly interpolated between the values of the pure components. The Fermi wave number k_F is calculated with the average valence of the alloy and the atomic volume defined previously.

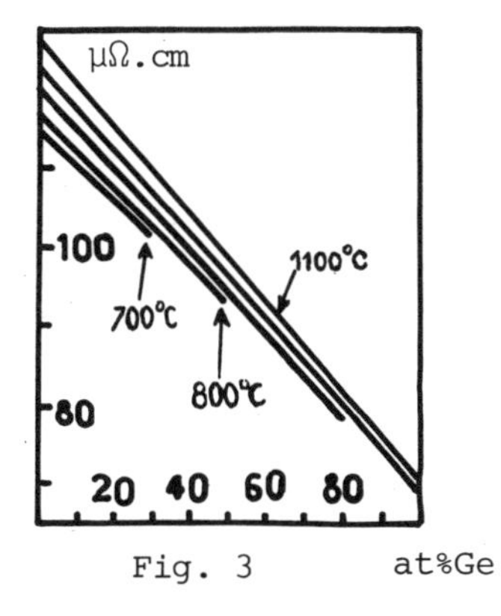

Fig. 3 at%Ge

Isotherms of the resistivity

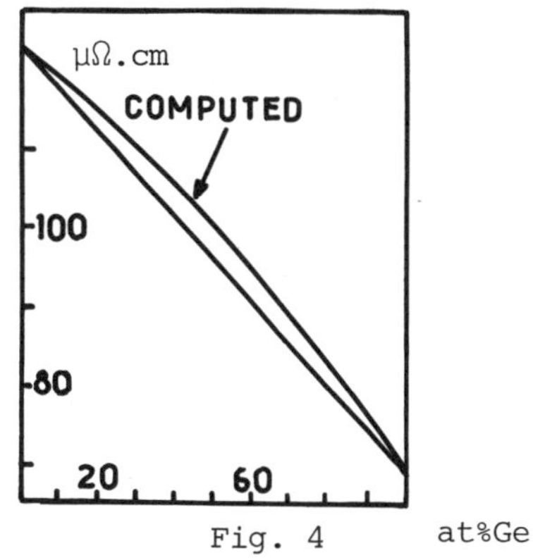

Fig. 4 at%Ge

Computed and measured isotherms

Experimental and calculated resistivities as a function of concentration are reported on fig. 4.
The agreement of these results is surprisingly good in regard to the crudeness of the approximations. The behaviour is the same as the Bi - Sn alloy (Verhoeven - Lieu 1965) and is well explained qualitatively in Faber's book p 456-457.
But we cannot appeal to the Faber - Ziman theory to explain the negative temperature coefficient of the alloy near the eutectic composition interm of displacement of $2k_F$ through the main peak of the interference functions as for alloys of monovalent with polyvalent metals.
A possible explanation could be a modification of short range order of atoms, but this must be confirmed by other experiments sensitive to a structural rearrangement.

ASHCROFT N.W. ; (1966) Phys. Lett. 23, 48
(1968) I Phys. C 2, 232
ASHCROFT N.W. ; LANGRETH J.C. ; (1967) Phys. Rev. 159, 500
FABER T.E. ; ZIMAN J.M. ; (1965) Phil. Mag. 11, 153
FABER T.E. ; Introduction to the theory of liquid metals. Cambridge At the University Press (1972)
HAISTY R.W. ; KREBS H. ; J. NON CRYST. SOLIDS (1969) 1, 5, 399
VERHOEVEN J.D. ; LIEU F.Y. ; (1965) Acta Mettallurgica 13, 927
WASEDA Y. ; (1976) Third int. conf. on liquid metals p. 231 Edited by R. Evans and D.A. Greenwood. Conference series Nr. 30. The Institute of Physics. Bristol and London.
ZIMAN J.M. ; (1961) Phil. Mag. 6, 1013.

ELECTRICAL RESISTIVITY OF LIQUID $(CU_{1-x}GA_x)_{.98}Fe_{.02}$ ALLOYS

J. Walter, G. Schubert, H. Coufal, S. Sotier, E. Lüscher

ABSTRACT. The electrical resistivity of liquid $Cu_{1-x}Ga_x$ alloys with 2 at. % of Fe-impurities has been measured up to temperatures of 1450 K. An increase is observed of the resistivity of these alloys compared with the pure $Cu_{1-x}Ga_x$ alloys. This increase can be understood in terms of the Friedel model of virtual bound states with a Fermi-energy varying from 7 eV up to 10 eV from pure Cu to pure Ga.

It has been known for a long time that Fe-impurities in Cu show a localized magnetic moment (see for example /1/). This can be seen in the low temperature magnetic susceptibility, NMR, specific heat, electrical resistivity - Kondo effect - and other physical properties. On the other hand, Fe-Ga alloys show only a very weak temperature dependence of the magnetic susceptibility that can be interpreted by assuming nonmagnetic alloys with a spinfluctuation temperature of about 1450 K /2/. Obviously $Cu_{1-x}Ga_x$ alloys with Fe-impurities would be a model-system to study the transition from a magnetic - $x = 0$ - to the nonmagnetic state - $x = 1$. In the liquid state the Ga-concentration x can be varied continously. The most direct information could be obtained by susceptibility measurements, but experiments on the electrical resistivity can also give some insight into the mechanism of the transition. The same type of experiment has been reported for the liquid $Cu_{1-x}Ga_x$ alloys with Mn impurities /3/. The experimental details of the four electrode DC method at temperatures up to 1450 K has been described elsewhere /4,5/.

Figure 1 shows the electrical resistivity of the pure host alloy together with the data for $(Cu_{1-x}Ga_x)_{.98}F_{.02}$ at 1450 K. Measurements at low Ga concentrations were exactly reproducible.

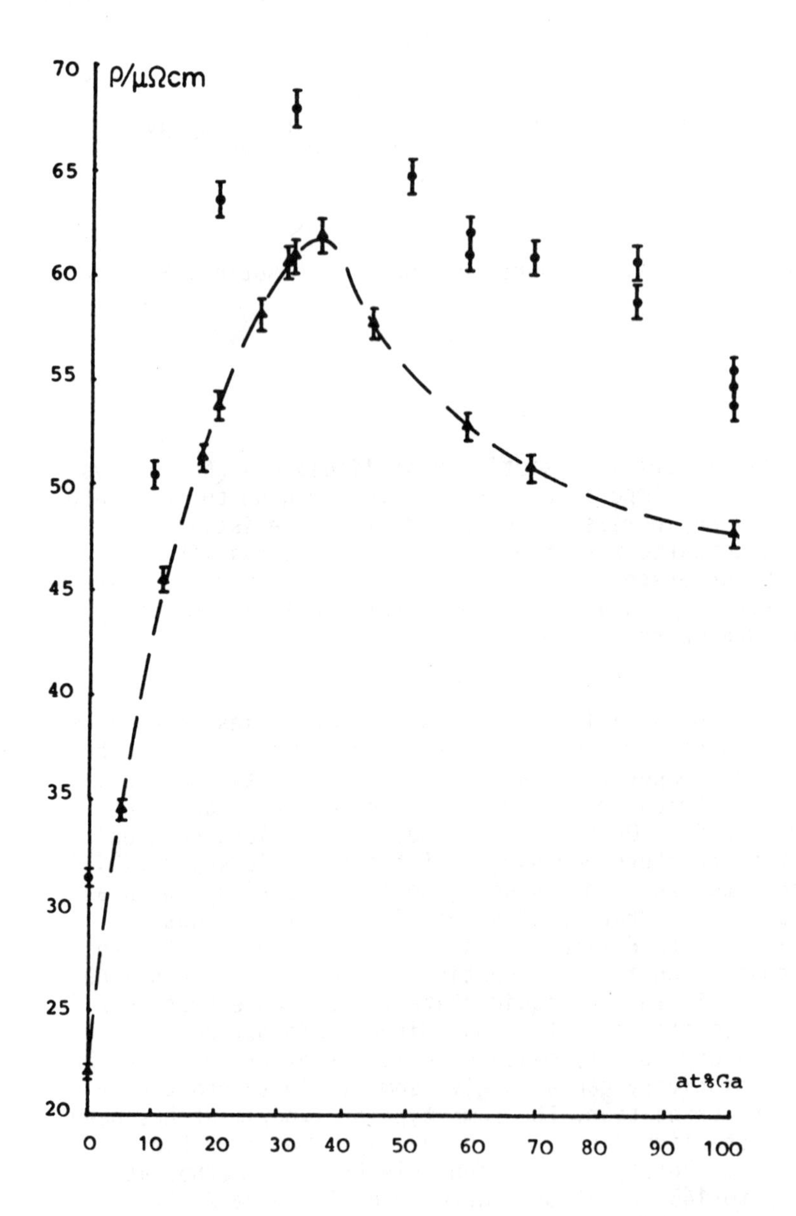

Fig. 1 Electrical resistivity ρ of $Cu_{1-x}Ga_x$ (▲) and $(Cu_{1-x}Ga_x)_{.98}Fe_{.02}$ (•) alloys versus Ga concentration at 1450 K.

At higher Ga concentrations, however, the observed resistivities show a certain spread that is probably due to the low solubility of Fe in Ga. The temperature coefficient $d\varrho/dT$ at the melting point for both systems is shown in Figure 2.

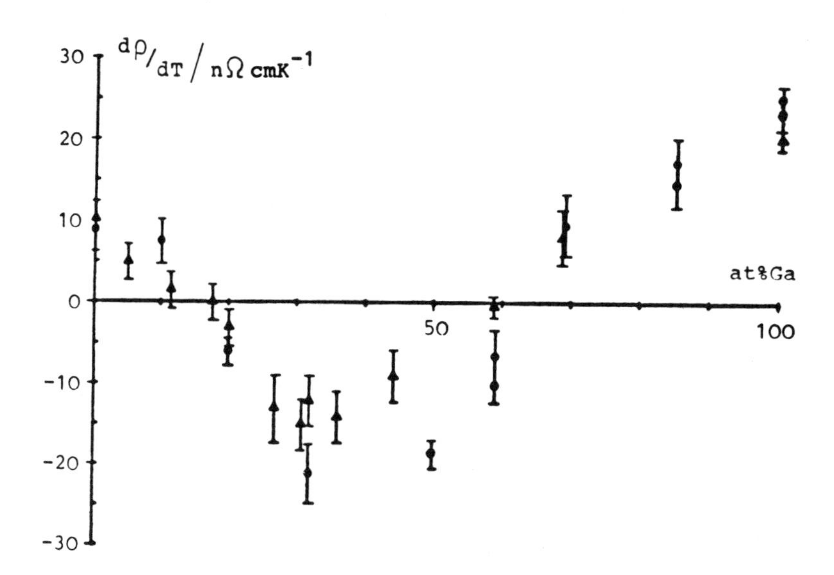

Fig. 2 Temperature coefficient $d\varrho/dT$ of the electrical resistivity at the melting point for $Cu_{1-x}Ga_x$ (▲) and $(Cu_{1-x}Ga_x)_{.98}Fe_{.02}$ alloys (•).

The results can be qualitatively described by using Friedel's scattering theory within this model. The increase of the resistivity $\Delta\varrho$ caused by a magnetic 3d-impurity can be written in the form

$$\Delta\varrho = \frac{4\pi c\,\hbar}{pe^2 k_F} \cdot \frac{5}{2}\,(\sin^2\eta_{2\uparrow} + \sin^2\eta_{2\downarrow}) \qquad (1)$$

when only d-wave phase shifts η_2 are considered. s- and p-wave scattering is usually neglected. In the nonmagnetic case $\eta_{2\uparrow} = \eta_{2\downarrow}$ leads to an even simpler form of equation (1).
Using Friedel's sum rule

$$Z_d = \frac{5}{\pi}\,(\eta_{2\uparrow} + \eta_{2\downarrow}) \qquad (2)$$

we find the following phase shifts η_2 and numbers Z_d of localized d-electrons for Fe-impurities in the pure host metals

Fe in	Cu	Ga
$\eta_{2\uparrow}$	π	$2.2\pm.2$
$\eta_{2\downarrow}$	$.75\pm.05$	$2.2\pm.2$
Z_d	$6.2\pm.1$	$7.0\pm.6$

The increase of $\Delta\varrho$ at Ga concentrations between 100 at. % and 85 at. % is due to the lowering of the Fermi level towards the maximum of the d-state density. This maximum occurs for $Cu_{.15}Ga_{.85}$ with $\eta_{2\uparrow} = \eta_{2\downarrow} = \pi/2$, where equation (1) leads to $\Delta\varrho = 11.9$ μΩcm compared to the measured $\Delta\varrho_{exp} = (10.9 \pm .9)$ μΩcm. For the evaluation of equation (1) a linear variation of the phase shifts with the concentration has been assumed.

For a more quantitative description of the very smooth magnetic-nonmagnetic transition additional information on the phase shifts, as for example from magnetic susceptibility measurements, is needed.

REFERENCES

1. G. Grüner, A. Zawadowski, 1974, Reports on Progress in Physics 37, 1497
2. P.L. Camwell, R. Dupree, C.J. Ford, 1973, J.Phys. F. Metal Phys. 3, 1015
3. J. Brunnhuber, Ch. Holzhey, H. Coufal, S. Sotier, 1979, Z.Physik B 33, 125
4. K. Thörmer, H. Coufal, G. Fritsch, H. Diletti, 1976, Phys. Lett. 56A, 489
5. Ch. Holzhey, H. Coufal, G. Schubert, J. Brunnhuber, S. Sotier 1979, Z.Physik B 33, 25

4.3 Amorphous metals

ELECTRON-PHONON INTERACTION IN AMORPHOUS METALS

K. Froböse, J. Jäckle

Fakultät für Physik, Universität Konstanz,
D - 7750 Konstanz, Germany

For simple amorphous metals the temperature dependence of the electrical resistivity is considerably weaker than for their crystalline counterparts. At higher temperatures ($T \gtrsim \Theta_D/2$) the temperature dependence becomes linear, and both positive and negative temperature coefficients are observed. This reduction of the effect of the electron-phonon interaction can be explained within the free electron model [1,2]. On the other hand, however, the experimental fact that the superconductiing transition temperatures are higher in the amorphous state than in the crystalline state indicates an enhancement of the electron-phonon interaction. We shall show that both effects can consistently be described by the same model.

ELECTRICAL RESISTIVITY

For amorphous metals the resistivity due to the scattering of the electrons by the disordered array of the ions is given by the Baym formula

$$\rho = \frac{m^2}{12\pi^3 Z^2 n_i e^2 \hbar^3} \int_0^{2k_F} dk\, k^3 |v(k)|^2 S'(k) \qquad (1)$$

where $v(k)$ is the pseudopotential, Z is the valency of the ions, and n_i is the ionic density. $S'(k)$ is a moment of the dynamic structure factor $S(k,\omega)$ containing the Bose function $n(\omega)$

$$S'(k) = \int \frac{d\omega}{2\pi} \beta h\omega n(\omega)\, S(k,\omega) \qquad (2)$$

The temperature dependent part of the resistivity consists of two competing contributions: the first one with a positive temperature coefficient originates from inelastic scattering and the second one is due to the reduction of the strong elastic scattering by the Debye-Waller factor. Since both contributions are of comparable magnitude, but opposite in sign, only a weak temperature dependence results. It has been shown [1] that the temperature dependence of the electrical resistivity for the Cu_xSn_{1-x} system can be well reproduced by taking into account only the incoherent part of the dynamic structure factor for the inelastic scattering. For a harmonic lattice this approximation means that the correlation length of the normal modes [2] is very short. Of course, in the low frequency regime this approximation is not applicable because in this limit the lattice vibrations are coherent plane waves. In the one-phonon approximation we obtain

$$S_{inel}^{incoh}(k,\omega) = \pi\hbar/(3Mn_i)k^2 \; (1+n(\omega)) \; F(\omega)/\omega \tag{3}$$

where $F(\omega)$ is the density of states and M is the ionic masse. In the high temperature limit which applies for $T \gtrsim \Theta_D/2$ we obtain for the temperature dependent part of the resistivity a linear term

$$\rho_T = \frac{m^2 (2k_F)^6 \; k_B T}{36\pi^3 n_i^4 e^2 \hbar M} (J_5 - J_5^{S_o}) \int d\omega F(\omega)/\omega^2 \tag{4}$$

where the averages of the pseudopotentials J_5 (inelastic) and $J_5^{S_o}$ (elastic) are defined by

$$J_n = \int_0^1 dt \; t^n |v(t \cdot 2k_F)|^2, \quad J_n^{S_o} = \int_0^1 dt \; t^n |v(t \cdot 2k_F)|^2 S_o(k)$$

Here $S_o(k)$ is the structure factor of the equilibrium positions of the ions. The temperature dependence of the electrical resistivity can be calculated when the structure factor, the pseudopotential, and the moment of the vibrational density of states $\int d\omega F(\omega)/\omega^2$ are known.

ELIASHBERG FUNCTION

We derive a relation between the temperature coefficient of the resistivity given by formula (4) and the Eliashberg function $\alpha^2 F(\omega)$ which characterizes the electron-phonon interaction and can be determined independently from superconduction tunnelling experiments. The Eliashberg function

$$\alpha^2 F(\omega) = \frac{1}{(2\pi)^3 \hbar^2 \; n_i \; V_F} \int_0^{2k_F} dk \; k |v(k)|^2 \frac{S_{inel}(k,\omega)}{1+n(\omega)} \tag{5}$$

describes the rate of inelastic electron scattering per unit of energy transfer $\hbar\omega$ averaged over all initial states on the Fermi surface. With the same approximation for $S_{inel}(k,\omega)$ as above we obtain

$$\alpha^2 F(\omega) = \frac{2 \; Z \; m}{n_i \hbar^2 M} \; J_3 \; F(\omega)/\omega \tag{6}$$

Combined with a Debye density of states we get a linear ω-dependence for $\alpha^2 F(\omega)$ which is indeed observed in the low frequency regime for amorphous metals. This is in contrast to crystalline materials where the variation is quadratic. The main reason for this relative enhancement is the lack of momentum conservation in disordered materials which increases drastically the number of possible scattering events [3]. The superconducting transition is mainly determined by the electron-phonon coupling constant

$$\lambda = \int d\omega\, \alpha^2 F(\omega)/\omega \tag{7}$$

In this integral the factor $1/\omega$ increases the relative weight of the low frequency region where the enhancement of $\alpha^2F(\omega)$ is observed. This explains the larger value of λ for the amorphous state. Combining (4) and (7) we find the following relation between the electron-phonon coupling constant λ and the temperature coefficient of the resistivity

$$\lambda = \frac{e^2\hbar\, Z\, n_i}{4\pi k_B\, m}\;\frac{J_3}{(J_5 - J_5 S_o)}\;\left.\frac{d\rho}{dT}\right|_{high} \tag{8}$$

TEST OF THE THEORY

The validity of the relation (8) can be tested for Ga for which most of the required experimental data are available. Unfortunately the temperature coefficient is only measured up to 40 K ($0.59\cdot10^{-12}$ $\mu\Omega$cm/k) [4] so that we have to assume that the linear temperature dependence around 40 K persists to higher temperatures of order $\Theta_D/2$. The static structure factor which is nearly equal to $S_o(k)$ was determined by X-ray measurements [5]. Four different pseudopotentials [6(twice),7,8] were taken from the literature. The corresponding calculated values of λ = 1.45, 1.76, 1.71, and 2.92 respectively, are scattered around the experimental values of λ_{exp} = 2.25 and 1.94 [9,10] determined by tunnelling spectroscopy. From this consistency we conclude that the model of free electrons and lattice vibrations with short correlation length is appropriate to the description of transport and superconducting properties of simple amorphous metals.

REFERENCES

1. K. Froböse, J. Jäckle, J.Phys.F Metal Phys.7, 2331 (1977)
2. J. Jäckle, K. Froböse, J.Phys.F Metal Phys.9, 967 (1979)
3. G. Bergmann, Phys.Rev.B3, 3797 (1971)
4. D. Korn, W. Mürer, G. Zibold, Z.Phys.260, 351 (1973)
5. A. Bererhi, L. Bosio, R. Cortès, J.Non-Cryst.Sol.30, 253 (1979)
6. W.A. Reed, Phys.Rev.188, 1184 (1969)
7. M. Appapilai, A.R. Williams, J.Phys.F Metal Phys.3, 759 (1973)
8. E. Carruthers, P.J. Lin-Chung, Phys.Rev.B17, 2705 (1978)
9. J.D. Leslie, J.T. Chen, T.T. Chen, Can.J.Phys.48, 2783 (1970)
10. J.E. Jackson, C.V. Briscoe, H. Wühl, Physica 55, 447 (1971)

TRIPLET CORRELATION IN POLK-TYPE MODEL OF $Fe_{80}B_{20}$ METALLIC GLASS

J. U. Madsen and R. M. J. Cotterill

Department of Structural Properties of Materials
The Technical University of Denmark
Building 307, DK-2800 Lyngby, Denmark

Introduction

The Field-Ion Microscopy (FIM) technique has been applied to the study of amorphous $Fe_{80}B_{20}$, see (1) for further details. It has been possible to derive a bond angle distribution function from the FIM pictures. A model for a-$Fe_{80}B_{20}$ has been constructed in the spirit of the Polk idea (2). In this communication a brief account is given of the principles of the model and of its correlation with experimental results.

The Model

The 84 largest cavities in a Lennard-Jones model glass (3), consisting of 336 particles within periodic boundaries, were identified. The 336 particles were taken to represent Fe atoms, 84 B atoms were inserted in the cavities, and the model was slightly expanded (2.1% linearly). The interactions were defined as the Lennard-Jones potential truncated at $r/r_o = 1.5$ for the Fe-Fe potential, V_{FeFe}, the Fe-B and B-B potentials being $V_{FeB}(r) = V_{FeFe}(1.25\ r)$, $V_{BB}(r) = V_{FeB}(1.25\ r)$. Thus no chemical effects were included in the potentials, only in the starting point. This, however, was sufficient to all but exclude the occurrence of B-B near neighbors, so the details of V_{BB} were in fact of no consequence. The ratio of the Fe-B to the Fe-Fe near neighbor distances (0.8) is consistent with the results of EXAFS studies (4). The model was relaxed using a restricted Monte-Carlo scheme for the minimization of the potential energy.

Results and Conclusion

The coordination numbers were found to be: Fe around Fe: 12.36; Fe around B: 6.67; hence B around Fe: 1.67; B around B: 0.21. In all cases the coordination numbers were calculated including neighbors within the first minimum of the relevant pair correlation function. For the Fe-Fe coordination number the value is consistent with the experimental results (5). The pair correlation functions were calculated; the Fe-Fe and Fe-B pair correlation functions are shown in Figures 1 and 2. For reference, see the experimental reduced radial distribution function of Waseda (6). The contribution of the Fe-B correlation is about 10% of that of the Fe-Fe correlation in an X-ray scattering experiment. The agreement is deemed satisfactory.

To elucidate the triplet correlation, which is experimentally accessible only by the FIM technique, the bond angle distribution functions were calculated, these being defined as the distribution functions for the angles between directions to near neighbors. The near neighbors were here defined as atoms lying within 1.2 r_{FeFe} from an atom. The distribution functions for FeFe and FeB bond angles at Fe atoms are displayed in Figures 3 and 4. The positions of the peaks in the FeFeFe function correlate with the predictions of the Kirkwood superposition approximation (7) applied to the FeFe pair correlation function; however, it remains to be investigated whether there is quantitative agreement. The FIM bond angle distribution function after smoothing is shown in Figure 5; the shoulder on the low-angle side of the 60° peak has been resolved and identified with a contribution from FeFeB triplets. From the ratio of the contents of the 60° peak to the low-angle peak one arrives at an estimate of 0.7 for the ratio of the imaging probability for B relative to Fe. Using this ratio the average bond angle distribution function appropriate to the FIM experiment has been calculated and is shown in Figure 6. It should be noted that in Figure 6 no allowance is made for instrumental resolution and projection errors and for possible surface reconstruction in the specimen.

The contributions from the Danish Research Councils, P. Jacobæus, F. Kragh, and W. Damgaard Kristensen are appreciated.

(1) P. Jacobæus, J. U. Madsen, F. Kragh, and R. M. J. Cotterill, Phil. Mag. (in press)
(2) D. E. Polk, Scripta Met. 4 (1970) 117; Acta Met. 20 (1972) 485
(3) W. Damgaard Kristensen, J. Non-Cryst. Sol. 21 (1976) 303
(4) R. Haensel, private communication
(5) T. Fukunaga, M. Misawa, K. Fukamichi, T. Masumoto, and K. Suzuki, in B. Cantor ed.: Rapidly Quenched Metals III, Vol. 2, p. 325 (Metals Society, London, 1978)
(6) Y. Waseda, ibid., p. 352
(7) J. G. Kirkwood, J. Chem. Phys. 3 (1935) 300

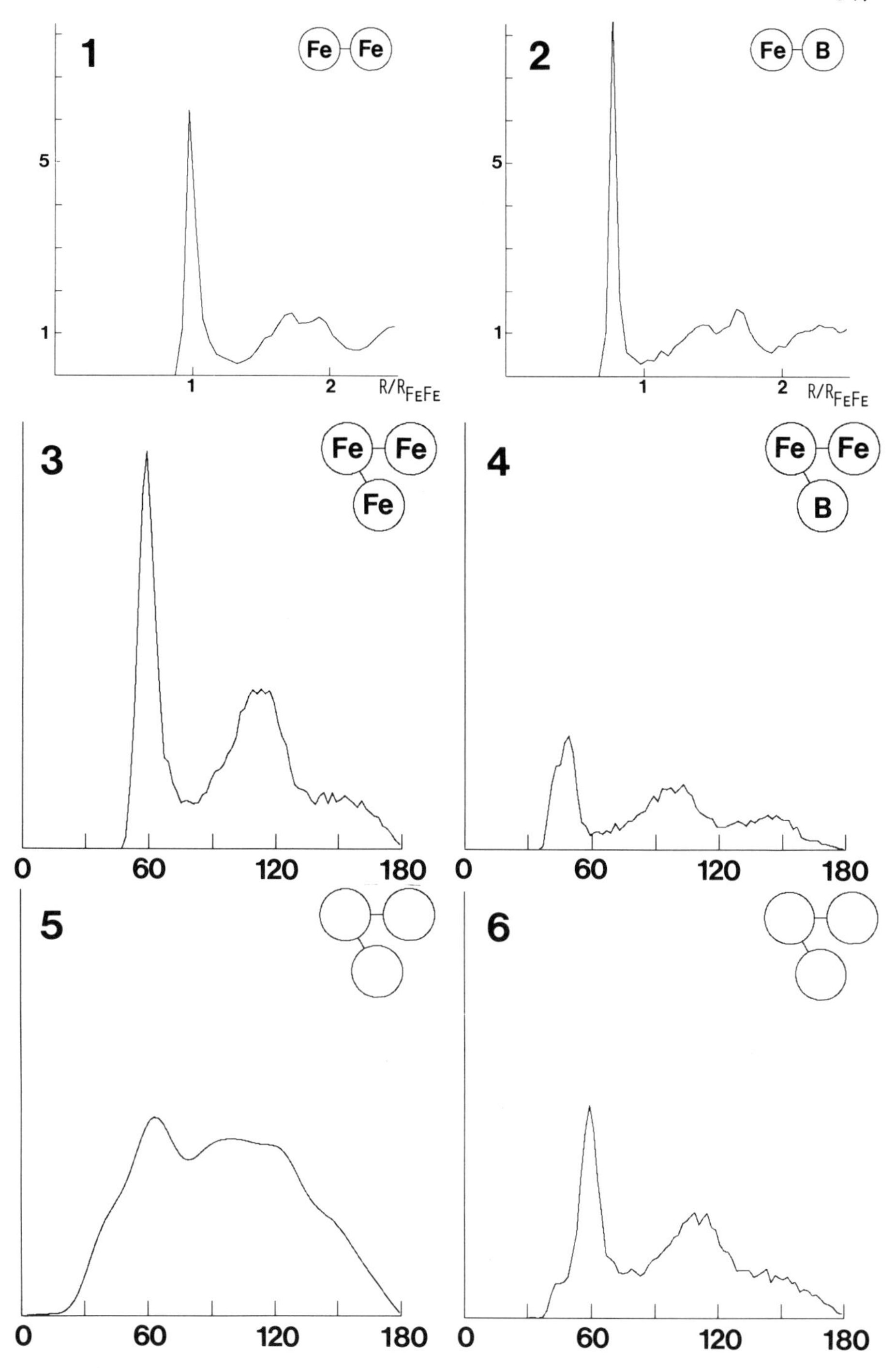

Figures 1 to 6. For explanation, see the text.

DYNAMICAL STRUCTURE FACTOR AND FREQUENCY DISTRIBUTION OF AMORPHOUS $Cu_{.46}Zr_{.54}$

J.-B. Suck[x], H. Rudin[xx], H.-J. Güntherodt[xx], H. Beck[+], J. Daubert[++], W. Gläser[++]

[x] Institut Laue-Langevin, 156X, F-38042 Grenoble Cedex
[xx] Institut für Physik, Universität CH-4056 Basel
[+] Institut de Physique, Université CH-2000 Neuchâtel
[++] Physik Department, T.U. München, D-8046 Garching.

The amorphous metal Cu_xZr_{1-x} has attracted much interest in recent years [1,2,3]. Its good glass forming ability makes it a preferred substance for experiments where a large amount of sample is required. We are reporting here first results from a measurement of the dynamical structure factor (scattering law) and the frequency distribution of amorphous $Cu_{.46}Zr_{.54}$ and the same sample after crystallization at 920 K.

The experiment was performed at the thermal neutron time-of-flight spectrometer IN4 of the I.L.L. in Grenoble with an incident energy E_o of 60.8 meV (v_o = 3410 m/sec). 60 spectra were recorded for scattering angles between 9 and 97 degrees. The energy resolution varied between 3 and 1.5 meV between 0 and 40 meV in the energy loss spectra. The measurements were done at room temperature. The data presented here were corrected for background (container) scattering, detector efficiency and self-absorption in sample and container. No corrections were applied for multiple scattering of the neutrons in the sample and for the broadening of the experimental results due to the finite energy and momentum resolution of the spectrometer. The scattering law was multiplied by $\exp(-\hbar\omega/2k_BT)$.

As an example the dynamical structure factor of amorphous $Cu_{.46}Zr_{.54}$ and of the polycrystalline sample at Q = 2.05 $Å^{-1}$ is shown in Fig. 1 for energy transfers between -45 meV (neutron energy gain) and 45 meV (energy loss spectra). For energy transfers between 10 and 30 meV the two spectra are very similar. The main difference is found for energies below 10 meV where a strong increase of scattered intensity is found in the dynamical structure

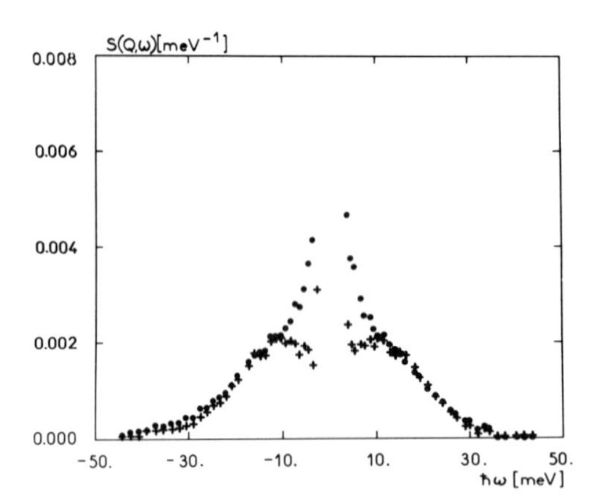

Fig. 1. The symmetrized dynamical structure factor of amorphous $Cu_{.46}Zr_{.54}$ (full circles) and of the same sample after crystallization (crosses) at a momentum transfer of Q = 2.05 $Å^{-1}$. The peak of the elastically scattered neutrons is not shown to enlarge the spectra of the inelastically scattered neutrons.

factor of the amorphous sample. These low frequency modes do not show predominant collective behaviour in the momentum range of this experiment. Generally, for both samples coherent scattering can be detected up to Q = 5 $Å^{-1}$. For higher momentum transfers the scattering law reflects mainly the single particle motion of the atoms.

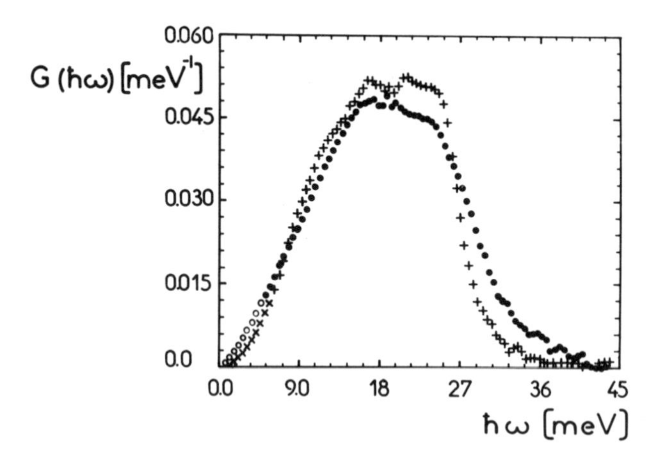

Fig. 2. Normalized frequency distribution of amorphous $Cu_{.46}Zr_{.54}$ (full circles) and of the polycrystalline sample (crosses).

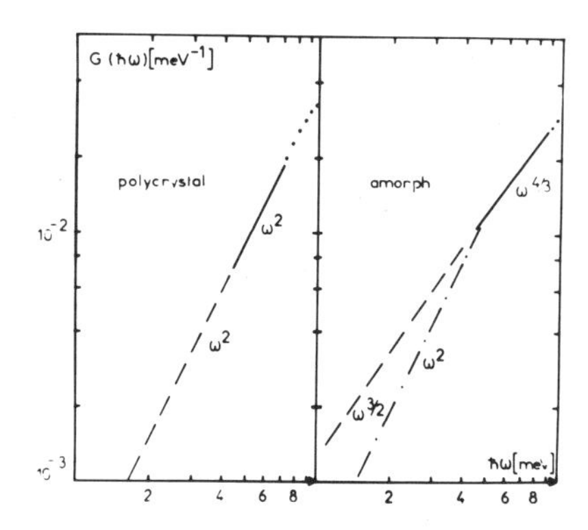

Fig. 3. Log-Log-plot of the frequency distributions of the polycrystalline and amorphous sample for energies between 1 and 10 meV. The experimentally obtained values are given as continuous line and full circles. The inserted spectra at low frequencies are given as dashed and dashed-dotted lines.

Fig. 2 shows the normalized frequency distribution of amorphous $Cu_{.46}Zr_{.54}$ and of the polycrystalline sample determined from the weighted sum of the neutron energy loss spectra [4]. The multiphonon contributions to the scattered intensities were corrected in the harmonic and incoherent approximations [5]. In the case of the polycrystalline sample one finds a broad, nearly structureless distribution. The optic modes are not separated from the acoustic modes. This lack of structure is most likely due to the superposition of the frequency spectra of several polycrystalline phases of Cu_xZr_{1-x} which developed during the crystallization process. The presence of several polycrystalline phases could be confirmed using X-ray diffraction after this experiment. Compared with this frequency distribution the spectrum of the amorphous sample is even broader and has less structure. It extends to higher energies than the frequency spectrum of the polycrystalline sample. These high energy modes are most likely due to the dynamic of atoms trapped at a shorter distance to their neighbours during the rapid quenching process than they would have in their crystalline phases. No values could be obtained for energies between 0 and 4.7 meV because of the resolution broadening of the peak of the elastically scattered neutrons. For the iterative and self-consistent multiphonons corrections a Debye spectrum was fitted to the measured distribution of the polycrystalline sample in this energy region. This was not possible in the case of the amorphous sample and a $\omega^{3/2}$-spectrum was arbitrarily chosen in this case. It was verified that this choice did not influence the final result of the iterative multiphonon correction. The most striking deviation between the two frequency distributions is found at energies between 4.7 and 7 meV. This is shown in more details in Fig. 3 where a log-log plot of the frequency spectra between 1 and 10 meV is shown. In the spectrum of the polycrystalline sample the inserted Debye spectrum at low energies is smoothly continued by the experimentally obtained values up to 7 meV. However for the amorphous sample, one finds a slope proportional to $\omega^{4/3}$ for energies between 4.7 and 7 meV. The energy region below 4.7 meV cannot be bridged by a Debye spectrum without producing an unphysical kink in the slope of the frequency distribution at 4.7 meV. The choice of the $\omega^{3/2}$-spectrum is better but still leads to a slight discontinuity in the derivative of the frequency distribution.

Experiments aiming to determine the frequency distribution of an amorphous metal down to much lower energies are in progress.

REFERENCES

1. T. Mizoguchi, S.v. Molnar, G.S. Cargill III, T. Kudo, N. Shiotani, H. Sekizawa : Amorphous Magnetism II, p.513 Plenum Press 1977.
2. Y. Waseda, T. Masumoto, S. Tamaki in "Liquid Metals III" Inst. Phys. Conf. Ser. No. 30, Bristol and London (1977).
3. T. Kudo, T. Mizoguchi, N. Watanabe, N. Niimura, M. Misawa, K. Suzuki, J. Phys. Soc. Jap. Letters 45, 1773 (1978).
4. V.S. Oskotskii, Sov. Phys. Sol. State 9, 420 (1967).
5. J.-B. Suck, Report KFK 2231, p. 50, Kernforschungzentrum Karlsruhe (1975).

EFFECT OF MAGNETIC FIELD ON THE ELECTRICAL RESISTIVITY OF AMORPHOUS TRANSITION METAL - METALLOID ALLOYS†

H. Gudmundsson[1], K.V. Rao[2] and A.C. Anderson[2]

[1]Dept. of Solid State Physics, Royal Institute of Technology, S-100 44 Stockholm 70

[2]Dept. of Physics and Material Research Laboratory, University of Illinois, Urbana, Illinois 61801

ABSTRACT

The electrical resistivity, ρ, has been measured from 0.1 to 300 K in fields of 0-22 T, for the amorphous alloy systems $Fe_yNi_{80-y}P_{14}B_6$+ and $(Fe,Co_{1-x}Ni_x)_{75}P_{16}B_6Al_3$++. A magnetic contribution to ρ is indicated by a rise in ρ at low temperatures which correlates with the magnetic state of the system (as characterized by the ordering temperature, T_c), and by a negative magnetoresistance both above and below T_c. Anisotropy is observed in the ferromagnetic state at $T \ll T_c$.

INTRODUCTION

A minimum in the electrical resistivity, ρ, at a characteristic temperature, T_{min}, with a logarithmic rise in ρ below it, and a T^2-dependence above it, are well known features of amorphous transition metal-metalloid alloys. The origin of this has been a subject of intense discussions during the last years. To help clarify this problem we have made systematic studies on such alloy systems containing Fe, Co, Ni, Mn and Cr, studying the dependence of ρ on temperature, magnetic field and alloy composition. We present here some results for the systems $Fe_yNi_{80-y}P_{14}B_6$ and $(Fe,Co_{1-x}Ni_x)_{75}P_{16}B_6Al_3$ in magnetic fields of 0-22 T in the temperature range 0.1-300 K. The samples were ribbons, typically ~30μm

---

†Funded by NSF grant DMR 77.0859 and The Swedish NFR.
+Samples made available by T.Egami, Univ.of PA, Philadelphia, PA.
++Samples made available by H.S.Chen, Bell Labs, Murray Hill, N.J.

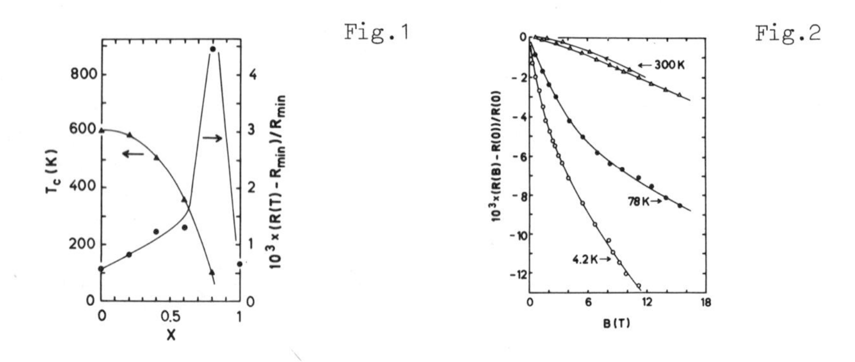

Fig.1 The concentration dependence of T_c and $(R(T)-R_{min})/R_{min}$ at $T=0.2xT_{min}$ for $(Fe_{1-x}Ni_x)_{75}P_{16}B_6Al_3$.

Fig.2 The MR as a function of the applied field for $(Fe_{.2}Ni_{.8})_{75}P_{16}B_6Al_3$ at three different temperatures.

thick and ~1 mm wide, produced by the meltspinning technique. The resistance was measured by standard four probe techniques to the relative accuracy of 1-10 ppm. The studies in high magnetic fields were carried out at the Francis Bitter National Magnetic Laboratory, MIT, Cambridge.

RESULTS AND DISCUSSIONS

It has been shown for these alloys that Fe and Co carry magnetic moment while Ni does not. Thus replacing Fe or Co by Ni can be considered as a dilution of the magnetic matrix and hence the ferromagnetic ordering temperature, T_c, decreases with increasing Ni concentration (1). A typical magnetic phase diagram is shown in figure 1. A spinglass region has been observed for low Fe concentrations (2). Figure 1 shows also a typical concentration dependence of the relative rise in ρ below T_{min}, $(R(T)-R_{min})/R_{min}$, taken at a fixed fraction of T_{min}. As we have previously pointed out (3), this shows a clear correlation between the low temperature anomaly in ρ and the magnetic state of the alloy as characterized by T_c, the rise being largest as $T_c \to 0$. Figure 2 shows a typical field dependence, in longitudinally applied field $(\bar{B}//\bar{J})$, of the magnetoresistance (MR), $(R(B)-R(0))/R(0)$, for an alloy in the magnetically dilute, Ni-rich, concentration range. The MR is negative and strongly field dependent in the whole temperature region, both below and above T_c. No saturation effects are seen at the highest fields. It may be pointed out that the T_{min} of this sample decreases by about 2 K as low a field as 0.56 T. The temperature dependence of the MR in a longitudinal field $(\bar{B}//\bar{J})$ of 0.56 T is shown for some of the alloys in figure 3. The MR is negative and somewhat temperature

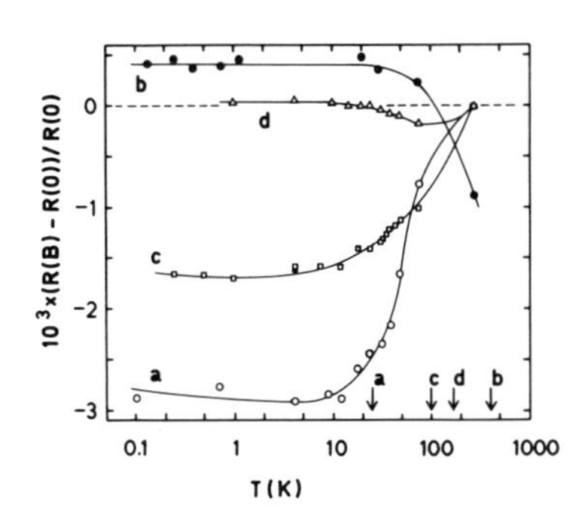

Fig.3

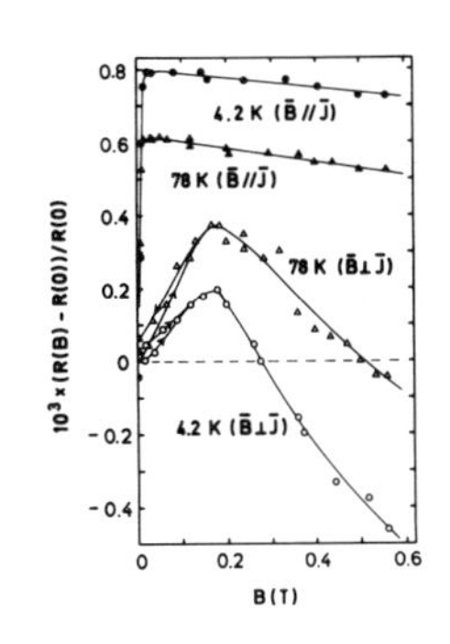

Fig.4

Fig. 3 The temperature dependence of the MR in longitudinal field of 0.56 T ($\bar{B}//\bar{J}$) for $Fe_yNi_{80-y}P_{14}B_6$, y=7 (a) and y=27 (b), $(Fe_{.2}Ni_{.8})_{75}P_{16}B_6Al_3$ (c) and $(Co_{.5}Ni_{.5})_{75}P_{16}B_6Al_3$ (d). The arrows mark the T_c of the alloys.

Fig. 4 The anisotropy of the MR for $(Fe_{.4}Ni_{.6})_{75}P_{16}B_6Al_3$ at T=4.2 K and 78 K.

dependent in a large temperature range around T_c for all the alloys. For the alloys where T_c is relatively high the MR changes sign and is positive at $T \ll T_c$. This positive MR is small and saturates at very low fields. However, if the applied field is transverse to the current ($\bar{B} \perp \bar{J}$) and in the plane of the ribbon, the MR may first pass through a weak maximum at low fields and then become negative as the field is increased, thus revealing a significant anisotropy in the ferromagnetic state, as illustrated in figure 4.

In conclusion we suggest that the negative field dependence of the MR below and above T_c and the correlation of the low temperature rise in ρ to the concentration of the elements carrying a magnetic moment indicate that there is a significant magnetic contribution to the resistivity in these alloys. Also the orientation of the magnetic field to the sample geometry is of crucial importance when investigating the MR in amorphous ferromagnetic alloys. Detailed analysis of these results will be published elsewhere.

REFERENCES

(1) E.M. Gyorgy, H.J. Leamy, R.C. Sherwood and H.S. Chen, AIP Conf. Proc. 29, 198 (1975).
(2) D. Onn, T.H. Antoniuk, T.A. Donelly, W.D. Johnson, T. Egami, J.T. Prater and J. Durand, J.Appl.Phys. 49, 1790 (1978).
(3) K.V. Rao, H. Gudmundsson, H.U. Åström and H.S. Chen, J.Appl. Phys. 50 (3), 1592 (1979)

ULTRASONIC BEHAVIOR OF SUPERCONDUCTING AMORPHOUS METAL PdZr AT LOW TEMPERATURES

G. Weiss and W. Arnold[+]

Max-Planck-Institut für Festkörperforschung
Heisenbergstrasse 1, 7 Stuttgart 80, FRG

At low temperatures the thermal and acoustic behavior of glasses is greatly influenced by the presence of elastic low energy states. These low energy states are responsible for the observed linear behavior of the specific heat below 1 K. The saturation of the ultrasonic attenuation reveales that these excitations cannot be harmonic oscillators and thus strongly support the original idea that the low energy states arise from quantum mechanical tunneling of a group of atoms from one configurational state to another /1/ which can be regarded as a two level system in a simplified form. Until presently there is no detailed microscopic picture of the tunneling particle despite the wealth of information obtained from specific heat, thermal conductivity, ultrasonic and dielectric measurements.

For the case of glassy metals it is not obvious that these materials should also contain tunneling states. Measurements of the relative variation of the sound velocity with temperature in PdSiCu /2/ indeed showed the characteristic logarithmic T dependence due to resonant interaction of phonons with tunneling states /3/ which therefore gives a clear indication of their existence in glassy metals.

Saturation of the ultrasonic absorption in PdSiCu at very low temperatures has also been reported /4,5/. However, much higher power level was required in these experiments because the lifetimes of the tunneling states are apparently four orders of magnitude shorter than those in insulating glasses. It has been proposed that the conduction electrons are responsible for this enhanced relaxation /5/. In a superconductor the electronic contri-

---
+ Present address: Brown Boveri Research Center, Baden, Switzerland

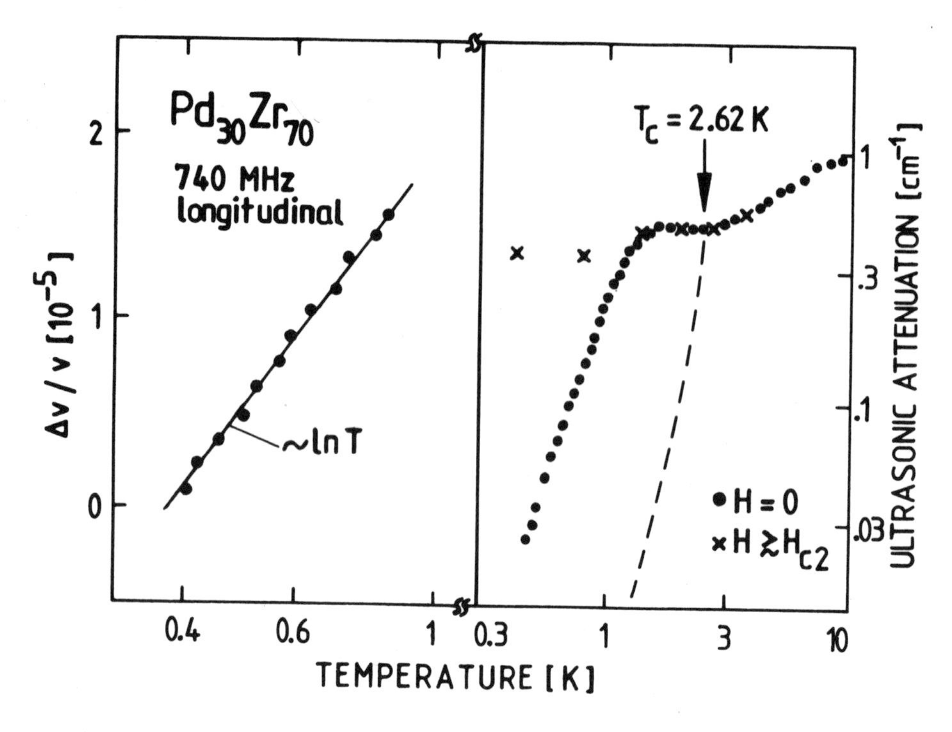

bution should drop drastically when cooling below T_c, as studied in detail in a recently depeloped theory /6/. This effect should be observable in the ultrasonic attenuation.

We have therefore performed ultrasonic measurements in the superconducting amorphous metal $Pd_{30}Zr_{70}$. (For experimental details see /7/.) Fig. 1a shows the relative variation of the sound velocity as a function of T below 0.8 K, whereas Fig. 1b shows the ultrasonic attenuation as 740 MHz and for $T \leq 10$ K. The logarithmic dependence of $\Delta v/v$ indeed shows that tunneling sites are also present in this material. From our measurements we deduce $nM^2 = 1.4 \cdot 10^7$ erg/cm^3 which results in a resonant absorption $\alpha = 1.3 \cdot 10^{-2}$ cm^{-1} /3/ (T= 0.4 K, f = 740 MHz), far too low to explain the attenuation results shown in Fig. 1b. Furthermore, the observed attenuation cannot be explained by the classical absorption mechanisms arising from the viscosity of the electron gas, because the mean free path of the electrons is only on the order of 10 Å in amorphous metals. We are therefore left with the relaxational absorption arising from strain modulation of the energy levels of the tunneling sites. In this process the temperature dependence of α is mainly determined by the temperature dependence of T_1 since, as usual, $\alpha \propto (A/(A^2+1))$ where $A = \omega T_1$. As pointed out above, in a superconductor below T_c only the thermally excited quasi-particles promote the relaxation rate /6/:

$$T_1^{-1} \sim (\rho V_\perp)^2 kT(\exp\Delta/kT + 1)^{-1} \qquad (1)$$

Here 2Δ is the energy gap and $\rho V_\perp$ is an effective coupling constant for the electron-tunneling state interaction. In Fig. 1b the dashed line shows the theoretical prediction /6/ using the density of states of the tunneling model and $\rho V_\perp=0.2$, both deduced from ultrasonic measurements in other metallic glasses. Obviously the steep fall of the absorption of T_c does not agree with our experimental results, because α drops drastically only for $T \lesssim 1.5$ K. Below 1.5 K the absorption also becomes strongly magnetic field dependent clearly indicating that electrons are involved in the attenuation process.

It might be possible to get better agreement by assuming a distribution of relaxation times and by taking into account the contribution of the thermal phonons to the relaxation rate. In fact, preliminary numerical calculations indicate that reasonable agreement between theory and experiment can be achieved by assuming $M \sim .3$ eV, where M describes the coupling tunneling state-phonons. However this assumption is only meaningful if there exists independent experimental verification of the magnitude of M and therefore further experiments are highly desirable in order to clarify this situation.

We thank our colleagues for helpful discussions and K. Dransfeld for stimulating and encouraging support. We are particularly grateful to H.J. Güntherodt for the generous supply of the samples.

REFERENCES

/1/ W.A. Phillips, J. Non-Cryst. Solids 31:267 (1978), and ref. contained therein.

/2/ G. Bellessa and O. Bethoux, Phys. Lett. 62A:125 (1977).

/3/ S. Hunklinger and W. Arnold, Physical Acoustics 12:155, eds. W.P. Mason and R.N. Thurston (Academic Press, New York, 1976), and ref. contained therein.

/4/ P. Doussineau, P. Legros, A. Levelut, A. Robin, J. de Phys. 39:L-265 (1978).

/5/ B. Golding, J.E. Graebner, A.B. Kane and J.L. Black, Phys. Rev. Lett. 41:1487 (1978).

/6/ J.L. Black and P. Fulde, Phys. Rev. Lett. 43:453 (1979)

/7/ G. Weiss, W. Arnold, K. Dransfeld and H.J. Güntherodt, to be published.

POSITRON ANNIHILATION IN IRON-BASED AMORPHOUS ALLOYS

G. Kögel

Zentrale Wissenschaftliche Einrichtung Physik,
Hochschule der Bundeswehr München
8014 Neubiberg, Germany

ABSTRACT. Results on positron lifetimes and Doppler-broadened annihilation lineshapes in Fe-based glassy amorphous metals are presented. From a simple model and by comparison with well annealed crystalline alloys we find strong evidence for positron trapping at large empty Bernal holes in the glassy alloy.

1. EXPERIMENTS

We have investigated the lifetimes and Doppler-broadened annihilation lines of four Fe-based glassy metals($Fe_{80}B_{20}$, $Fe_{78}Mo_2B_{20}$, $Fe_{40}Ni_{40}P_{14}B_6$, $Fe_{32}Ni_{36}Cr_{14}P_{12}B_6$) supplied by Allied Chemical Corp. Standard experimental setups were used/1/. The statistical accuracy of the lifetimes is better than 1 ps if not stated otherwise. Results at ambient temperature are presented in table 1 and fig. 1a. The lifetime spectra for the glassy metals always show a simple exponential decay envolving only a single lifetime. Therefore all positrons annihilate from the same state/1/. The temperature dependence of the annihilation characteristics (cf fig. 1b) is very strong in glassy metals when compared with crystalline transitional metals.

Table 1 : observed lifetimes (ps)

Fe bulk	Fe[a] vacancy	$Fe_{80}B_{20}$	$Fe_{40}Ni_{40}P_{14}B_6$	$Fe_{32}Ni_{36}Cr_{14}P_{12}B_6$	$Fe_{78}Mo_2B_{20}$ glass	$Fe_{78}Mo_2B_{20}$ cryst.[b]	$Fe_{78}Mo_2B_{20}$ cryst.[c]
110	170	143	157	160	143	137	116

[a]) /2/ ; [b]) immediately after crystallization at 430°C

[c]) 24 h annealed at 850°C

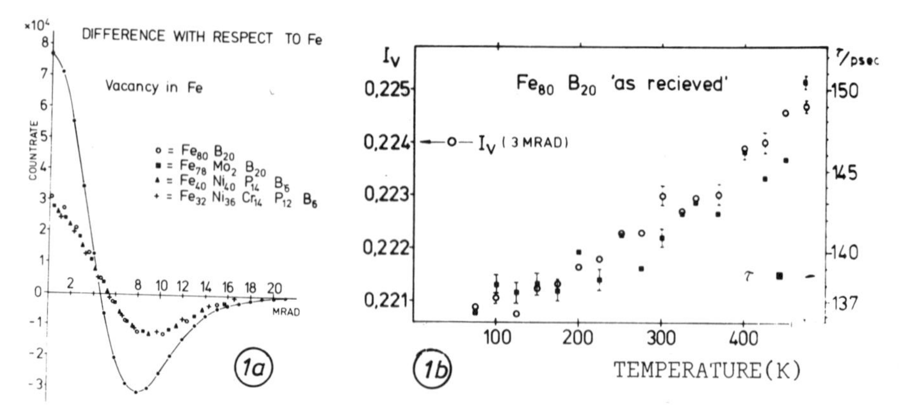

Fig.1a: Annihilation line shapes. The differences of normalized lines with respect to Fe are plotted. Vacancy lineshape from/3/.
Fig.1b: Temperature dependence of annihilation characteristics. I_V is the fraction of counts in the region -1.5 to +1.5 mrad.

2. SIMPLE ESTIMATES

At present there is no theory of positron annihilation in amorphous metals. Simple estimates are only possible for the positron lifetime. The positrons mainly occupy the interstitial cells/1/. In crystals there is one type of interstitial cell with an effective electron density, n_{cr}, yielding a lifetime τ_{cr} of the positrons. In the dense random packing model a metallic glass consists of different interstitial cells (Bernal holes)/4/ with different effective densities, n_i, experienced by the positrons. So in a state where the positron occupies the cells of type i with probability x_i the relation between effective electron density and positron lifetime τ/5/ may be rewritten as

$$1/\tau = (\Sigma_i x_i n_i / n_{cr})/\tau_{cr} = \Sigma_i x_i/\tau_i \quad . (1)$$

In this expression the effective densities contain all problems. For Fe and B we expect approximately the same depletion of positrons at the ion cores and also the effective radii of the 3d-electrons in Fe and the 2s,2p-electrons in B are similar. Therefore as a crude approximation in the Fe-B alloys we may replace n_i/n_{cr} by the ratio of the number densities. Using this assumption the expected lifetimes τ_i for positrons in the different canonical holes /4/ of glassy $Fe_{80}B_{20}$, in a vacancy and in crystalline Fe_3B, Fe_2B are calculated from the bulk iron value. The results are presented in table 2. The parameters of the canonical holes in a dense random packing of Fe-atoms were taken from /4/. V_{free} is the difference of the cell and the atomic volumes in units of the atomic volume Ω of bulk Fe.
In spite of the crude approximations made the calculated values are in good agreement with the experimental values available.

Table 2 : calculated lifetimes τ_i(ps)

cell type	empty cell n_i/n_{cr}	τ_i	$\tau_{obs.}$	V_{free}	cell contains B-atom n_i/n_{cr}	τ_i	$\tau_{obs.}$
Fe(bulk)	1	-	110	0	-	-	-
Fe(vacancy)	0.67	164	170	1	-	-	-
tetrahedron	0.95	116	-	0.01	2.97	37	-
octahedron	0.88	125	-	0.09	1.39	79	-
dodecahedron	0.86	128	-	0.19	1.32	83	-
trig.prism	0.82	134	-	0.31	1.03	107	-
Fe_3B	-	-	-	-	0.99	111	110±10
Archimedian antiprism	0.83	133	-	0.38	1.00	110	-
Fe_2B	-	-	-	-	1.01	109	110±10

3.INTERPRETATION

For the glassy alloys the annihilation lineshapes and the lifetimes presented in this work clearly indicate annihilation from a definite state between bulk and vacancy.Similar facts have been reported for PdSi glasses/6/.From table 2 we conclude that the annihilation takes place in empty trig.prisms and Archimedian antiprisms.At 20% B content only a very small fraction of these holes may be empty.Therefore the positrons have to be trapped into bound states at these empty holes.From a square well potential and taking into account the positron pseudopotential of 5.8 eV for iron/7/we have estimated a minimal free hole volume of 0.5Ω to bind a positron.So binding is only expected for slightly expanded holes or at accidental neighbours.In both cases the binding energy will be small.So at higher temperatures thermal detrapping from the smaller holes with shorter positron lifetimes should occur.This may explain the observed strong temperature dependence (cf fig.1b).For nonlocalised positrons one should expect a weak temperature dependence because $Fe_{80}B_{20}$ shows essentially no thermal expansion/8/.Our interpretation is further supported by the fact that the annihilation characteristics remain nearly unchanged by crystallization.In the highly defective state following crystallization positrons annihilate at grain boundaries. The shorter bulk lifetimes are only observable in well annealed crystalline alloys (cf table 1).

4.REFERENCES

/1/ W.Triftshäuser,this volume
/2/ P.Hautojärvi et al.,to be published in Solid State Commun.
/3/ M.Weller et al.,Solid State Commun. 17,1223 (1975)
/4/ G.S.Cargill III,Solid State Physics 30,227 (1975)
/5/ R.N.West,Adv.Physics 22,263 (1973)
/6/ H.S.Cheng,S.Y.Chuang,Appl.Phys.Lett. 27,316 (1975)
/7/ R.Nieminen,C.H.Hodges,Solid State Commun. 18,1115 (1976)
/8/ K.Fukamichi et al.,Solid State Commun. 23,955 (1977)

CHANGES OF SUPERCONDUCTING PROPERTIES AND ELECTRICAL RESISTIVITY OF AMORPHOUS COMPOUNDS BY MEANS OF LOW TEMPERATURE HEAVY ION IRRADIATION

J. Bieger, G. Saemann-Ischenko, H. Adrian, M. Lehmann,
P. Müller, L. Söldner
Physikalisches Institut der Universität Erlangen-Nürnberg,
Western Germany

E.L. Haase
Institut für angewandte Kernphysik, KFK Karlsruhe,
Western Germany

C.C. Tsuei
IBM Thomas J. Watson Research Center, Yorktown Heights,
New York

ABSTRACT. Thin Films of $Nb_{80}Si_{20}$ and $Nb_{75}Ge_{25}$ were irradiated ($T_{irr} \leq 20K$) with 20 MeV sulphur ions up to fluences of 10^{16} cm^{-2} and subsequently annealed up to room temperature. Irradiation enhances T_c by about 0.5K. In the case of $Nb_{75}Ge_{25}$ ϱ decreases by 1.5% and dH_{c2}/dT by 15%. Annealing at room temperature nearly restores the initial values of T_c and ϱ. The dH_{c2}/dT vs. T_c curves for irradiation and annealing are almost identical for weakly irradiated samples, whereas a pronounced difference is observed after high dose irradiation. Comparison is made with heavy ion irradiation of crystalline A15 - Nb_3Ge.

Heavy ion irradiation of crystalline superconductors produces lattice defects which cause changes of many properties of the normal and the superconducting state. To investigate, how amorphous superconductors are affected by irradiation, evaporated films of $Nb_{80}Si_{20}$ (thickness $\approx$ 200 nm) and sputtered films of $Nb_{75}Ge_{25}$ (thickness $\approx$ 300 nm) were irradiated at temperatures below 20 K with 20 MeV sulphur ions up to fluences of 10^{16} particles/cm². The range of the projectiles is $\approx 3\mu m$, therefore no sulphur was implanted. After irradiation an isochronal annealing program was performed up to room temperature (holding time 10 min.).

The effect of irradiation on the superconducting critical temperature T_c of $Nb_{75}Ge_{25}$ is shown in Fig.1a. One observes a monotonous increase of T_c up to $\Phi t \approx 5 \cdot 10^{14}$ cm^{-2}, followed by a slight

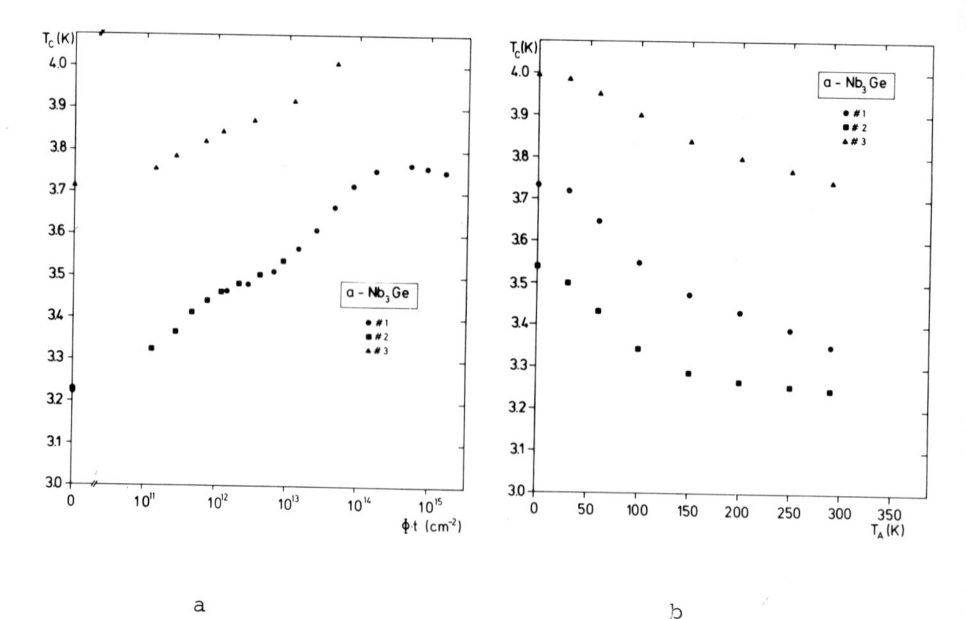

Fig.1 Superconducting critical temperature T_c as a function of irradiation fluence Φt (a) and annealing temperature T_A (b).

decrease. Although different samples have different starting values of T_c, they behave in a similar way. For $Nb_{80}Si_{20}$, T_c rises from 4.7 K before irradiation to 5.3 K at $\Phi t=10^{15}$ cm^{-2}. The recovery of T_c of $Nb_{75}Ge_{25}$ during annealing is shown in Fig.1b, indicating that irradiation produces structural changes which are stable only at low temperatures and anneal without pronounced steps. The same is true for $Nb_{80}Si_{20}$. For a discussion of the behaviour of T_c it is convenient to use a simplified McMillan-type formula:

$$T_c \approx \frac{\langle\omega\rangle}{1,20} \cdot \exp\left[-\frac{1+\lambda}{\lambda-\mu^*}\right] \text{ with } \lambda = \frac{N(E_F)\cdot\langle J^2\rangle}{M\cdot\langle\omega^2\rangle} \quad \text{and } \mu^* = 0.11$$

In the case of crystalline A15 - Nb_3Ge the decrease of T_c upon irradiation can be explained by a decrease in the electronic density of states $N(E_F)$, determined from the slope of the upper critical field $H'_{c2}=\frac{dH_{c2}}{dT}(T_c)$ and the residual resistivity ϱ via the relation $H'_{c2}\propto N(E_F)\cdot\varrho$ (dirty limit) [1,2]. From Fig.2 it can be seen that, although T_c increases, H_{c2} decreases by 15% and ϱ by 1,5%. Thus $\frac{H'_{c2}}{\varrho}$ decreases upon irradiation and cannot explain the increasing T_c. We suggest that irradiation changes the atomic short range order in such a way that the first peak in the structure factor S(k) is smeared out. The increase of S(k) for small k is expected to lead to an enhancement of $\langle J^2\rangle$ [3] and therefore of T_c. The mechanism of smoothing S(k) also explains the observed decrease of ϱ. The diffraction model [4] yields $\varrho\propto S(2k_F)$. From the negative

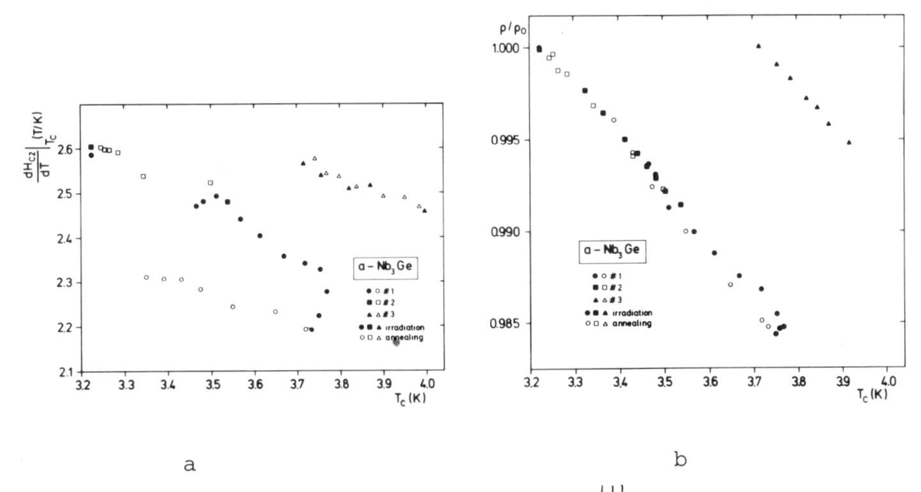

a b

Fig.2 Slope of the upper critical field $\frac{dH_{c2}}{dT}(T_c)$ (a) and normalized resistivity ρ/ρ_0 (b) vs. T_c during irradiation (full symbols) and annealing (open symbols).

temperature coefficient $\alpha = \frac{1}{\rho} \cdot \frac{d\rho}{dT}$ we conclude that $2k_F$ lies near the first peak of S(k). Then irradiation reduces $S(2k_F)$ similar as temperature does, and therefore ρ decreases. Fig.2b shows that the ρ vs. T_c correlation is the same for irradiation and annealing. For H'_{c2} this is only true for samples #2 and #3, which were not exposed to very high fluences, whereas the high dose irradiated sample #1 shows an irreversible behaviour in the sense that T_c recovery is 80% of the irradiation effect and H'_{c2} recovery only 30% after annealing at room temperature. This irreversibility is also reflected in the width of the superconducting transition T_φ, defined as the temperature range between 10% and 90% of normal state resistivity. For small fluences T_φ does not change very much, but it drops from 35 mK to 7 mK for high fluences and retains this value even after annealing at room temperature.

REFERENCES

1. P. Müller, G. Saemann-Ischenko, H. Adrian, J. Bieger, M. Lehmann, E.L. Haase, Proc. 3rd Int. Conf. on Superconductivity in d- and f-Band Metals, San Diego, 1979
2. R. Koepke, G. Bergmann, Solid State Commun. 19, 435 (1976)
3. K. Bennemann, private communication
4. L.V. Meisel, P.J. Cote, Phys.Rev. B17, 4652 (1978)

LIQUID METALS AND SUPERCONDUCTIVITY LAUNCH A NEW GENERATION OF ELECTRIC MACHINES*

J.-T. Eriksson°, A. Arkkio, P. Berglund, J. Luomi,
and M. Savelainen
Technical Research Centre of Finland,
02150 Espoo 15, Finland

1. THE SUMO MOTOR

1.1 Why and how

It has been established that using superconductivity for the creation of high density magnetic fields one can achieve radical savings in weight and volume of electric machines.

Liquid metals for current collection have always been very attractive to machine designers. Mostly due to expected high current capacity and low loss level. Metallurgic and hydrodynamic factors, however, have suppressed any success in arranging a reliable collector set so far.

A project aiming at the construction of a superconducting homopolar machine was started in August 1977. The dead line for running the motor was scheduled to May 1979. The first and successful full load test took place on May 24, 1979.

The 50 kW motor is of the drum type. Two superconducting solenoids are placed in a nonrotating cryostat inside the rotor frame. The main current circuit, rated 10,000 A, consists of copper bars on the stator and copper shells on the rotor. The

---

* This project has been undertaken by the Technical Research Centre of Finland in cooperation with the Helsinki University of Technology. It is partly supported by the Ministry of Trade and Industry, and Oy Strömberg Ab.

° Present address: Laboratory of electromechanics, Helsinki University of Technology, SF-02150 Espoo 15, Finland.

Fig. 1. Test bench installation. The SUMO motor to the right. The loading generators are rated 37 kW each at 1200 rpm.

liquid metal current collection system comprises two intermediate rotors which rotating at half nominal speed reduce the risk of hydrodynamic instability and on the same time establish the necessary centrifugal force to keep the liquid metal in the collector grooves. This arrangement makes it possible to regulate the shaft speed over the whole range zero to nominal speed.

1.2 Specifications

The following dates are based on test results. In general, they are in close agreement with design values.

Power output	51 kW	Efficiency	92 %
Rotational speed	20 s^{-1}	Current density in liquid metal	1.7 MA/m^2
Voltage	5.77 V		
Main current	10 kA	Collector losses	1.1 kW
Field current	148 A	Collector voltage drop (resistive)	4 mV

Best achievement is a 1 hour run at 75 kW power output. The field current was 170 A and the armature current 12 kA. The efficiency was at best (before warm-up) 96%. No indications of overheating was registrated.

Long run tests at rated power have been performed twice. Both lasting about 16 hours. Interruption was caused by the formation of a cover liquid/liquid metal emulsion, where the nonconducting cover liquid occupies the outer phase.

2. CURRENT COLLECTION

2.1 Liquid versus solid brushes

A comparison of the most dominant factors governing the current collection process gives a straight answer why liquid metals are so attractive:

	Solid brush	Liquid metal
Current capacity, MA/m^2	0.3	5
Contact voltage drop, mV	500	0.5
Shear velocity, m/s	30	100

Homopolar motors working with very high armature currents are simply not realistic if the current has to be spread out over a sufficient area to fulfill the density requirements of solid brushes.

2.2 Gallium-indium

The eutectic of gallium-indium (76/24 w/o) has been chosen for two main reasons:

1. In a superconducting machine where magnetic stray fields are considerable Ga-In is favored before mercury and NaK on a loss basis.
2. Both gallium and indium are considered nontoxic in small amounts. Accordingly Ga-In is easy to handle. This is particularly important in a research activity where progress is dependent on the number of experiments. Simple equipment makes visual observations possible.

The melting point is 15.7°C, conveniently below room temperature.

2.3 Hydrodynamics

The hydrodynamics of an open channel MHD-driven Couette flow is quite complicated. There is always a risk that the liquid is thrown out or at least continuously decreasing due to aerosol effects. Under certain circumstances complete instability is developed. The origin is a combination of strong turbulence, secondary flows, adhesion and mechanical irregularities. A first measure against unwanted effects is to give the channel a proper design. The introduction of a viscous cover liquid offers several advantages. The liquid acts as a damper against instabilities. It can be used to remove impurities and oxides and to work as a cooling agent. Given the right chemical composition the cover liquid maintain a high surface tension between the liquid metal and the channel wall.

2.4 Interfacial phenomena

Gallium is known to be very reactive with most metals, including good electric conductors like copper, aluminium and silver. Metals from groups VI and VIII obtain acceptable or even good corrosion resistance. However, very few dates concerning solubility and corrosion in the temperature range 15...100 °C are available.

In a current collector the contact resistance between electrode and liquid metal must be very low. According to our experience it should not exist at all. Tunneling through thin impurity or oxide films usually results in high voltage drops and moreover, tend to increase with time. Good collecting properties ask for good wetting i.e. ideal metal to metal contact. As expected wetting and contact resistance are closely related to each other. Unfortunately good wetting often means interfacial reactions.

2.5 Corrosion experiments

In order to establish corrosion dates equivalent to the conditions occurring in a homopolar machine, liquid metal has been pumped by MHD means in a circular closed duct, with a 3 x 3 mm cross section and an average diameter of 12 mm. Materials under investigation are copper, nickel, chromium, and rhodium. Even lead was the subject of our interest, but within a week the solded surface broke up completely thus exposing the copper base metal.

The temperature is held at 70 °C. An average current density of 1 MA/m^2 and a flux density of 0.25 T result in a mean flow velocity of 2.3 m/s. After a test period of 1176 hours copper dissoluted to a saturation level of 54 ppm. A severe transport effect occurred due to electric field. Copper dissolved at the anode and a solid gallium copper compound was formed on the cathode. Electrodeposited nickel did not show any visible signs of corrosion. The nickel concentration in the liquid metal was less than 27 ppm. Electrodeposited chromium on nickel and copper did not indicate corrosion, Cr concentration was less than 2.3 ppm. Electrodeposited rhodium on copper resulted in a severe copper dissolution with the same transport phenomenon as in the pure copper case. Rhodium necessitates a nickel middle coating. As a conclusion we may consider chromium, like rhodium, as a potential alternative, but a strong oxide layer and (hence) poor wetting increase the risk of undesired contact failures. Nickel seems to present an acceptable solution as long as the temperature is held within moderate limits. Nickel can easily be wetted by mechanical means.